STOCHASTIC PROCESSES

STOCHASTIC PROCESSES

SECOND EDITION

J. MEDHI

Professor Emeritus
Gauhati University

JOHN WILEY & SONS

NEW YORK • CHICHESTER • BRISBANE • TORONTO • SINGAPORE

First Published in 1994 by
WILEY EASTERN LIMITED
4835/24 Ansari Road, Daryaganj
New Delhi 110 002, India

Distributors:

Australia and New Zealand :
JACARANDA WILEY LIMITED
PO Box 1226, Milton Old 4046, Australia

Canada :
JOHN WILEY & SONS CANADA LIMITED
22 Worcester Road, Rexdale, Ontario, Canada

Europe and Africa :
JOHN WILEY & SONS LIMITED
Baffins Lane, Chichester, West Sussex, England

South East Asia :
JOHN WILEY & SONS (PTE) LIMITED
05-04, Block B, Union Industrial Building
37 Jalan Pemimpin, Singapore 2057

Africa and South Asia :
WILEY EASTERN LIMITED
4835/24 Ansari Road, Daryaganj
New Delhi 110 002, India

North and South America and rest of the world :
JOHN WILEY & SONS INC.
605, Third Avenue, New York, NY 10158, USA

Library of Congress Cataloging-in-Publication Data

ISBN 0-470-22053-8 John Wiley & Sons Inc.
ISBN 81-224-0549-5 Wiley Eastern Limited

Printed in India at Chaman Offset Printers, New Delhi.

Kālo hi eko nityaśca Vȳapakaḥ
—Bhāgavad-Purāna

To

My Mother

and to

The Memory of My Father

Preface to the Second Edition

It is more than 10 years that the first edition appeared. The favourable reviews (one of which marked the book as *the clear choice*) of the first edition were very encouraging indeed. The book received a very welcome response from the intended audience—ranging over a wide range of disciplines. The fact that the book was referred to in papers, not only in statistics, operations research, engineering and management journals, but also in leading journals in other disciplines, such as, *Physical Review, J. Math. Physics, Transportation Science, Arch. G. Psychology, J. Theoretical Biology, Water Resource Res., Forest Science* and so on speaks of the wide acceptability, of the book on the one hand, as well as of the growing application of stochastic processes in diverse areas, on the other hand.

I have received several favourable communications from interested readers from the world over, from faculty and students to researchers, using methods of stochastic processes in a wide range of areas. John Wiley and Sons, New York have kindly got the book reviewed thoroughly by a learned scholar (whose identity was not known to me); detailed comments and constructive suggestions obtained from the anonymous referee were very useful in carrying out revision of the book. All these have served as valuable feedbacks and have been helpful in my revision work.

In order to facilitate use of the second edition by those who are already familiar with the 1982 edition, I have avoided a drastic change in the basic structure; the original arrangement of matter has been adhered to. The number of chapters has been kept at 10, as before, while considerable additions and alterations have been made in the broad topics covered in the Chapters.

The first Chapter has been entirely devoted to probability distributions and their properties, which are later used in the models. The mathematical preliminaries covered are put in the Appendix; this would be helpful for quick reference of the results. Chapter 10, which has been substantially enlarged and made up-to-date could be used for a one-semester course in Queueing Theory. Many sections have been thoroughly revised and new material has been added. Mention may be made of inclusion of topics, such as, martingales in Chapter 2, reducible chains in Chapter 3, randomization in Chapter 4, renewal theory in discrete time and regenerative inventory systems in Chapter 6, generalised Galton Watson processes in Chapter 9, network of Markovian queues, $GI/G/1$ model, and queues with vacations in Chapter 10. Apart from inclusion of new topics, new examples and new exercises, there has been considerable expansion of the discussion of the topics through notes, remarks, and so on. The references have been made up-to-date. The additions and alterations have resulted in increase of the volume by over 50%, from less than 400 pages to about 600 pages. In fact, this enlarged and revised edition, while retaining the structure and following the objective as well as the philosophy of the 1982 edition, removes the deficiencies, updates the material and references and aims at a broader perspective and more elaborate treatment of the subject matter.

My thanks are due to many colleagues and other users whose favourable comments served as feedbacks as well as encouragement. I am also thankful to the anonymous referee in U.S.A. for his valuable comments, to John Wiley and Sons, New York for getting the book reviwed, to Mr. A. Machwe of Wiley Eastern Ltd, New Delhi for his interest and deep appreciation in my works and his willingness to render me any help that I needed, and to Dr. A. Subramaniyam, IIT, Bombay for his help in several ways.

My elder son Deepankar Medhi, University of Missouri-Kansas City (formerly of Bell Labs) and my elder daughter Shakuntala Choudhury of A.T. & T, Technology Systems, NJ rendered immense technical assistance.

May 1993   JYOTIPRASAD MEDHI

U.N. Bezbarua Rd
Silpukhuri West
Guwahati 781 003
Assam, India
    Tel : (0361)542959

Preface to the First Edition

"Currently in the period of dynamic indeterminism in science, there is hardly a serious piece of research, which, if treated realistically, does not involve operations on stochastic processes" (Neyman, *J. Am. Stat. Ass.* (1960) p. 625). It is thus natural to find a growing awareness and interest in this subject in scientific and technological studies everywhere.

It is felt that, unlike statistical theory, a large variety of introductory and intermediate level texts dedicated to stochastic processes are not available, and that there is need for more books at this level. This book which aims at the level between that of elementary probability texts and of the excellent advanced works on stochastic processes, has been written keeping in view students of diverse interests and background. The book deals mainly with methods and applications.

The prerequisites for using this book are a course on elementary probability theory and statistics and a course on advanced calculus. Some of the basic mathematical concepts and techniques used (in the sequel) have been treated in some details in Chapter 1. The theoretical results developed have been followed by a large number of illustrative examples. These have been supplemented by a large number of exercises, answers to most of which are also given. Lists of references are provided for the benefit of those who need more details or wish to pursue the subject further.

It is expected that the book would be suitable as a text book for advanced undergraduate, post graduate and research level courses in Statistics, Applied Mathematics, Operations Research, Computer Science, Management Science etc. as also for graduate and beginning research level courses of study in Physics, Life Sciences and different branches of Engineering. The course could be covered in one or two semesters depending on the interest of the users and on the extent or coverage desired. Researchers who need knowledge of applied probability and applied stochastic processes would also profit from using the book. The illustrative examples spread over areas, such as, biological sciences to management sciences would lend a special appeal to not only those who wish to pursue the subject but also to those who wish to go for applications. Some materials, specially on applications from some of the most recent papers are also included. There is a somewhat exhaustive chapter on stochastic processes in Queues, and this portion could be used also for a course on queueing theory. The material presented in the book would, it is hoped, be suitable, in general, for students having some mathematical maturity who wish to gain an understanding of the subject and to see how the theory works.

The book has grown out of the author's long 30 years' experience of teaching and research. The suggestion of writing a book on the lines of his lecture notes, was first made to him by groups of students who took such a course given by him at the University of Montreal, Canada.

I look forward to receive comments and suggestions on the work from students, teachers and researchers.

I remain intellectually indebted to all those workers whose works have stimulated my interest in the subject. In the present venture, encouragements and suggestions have been received from a number of learned scholars both in this country and abroad. I am thankful to them all.

I am indeed most grateful to Professor N.U. Prabhu, School of Operations Research, Cornell University U.S.A. and to Professor R. Fortet, Université de Paris, France, both of whom kindly read the manuscript thoroughly and made very valuable comments and offered constructive suggestions. I recall also the appreciative interest shown and the constant encouragement given by Professor Prabhu.

I am indebted to Professor M.B. Priestley, University of Manchester Institute of Science & Technology, U.K. and to.Professor R.M. Loynes, University of Sheffield, U.K. for having glanced through the manuscript within a short time and for making useful comments, specially with regard to coverage.

I received considerable assistance with proofreading from some of my former research students and present colleagues, namely, Dr. S.B. Nandi, Dr. A. Borthakur, Dr. H.K. Baruah and Mr. D.C. Jain as well as from my elder son, Deepankar; and help in many other ways from my younger son, Shubhankar and two daughters, Shakuntala and Alakanandaa. Acknowledgement is also made of the excellent typing by S. Borah and A. Choudhury.

JYOTIPRASAD MEDHI

Gauhati, India
August 1981

Contents

CHAPTER 1

Probability Distributions

1.1 GENERATING FUNCTIONS

1.1.1 Introduction

In dealing with integral-valued random variables, it is often of great convenience to apply the powerful tool of generating functions. Many stochastic processes that we come across involve non-negative integral-valued random variables, and quite often we could use generating function in their studies. The principal advantage of its use is that a single function (generating function) may be used to represent a whole set of individual items.

Definition. Let $a_0, a_1, a_2, \ldots$ be a sequence of real numbers. Using a variable s, we may define a function

$$A(s) = a_0 + a_1 s + a_2 s^2 + \cdots = \sum_{k=0}^{\infty} a_k s^k. \tag{1.1}$$

If this power series converges in some interval $-s_0 < s < s_0$, then $A(s)$ is called the generating function (or the *s-transform*) of the sequence, $a_0, a_1, a_2, \ldots$. The variable s itself has no particular significance. Here we assume s to be real but a generating function with a complex variable z is also sometimes used. It is then called a z-transform. The power series (1.1) can be differentiated any number of times, also $\frac{d}{ds} \Sigma = \Sigma \frac{d}{ds}$ holds, within its radius of convergence.

Differentiating (1.1) k times, putting $s = 0$ and dividing by $k!$, we get a_k, i.e.

$$a_k = \frac{1}{k!} \left[\frac{d^k A}{ds^k} \right]_{s=0}. \tag{1.2}$$

1.1.2 Probability Generating Function : Mean and Variance

Suppose that X is a random variable that assumes non-negative integral values $0, 1, 2, \ldots$, and that

$$\Pr\{X = k\} = p_k, \quad k = 0, 1, 2, \ldots, \sum_k p_k = 1. \tag{1.3}$$

If we take a_k to be the probability p_k, $k = 0, 1, 2, \ldots$ then the corresponding generating function $P(s) = \Sigma p_k s^k$ of the sequence of probabilities $\{p_k\}$ is known as the *probability generating function* (p.g.f.) of the random variable X. It is sometimes also called the *s-transform* (or *geometric transform*) of the r. v. X.

We have $P(1) = 1$; the series $P(s)$ converges for at least $-1 \leq s \leq 1$ and is infinitely differentiable. The function $P(s)$ is defined by $\{p_k\}$ and in turn defines $\{p_k\}$ uniquely, i.e. a p.g.f. determines a distribution uniquely. Again

$$P(s) = \sum_{k=0}^{\infty} \Pr\{X = k\} s^k = E(s^X), \qquad (1.4)$$

where $E(s^X)$ is the expectation of the function s^X (a random variable) of the random variable X.

The first two derivatives of $P(s)$ are given by

$$P'(s) = \sum_{k=1}^{\infty} k p_k s^{k-1}, -1 < s < 1 \qquad (1.5a)$$

and

$$P''(s) = \sum_{k=1}^{\infty} k(k-1) p_k s^{k-2}. \qquad (1.5b)$$

The expectation $E(X)$ is given by

$$E(X) = \sum_{k=1}^{\infty} k p_k = \lim_{s \to 1} P'(s) = P'(1). \qquad (1.6)$$

We have

$$E\{X(X-1)\} = \sum_{k} k(k-1) p_k = \lim_{s \to 1} P''(s) \qquad (1.7)$$

and

$$E(X^2) = E\{X(X-1)\} + E(X) = P''(1) + P'(1).$$

Hence

$$\mathrm{var}\,(X) = E(X^2) - [E(X)]^2$$

$$= P''(1) + P'(1) - [P'(1)]^2 \qquad (1.8)$$

If, as $s \to 1$, $\Sigma\, k p_k$ diverges then we say that $E(X) = P'(1) = \infty$ and if $\Sigma\, k(k-1)\, p_k$ diverges then $P''(1) = \infty$ and $\mathrm{var}\,(X) = \infty$.

The relation (1.6) gives the mean and (1.8) the variance of X in terms of the p.g.f. of X. In fact, moments, cumulants etc. can be expressed in terms of generating functions. More generally, the kth factorial moment of X is given by

$$E\{X(X-1)\ldots(X-k+1)\} = \frac{d^k}{ds^k} P(s)\big|_{s=1} \quad \text{for } k = 1, 2, \ldots$$

Note that $P(e^t)$ is the moment generating function and $P(1+t)$ is the factorial moment generating function. Further, the p.g.f. $P(s) = E(s^X)$ is a special case of the characteristic function $E(e^{itX})$ or the Fourier transform of X.

Note 1. Assuming that term by term integration is valid, we get

$$\int_0^1 P(s)ds = E\left(\frac{1}{X+1}\right).$$

See also Exercise 1.6.

Note 2. Let $q_k = \Pr(X > k) = p_{k+1} + p_{k+2} + \ldots, \quad k = 0, 1, 2 \ldots$

and $\qquad r_k = \dfrac{\Pr(X > k - 1)}{k}, \quad k = 1, 2, \ldots$

Then

$$Q(s) = \sum_{k=0}^{\infty} q_k s^k = \frac{1 - P(s)}{1 - s}, \quad s \neq 1$$

and $\qquad R(s) = \sum_{k=1}^{\infty} r_k s^k = \int_0^s Q(y)dy.$

Example 1(a). *Bernoulli distribution* : Let X be a r.v. that assumes only two values 0 and 1 with

$$P\{X = 1\} = p, P\{X = 0\} = q = 1 - p \ (0 < p < 1).$$

The p.g.f. of the r.v. X is

$$P(s) = E[s^X] = q + sp.$$

We have $E(X) = p$ and var $(X) = pq$. Independent trials with only two possible outcomes ("success" and "failure") at each trial are known as Bernoulli trials. Denote "success" by 1 and "failure" by 0 and their probabilities by p and q respectively; we then get Bernoulli distribution and X is called a Bernoulli r.v. with parameter p.

Example 1(b): *Binomial distribution*: Let X be a r.v. denoting the number of successes (or failures) in a fixed number n of Bernoulli trials. Then X has binomial distribution having p.m.f.

$$p_k = \Pr(X = k) = \binom{n}{k} p^k q^{n-k}, \quad k = 0, 1, 2, \ldots, n$$

The p.g.f. of X is

$$P(s) = (q + sp)^n$$

with $E(X) = np$ and var $(X) = npq$.

The binomial r.v. has parameters n and p and is the sum of n independent Bernoulli r.v.'s each with parameter p.

Example 1(c). *Poisson distribution*: Let X be a Poisson variate with p.m.f.

$$p_k = \Pr\{X = k\} = \frac{\exp(-\lambda)\lambda^k}{k!}, \quad k = 0, 1, 2, \ldots$$

The p.g.f. of Poisson distribution (or of the random variable X) is given by

$$P(s) = \sum_{k=0}^{\infty} p_k s^k = \sum_{k=0}^{\infty} \frac{\exp(-\lambda)\lambda^k}{k!} s^k$$

$$= \exp(-\lambda)\exp(\lambda s)$$

$$= \exp\{\lambda(s - 1)\}.$$

Since $P'(s) = \lambda \exp\{\lambda(s-1)\}$ and $P''(s) = \lambda^2 \exp\{\lambda(s-1)\}$, we have, from (1.6) and (1.8)

$$E(X) = \lambda \quad \text{and} \quad \text{var}(X) = \lambda.$$

Example 1(d). *Geometric distribution*: Let X be a random variable with geometric distribution $\{p_k\}$, where

$$p_k = \Pr\{X = k\} = q^k p, \quad k = 0, 1, 2 \ldots$$

$$0 < q = 1 - p < 1.$$

The p.g.f. of X is

$$P(s) = \sum_{k=0}^{\infty} p_k s^k = p \sum_{k=0}^{\infty} (qs)^k = \frac{p}{1 - qs}.$$

We have

$$P'(s) = \frac{pq}{(1-qs)^2} \;,\quad P''(s) = \frac{2pq^2}{(1-qs)^3} \;.$$

Using (1.6) and (1.8) we have

$$E(X) = P'(1) = \frac{pq}{(1-q)^2} = \frac{q}{p}$$

and

$$\text{var}\,(X) = P''(1) + P'(1) - [P'(1)]^2 = \frac{2q^2}{p^2} + \frac{q}{p} - \left(\frac{q}{p}\right)^2 = \frac{q}{p^2} \;.$$

An example of a random variable X that follows geometric distribution is given below. Consider a sequence of Bernoulli trials with probability of success p and that of failure $q = 1 - p$ at each trial. Then the number X of failures preceding the first success has geometric distribution with $\Pr\{X = k\} = q^k p$ $(k = 0, 1, 2, \ldots)$. The r.v. X is also called the waiting time for the first success.

Non-ageing property of geometric distribution:

For a geometric r.v. X, we have

$$\Pr\{X = s + r \mid X \geq s\} = \frac{q^{s+r}p}{q^s}$$

$$= q^r p = \Pr\{X = r\}.$$

This property, called non-ageing (or memoryless) property, characterizes geometric distribution among all distributions of discrete non-negative integral r.v.'s.

Example 1(e). *Logarithmic series distribution*:
The r.v. X has logarithmic series distribution if

$$p_k = \Pr\{X = k\} = \frac{\alpha q^k}{k}, \quad k = 1, 2, 3, \ldots$$

$$\alpha = -1/(\log p)$$

$$0 < q = 1 - p < 1.$$

The p.g.f. of X is

$$P(s) = \sum_{k=1}^{\infty} p_k s^k = \sum_{k=1}^{\infty} \frac{\alpha q^k}{k} s^k$$

$$= -\alpha \log(1 - sq) = \frac{\log(1 - sq)}{\log(1 - q)}.$$

We have

$$E(X) = \frac{\alpha q}{1 - q}$$

and

$$\text{var}(X) = \frac{\alpha q(1 - \alpha q)}{(1 - \alpha)^2}.$$

Example 1 (f). Let X be a random variable with p.g.f. $P(s)$. To find the p.g.f. of the random variable $Y = mX + n$, where m, n are integers and $m \neq 0$.

Let

$$P_X(s) \text{ and } P_Y(s)$$

be the generating functions of X and Y respectively. We have

$$P_Y(s) = E\{s^Y\} = E\{s^{mX+n}\}$$

$$= E\{s^{mX} \cdot s^n\} = s^n \cdot E\{(s^m)^X\}$$

$$= s^n P_X(s^m).$$

1.1.2.1 Determination of $\{p_k\}$ from a given $P(s)$

From the above examples, we see how a single generating function $P(s)$ may be used to represent a whole set of probabilities

$$p_k = \Pr\{X = k\}, \quad k = 0, 1, 2, \ldots$$

In these examples we were concerned with the problem of finding $P(s)$ for a given set of p_k's. In many cases the reverse problem arises: to determine p_k from a given p.g.f. $P(s)$. Many situations arise, where it is easier to find the p.g.f. $P(s)$ of a variable rather than the probability distribution $\{p_k\}$ of the variable. One proceeds to find first the p.g.f. $P(s)$ and then to find the probability p_k from the function $P(s)$. Even without finding the p_k's one can find the moments of the distribution from $P(s)$.

Again, p_k can be (uniquely) determined from $P(s)$ as follows:

p_k can be found from $P(s)$ by applying (1.2), i.e.

$$p_k = \frac{1}{k!} \left[\frac{d^k P(s)}{ds^k} \right]_{s=0};$$

p_k is also given by the coefficient of s^k in the expansion of $P(s)$ as a power series in s.

When $P(s)$ is of the form $P(s) = U(s) / V(s)$, it may be convenient to expand $P(s)$ in a power series in s first by decomposing $P(s)$ into partial fractions. Suppose that $s_1, \ldots, s_r$ are the *distinct roots* of $V(s)$, i.e. $V(s) = (s - s_1) \ldots (s - s_r)$ (apart from a constant factor c, which, for simplicity, we take to be equal to 1), then $P(s)$ can be decomposed into partial fractions as

$$P(s) = \frac{a_1}{s_1 - s} + \cdots + \frac{a_r}{s_r - s},$$

where a_i's can be determined. It may be verified that

$$a_i = -U(s_i)/V'(s_i).$$

Now

$$\frac{a_i}{s_i - s} = \frac{a_i}{s_i(1 - s/s_i)} = \frac{a_i}{s_i}\left(1 - \frac{s}{s_i}\right)^{-1}$$

$$= \frac{a_i}{s_i}\left\{1 + \frac{s}{s_i} + \left(\frac{s}{s_i}\right)^2 + \cdots\right\} \text{ for } |s| < |s_i|$$

$$= a_i \sum_{k=0}^{\infty} \frac{s^k}{s_i^{k+1}}.$$

Hence $p_n \equiv$ coefficient of s^n in $P(s)$

$$= \frac{a_1}{s_1^{n+1}} + \cdots + \frac{a_i}{s_i^{n+1}} + \cdots \frac{a_r}{s_r^{n+1}}.$$

The above is an exact expression for p_n. The calculation of p_n by the above expression involves determination of all the roots of $V(s)$. A simple approximation can be had as follows.

Suppose that s_1 is a root which is smaller in absolute value than all the other roots. Then as n increases, the first term becomes the dominant term, and the contributions of the other terms become negligible. Thus, as $n \to \infty$

$$p_n \cong \frac{a_1}{s_1^{n+1}}$$

The case when $V(s)$ has *multiple roots* can be treated in a similar manner. Suppose that s_1 is a root of order m; then we shall have

$$a_1 = \lim_{s \to s_1} [(s - s_1)^m P(s)]$$

$$= \frac{(-1)^m m! U(s_1)}{V^m(s_1)} .$$

If s_1 is of smaller absolute value than the other roots of $V(s)$, we shall get as approximation, as $n \to \infty$,

$$p_n \cong \frac{a_1}{s_1^{n+m}} \binom{n+m-1}{m-1}.$$

1.1.3 Sum of (a Fixed Number of) Random Variables

Let X and Y be two independent non-negative integral-valued random variables with probability distributions given by

$$\Pr\{X = k\} = a_k, \quad \Pr\{Y = j\} = b_j .$$

The sum $Z = X + Y$ is a random variable. The event $\{Z = r\}$ can happen in the following mutually exclusive ways with corresponding probabilities:

$(X = 0 \text{ and } Y = r)$ with probability $a_0 b_r$

$(X = 1 \text{ and } Y = r - 1)$ with probability $a_1 b_{r-1}$

$\ldots$ $\qquad\qquad\qquad\qquad$ $\ldots$

$(X = t \text{ and } Y = r - t)$ with probability $a_t b_{r-t}$

$\ldots$ $\qquad\qquad\qquad\qquad$ $\ldots$

$(X = r \text{ and } Y = 0)$ with probability $a_r b_0.$

Hence the distribution of Z is given by

$$c_r = \Pr\{Z = r\} = a_0 b_r + a_1 b_{r-1} + \cdots + a_r b_0 = \sum_{t=0}^{r} a_t b_{r-t}. \qquad (1.9)$$

This new sequence $\{c_r\}$ that results from a combination of the two sequences $\{a_k\}$, $\{b_j\}$ in this particular way indicated by (1.9) is called the *convolution* of $\{a_k\}$ and $\{b_j\}$ and is denoted by

$$\{c_r\} = \{a_r\} * \{b_r\}.$$

Let $A(s), B(s), C(s)$ be the p.g.f. of X, Y and Z respectively.
Then from (1.9) it follows that

$$C(s) = A(s) \cdot B(s) \qquad (1.9a)$$

This also follows from the fact that

$$C(s) = E(s^{X+Y}) = E(s^X)E(s^Y),$$

$$= A(s) \cdot B(s),$$

because of independence of X, Y.

We have thus the following theorem:

Theorem 1.1. *The p.g.f. of the sum of two independent random variables X and Y is the product of the p.g.f. of X and that of Y.*

The result will also hold good in the case of the sum S_n of n non-negative independent integral-valued random variables $X_1, \ldots, X_n$, i.e. the p.g.f. of S_n is the product of the p.g.f.'s of $X_1, X_2, \ldots, X_n$.

In particular, we have the following:

Theorem 1.2. The sum $S_n = X_1 + \cdots + X_n$ of a *fixed* number n of identically and independently distributed (i.i.d.) random variables X_i has the p.g.f. $\{P(s)\}^n$, each X_i having p.g.f. $P(s)$.

The results stated above are of considerable importance. These can be employed to find the distribution of the sum of a number of independent random variables.

It can be easily verified that if X_i are i.i.d. Bernoulli r.v.'s, then S_n has Binomial distribution.

Example 1(g). *Sum of independent Poisson variates*: Let X_1, X_2 be two independent Poisson variates with parameters (means) λ_1, λ_2 respectively. To find the distribution of the sum $Z = X_1 + X_2$.

The p.g.f. of X_i is exp $\{\lambda_i(s - 1)\}$, $i = 1, 2$.

Using (1.9a) the p.g.f. of Z is obtained as

$$[\exp\{\lambda_1(s - 1)\}] [\exp\{\lambda_2(s - 1)\}]$$

$$= \exp\{(\lambda_1 + \lambda_2)(s - 1)\}.$$

But this is the p.g.f. of a Poisson variate with parameter $\lambda_1 + \lambda_2$. Hence the sum of two independent Poisson variates is a Poisson variate. More generally, if X_i, $i = 1, 2$, $\ldots, n$ are n independent Poisson variates with parameters λ_i, then the sum $S = X_1 + X_2 + \cdots + X_n$ is a Poisson variate with parameter $\sum_{i=1}^{n} \lambda_i$.

Example 1 (h). Let X_1, X_2 be two independent random variables each with the same geometric distribution, $\Pr(X_i = k) = q^k p$, $k = 0, 1, 2, \ldots$.

The p.g.f. of X_i is $p / (1 - qs)$ for $i = 1, 2,$. Thus, the sum $Z = X_1 + X_2$ has the p.g.f.

$$P(s) = \left(\frac{p}{1-qs}\right)^2 = p^2 \sum_{r=0}^{\infty} \binom{r+1}{r} q^r s^r.$$

Hence

$$\Pr\{Z = r\} \equiv \text{coefficient of } s^r \text{ in } P(s)$$

$$= \binom{r+1}{r} q^r p^2 = (r+1)q^r p^2, \quad r = 0, 1, 2, \ldots$$

This gives the distribution of the sum Z. If X_1 denotes the number of failures preceding the first success and X_2 the number of failures following the first success and preceding the second success, then $Z = X_1 + X_2$ denotes the number of failures preceding the second success.

The sum $S_r = X_1 + \cdots + X_r$ of r i.i.d. geometric variates denotes the number of failures preceding the rth success. It has the p.g.f. $P(s) = \left(\frac{p}{1-sq}\right)^r$. It follows that

$$\Pr\{S_r = k\} \equiv \text{coefficient of } s^k \text{ in } P(s)$$

$$= \binom{r+k-1}{k} p^r q^k, \quad k = 0, 1, 2, \ldots \qquad (1.10)$$

and

$$\sum_k \binom{r+k-1}{k} p^r q^k = 1. \qquad (1.10a)$$

Here r is a positive integer. Now the quantity given by the r.h.s. of (1.10) is non-negative and (1.10a) holds for any positive r. The distribution (1.10) for $r > 0$ is known as *negative binomial distribution* with *index* r, and mean rq/p. In particular, when r is a positive integer, it is known as *Pascal distribution*.
(See Exercise 1.9 for distribution of the sum of two independent geometric variables).

Example 1(i). If X, Y are independent r.v.'s with generating functions $A(s)$, $B(s)$ respectively, then the generating function $D(s)$ of $W = X - Y$ is given by

$$D(s) \equiv E\{s^{X-Y}\} = E\{s^X(s^{-1})^Y\} = E\{s^X\} E\{(1/s)^Y\},$$

because of independence of X, Y; thus $\Pr\{X - Y = k\}$ is the coefficient of s^k, $k = 0$, $\pm 1, \pm 2, \ldots$ in the expansion of

$$D(s) = A(s)B(1/s).$$

1.1.4 Sum of a Random Number of Discrete Random Variables

In Section **1.1.3** we considered the sum $S_n = X_1 + \cdots + X_n$ of a fixed number of mutually independent random variables. Sometimes we come across situations when we have to consider the sum of *a random number N* of random variables. For example, if X_i denotes the number of persons involved in the *i*th accident (in a day in a certain city) and if the number N of accidents happening on a day is a random variable, then the sum $S_N = X_1 + \cdots + X_N$ denotes the total number of persons involved in accidents on a day.

We have the following theorem for the sum S_N.

Theorem 1.3. Let $X_i, i = 1, 2, \ldots$, be identically and independently distributed random variables with Pr $\{X_i = k\} = p_k$ and p.g.f.

$$P(s) = \sum_k p_k s^k \text{ for } i = 1, 2, \ldots \qquad (1.11)$$

Let

$$S_N = X_1 + \ldots + X_N, \qquad (1.12)$$

where N is a random variable independent of the X_i's. Let the distribution of N be given by Pr $\{N = n\} = g_n$ and the p.g.f. of N be

$$G(s) = \sum g_n s^n. \qquad (1.13)$$

Then the p.g.f. $H(s)$ of S_N is given by the compound function $G(P(s))$, i.e.

$$H(s) = \sum_j \Pr\{S_N = j\} s^j = G(P(s)). \qquad (1.14)$$

Proof: Since N is a random variable that can assume values $0, 1, 2, 3, \ldots$, the event $S_N = j$ can happen in the following mutually exclusive ways: $N = n$ and $S_n = X_1 + \ldots + X_n = j$ for $n = 1, 2, 3, \ldots$ To meet the situation $N = 0$ let us define $X_0 = S_0 = 0$ so that

$$X_0 + X_1 + \ldots + X_N = S_N, \quad n \geq 0.$$

We have $h_j = \Pr\{S_N = j\} = \sum_{n=0}^{\infty} \Pr\{N = n \text{ and } S_n = j\}.$

Since N is independent of X_i's and therefore of S_n,

$$h_j = \sum_{n=0}^{\infty} \Pr\{N = n\} \Pr\{S_n = j\} \qquad (1.15)$$

$$= \sum_{n=0}^{\infty} g_n \Pr\{S_n = j\}. \tag{1.16}$$

(The inclusion of the value 0 for n can be justified since Pr $\{S_0 = 0\} = 1$ and Pr $\{S_0 > 0\} = 0$.)

The p.g.f. of the sum $S_N = X_1 + \cdots + X_N$ is thus given by

$$H(s) = \sum_j h_j s^j = \sum_j \Pr\{S_N = j\} s^j \tag{1.17}$$

$$= \sum_{j=0}^{\infty} \left[\left\{ \sum_{n=0}^{\infty} g_n \Pr\{S_n = j\} \right\} \right] s^j \tag{1.18}$$

$$= \sum_n \left[\sum_j \Pr\{S_n = j\} s^j \right] g_n$$

$$= \sum_n g_n [P(s)]^n = G(P(s)),$$

since the expression $\{P(s)\}^n$ is the p.g.f. of the sum S_n of a fixed number n of i.i.d. random variables (Theorem 1.2).   ▲

As $P(s)$ and $G(s)$ are p.g.f.'s, so also is $G(P(s))$; $G(P(s)) = 1$ for $s = 1$.

Since $G(P(s))$ is a compound function, the corresponding distribution is known as *compound distribution*. It is also known as *random sum distribution*.

Mean and Variance of S_N

We have

$$E(S_N) = H'(s)\,|_{s=1} = [P'(s)G'(P(s))]_{s=1}$$

$$= P'(1)G'(P(1)) = P'(1)G'(1)$$

so that

$$E\{S_N\} = E\{X_i\} E\{N\}. \tag{1.19}$$

Again,

$$H''(s) = P''(s)G'(P(s)) + G''(P(s))[P'(s)]^2$$

Thus

$$E(S_N^2) - E(S_N) = H''(s)\,|_{s=1}$$

$$= [E(X_i^2) - E(X_i)] E(N)$$

$$+ [E(N^2) - E(N)] [E(X_i)]^2$$

Using (1.19) we get

$$E(S_N^2) = E(X_i^2)E(N) - E(N)[E(X_i)]^2$$

$$+ E(N^2)[E(X_i)]^2$$

$$= E(N)\,\text{var}\,(X_i) + E(N^2)[E(X_i)]^2$$

Finally,

$$\text{var}\,(S_N) = E(S_N^2) - [E(S_N)]^2$$

$$= E(N)\text{var}\,(X_i) + [E(X_i)]^2\text{var}\,(N). \qquad (1.20)$$

Note: The result (1.20) can also be obtained by using the relation

$$\text{var}\,(X) = E\,[\text{var}\,(X\,|\,Y)] + \text{var}\,(E(X\,|\,Y)) \qquad (a)$$

and taking $X = S_N$ and $Y = N$. (See Sec. 1.1.5 for conditional expectation).

We have, $E(S_N\,|\,N = n) = nE(X_i)$ with probability $\Pr\,(N = n) = g_n$

so that

$$E(S_N\,|\,N) = \sum_n nE(X_i)g_n$$

$$= E(X_i)E(N)$$

and

$$[E(S_N\,|\,N)]^2 = \sum_n [nE(X_i)]^2 g_n$$

$$= [E(X_i)]^2 E(N^2).$$

Thus

$$\text{var}\,(S_N\,|\,N) = [E(X_i)]^2\,\text{var}\,(N). \qquad (b)$$

Again

$$\text{var}\,(S_N\,|\,N = n) = n\,\text{var}\,(X_i) \text{ with probability } g_n$$

so that

$$E[\text{var}(S_N \mid N)] = \sum_n n \text{var}(X_i)g_n = [\text{var}(X_i)] \, E(N). \tag{c}$$

Using (b) and (c) in (a) we get (1.20).

Example 1(j). *Compound Poisson distribution*: If N has a Poisson distribution with mean λ, then $G(s) = \sum g_n s^n = \exp\{\lambda(s-1)\}$ and hence the sum $X_1 + \ldots + X_N$ has p.g.f.

$$H(s) = G(P(s)) = \sum g_n [P(s)]^n = \exp\{\lambda[P(s)-1]\}.$$

The distribution having a generating function of the form $\exp\{\lambda[P(s)-1]\}$, where $P(s)$ is itself a generating function, is called *compound Poisson* distribution.
   The mean is given by

$$E\{S_N\} = E(X_i)E\{N\} = \lambda E(X_i).$$

Here $g_n = e^{-\lambda}\lambda^n / n!$. Similarly, taking $g_n = (1-p)p^n$, one gets *compound geometric* distribution, with $H(s) = pP(s)/[1-qP(s)]$.

1.1.5 Generating Function of Bivariate Distribution

Bivariate Distribution: Suppose that X, Y is a pair of integral-valued random variables with joint probability distribution given by

$$\Pr\{X = j, Y = k\} = p_{jk}, \quad j,k = 0,1,2,\ldots, \quad \sum_{j,k} p_{jk} = 1. \tag{1.21}$$

The marginal distributions are given by

$$\Pr\{X = j\} = \sum_{k=0}^{\infty} p_{jk} = f_j, \qquad j = 0,1,2,\ldots \tag{1.22}$$

$$\Pr\{Y = k\} = \sum_{j=0}^{\infty} p_{jk} = g_k, \qquad k = 0,1,2,\ldots \tag{1.23}$$

If (j, k) is any point at which $f_j \neq 0$ (i.e. $f_j > 0$), then the conditional distribution of Y given $X = j$ is given by

$$\Pr\{Y = k \mid X = j\} = \frac{\Pr\{Y = k, X = j\}}{\Pr\{X = j\}} = \frac{p_{jk}}{f_j}, \quad k = 0,1,2,\ldots \tag{1.24}$$

Conditional Expectation

The conditional expectation $E\{Y \mid X = j\}$ is given by

$$E\{Y \mid X = j\} = \sum_k k \Pr\{Y = k \mid X = j\}$$

$$= \sum_k k \frac{p_{jk}}{f_j} = \frac{\sum_k k p_{jk}}{f_j}, \quad j = 0, 1, 2, \ldots \quad (1.25)$$

For X given but unspecified, we note that $E\{Y \mid X\}$ is a random variable that assumes the value $[\sum_k k p_{jk}/f_j]$ with $\Pr\{X = j\} = f_j > 0$.

Hence the expectation of the random variable $E\{Y \mid X\}$ is given by

$$E[E\{Y \mid X\}] = \sum_j E\{Y \mid X = j\} \Pr(X = j)$$

$$= \sum_j \left\{ \frac{\sum_k k p_{jk}}{f_j} \right\} f_j$$

$$= \sum_j \sum_k k p_{jk}$$

$$= \sum_k k \left\{ \sum_j p_{jk} \right\} = \sum_k k \Pr\{Y = k\} = E(Y). \quad (1.26)$$

In the same way, we can prove that

$$E[E\{Y^2 \mid X\}] = E[Y^2]; \quad (1.27)$$

more generally, for any function $\phi(Y)$ whose expectation exists,

$$E[E\{\phi(Y) \mid X\}] = E[\phi(Y)]. \quad (1.28)$$

Note (1). The results (1.26–1.28) which are given here for discrete random variables X, Y will also hold, *mutatis mutandis*, for continuous random variables.
Note (2). The result (1.26) holds for all r.v.'s provide $E(Y)$ exists.
Enis (*Biometrika* (1973) 432) cites an example where $E[E(Y \mid X)] = 0$ but $E(Y)$ does not exist and consequently (1.26) does not hold.

Bivariate Probability Generating Function

Definition. The probability generating function (bivariate p.g.f.) of a pair of random variables X, Y with joint distribution given by (1.21) is defined by

$$P(s_1, s_2) = \sum_{j,k} p_{jk} s_1^j s_2^k, \quad (1.29)$$

s_1, s_2 being dummy positive real variables chosen so as to make the double series convergent. We have the following results.

Theorem 1.4. (a) The p.g.f. A (s) of the marginal distribution
of X is given by A $(s) = P$ $(s, 1)$.
(b) The p.g.f. B (s) of Y is given by
B $(s) = P$ $(1, s)$.
(c) The p.g.f. of $(X + Y)$ is given by P (s, s).

Proof: Since the convergent double series (1.29) consists of positive real terms, the change of order of summation is justified.

(a) We have from (1.29)

$$P(s,1) = \sum_{j,k} p_{jk}s^j = \sum_{j=0}^{\infty}\sum_{k=0}^{\infty} p_{jk}s^j$$

$$= \sum_{j=0}^{\infty}\left\{\sum_{k=0}^{\infty} p_{jk}\right\}s^j = \sum_{j=0}^{\infty} \Pr\{X = j\}s^j, \text{ from (1.22)}$$

$$= A \ (s) \ \text{(the p.g.f of } X).$$

(b) It can be proved in the same way.

(c) We have

$$P(s,s) = \sum_{j,k=0}^{\infty} p_{jk}s^{j+k} = \sum_{m=0}^{\infty}\sum_{r=0}^{m} p_{r\,m-r}s^m.$$

as can be easily verified.

Now

$$\Pr\{X+Y = m\} = \sum_{r=0}^{m} \Pr\{X = r, Y = m-r\} = \sum_{r=0}^{m} p_{r\,m-r}.$$

Hence

$$P(s,s) = \sum_{m=0}^{\infty} [\Pr\{X+Y = m\}]s^m,$$

which shows that P (s, s) is the p.g.f. of $X + Y$.   ▲

Remarks: (1) If X, Y are independent,

$$p_{jk} = \Pr\{X = j, Y = k\} = \Pr\{X = j\} \Pr\{Y = k\}$$

and then

$$P(s_1, s_2) = A(s_1)B(s_2) = P(s_1, 1)P(1, s_2)$$

and conversely, from this relation follows the independence of X and Y.

(2) The probabilities p_{jk} can be (uniquely) determined from $P\,(s_1, s_2)$ as follows:

$$p_{jk} = \frac{1}{j!k!}\left[\frac{\partial^{j+k}P(s_1,s_2)}{\partial s_1^j \partial s_2^k}\right]_{s_1 = s_2 = 0}$$

Example 1(k). Consider a series of Bernoulli trials with probability of success p. Suppose that X denotes the number of failures preceding the first success and Y the number of failures following the first success and preceding the second success. The sum $(X + Y)$ gives the number of failures preceding the second success.

The joint distribution of X, Y is given by

$$p_{jk} = \Pr\{X = j, Y = k\} = q^{j+k}p^2, \quad j,k = 0,1,2,\ldots;$$

and the bivariate generating function is given by

$$P(s_1,s_2) = \sum_{j,k=0}^{\infty} p_{jk}s_1^i s_2^k = \sum_{j,k} q^{j+k}p^2 s_2^j s_1^k$$

$$= p^2\left\{\sum_{j=0}^{\infty}(qs_1)^j\right\}\left\{\sum_{k=0}^{\infty}(qs_2)^k\right\}$$

$$= \frac{p^2}{(1-s_1 q)(1-s_2 q)}.$$

The p.g.f. of X is given by

$$A(s) = P(s,1) = \frac{p^2}{(1-sq)(1-q)} = \frac{p}{1-qs}.$$

The p.g.f. of $X + Y$ is given by

$$P(s,s) = \left(\frac{p}{1-sq}\right)^2.$$

(See Example 1(h)).

1.2 LAPLACE TRANSFORMS

1.2.1. Introduction

Laplace transform is a generalization of generating function. Laplace transforms serve as very powerful tools in many situations. They provide an effective means for the solution of many problems arising in our study. For example, the transforms are very effective for solving linear differential equations. The Laplace transformation reduces a linear differential equation to an algebraic equation. In the study of some probability distributions, the method could be used with great advantage, for it happens quite often that it is easier to find the Laplace transform of a probability distribution rather than the distribution itself.

Definition. Let $f(t)$ be a function of a positive real variable t. Then the Laplace transform (L.T.) of $f(t)$ is defined by

$$\bar{f}(s) = \int_0^\infty \exp(-st)f(t)dt \tag{2.1}$$

for the range of values of s for which the integral exists.

We shall write $\bar{f}(s) = L\{f(t)\}$ to denote the Laplace transform of $f(t)$.

Example 2(a).

(i)   If $f(t) = c$ (const.), $\bar{f}(s) = \int_0^\infty \exp(-st)cdt = c/s$ $(s > 0)$

If $f(t) = t$, $\bar{f}(s) = \int_0^\infty t\exp(-st)dt = 1/s^2$ $(s > 0)$

If $f(t) = t^n$, $\bar{f}(s) = \int_0^\infty t^n \exp(-st)dt = \Gamma(n+1)/s^{n+1}$ $(s > 0)$

Though t^n $(0 > n > -1)$ is infinite at $t = 0$, the result holds for $n > -1$. In particular, if $f(t) = t^{-1/2}$, then, $\bar{f}(s) = \Gamma\left(\frac{1}{2}\right) / \sqrt{s} = \sqrt{\dfrac{\pi}{s}}$.

(ii)   Let $f(t) = e^{at}$, then $\bar{f}(s) = \int_0^\infty \exp(-st)\exp(at)dt$

$$= 1/(s-a)   (s > a)$$

(iii)   Let $f(t) = \sin t$, then $\bar{f}(s) = \int_0^\infty \exp(-st)\sin t\,dt$

$$= 1/(s^2 + 1)   (s > 0)$$

(iv)   Let $f(t) = e^{-t}t^n$, then $\bar{f}(s) = \int_0^\infty \exp(-st)\exp(-t)t^n dt$

$$= \int\limits_{0}^{\infty} \exp\{-t(s+1)\}\, t^n dt$$

$$= \frac{\Gamma(n+1)}{(s+1)^{n+1}} \quad (n > -1).$$

Example 2(b). Dirac-delta (or Impulse) function located at a is defined as

$$\delta(t-a) = 1, \quad t = a$$

$$= 0, \quad t \neq a$$

The L.T. of $\delta (t - a)$ equals

$$\int\limits_{0}^{\infty} e^{-st}\delta(t-a)dt = e^{-sa}.$$

Unit step function (at a) is defined as

$$u_a(t) = \begin{Bmatrix} 0, t < a \\ 1, t > a \end{Bmatrix}$$

The L.T. of $u_a(t)$ is e^{-as}/s.

1.2.2 Some Important Properties of Laplace Transforms : see Appendix A1

1.2.3 Inverse Laplace Transform

Definition. If $\bar{f}(s)$ is the L.T. of $f(t)$, i.e. $L\{f(t)\} = \bar{f}(s)$, then $f(t)$ is called the inverse Laplace transform $\bar{f}(s)$. For example, the inverse Laplace transform of $\bar{f}(s) = 1/s^2$ is $f(t) = t$. There is an inversion formula which gives $f(t)$ in terms of $\bar{f}(s)$; if $\bar{f}(s)$ exists, then $f(t)$ can be *uniquely* determined subject to certain conditions satisfied by $f(t)$. In particular, if two continuous functions have the same transform, they are identical.

Extensive tables of Laplace transforms are also available and reference may be made to them, whenever necessary, to obtain either the L.T. of a function $f(t)$ or the inverse L.T. of a function $\bar{f}(s)$. Techniques for numerical inversion of Laplace transforms are discussed in Bellman et al. (1966).

1.3   LAPLACE (STIELTJES) TRANSFORM OF A PROBABILITY DISTRIBUTION OR OF A RANDOM VARIABLE

1.3.1 Definition

Let X be a non-negative random variable with distribution function

$$F(x) = \Pr\{X \leq x\}.$$

The Laplace (Laplace-Stieltjes) transform F^* (s) of this distribution is defined, for $s \geq 0$, by

$$F^*(s) = \int_0^\infty \exp(-sx)dF(x) \tag{3.1}$$

We shall say that (3.1) also gives the ''Laplace (Laplace-Stieltjes) transform of the random variable X''.
We have

$$F^*(s) = E\{\exp(-sX)\} \tag{3.1a}$$

and

$$F^*(0) = 1. \tag{3.2}$$

Suppose that X is a continuous variate having density $f(x) = F'(x)$. Then form (3.1),

$$F^*(s) = \int_0^\infty \exp(-sx)f(x)dx \tag{3.3}$$

(this is the ''ordinary'' Laplace transform $L\{f(x)\} \equiv \bar{f}(s)$ of the density function $f(x)$).

By a Laplace transform of a r.v. X, we shall mean the L.S.T. of the distribution function $F(\cdot)$ of X; this is equal to the ordinary L.T. of the density function of X when this exists. We have

$$F^*(s) = \bar{f}(s).$$

In case X is an integral-valued random variable with distribution $p_k = \Pr\{X = k\}$, $k = 0, 1, 2, \ldots$ and p.g.f. $P(s) = \sum_k p_k s^k$, we can stretch the language and define the L.T. of X by

$$F^*(s) = E\{\exp(-sX)\} = P\{\exp(-s)\}. \tag{3.4}$$

Thus in case of a discrete random variable assuming non-negative values 0, 1, 2, 3, . . . , the L.T. of the variable differs from its p.g.f. only by a change of variable: $\exp(-s)$ in the former replaces s in the latter. Thus there is a close analogy between the properties of Laplace transforms and those of generating functions or s-transforms.

Note. When it exists, the function

$$M(t) = E\{e^{tX}\} = F^*(-t)$$

is called the *moment generating function* of the r.v. X (it is the generating function of the sequence $\mu_n / n!$, where $\mu_n = E\{X^n\}$ is the nth central moment of X).

The function

$$\phi(t) = E\{e^{itX}\}$$

defined for every r.v. X is called the *characteristic function* of X.

When X is non-negative and integral valued, then its p.g.f. is considered; so is its L.T. when X is non-negative and continuous as these are easier to handle. The characteristic function, defined for all r.v.'s, is a more universal tool.

1.3.2 The Laplace Transform of the Distribution Function in Terms of that of the Density Function

Let X be a continuous (and non-negative) r.v. having density function $f(x)$ and distribution function

$$\Pr\{X \leq x\} = F(x) = \int_0^x f(t)dt$$

The (ordinary) Laplace transform of the distribution function $F(x)$ is

$$L\{F(x)\} = \int_0^\infty \exp(-sx)F(x)dx$$

We have from **A.6** (Appendix)

$$\overline{F}(s) = L\{F(x)\} = \int_0^\infty \exp(-sx)\left\{\int_0^x f(t)dt\right\}dx$$

$$= L\left\{\int_0^x f(t)dt\right\} = \overline{f}(s)/s.$$

The relation can also be obtained by integrating by parts the relation (3.1). Thus we get

$$F^*(s) = \overline{f}(s) = s\overline{F}(s). \tag{3.5}$$

1.3.3 Mean and Variance in Terms of (Derivatives of) L.T.

We note here that differentiation under the integral sign is valid for the L.T. given by (3.1), since the integrand is bounded and continuous.

Differentiating (3.1) with respect to s, we get

$$\frac{d}{ds}F^*(s) = -\int_0^\infty x\exp(-sx)dF(x)$$

$$\frac{d^2}{ds^2}F^*(s) = (-1)^2\int_0^\infty x^2\exp(-sx)dF(x)$$

and, in general, for $n = 1, 2, \ldots$

$$\frac{d^n}{ds^n}F^*(s) = (-1)^n\int_0^\infty x^n\exp(-sx)dF(x). \tag{3.6}$$

The differentiation under the integral is valid since the new integrands are continuous and bounded.

We can use the above relation to find $E(X^n)$, the nth moment of X when it exists; we have

$$(-1)^n\left[\frac{d^n}{ds^n}F^*(s)\right]_{s=0} = \int_0^\infty \{\exp(-sx)\}x^n dF(x) \text{ for } s = 0$$

$$= \int_0^\infty x^n dF(x) = E(X^n), \quad n = 1, 2, \ldots$$

when the l.h.s. exists, i.e. the r.v. X possesses a finite nth moment *iff* $\lim \frac{d^n}{ds^n}F^*(s)]_{s=0}$ exists.

We have

$$E(X) = -\left[\frac{d}{ds}F^*(s)\right]_{s=0} \tag{3.7}$$

$$E(X^2) = \left[\frac{d^2}{ds^2}F^*(s)\right]_{s=0} \tag{3.8}$$

and
$$\text{var}(X) = \left[\frac{d^2}{ds^2}F^*(s)\right]_{s=0} - \left[\left\{-\frac{d}{ds}F^*(s)\right\}_{s=0}\right]^2. \tag{3.9}$$

1.3.4 Some Important Distributions

1.3.4.1 A special kind of discrete distribution:

Suppose that X is a random variable whose whole mass is concentrated in one single point, say, point a. This implies that the variable is 'almost always' equal to a, i.e. $\Pr(X = a) = 1$ and $\Pr(X \neq a) = 0$. Its distribution function is given by

$$F(x) = \Pr\{X \le x\} = 0 \quad \text{for } x < a$$

$$= 1 \quad \text{for } x \ge a.$$

We have $E(X) = a$ and $\text{var}(X) = E(X^2) - (E(X))^2 = a^2 - a^2 = 0$.
The L.T. of the distribution is given by

$$F*(s) = E\{\exp(-sX)\} = \exp(-sa)\Pr(X = a) = \exp(-sa). \qquad (3.10)$$

This is the L.T. of Dirac-delta function located at a.
The distribution is known as *degenerate distribution*.

1.3.4.2 Poisson distribution:

Let X have Poisson distribution with mean λ. Its p.g.f. is $P(s) = \exp\{\lambda(s-1)\}$.
From (3.4), its L.T. is given by

$$F*(s) = P\{\exp(-s)\} = \exp[\lambda\{\exp(-s) - 1\}]. \qquad (3.11)$$

1.3.4.3 Negative exponential distribution:

Let X have the negative exponential distribution with parameter λ. Its density function is

$$f(x) = \lambda\exp(-\lambda x), \quad \lambda > 0, \quad 0 \le x < \infty, \qquad (3.12)$$

$$= 0, \qquad\qquad\qquad x < 0.$$

The distribution function $F(x) = \Pr[X \le x] = 1 - e^{-\lambda x}$ and $\Pr[X \ge x] = e^{-\lambda x}$.

The L.T. $F*(s)$ of the r.v. X is

$$F*(s) = \int_0^\infty \{\exp(-sx)\}\{\lambda\exp(-\lambda x)\}\, dx = \frac{\lambda}{(s+\lambda)}.$$

We have

$$\frac{d}{ds}F*(s) = -\frac{\lambda}{(s+\lambda)^2}, \qquad \frac{d^2}{ds^2}F*(s) = \frac{2\lambda}{(s+\lambda)^3}.$$

Hence from (3.7) and (3.8)

$$E(X) = -\left[-\frac{\lambda}{\lambda^2}\right] = \frac{1}{\lambda}, \qquad E(X^2) = \frac{2\lambda}{\lambda^3} = \frac{2}{\lambda^2} \qquad (3.13)$$

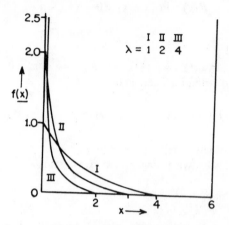

Fig. 1.1 The negative exponential density function for some values of λ

and
$$\text{var}(X) = \frac{2}{\lambda^2} - \frac{1}{\lambda^2} = \frac{1}{\lambda^2}.$$
(3.14)

Further, $E(X^r) = \dfrac{r!}{\lambda^r}$, $r = 1, 2, \ldots$

The distribution is known also simply as *exponential distribution*.

Properties of exponential distribution

(a) *Non-ageing (or memoryless or Markov) property*

For a r.v. X having exponential distribution with parameter λ, we have

$$\Pr\{X \geq x + y \mid X \geq x\} = \frac{e^{-\lambda(x+y)}}{e^{-\lambda x}}$$

$$= e^{-\lambda y} = \Pr\{X \geq y\}.$$

This implies that if the duration X of a certain activity has exponential distribution with parameter λ and if the activity is observed after a length of time x after its commencement, then the *remaining duration* of the activity is independent of x and is also distributed as an exponential r.v. with the same parameter λ.

Again, if X is a non-negative continuous r.v. with non-ageing property, then it can be shown that X must be exponential.

This non-ageing or memoryless property characterizes exponential distribution among all distributions of *continuous non-negative r.v.'s*.

(b) *Minimum of two independent exponential distributions*:

Suppose that X_1, X_2 have independent exponential distributions with parameters λ_1, λ_2 respectively. Let

$$Z = \min(X_1, X_2).$$

We have

$$\Pr\{Z \geq x\} = \Pr\{X_1 \geq x\}\Pr\{X_2 \geq x\}$$

$$= e^{-\lambda_1 x} e^{-\lambda_2 x}$$

$$= e^{-(\lambda_1 + \lambda_2)x}$$

so that Z is exponential with parameter $(\lambda_1 + \lambda_2)$.

This implies that if the durations X_1 and X_2 of two activities A_1 and A_2 have independent exponential distributions with parameters λ_1, λ_2 respectively and these activities are observed when neither has been completed, then the duration of the interval Z upto the first completion of one of the activities has also exponential distribution with parameter $\lambda_1 + \lambda_2$.

The probability that the activity A_1 will be completed earlier than the activity A_2 is given by

$$\Pr\{X_1 < X_2\} = \int_0^\infty \Pr\{x \leq X_1 < x + dx\}\Pr\{X_2 > x\}$$

$$= \int_0^\infty \lambda_1 e^{-\lambda_1 x} dx\, e^{-\lambda_2 x}$$

$$= \frac{\lambda_1}{\lambda_1 + \lambda_2}.$$

Similarly,

$$\Pr\{X_2 < X_1\} = \frac{\lambda_2}{\lambda_1 + \lambda_2}.$$

These results can be generalised to a number of variables. For example, if there are n independent exponential variables X_i, with parameter λ_i, $i = 1, 2, \ldots, n$ then

$Z = \min\{X_1, X_2, \ldots, X_n\}$ is exponential with parameter $\lambda = \sum_{i=1}^n \lambda_i$.

Further, of the n activities $A_1, \ldots, A_n$ whose durations are independent exponential r.v.'s $X_1, \ldots, X_n$, the probability that the activity A_1 will be completed earlier than any of the other activities equals

$$\frac{\lambda_1}{\sum \lambda_i}.$$

(c) *Characterization of exponential distribution*:

Several characterizations have been obtained. We mention the following.
(i) X has exponential distribution *iff*

$$E(X \mid X > y) = y + E(X) \text{ for all } y$$

or

$$E(X - y \mid X > y) = E(X) \quad \text{for all } y.$$

(ii) Two independent continuous variables X_1, X_2 are exponential *iff*

$$Z = \min(X_1, X_2) \quad \text{and} \quad W = X_1 - X_2$$

are independent.

1.3.4.4 Uniform (rectangular) distribution:

Let X have uniform distribution in $(0, 1)$. The density function of X is

$$f(x) = 1, \qquad 0 \le x \le 1$$

$$= 0, \qquad \text{otherwise.}$$

The L.T. $F^*(s)$ of the r.v. X is given by

$$F^*(s) = \int_0^\infty \exp(-sx) f(x) dx = \int_0^1 \exp(-sx) dx = \{1 - \exp(-s)\}/s.$$

The mean of X is $\frac{1}{2}$ and the variance is $1/12$.

1.3.4.5 Gamma distribution:

Let X have a two parameter gamma distribution with parameters λ, k ($\lambda > 0$ is the *scale* parameter and $k > 0$ is the *shape* parameter).
The density function of the r.v. X is

$$f_{\lambda,k}(x) = \frac{\lambda^k x^{k-1} \exp(-\lambda x)}{\Gamma(k)}, \quad x > 0$$

$$= 0, \quad x < 0. \tag{3.15}$$

The L.T. $F^*(s)$ is given by

$$F^*(s) = \int_0^\infty \exp(-sx) f_{\lambda,k}(x) dx$$

$$= (\lambda^k/\Gamma(k)) \int_0^\infty [\exp\{-(s+\lambda)x\}]x^{k-1}dx$$

$$= (\lambda^k/\Gamma(k)) \int_0^\infty \frac{\exp(-t)t^{k-1}}{(s+\lambda)^k}dt, \qquad (\text{putting } (s+\lambda)x = t)$$

$$= \left(\frac{\lambda}{s+\lambda}\right)^k. \tag{3.16}$$

We have

$$\frac{d}{ds}F^*(s) = -\frac{k\lambda^k}{(s+\lambda)^{k+1}},$$

$$\frac{d^2}{ds^2}F^*(s) = \frac{k(k+1)\lambda^k}{(s+\lambda)^{k+2}}.$$

Hence from (3.7) and (3.8)

$$E(X) = -\left[\frac{-k\lambda^k}{\lambda^{k+1}}\right] = \frac{k}{\lambda}, \tag{3.17}$$

$$E(X^2) = \frac{k(k+1)\lambda^k}{\lambda^{k+2}} = \frac{k(k+1)}{\lambda^2},$$

and

$$\text{var}(X) = \frac{k(k+1)}{\lambda^2} - \left(\frac{k}{\lambda}\right)^2 = \frac{k}{\lambda^2}. \tag{3.18}$$

The coefficient of variation $= \dfrac{\text{s.d. }(X)}{E(X)} = 1/\sqrt{k}$: it is less (greater) than 1 according as k is greater (less) than 1.

Further, $E(X^r) = [k(k+1)\ldots(k+r-1)]/\lambda^r, r \geq 1$.

When $\lambda = 1$, we get single parameter gamma variate with density

$$\frac{x^{k-1}\exp(-x)}{\Gamma(k)}(x > 0);$$

and its Laplace transform is $1/(s+1)^k$.

When $k = 1$, the gamma distribution becomes the negative exponential distribution with density $\lambda \exp(-\lambda x)$ and having Laplace transform $\lambda/(s+\lambda)$. Another special case is $\lambda = 1/2$ and $k = n/2$, where n is a positive integer. The density then becomes (for $x > 0$)

$$\frac{x^{n/2-1}\exp(-x/2)}{2^{n/2}\Gamma(n/2)}$$

and the random variable is then said to have a *chi-square* (χ^2) distribution with n degrees of freedom. The L.T. of the density is $(1 + 2s)^{-n/2}$.

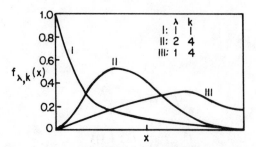

Fig. 1.2 The gamma density $f_{\lambda,k}(x)$ for some values of λ, k

The gamma distribution has considerable scope for practical application; many empirical distributions can be represented roughly by a suitable choice of the parameters λ, k. Burgin (*Opnl. Res. Qrly.* **26** (1975) 507–525) discusses its application in inventory control.

1.3.4.6 *Erlang distribution*:

Writing λk for λ, we have a gamma distribution with parameters $\lambda k, k$ having density function

$$f_{\lambda k,k}(x) = \frac{(\lambda k)^k x^{k-1}\exp(-k\lambda x)}{\Gamma(k)}, \quad x > 0 \tag{3.19}$$

When k is a positive integer, the variable having this for its density function is said to have *Erlang* (or *Erlang–k*) distribution and is denoted by E_k. In fact, when k is a positive integer, gamma distribution with density $f_{\lambda k,k}(x)$ is known as *Erlang–k* distribution. *Erlang–k* density is shown in Fig. 1.6.

The L.T. of *Erlang–k* distribution is

$$\left(\frac{\lambda k}{s + \lambda k}\right)^k; \tag{3.20}$$

We have $E(E_k) = k/\lambda k = 1/\lambda$; var $(E_k) = k/(\lambda k)^2 = 1/k\lambda^2$; $\tag{3.21}$

The coefficient of variation equals $1/\sqrt{k}$ which is < 1 for all k (> 1). Further

$$m_r = E [E_k^r], \quad r \geq 2$$

$$= \frac{(k + 1)(k + 2)...(k + r - 1)}{k^{r-1} \lambda^r}.$$

1.3.4.7 Hyper-exponential distribution:

A mixture of k (≥ 2) independent exponential distributions having pdf

$$f(x) = \sum_{i=1}^{k} a_i \lambda_i e^{-\lambda_i x}, \quad 0 \leq a_i \leq 1, \sum a_i = 1, \quad (x \geq 0)$$

is called a k-stage hyper-exponential distribution. We have, for its L.T.

$$F^*(s) = \sum_{i=1}^{k} a_i \left(\frac{\lambda_i}{s + \lambda_i} \right)$$

with
$$E(X) = \text{mean} = \sum_{i=1}^{k} \frac{a_i}{\lambda_i}$$

and
$$\sigma_X^2 = \text{variance} = 2 \sum_{i=1}^{k} \frac{a_i}{\lambda_i^2} - \left(\sum_{i=1}^{k} \frac{a_i}{\lambda_i} \right)^2.$$

The coefficient of variation $\sigma_X / E(X)$ is always greater than 1.

It corresponds to a mixed-exponential distribution with k-stages in parallel: the stage i is traversed with probability a_i, the time taken having exponential distribution with mean $1/\lambda_i$. It is also known as *mixed-exponential*.

A hyper-exponential distribution with k-stage is denoted by H_k.

1.3.4.8 Hypo-exponential distribution:

A sum of k (≥ 2) independent exponential distributions $Z = \sum_{i=1}^{k} X_i$ where X_i are exponential with parameters λ_i ($\lambda_i \neq \lambda_j$, $i \neq j$) is called a k-stage hypo-exponential distribution.

While k-stage hyper-exponential corresponds to k-stages in *parallel*, k-stage hypo-exponential corresponds to k-stages in *series*, the time taken to traverse stage i being exponential with mean $1/\lambda_i$. See Fig. 1.3.

For $k = 2$, the pdf is given by

$$f(x) = a_1 \lambda_1 e^{-\lambda_1 x} + a_2 \lambda_2 e^{-\lambda_2 x}$$

where
$$a_1 = \frac{\lambda_2}{\lambda_2 - \lambda_1}, a_2 = \frac{\lambda_1}{\lambda_1 - \lambda_2}.$$

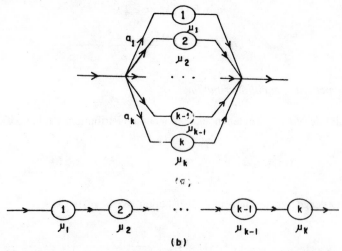

Fig. 1.3 Schematic represention of (a) hyper-exponential (b) hypo-exponential distributions, each of k stages

The pdf of k-stage hypo-exponential Z is given by

$$f(x) = \sum_{i=1}^{k} a_i \lambda_i e^{-\lambda_i x}$$

where

$$a_i = \prod_{\substack{j=1 \\ j \neq i}}^{k} \frac{\lambda_j}{\lambda_j - \lambda_i}, \quad i = 1, 2, ..., k.$$

The L.T. of the distribution is

$$F^*(s) = \sum_{i=1}^{k} a_i \left(\frac{\lambda_i}{s + \lambda_i} \right)$$

with

$$E(Z) = \sum_{i=1}^{k} \frac{1}{\lambda_i}$$

and

$$\sigma_Z^2 = \sum_{i=1}^{k} \frac{1}{\lambda_i^2}.$$

The coefficient of variation $\sigma_Z / E(Z)$ is always less than 1.

Note that, $\sum_{i=1}^{k} a_i = 1$, since $f(x)$ is pdf. Though the form of the pdf appears to resemble that of a hyper-exponential distribution, it is different; it is of a particular form in case of the hypo-exponential distribution.

Note: The prefix hyper(hypo) can be linked to the fact that the coefficient of variation [c.v.] is greater (less) than unity, the c.v. of exponential distribution.

1.3.4.9 Coxian Distribution:

Exponential, Erlang, hyper- and hypo-exponential distributions all have their L.T.'s as rational functions of s. The poles are on the negative real axis of the complex s-plane. A family of distributions is obtained by a generalisation of an idea (due to Erlang) by Cox (1955). The family of distributions has as L.T. a rational function of s with the degree of polynomial in the numerator less than or equal to that of the denominator. This family has been increasingly considered in the literature as a more general service time distribution (in queueing context) and repair time distribution (in reliability context). Extensive references to Coxian distributions appear also in computer science and telecommunication literature.

Consider a service facility, with m phases of service channels (nodes); the system is a subnetwork of m nodes. The service time distribution in the node (phase) i, $i = 1, 2, \ldots, m$ is exponential with rate μ_i, service time in any node being independent of the service time in other nodes. The job (or the customer) needing service can be at one of the m stages at a given time and no other job can be admitted for service until the job receiving service at one of the nodes (phases) has completed his service and departs from the system.

A job enters from the left and moves to the right. After receiving service at node i, the job may leave the system with probability b_i or move for further service to the next node $(i + 1)$ with probability a_i, $a_i + b_i = 1$, $i = 1, \ldots m$; we can include $i = 0$ such that $a_0 = 0$ indicates that the job does not require any service from any of the nodes and departs from the system without receiving any service from any node, whereas $a_0 = 1$ indicates that it needs service at least from the first node. After receiving service at the last node m, if it reaches that node, the job departs from the system, so that $b_m = 1$.

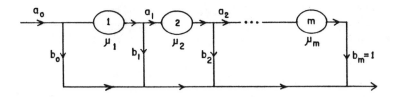

Fig. 1.4 Schematic representation of Coxian distribution with m phases

The distribution is denoted by K_m (or C_m as a tribute to Cox). The probability that a job receives (requires) services at nodes $1, 2, \ldots, k$ ($k \leq m$) and departs from the system after service completion at the facility equals $A_k b_k$, where $A_k = a_0 a_1 \ldots a_{k-1}$. We have

$$b_0 + \sum_{k=1}^{m} A_k b_k = 1.$$

Let γ be the total service time of a job. The L.T. of the r.v. γ is given by

$$F^*(s) = b_0 + \sum_{k=1}^{m} A_k b_k \left\{ \prod_{i=1}^{k} \frac{\mu_i}{s + \mu_i} \right\}$$

$$= b_0 + A_1 b_1 \left(\frac{\mu_1}{s + \mu_1} \right) + \sum_{k=2}^{m} \{a_1 a_2 \ldots a_{k-1}\} b_k \left[\prod_{i=1}^{k} \frac{\mu_i}{s + \mu_i} \right].$$

The last term can also be written as

$$\sum_{k=0}^{m} \frac{b_k \mu_k}{s + \mu_k} \left[\prod_{i=0}^{k-1} (1 - b_i) \frac{\mu_i}{s + \mu_i} \right].$$

In general, we have $b_0 = 1$, then $A_1 = 1$; the service is received at least at the first node. We have

$$F^*(0) = 1,$$

$$E(\gamma) = \sum_{k=1}^{m} A_k b_k \left\{ \sum_{i=1}^{k} \frac{1}{\mu_i} \right\},$$

and that the coefficient of variation $\sigma_\gamma / E(\gamma)$ is not less that $1 / \sqrt{m}$, that is, it ranges from $1/\sqrt{m}$ to ∞, depending on the values of the parameters involved.

Particular Case, $m = 2$ (Two-phase facility): K_2-distribution.

Suppose that a job needs service at node 1 and then leaves the system with probability q or (after receiving service at node 1) goes to node 2 with probability p ($= 1 - q$) for receiving further service before departing from the system. The service times at nodes 1 and 2 are independent exponential variables with parameters μ_1 and μ_2 respectively. The total time that the job requires is a random variable (X) which has K_2 distribution.

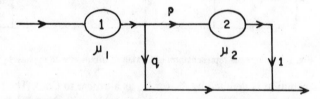

Fig. 1.5 Schematic representation of two-phase Coxian distrubution (K_2)

The configuration here is like a 2-stage parallel exponential system as described below.

(A): The job receives service with probability q and then departs from the system;

(B): or the job receives service with probability p at the both the facilities sequentially and then departs from the system.

The r.v. X can be written as

$$X = X_1, \text{ with probability } q$$

$$= X_1 + X_2, \text{ with probability } p$$

where X_1, X_2 are independent exponential variables with parameters μ_1 and μ_2 respectively. Thus we can at once write down the L.T. $F*(s)$ as

$$F*(s) = \frac{q\mu_1}{s + \mu_1} + (1 - q) \prod_{i=1}^{2} \alpha_i \left(\frac{\mu_i}{s + \mu_i} \right)$$

where

$$\alpha_i = \prod_{\substack{j=1 \\ j \neq i}}^{2} \frac{\mu_j}{\mu_j - \mu_i}.$$

On simplification, we get

$$F*(s) = \frac{q\mu_1}{s + \mu_1} + \frac{(1 - q)\mu_1\mu_2}{(s + \mu_1)(s + \mu_2)}$$

For $q = 1$, K_2 reduces to an exponential distribution and for $p = 1$, $\mu_1 = \mu_2$, K_2 reduces to Erlang-2 distribution. Hyper-exponential (H_2) is also a particular case of K_2-distribution as can be seen below.

Consider a H_2 distribution having L.T.

$$G*(s) = \frac{\theta v_1}{s + v_1} + \frac{(1 - \theta)v_2}{s + v_2}, \quad (v_1 > 0, v_2 > 0, \ 0 < \theta < 1)$$

and let

$$\mu_1 = \max(v_1, v_2), \qquad \mu_2 = \min(v_1, v_2)$$

and

$$q = \frac{\theta v_1 + (1 - \theta)v_2}{\mu_1};$$

then it can be seen that $G*(s) = F*(s)$ (for K_2 distribution). Thus a K_2 distribution becomes a H_2 distribution. A necessary condition for a K_2 distribution to be a H_2 distribution is that

$$(\mu_2/\mu_1) < q;$$

it follows since

$$\min(v_1, v_2) < \theta v_1 + (1 - \theta) v_2.$$

The first three moments of K_2 distribution can be obtained by taking derivatives of $F^*(s)$ or alternatively (and simply) by considering X in terms of X_1 and $X_1 + X_2$ and taking moments of X_1 and $X_1 + X_2$. Using

$$E(X_i) = 1/\mu_i, \quad E(X_i^2) = 2/\mu_i^2,$$

$$E(X_i^3) = 6/\mu_i^3, \quad i = 1, 2,$$

we get

$$m_1 = \frac{q}{\mu_1} + p\left(\frac{1}{\mu_1} + \frac{1}{\mu_2}\right) = \frac{1}{\mu_1} + \frac{p}{\mu_2}$$

$$m_2 = \frac{2q}{\mu_1^2} + 2p\left(\frac{1}{\mu_1^2} + \frac{1}{\mu_2^2} + \frac{1}{\mu_1 \mu_2}\right)$$

$$= \frac{2}{\mu_1^2} + \frac{2p(\mu_1 + \mu_2)}{\mu_1 \mu_2^2}$$

$$m_3 = \frac{6q}{\mu_1^3} + 6p\left(\frac{1}{\mu_1^3} + \frac{1}{\mu_2^3} + \frac{1}{\mu_1^2 \mu_2} + \frac{1}{\mu_1 \mu_2^2}\right)$$

$$= \frac{6}{\mu_1^3} + 6p\left(\frac{\mu_1^2 + \mu_1 \mu_2 + \mu_2^2}{\mu_1^2 \mu_2^3}\right).$$

From the above, the squared coefficient of variation

$$V_s^2 = \frac{\text{variance}(K_2)}{[E(K_2)]^2}$$

$$= \frac{\mu_2^2 + p(1+q)\mu_1^2}{(\mu_2 + p\mu_1)^2}$$

and

$$V_s^2 \geq \frac{1}{2} \Rightarrow (\mu_2 - p\mu_1)^2 \geq 4p\mu_1^2(p-1)$$

which is obvious, since the r.h.s. of the inequality is a negative quantity, p being less than 1. Thus the squared coefficient of variation of K_2 distribution lies between $\frac{1}{2}$ and ∞. Further, $V_s^2 = \frac{1}{2}$ when $p = 1$, $\mu_1 = \mu_2$ (in case of E_2 distribution).

Note that $V_s \geq 1$ when $(\mu_2/\mu_1) < q$, i.e. the distribution is H_2.

K_2-distribution has three parameters μ_1, μ_2 and q (or $p = 1 - q$). Consider fitting a given distribution with moments m_1, m_2 and m_3 by a K_2-distribution.

Denote

$$\mu_1 + \mu_2 = u, \quad \mu_1 \mu_2 = v \, ;$$

then it can be seen that

$$u = (6m_1 m_2 - 2m_3) / (3m_2^2 - 2m_1 m_3)$$

$$v = (12m_1^2 - 6m_2) / (3m_2^2 - 2m_1 m_3)$$

and that the three parameters of K_2-distribution can be written as

$$\mu_1, \mu_2 = \frac{1}{2} [u \pm (u^2 - 4v)^{1/2}]$$

$$p = \mu_2 (m_1 \mu - 1) / \mu_1.$$

Thus μ_1, μ_2 and p are expressed in terms of the three moments of the distribution given.

A three-moment fit of the given distribution by a K_2-distribution is thus feasible provided the above relations yield a set of real numbers μ_1, μ_2 and p (parameters of K_2-distribution) such that $\mu_1 > 0$, $\mu_2 > 0$ and $0 < p < 1$.
[Yao and Buzacott (1985)].

Remarks

(1) Any distribution having as its L.T. a rational function (i.e. having RLT) $F^*(s) = P(s) / Q(s)$ such that the degree of $P(s)$ is less than or equal to that of $Q(s)$ can be represented as a (phase-type) Coxian distribution with number of phases equal to the degree say m, of the polynomial $Q(s)$ (a subnetwork of m nodes). The distribution is denoted by K_m (and also by C_m).
(2) Cox shows that no further generalisation of this type of phase-type distribution is possible by considering feed–forward and feed-backward concepts. One may consider that a job instead of departing from node i or going to next node $(i + 1)$ on the right for further service, may go from node i to any node in the system (not necessarily to node $i + 1$), if it does not depart from the system after receiving service at node i. However, this does not lead to further generalisation.
(3) Any pdf f (with df F) can be approximated arbitrarily closely by the pdf (df) of a distribution having RLT, that is, of a Coxian distribution.
(4) As the coefficient of variation of K_m lies between $1/\sqrt{m}$ to ∞, this gives a wider choice of distributions
(5) Any linear combination of Coxian distribution is again a Coxian distribution.

In view of the above, Coxian distribution has been considered to be of great versatility and has assumed importance. For an account and properties of Coxian distribution, reference may be made to Cox (1955), Gelenbe and Mitrani (1980), Yao and Buzacott (1985), Botta et al. (1987) and Bertsimas & Papaconstantinou (1988).

1.3.4.10 Normal Distribution:

The pdf of normal distribution is given by

$$f(x) = \frac{1}{\sigma\sqrt{(2\pi)}} \exp\left\{-\frac{1}{2}\left(\frac{x-\mu}{\sigma}\right)^2\right\}, -\infty < x < \infty.$$

The parameters are μ and σ, the mean and the s.d. respectively of the distribution. The r.v. X with the above pdf is denoted by $N(\mu, \sigma)$.

The r.v. takes positive as well as negative values. The characteristic function is given by

$$\phi(t) = E\{e^{itX}\} = \exp\left\{it\mu - \frac{t^2\sigma^2}{2}\right\}$$

The pdf is bell shaped and is symmetrical about μ with

$$\Pr\{\mu - \sigma < X < \mu + \sigma\} = 0.683$$

$$\Pr\{\mu - 2\sigma < X < \mu + 2\sigma\} = 0.956$$

$$\Pr\{\mu - 3\sigma < X < \mu + 3\sigma\} = 0.997$$

The normal variate $N(0, 1)$ with $\mu = 0$ and $\sigma = 1$ is called the standard normal variate. Its d.f. is given by

$$\Phi(x) = \frac{1}{\sqrt{2\pi}} \int_{-\infty}^{x} \exp\left\{-\frac{t^2}{2}\right\} dt$$

and $\Phi(-x) = 1 - \Phi(x)$.

Extensive tables of the pdf and d.f. of $N(0, 1)$ are available.

Normal distribution plays a very important role in statistical analysis.

A key result is the *Central Limit Theorem* (CLT) which states:

Under certain very general conditions, the sum of n independent r.v.'s having finite mean and variance, is asymptotically normally distributed as $n \to \infty$. A simple version of CLT for equal components is as follows.

If $X_1, \ldots, X_i, \ldots$ are i.i.d. random variables, with $E(X_i) = \mu$, var $(X_i) = \sigma^2$ (both finite) and $S_n = X_1 + \ldots + X_n$, then as $n \to \infty$,

$$\Pr\left\{\frac{S_n - n\mu}{\sigma\sqrt{n}} < \alpha\right\} \to \Phi(\alpha)$$

where $\Phi(\cdot)$ is the d.f. of $N(0, 1)$.

1.3.5 Three Important Theorems

We state, without proof, three theorem concerning the L.T. of probability distributions

Theorem 1.5.. *Uniqueness Theorem*: Distinct probability distributions on $[0, \infty)$ have distinct Laplace transforms.

It has the important implication that a probability distribution is recognizable by its transform. If the L.T. of a random variable X is known, then by identifying this with the form of L.T. of a distribution, one can conclude that X has the corresponding distribution; if this form does not resemble with the form of a L.T. of any standard distribution one can proceed to find the inverse of the L.T. (for which numerical methods have also been developed) to get the distribution. Even without finding the inverse, i.e. the form of the distribution, one can compute the moments of the distribution. Because of this theorem, perhaps, L.T. has the role it now plays in the study of probability distributions.

Theorem 1.6. *Continuity Theorem*: Let $\{X_n\}$, $n = 1, 2, \ldots$ be a sequence of random variables with distribution functions $\{F_n\}$ and L.T.'s $\{F_n^* (s)\}$. If as $n \rightarrow \infty$, F_n tends to a distribution function F having transform $F^* (s)$, then as $n \rightarrow \infty$, $F_n^* (s) \rightarrow F^* (s)$ for $s > 0$, and conversely.

This theorem can be used to obtain the limit distribution of a sequence of random variables.

Theorem 1.7. *Convolution Theorem*: The Laplace transform of the convolution of two independent random variables X, Y is the product of their transforms.

The integral

$$\int_0^x g(x - y)f(y)dy \qquad (3.22a)$$

(denoted by $f* g$) is called the convolution of the two functions f and g. The convolution U of F and G is given by

$$U(x) = \int_0^x G(x - y)dF(y) \qquad (3.22b)$$

and

$$L\{U(x)\} = L\{G(x)\}L\{F(x)\}. \qquad (3.22c)$$

In case of discrete random variables the result is essentially the same as that stated in (1.10) for generating function of the convolution of two random variables. In case of continuous random variables we have now the analogous result.

The above result is equivalent to the assertion that if X, Y are two independent random variables, then

$$E\{e^{-s(X+Y)}\} = E\{\exp(-sX)\exp(-sY)\}$$

$$= E\{\exp(-sX)\}E\{\exp(-sY)\}. \tag{3.23}$$

(Note that the *converse* is *not* true).
The result can also be stated as:

Theorem 1.8. The L.T. of the sum of two independent variables is the product of L.T.'s of the variables.

We can use the result to find the distribution of the sum of two or more independent random variables, when the variables are continuous or discrete. The method of generating functions is applicable only when the variables are non-negative integral-valued.

The following result immediately follows:

Theorem 1.9. The sum $S_n = X_1 + \ldots + X_n$ of n (a *fixed* number) identically and independently distributed random variables X_i has the L.T. equal to $[F^*(s)]^n$, $F^*(s)$ being the L.T. of X.

As an application we consider the following result:

Theorem 1.10. The sum of k identical and independent negative exponential distributions with parameters λ follows gamma distribution with parameters λ, k.

Let

$$X_i, i = 1, 2, \ldots 1, 2, \ldots k, \quad \text{have densities}$$

$$f_i(x) = \lambda \exp(-\lambda x), x \geq 0 \quad \text{for all } i.$$

From Example 3(c), the L.T. of X_i is

$$F_i^*(s) = \lambda/(s + \lambda)$$

and so the L.T. of $S = X_1 + X_2 + \ldots + X_k$ is $(\lambda/s + \lambda))^k$.
But this is the L.T. of the gamma distribution with density $f_{\lambda,k}(x)$ given in (3.15) (Example 3(e)). Hence the theorem.

We can easily obtain the mean and variance of the gamma distribution as follows:

$$E(S) = E(\Sigma X_i) = \Sigma E(X_i) = k/\lambda$$

$$\text{var}(S) = \Sigma \text{var}(X_i) = k/\lambda^2 \text{ (see (3.17) and (3.18))}.$$

Limiting form of Erlang–k distribution: As $k \to \infty$, Erlang-k distribution tends to a discrete distribution concentrated at $1/\lambda$. Let $\{E_k\}$ be a sequence of r.v.'s, E_k having Erlang–k distribution with density

$$f_{\lambda k, k}(x) = \frac{\{(\lambda k)^k x^{k-1} \exp(-k\lambda x)\}}{\Gamma(k)}, \quad k = 1, 2, 3, \ldots$$

Its L.T. is given by $F_k^*(s) = \left(\dfrac{\lambda k}{s + \lambda k}\right)^k = \left(1 + \dfrac{s}{\lambda k}\right)^{-k}$

As $k \to \infty$, $F_k^*(s) \to \exp(-s/\lambda)$. $\hspace{3cm}$ (3.24)

But $\exp(-s/\lambda)$ is the L.T. of the discrete distribution concentrated at $a = 1/\lambda$ (Example 3 (a)). Therefore, from the continuity theorem it follows that, as $k \to \infty$, the distribution of E_k tends to that variable whose whole mass is concentrated at $1/\lambda$.

This can also be seen from the fact that, as $k \to \infty$, the mode $(k - 1)/k\lambda$ of E_k moves to the right to $1/\lambda$ and that as var $(E_k) = 1/k\lambda^2 \to 0$, the whole mass of E_k tends to concentrate at the single point $1/\lambda$.

We thus have exponential distribution for $k = 1$ and the degenerate distribution (deterministic case) for $k \to \infty$. A suitable value of k $(1 < k < \infty)$ gives a good fit for many empirical distributions arising in practice.

An idea of how the density varies as k increases can be seen from Fig. 1.6.

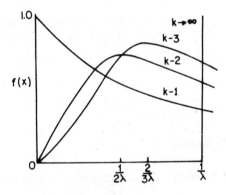

Fig. 1.6 The density function of Erlang–k distribution for some values of k

1.3.6 Geometric and Exponential Distributions

The geometric and exponential distributions are unique in the sense that these are the only distributions with memoryless or non-aging property. In fact, there is a close connection between these two distributions. Exponential distribution is the continuous analogue of geometric distribution and discretization of exponential distribution leads to geometric distributions.

We show below how one can get the exponential distribution from the geometric distribution.

Suppose that the occurrence (success) or non-occurence (failure) of an event E may happen only at the discrete time points $\delta, 2\delta, 3\delta, \ldots$ each subinterval being of length δ, the probability of occurrence being $p = \lambda\,\delta$ and that of non-occurrence being $q = 1 - \lambda\delta$.

The number of failures N preceding the first success has a geometric distribution

$$\Pr\{N = k\} = q^k p, \quad k = 0, 1, 2, \ldots$$

Now the event that the first success occurs at trial number $(k + 1)$ or beyond has

the probability $\sum_{r=k}^{\infty} q^r p = q^k p / (1 - q) = q^k$, and therefore the probability that the time

T required for the first success exceeds $k\delta$ is given by

$$\Pr\{T > k\delta\} = \Pr\{N \geq k\} = \sum_{r=k}^{\infty} q^r p = q^k = (1 - \lambda\delta)^k.$$

Now suppose that $\delta \to 0$ as $k \to \infty$, such that $k\,\delta = t$ remains fixed, then as $k \to \infty$.

$$\Pr(T > t) = (1 - \lambda t/k)^k = \{(1 - \lambda t/k)^{k/\lambda_t}\}^{\lambda t} \to \{e^{-1}\}^{\lambda t} = e^{-\lambda t}.$$

Thus T has an exponential distribution with mean

$$\frac{1}{\lambda} = \frac{\delta}{p}.$$

1.3.7 Sum of a Random Number of Continuous Random Variables

In Sec. 1.1.4 we considered the sum S_N of a random number N of random variables X_i. Theorem 1.3 is stated in terms of the p.g.f. of X_i on the assumption that the X_i's are discrete variables (with non-negative integral values). Here we assume that the X_i's are mutually independent and identically distributed non-negative random variables which may be continuous as well.

Replacing the p.g.f. $P(s)$ of X_i by its L.T. $F^*(s)$, we can get an analogous result. This may be stated as follows:

Theorem 1.11. If X_i are identically and independentally distributed random variables having for their L.T.'s

$$F^*(s) = E\{\exp(-sX_i)\} \quad \text{for all } i \tag{3.25}$$

and if N (independent of X_i's) is a random variable having p.g.f.

$$G(s) = \sum_n \Pr\{N = n\} s^n = \sum_n g_n s^n,$$

then the L.T. $F_{S_N}^*(s)$ of the sum $S_N = X_1 + \cdots + X_N$,

is given by
$$F_{S_N}^*(s) = G(F^*(s)). \tag{3.26}$$

The proof is identical with that of Theorem 1.3. It also follows that

$$E\{S_N\} = E\{X_i\} E\{N\}. \tag{3.27}$$

It is to be noted that S_N is a continuous random variable.

Corresponding to $h_j = \Pr\{S_N = j\}$ we shall have density function $f(x)$ of S_N, so that

$$f(x)dx = \Pr\{x \le S_N < x + dx\} = \sum_{n=0}^{\infty} \Pr\{N = n \text{ and } x \le S_n < x + dx\}$$

(3.28)

$$= \sum_{n=0}^{\infty} \Pr\{N = n\} \Pr\{x \le S_n < x + dx\}$$

$$= \sum_{n} g_n \Pr\{x \le S_n < x + dx\}.$$

(3.29)

Example 3(h). *Random geometric sum (or geometric compounding) of exponential variables*:

If X_i, $i = 1, 2, \ldots$ have negative exponential distribution each with mean $1/\lambda$ and $S_N = X_1 + \ldots + X_N$, where N has geometric distribution given by $\Pr\{N = n\} = pq^{n-1}$, $n = 1, 2, \ldots$ then S_N has negative exponential distribution with mean $1/\lambda p$.

The p.g.f. of N is $G(s) = ps / (1 - qs)$

and the L.T. of the r.v. X_i is $F^*(s) = \lambda / (s + \lambda)$.

From the Theorem 1.11, it follows that the sum $S_N = X_1 + \ldots + X_N$ has L.T. equal to $G(F^*(s))$. Replacing s in $G(s)$ by $F^*(s) = \lambda / (s + \lambda)$, we have

$$G(F^*(s)) = \frac{p\lambda/(s+\lambda)}{1 - q\lambda/(s+\lambda)} = \frac{p\lambda}{s + p\lambda};$$

(3.30)

but this is the L.T. of the exponential distribution with mean $1/\lambda p$.

Note. The above example exhibits an interesting result of preservation of memoryless property.

Denote by $F_n^*(s)$ the L.T. of n-fold convolution of X_i i.e., (L.T. of $X_1 + X_2 + \ldots X_n$), then

$$F_n^*(s) = \left(\frac{\lambda}{s + \lambda}\right)^n$$

Thus we can write

$$G(F^*(s)) = \sum_{n=1}^{\infty} pq^{n-1} F_n^*(s)$$

$$= \frac{p\lambda}{s + p\lambda}$$

The random geometric sum of independent and identical exponential variables is again an exponential variable. This is a closure property.

Example 3(i). Let X_i have exponential distribution with mean $1/a$ and let N have Poisson distribution with mean c. We have $F^*(s) \equiv$ L.T. of $X_i = a/(s+a)$, and $G(s) = e^{c(s-1)}$ and hence the L.T. of $S_N = X_1 + \ldots + X_N$ is

$$G(F^*(s)) = \exp\left(c\left(\frac{a}{s+a} - 1\right)\right), \text{ replacing } s \text{ by } F^*(s) = a/(s+a)$$

$$= \exp(-cs/(s+a)).$$

One may be interested to find the probability density function of S_N.

1.4 CLASSIFICATION OF DISTRIBUTIONS

Let X be a non-negative continuous r.v. Let $F(x)$ be the d.f. of X and let $f(x) = F'(x)$ be the p.d.f. of X.

Let $R(x) = 1 - F(x) = \Pr\{X > x\}$ be the complementary d.f. of X; we have $f(x) = -R'(x)$. The Mean time to failure (MTTF) is given by

$$E\{X\} = \int_0^\infty tf(t)dt = \int_0^\infty R(t)dt.$$

In reliability literature the life time or time to failure of a component or unit can be denoted by X. Then $R(x)$ denotes the survival probability at age x or the reliability of the unit at time x. $R(x)$ is called the survivor function of X. It is also referred to as reliability function. The r.v. X may also represent service time or repair time or completed length of service (as in manpower studies) and so on.

1.4.1 Hazard (or Failure) Rate Function

Definition. Instantaneous failure rate (or hazard rate). The failure rate (or hazard rate) function of the r.v. X or of the distribution F is defined by

$$h(t) = \lim_{x \to 0} \frac{1}{x} \frac{F(t-x) - F(t)}{R(t)}$$

$$= \frac{f(t)}{R(t)} = \frac{-R'(t)}{R(t)}$$

for values of t for which $F(t) < 1$.
The probability element

$$h(t)dt = \frac{f(t)dt}{R(t)} \cong \frac{\Pr\{t \leq X < t+dt\}}{\Pr\{X > t\}}$$

gives the *conditional* probability that X lies in the interval $(t, t+dt)$ given that X exceeds t, whereas $f(t)\,dt$ gives the *unconditional* probability that X lies in the interval $(t, t+dt)$. In terms of lifetime, while $h(t)\,dt$ gives the conditional probability that it will fail in $(t, t+dt)$ given that it has survived until t, $f(t)\,dt$ gives the unconditional probability that the component will fail in $(t, t + dt)$.

The failure rate corresponds to age-specific death rate in demographic studies. Integrating

$$h(t) = -\frac{R'(t)}{R(t)}$$

we get

$$R(t) = R(0)\exp\left\{-\int_0^t h(x)dx\right\} .$$

The r.v. X or the distribution F is said to be an increasing (decreasing) failure rate or hazard rate distribution IFR or IHR (DFR or DHR) according as $h(t)$ is an increasing (decreasing) function of t for $t \geq 0$.

We list below some non-negative distributions along with their hazard rates.

Distribution	$f(t)$ (p.d.f.)	$h(t)$ (hazard rate function)
Exponential	ae^{-at} $(a > 0)$	a
Gamma	$\dfrac{a^\alpha t^{\alpha-1} e^{-at}}{\Gamma(\alpha)}$ $(a > 0, \alpha > 0)$	
Weibull	$a\alpha t^{\alpha-1} e^{-at^\alpha}$ $(a > 0, \alpha > 0)$	$a\alpha t^{\alpha-1}$
Hypo-exponential (2-stages)	$\dfrac{ab}{b-a}(e^{-at} - e^{-bt})$ $(a > 0, b > 0, a \neq b)$	$\dfrac{ab(e^{-at} - e^{-bt})}{be^{-at} - ae^{-bt}}$
Hyper-exponential (2-stages)	$pae^{-at} + (1-p)be^{-bt}$ $(a > 0, b > 0, 0 < p < 1)$	$\dfrac{pae^{-at} + (1-p)be^{-bt}}{pe^{-at} + (1-p)e^{-bt}}$
Modified extreme value	$\dfrac{1}{a}\exp\left\{-\dfrac{(e^t - 1)}{a} + t\right\}$ $(a > 0)$	$\dfrac{e^{at}}{a}$

The exponential distribution has constant hazard rate and is the only continuous distribution with this property. The modified extreme value distribution is an IFR distribution. The Weibull distribution is an IFR distribution for $\alpha > 1$ and DFR distribution for $0 < \alpha < 1$. The hypo-exponential distribution is an IFR distribution with failure rate increasing from 0 to min $\{a, b\}$. The hyper-exponential distribution is a DFR with failure rate decreasing from $pa + (1 - p)b$ to min $\{a, b\}$. The Gamma distribution is IFR for $\alpha > 1$ and DFR for $0 < \alpha < 1$. For $\alpha = 1$, gamma distribution as well as Weibull distribution degenerate to an exponential distribution. The normal distribution is IFR.

The above results can be had by considering the sign of $h'(t)$.

Definition: Hazard (or failure) rate average is defined by

$$\frac{1}{t}\int_0^t h(x)dx = \frac{1}{t}H(t)$$

$H(t)$ is called the cumulative hazard function. The r.v. X or the distribution F is called an increasing (decreasing) hazard or failure rate average IFRA or IHRA according as the hazard rate average is an increasing (decreasing) function of t, $t \geq 0$.

1.4.2 Mean Residual Life (MRL)

The quantity $E\{X - t \mid X > t\}$, $t \geq 0$ is called the *mean residual life (time)* for the r.v. X.

The r.v. X is said to have increasing mean residual life IMRL if $E(X) < \infty$

and

$$r(t) = E\{X - t \mid X > t\} = \frac{\int_t^\infty R(x)dx}{R(t)}$$

is increasing in $t > 0$.

Note that an exponential distribution with mean μ has a constant mean residual life, equal to μ and has a constant hazard rate equal to $1/\mu$.

Analogous definitions can be given for discrete r.v.'s.

Note that $h(t)$ and $r(t)$ are connected by the relation

$$h(t) = \frac{1 + r'(t)}{r(t)}.$$

Definition: Let X, the life time of a unit or component, have d.f. $F(\cdot)$ and complementary d.f. $R(x) = 1 - F(x)$. Then the component is said to be *New Better (Worse) than used NBU (NWU)* if

$$\{1 - F(x+y)\} \leq (\geq)[\{1 - F(x)\}\{1 - F(y)\}].$$

In other words, a unit is **NBU (NWU)** if the survival probability of a unit aged x is less (greater) than the corresponding survival probability of a new unit.

The component is said to be **New Better (Worse) than Used in Expectation NBUE (NWUE)** if

$$r(t) = \frac{\int_t^\infty R(x)dx}{R(t)} \leq (\geq)E(X) = r(0)$$

In other words, a unit is **NBUE (NWUE)** if the conditional expectation of the residual life of a unit at age t is less (greater) than the expectation of the life time of a new unit, that is, $r(t) \leq (\geq) r(0) = E(X)$.

Remarks: 1. 1(a) Suppose that life distribution of an item has non-decreasing $h(t)$ for $t > 0$, i.e. The distribution has IFR property. This implies that

1(b). It will have IFRA, i.e., $H(t)/t$ is nondecreasing for $t > 0$.
1(c). It implies that it is new better than used, i.e., $r(t) \leq r(0)$, $t \geq 0$.
1(d). It implies that it has decreasing mean residual life, i.e., $r(t)$ is decreasing for $t \geq 0$.

In other words,
$$1(a) \Rightarrow 1(b) \Rightarrow 1(c)$$
and
$$1(a) \Rightarrow 1(d) \Rightarrow 1(c).$$

2. The functions $f(t)$ (p.d.f.), $F(t)$ (d.f.), $R(t)$ (complementary d.f.) and $h(t)$ (hazard function) give mathematically equivalent specification of the distribution of a random variable.

3. One can have corresponding definition for discrete r.v. T. Suppose that the r.v. takes values t_1, t_2, with p.m.f. $p_j = P(T = t_j)$ then the hazard function is defined by

$$h(t_j) = p_j / P(T > t_j).$$

For example, hazard function of Poisson distribution is monotone increasing.

4. Block and Savits (*Ann. Prob.* **8**, 465 (1980)) obtain necessary and sufficient conditions in terms of Laplace transforms for characterizing various classes of life distributions such as IFR, DFR, IHRA, IDRA, NBU, NBUE etc. (See Exercise 1.11).

5. In Example 3(h) we find a closure property of random geometric sum of exponential variables. A more general property of random geometric sum is stated below.
If X_i are positive (discrete or continuous) r.v.'s with DFR then the random geometric sum of such variables is again a r.v. with DFR.

6. A discrete mixture of densities $f_k(x)$ is defined by

$$f(x) = \Sigma f_k(x)p_k, \quad p_k \geq 0, \quad \Sigma p_k = 1$$

where $\{p_k\}$ may not necessarily imply a standard probability distribution).

A discrete mixture of densities each having DFR is again a DFR density.

However, a discrete mixture of IFR densities is not necessarily an IFR density (see Proschan, *Techno.* **5**, 375 (1963) and Shanthikumar, *Ann. Prob.* **16**, 397, 1988).

1.4.3 Further Properties

The MRL function $r(t)$ is related to the survivor function $R(t)$ by the relations

$$r(t) = \frac{1}{R(t)} \int\limits_{t}^{\infty} R(x)dx$$

and

$$R(t) = \exp\left[-\int\limits_{0}^{t} \frac{1 + r'(x)}{r(x)} dx\right]$$

A set of necessary and sufficient conditions for a function $r(t)$, $t \geq 0$ to be an MRL function are:

(i) $r(t) \geq 0$

(ii) $r'(t) \geq -1$

(iii) $\lim\limits_{t \to 0} \left\{ \frac{r(t)}{t \ln t} \right\} = 0.$

[For proof refer to Swartz, G.B., The mean residual lifetime function, *IEEE Trans. Rel.* **R-22**, 108-9 (1973)].

Both the hazard rate function $h(t)$ and the mean residual life function $r(t)$ are indicators of aging; though they look similar, there are essential differences.

While the hazard rate function takes into account the immediate future (in assessing component failure or service completion), the mean residual life takes into account the complete future life. This can be seen from the following.
We have

$$h(t) R(t) = -\frac{d}{dt} R(t)$$

$$= f(t), \text{ when the distribution has p.d.f. } f(\cdot)$$

and

$$r(t) R(t) = \int\limits_{t}^{\infty} R(x)dx.$$

The r.h.s. of the former relation depends on the probability law at the point t only, whereas the r.h.s. of the latter relation depends on the probability law at all points in the interval (t, ∞). For example, consider as lifetime distribution the uniform distribution in $[a, b]$. Then

$$f(x) = \frac{1}{b-a}, \quad a \leq x \leq b$$

$$= 0, \quad \text{otherwise.}$$

We get

$$h(t) = 0, \quad 0 \leq t \leq a$$

$$= \frac{1}{b-t}, \quad a < t \leq b$$

and
$$r(t) = \frac{a+b}{2} - t, \quad 0 \leq t \leq a$$

$$= \frac{1}{2}(b-t), \quad a \leq t \leq b.$$

In the interval $[0, a]$, the hazard rate function $h(t)$ does not give any indication of wearout; the actual age cannot be obtained from the hazard rate function prior to the point a. The MRL function $r(t)$ gives a more descriptive measure of the process of aging than the hazard rate function.

It has been shown that white IHF implies DMRL, the latter does not imply the former, so that the class of IHF distributions forms a proper subset of the class of DMRL distributions. For further results, see Muth (1977, 1980).

EXERCISES

1.1 (a) Find the generating function of the *Fibonacci numbers* $\{f_n\}$ defined by
$f_n = f_{n-1} + f_{n-2}, n \geq 2, f_0 = 0, f_1 = 1$.

(b) Let X be a random variable denoting the number of tosses required to get two consecutive heads (say, event E_2) when a fair coin is tossed. Show that the p.g.f. of X is $[(s^2/4)\{1 - s/2 - (s/2)^2\}^{-1}]$.

(c) Show further that, if a fair coin is tossed indefinitely, the event E_2 is certain to occur and that the expected number of trials needed is 6.
Examine the case when the coin is biased.

(d) Find the probability that *two* or *more* consecutive heads will *not* occur if a fair coin is tossed n times.

1.2 Let Y have the distribution of the geometric form (modified geometric or decapitated geometric) given by

$$\Pr\{Y = k\} = q^{k-1}p, \quad k = 1, 2, 3, \ldots$$

Show that the p.g.f. of Y is $ps / (1 - sq)$, and that

$$E(Y) = 1/p \quad \text{and var } (Y) = q/p^2$$

(The number of trials Y upto and including the first success in a sequence of Bernoulli trials with probability p of success has the above distribution;

again $Y = X + 1$, where X has geometric distribution considered in Example 1 (d).)

1.2.1 Application to a population model (Keyfitz, 1969).

(a) If couples continue to have children till they have one *male* child, then the distribution of the family size (i.e. number of children in the family) has the same distribution as Y (with p, the probability of having a boy). The expected number of children in the family is then $1/p$; this becomes 2 when $p = 1/2$.

For
$$p = \frac{1}{2} + \varepsilon, \quad \frac{1}{p} \cong 2 - 4\varepsilon.$$

[In case of USA, ε is estimated at $\hat{\varepsilon} = 0.0115$ for 1965].
Suppose that couples continue to have children till they have one boy and one girl. Find the p.g.f. of the number of children Z. Show that

$$E(Z) = \frac{1}{p} + \frac{1}{q} - 1$$

(b) Suppose that couples continue to have children till they have m *boys* and n *girls*. Let the number of children be N. Then show that

$$\Pr(N = m + n + k) = \binom{N-1}{n-1} p^{N-n} q^n$$

$$+ \binom{N-1}{m-1} p^m q^{N-m},$$

$$m \geq 1, n \geq 1, k \geq 0.$$

1.2.2 The decapitated geometric r.v. is a particular case of the shifted geometric variable W with

$$\Pr\{W = k\} = q^{k-m} p, \quad k = m, m+1, \ldots$$

Find the mean and variance of W.

1.3 (a) Let X be a positive discrete random variable such that

$$\Pr\{X = k\} = p_k, \quad \Pr\{X > k\} = q_k = \sum_{r=k+1}^{\infty} p_r, \quad k \geq 0,$$

and

$$P(s) = \sum_{k=0}^{\infty} p_k s^k \quad \text{and} \quad Q(s) = \sum_{k=0}^{\infty} q_k s^k.$$

Obtain a relation between $P(s)$ and $Q(s)$.

Show that $E(X) = Q(1) = \sum_{k=0}^{\infty} q_k$

Find an expression for $E(X^r)$ in terms of $\{q_k\}, r \geq 1$.

(b) If X is a continuous non-negative random variable with d.f.
$F(t) = \Pr\{X \leq t\}$, then show that

$$E\{X\} = \int_0^{\infty} [1 - F(t)] dt;$$

more generally, the rth moment equals

$$E\{X^r\} = r \int_0^{\infty} t^{r-1}[1 - F(t)] \, dt, \, r \geq 1.$$

(c) If X, Y are jointly continuous random variables, then show that

$$E(X) = \int_{-\infty}^{\infty} E[X \mid Y = y] f(y) dy, \text{ where } f(y) \text{ is the p.d.f. of } Y.$$

(d) Show that

$$\text{var}\left(\sum_{i=1}^{n} X_i\right) = \sum_{i=1}^{n} \text{var}(X_i) + 2 \sum_{1 \leq i < j \leq n} \text{cov}(X_i, X_j).$$

1.4. Find the mean and variance of the negative binomial (NB) distribution considered in Example 1(h). Obtain the p.g.f. of the sum of m i.i.d. NB variables and find its mean and variance.

1.5. (a) Let X have a zero-truncated (or decapitated) Poisson distribution with zero class missing, i.e.

$$p_k = \Pr\{X = k\} = \frac{(e^a - 1)^{-1} a^k}{k!}, \quad k = 1, 2, 3, \ldots$$

Show that the p.g.f. of X is given by $P(s) = (e^a - 1)^{-1} (e^{as} - 1)$.

Verify that $P(1) = \sum_{k=1}^{\infty} p_k = 1$; show that $E(X) = ae^a / (e^a - 1)$ and

$$\text{var}(X) = e^a [1 - a/(e^a - 1)] / (e^a - 1).$$

(b) If X, Y are two independent zero-truncated Poisson variates, find the p.g.f. of the sum $(X + Y)$ and the distribution $\Pr\{X + Y = n\}, n = 2, 3, \ldots$, when (i) X, Y are identical with the same parameter a, (ii) X, Y have parameters a_1, a_2 (see Medhi, *Metrika*, **20**, 215–218 (1973)).

(c) Find the p.g.f. of the sum $S_n = X_1 + \ldots + X_n$ of n independent and identical zero-truncated Poisson variates. Find $E(S_n)$ and $\Pr\{S_n = m\}, m = n, n + 1, n + 2 \ldots$

1.6. Negative moments of a positive random variable

Let X be a positive r.v. having p.g.f. $P(s)$. If r is a positive integer and $a > 0$, then show that

$$E\left[\frac{1}{(x+a)^r}\right] = \int_0^1 g_r(s)\,ds$$

where $g_r(s)$ can be found as follows

$$g_1(s) = E[s^{X+a-1}] = s^{a-1}P(s)$$

and

$$g_{k+1}(s) = \frac{1}{s}\int_0^s g_k(t)\,dt,$$

$$k = 1, 2, \ldots, r-1.$$

With $r = 1$

$$E\left[\frac{1}{(X+a)}\right] = \int_0^1 g_1(s)\,ds = \int_0^1 s^{a-1}P(s)\,ds$$

Further show that

$$E\left[\frac{1}{(X+a)(X+b)}\right] = \int_0^1 t^{b-a-1}\left\{\int_0^t s^{a-1}P(s)\,ds\right\}$$

$$b > a > 0.$$

(Chao and Strawderman, *J.Am. Stat. Ass.* **67** (1972) 429–431).

1.7   Let $S_N = X_1 + X_2 + \ldots + X_N$, where N has Poisson distribution with mean a.

(a)   If X_i have i.i.d geometric distributions with mean q/p, find (i) the p.g.f. of S_N, and (ii) $E(S_N)$ and var (S_N).

(b)   If X_i have i.i.d. Bernoulli distributions with Pr $\{X_i = 1\} = p$, and Pr $\{X_i = 0\} = 1 - p = q$, show that S_N has Poisson distribution with mean ap.

Show that the joint distribution of S_N and N has probability mass function

$$\Pr\{N = n, S_N = y\} = \frac{e^{-a}a^n p^y q^{n-y}}{y!(n-y)!}$$

for $n = 0, 1, 2, \ldots,$ $y = 0, 1, 2, \ldots, n$

and that the conditional distribution of N given $S_N = y$ has probability mass function

$$\Pr\{N = n \mid S_N = y\} = \frac{e^{-aq}(aq)^{n-y}}{(n-y)!}, \quad n \geq y.$$

Show that $\quad\quad\quad$ cov $(N, S_N) = ap$.

(c) If X_i have i.i.d. Poisson distributions with mean b, show that the joint distribution of S_N and N has mass function

$$\Pr\{N = n, S_N = y\} = \frac{\exp\{-(a + bn)\}a^n(nb)^y}{n!y!}$$

for $\quad\quad\quad\quad\quad\quad n = 0, 1, 2, ..., y = 0, 1....$

Show that $E(S_N) = ab$ and var $(S_N) = ab(1 + b)$.

(Accident models have been constructed and fitted to actual data involving traffic accidents (Leiter and Hamdan, *Int. Stat. Rev.* **41** (1973) 87–100) taking N as the number of accidents; X_i (in (b)) as the indicator variable indicating whether the ith accident results in a fatality or not; X_i (in (c)) as the number of persons fatally injured in the ith accident.)

(d) If X_i have i.i.d. binomial distributions with parameters m and p, then show that

$$\Pr\{N = n, S_N = y\} = \frac{e^{-a}a^n}{n!}\binom{mn}{y}p^y q^{mn-y}.$$

Find $E\{S_N\}$ and var $\{S_N\}$.
Show that

$$\text{cov}\{N, S_N\} = mpa;$$

and $\quad\quad\quad\quad\quad\quad\quad\text{corr}\{N, S_N\} = \left(\frac{mp}{mp+q}\right)^{1/2}$ (independent of a).

1.8 Let $Z = X_1 + \ldots + X_N$, where $\{N\}$ has zero-truncated Poisson distribution (defined in 1.5(a) above) (N is independent of X_i) and X_i have i.i.d. exponential distributions with mean $1/\lambda$. Show that the L.T. $F^*(s)$ of Z equals

$$F^*(s) = \{\exp(a\lambda/(\lambda + s)) - 1\}/(e^a - 1),$$

and that

$$E(Z) = ae^a/\{\lambda(e^a - 1)\}.$$

It may be further verified that

$$f(t) = \frac{\exp(-\lambda t)}{\exp(a) - 1} \sqrt{\left(\frac{a\lambda}{t}\right)} I_1(2\sqrt{(\lambda at)})$$

is the probability density function of S_N. (Show that L.T. of $f(t)$ is $F^*(s)$.)
$I_r(x)$ is the modified Bessel function of order r (≥ -1) given by

$$I_r(x) = \sum_{k=0}^{\infty} \frac{1}{k!\Gamma(k+r+1)}(x/2)^{2k+r}.$$

(The above result can be used in Queueing theory in a model of phase-type service, where the r.v. N denotes the number of phases associated with a customer's service, and the i.i.d. r.v. X_i denotes the duration of service time at the ith phase).

The Bessel function density $f(t)$ is quite similar, for the value of the parameter a of 10 or over, to the gamma density $f_{\lambda,a}(t)$. However, $f(t)$ gives a better fit to a large number of empirical distributions (see Gaver, *Opns. Res.* **2** (1954) 147).

1.9. Suppose that X_i, $i = 1, 2$, are two independent geometric variables with Pr $\{X_i = k\} = q_i^k p_i$, $p_i + q_i = 1$, $i = 1, 2$, $k = 0, 1, 2, \ldots$ Find the bivariate generating function $P(s_1, s_2)$ of the pair (X_1, X_2) and from the form of the generating function for the sum $S = X_1 + X_2$, verify that

$$\text{Pr}\{S = k\} = \sum_{r=0}^{k} q_1^r p_1 q_2^{k-r} p_2 = \frac{p_1 p_2}{(q_1 - q_2)}(q_1^{k+1} - q_2^{k+1}), \quad k = 0, 1, 2, \ldots$$

1.10 Find the Laplace transforms of:

(i) $\sin at$;   (ii) $a \exp(-2t)$;   (iii) $(t^2 + 1)^2$;   (iv) $t^3 \exp(-3t)$;

(v) $e^{-t} \cos 2t$;   (vi) $t^n e^{at}$ $(n > -1)$;

(vii) $\dfrac{\{\exp(-at)\}(at)^n}{n!}$ $(n = 0, 1, 2, \ldots)$;   (viii) $\dfrac{a \exp\{-(a^2/4t)\}}{2\sqrt{(\pi t^3)}}$;

(ix) $\sqrt{(t^r)} I_r(2\sqrt{t})$;

(x) $\{\exp(-t-1)\} \sqrt{(t^r)} I_r(2\sqrt{t})$;

(xi) $\dfrac{\exp(-\lambda t)}{e^a - 1} \sqrt{\left(\dfrac{\lambda a}{t}\right)} I_1(2\sqrt{(\lambda a t)})$.

1.11 *Laplace transforms of life distributions (Block & Savits, 1980)*
Let X be a non-negative r.v. with d.f. F and L.T.

$$F^*(s) = \int_0^{\infty} e^{-sx} dF(x)$$

Define

$$a_n(s) = \frac{(-1)^n}{n!} \frac{d^n}{ds^n}\left[\frac{1-F^*(s)}{s}\right], n \geq 0, s \geq 0$$

and

$$\alpha_n(s) = s^{n+1}a_n(s), n \geq 0, \quad \alpha_0(s) = 1, s \geq 0.$$

Verify that, for $n \geq 0, s \geq 0$

$$a_n(s) = \int_0^\infty \frac{u^n}{n!}e^{-su}\overline{F}(u)du, \quad \overline{F} = 1-F$$

$$\alpha_{n+1}(s) = s\int_0^\infty \frac{(su)^n}{n!}e^{-su}\overline{F}(u)du.$$

Show that
(1) X has IFR *iff* $\{\alpha_n(s), s \geq 1\}$ is log concave in n, for all $s \geq 0$
(2) X has IFRA *iff* $[\alpha_n(s)]^{1/n}$ is decreasing in $n \geq 1$, for all $s \geq 0$
(3) X has NBU distribution *iff* $\alpha_{n+m}(s) \leq \alpha_n(s)\,\alpha_m(s)$, $n, m > 0, s > 0$
(4) X has NBUE *iff*

$$\alpha_n(s)\sum_{k=0}^\infty \alpha_k(s) \geq \sum_{k=n}^\infty \alpha_k(s), \quad n > 0, s > 0.$$

1.12 Memory as a property of lifetime distribution

Define *virtual age* at time t of an equipment with lifetime distribution X by

$$v(t) = r(0) - r(t)$$

where $r(t) = E\{X - t \mid X > t\}$ is the expected residual lifetime (MRL).
Define interval memory over (t_1, t_2) by

$$m(t_1, t_2) = \frac{v(t_2) - v(t_1)}{t_2 - t_1}$$

The (lifetime) distribution X is said to have positive/perfect/negative/no memory according as:

        positive memory if $0 < m < 1$
        perfect memory if $m = 1$
        no memory if $m = 0$
        negative memory if $m < 0$.

[Negative memory implies decrease of virtual age with increase of actual age.]

Show that exponential and geometric distributions have *no memory* ($m (t_1, t_2)$ $= 0$ for all $t_1 > t_2$)) and that uniform distribution in (a, b) has positive memory in (a, b) and perfect memory in $(0, a)$. Further, the degenerate distribution with all mass at a point has perfect memory.

Refer to Muth (1980) for an account of the notion of using memory for classification of distributions.

1.13 If X and Y are independent exponential r.v.'s with parameters a and b, and $W = \max(X, Y)$, then $F(x) = \Pr\{W \le x\} = 1 - e^{-ax} - e^{-bx} + e^{-(a+b)x}$.

If $X_1, X_2, \ldots, X_n$ are i.i.d. exponential variables with mean 1, and

$$Y_n = \max\{X_1, X_2, \ldots, X_n\}$$

then show that

$$E\{Y_n\} = 1 + \frac{1}{2} + \ldots + \frac{1}{n};$$

and

$$\mathrm{var}\{Y_n\} = 1 + \frac{1}{2^2} + \ldots + \frac{1}{n^2}.$$

1.14 (a) The non-negative valued random variable X is said to have a *modified exponential distribution* with an non-zero probability mass at '0', if its distribution function is

$$F(x) = \Pr(X \le x) = 1 - k \exp(-\lambda x),$$

$$x \ge 0, \lambda > 0, 0 < k < 1.$$

Find (i) the frequency density of X, (ii) $E(X)$, and (iii) L.T. of X.
   (b) A r.v. T having pdf

$$f(t) = \lambda \exp\{-(\lambda - \mu)t\}, \qquad t \ge \mu$$

is said to be an exponential r.v with threshold μ. Find $E(T)$ and var (T).

1.15 (a) Show that the sum of two independent gamma variables having parameters λ, k and λ, l is a gamma variable with parameters $\lambda, k + l$.
   (b) A r.v. X is said to have a *generalised (3-parameter) gamma* distribution if its pdf is

$$f(x) = \frac{c}{\Gamma(k)} \lambda^{ck} x^{ck-1} e^{-(\lambda x)^c}, \qquad 0 \le x < \infty$$

$$(c > 0, k > 0, \lambda > 0).$$

Find (i) $E(X)$ and (ii) var (X).

When $k = 1$, the distribution of X reduces to *Weibull distribution*; when $k = 1$, $c = 2$, it reduces to *Rayleigh distribution*.

When $c = 1$, it reduces to (2-parameter) gamma distribution.

1.16   Find the Laplace transform of the variable X (having triangular distribution) with density function

$$f(x) = x, \qquad 0 \leq x \leq 1$$

$$= 2 - x, \qquad 1 \leq x \leq 2.$$

Hence show that the *triangular distribution* can be obtained as the sum of two independent random variables each with a rectangular distribution in $(0, 1)$.

REFERENCES

R. Bellman, R.E. Kalaba and J.A. Lockett. *Numerical Inversion of Laplace Transforms*, American Elsevier, New York (1966).

D. Bertsimas and X. Papaconstantinou. On the steady-state solution of the M/C_2 $(a, b)/s$ queueing system, *Transportation Science*, **22**, 125-138 (1988).

H.W. Block and T.H. Savits. Laplace transforms for classes of life distributions, *Ann. Prob.* **8**, 465-474 (1980).

———— Multivariate increasing failure rate average distribution, *Ann. Prob.* **8**, 793-801 (1980).

R.F. Botta, C.M. Harris and W.G. Marechal. Characterizations of generalized hyperexponential distribution functions, *Comm. in Statist.—Stochastic Models*, **3**, 115–148 (1987).

D.R. Cox. A use of complex probabilities in the theory of stochastic processes, *Proc. Comb. Phil. Soc.* **51**, 313–319 (1955).

W. Feller. *An Introduction to Probability Theory and Its Applications*, Vol. I, 3rd ed. (1968), Vol II (1966), Wiley, New York.

E. Gelenbe and I. Mitrani. *Analysis and Synthesis of Computer Systems*, Academic Press, London (1980).

V.B. Iverson and Nielsen. Some properties of Coxian distributions with applications: in *Modeling Techniques and Tools for Performance Analysis* (Ed. Abu El Ata), North Holland, Amsterdam (1985).

N.L. Johnson and S. Kotz. (1) *Discrete Distributions* (1969), (2) *Continuous Univariate Distributions* (1970), Houghton Mifflin, Boston MA.

N. Keyfitz. *Introduction to Mathematics of Population*, Addison-Wesley, Reading, MA (1969) [with revisions 1977].

J.F. Lawless, *Statistical Models and Methods of Lifetime Data*, Wiley, New York (1982).

E.J. Muth. Reliability models with positive memory derived from the mean residual life function: in *The Theory and Applications of Reliability*, Vol. II, 401-435, Academic Press, New York (1977).

———— Memory as a property of probability distributions, *IEEE Trans. on Rel.* Vol. **R-29**, 160-165 (1980).

M.F. Neuts, *Matrix Geometric Solutions in Stochastic Models*, Johns Hopkins Press, Baltimore (1981).

J.G. Shanthikumar. DFR property of first passage times and its preservation under geometric compounding, *Ann. Prob.* **16**, 397–407 (1988).

M.G. Smith. *Laplace Transform Theory*, Van Nostrand, London (1966).

H.M. Srivastava and B.R.K. Kashyap. *Special Functions in Queueing Theory and Related Stochastic Processes*, Academic Press, New York (1982).

S.K. Srinivasan and K.M. Mehata. *Probability and Random Processes*, 2nd ed., Tata-McGraw Hill, N. Delhi (1981).

K.S. Trivedi. *Probability and Statistics with Reliability, Queuing and Computer Science Applications*, Prentice-Hall International (1982). Prentice-Hall India Reprint (1988).

D.D. Yao and J.A. Buzacott. Queueing models for a flexible machining station, Part II: The methods of Coxian phases, *Euro. J. Opns. Res.* **19**, 241-252 (1985).

CHAPTER 2

Stochastic Processes: Some Notions

2.1 INTRODUCTION

Since the last century there have been marked changes in the approach to scientific enquiries. There has been greater realisation that probability (or non-deterministic) models are more realistic than deterministic models in many situations. Observations taken at different time points rather than those taken at a fixed period of time began to engage the attention of probabilists. This led to a new concept of indeterminism: indeterminism in dynamic studies. This has been called ''dynamic indeterminism'' (Neyman (1960), traces the transitions). The period of dynamic indeterminism began roughly with the work of Mendel (1822-1884). The physicists (see Chandrasekhar, 1943) played a leading role in the development of dynamic indeterminism. Many a phenomenon occurring in physical and life sciences are studied now not only as a random phenomenon but also as one changing with time or space. Similar considerations are also made in other areas, such as, social sciences, engineering and management and so on. The scope of applications of random variables which are functions of time or space or both has been on the increase.

Families of random variables which are functions of say, time, are known as stochastic processes (or random processes, or random functions). A few simple examples are given as illustrations.

Example 1(a). Consider a simple experiment like throwing a true die. (i) Suppose that X_n is the outcome of the nth throw, $n \geq 1$. Then $\{X_n, n \geq 1\}$ is a family of random variables such that for a distinct value of n (= 1, 2, . . .), one gets a distinct random variable X_n; $\{X_n, n \geq 1\}$ constitutes a stochastic process, known as Bernoulli process. (ii) Suppose that X_n is the number of sixes in the first n throws. For a distinct value of $n = 1, 2, \ldots$, we get a distinct binomial variable X_n; $\{X_n, n \geq 1\}$ which gives a family of random variables is a stochastic process. (iii) Suppose that X_n is the maximum number shown in the first n throws. Here $\{X_n, n \geq 1\}$ constitutes a stochastic process.

Example 1(b). Consider that there are r cells and an infinitely large number of identical balls and that balls are thrown at random, one by one, into the cells, the ball thrown being equally likely to go into any one of the cells. Suppose that X_n is the number of occupied cells after n throws. Then $\{X_n, n \geq 1\}$ constitutes a stochastic process.

Example 1(c). Consider a random event occurring in time, such as, number of telephone calls received at a switchboard. Suppose that $X(t)$ is the random variable which represents the number of incoming calls in an interval $(0, t)$ of duration t units. The number of calls within a fixed interval of *specified duration*, say, one unit of time, is a random variable $X(1)$ and the family $\{X(t), t \in T\}$ constitutes a stochastic process $(T = [0, \infty))$.

2.2 SPECIFICATION OF STOCHASTIC PROCESSES

The set of possible values of a single random variable X_n of a stochastic process $\{X_n, n \geq 1\}$ is known as its state space. The *state space* is discrete if it contains a finite or a denumerable infinity of points; otherwise, it is continuous. For example, if X_n is the total number of sixes appearing in the first n throws of a die, the set of possible values of X_n is the finite set of non-negative integers $0, 1, \ldots, n$. Here, the state space of X_n is discrete. We can write $X_n = Y_1 + \ldots + Y_n$, where Y_i is a discrete r.v. denoting the outcome of the ith throw and $Y_i = 1$ or 0 according as the ith throw shows six or not. Secondly, consider $X_n = Z_1 + \ldots + Z_n$, where Z_i is a continuous r.v. assuming values in $[0, \infty)$. Here, the set of possible values of X_n is the interval $[0, \infty)$, and so the state space of X_n is continuous.

In the above two examples we assume that the parameter n of X_n is restricted to the non-negative integers $n = 0, 1, 2, \ldots$. We consider the state of the system at distinct time points $n = 0, 1, 2, \ldots$, only. Here the word *time* is used in a wide sense. We note that in the first case considered above the "time n" implies throw number n.

On the other hand, one can visualise a family of random variables $\{X_t, t \in T\}$ (or $\{X(t), t \in T\}$) such that the state of the system is characterized at every instant over a finite or infinite interval. The system is then defined for a continuous range of time and we say that we have a family of r.v. in *continuous* time. A stochastic process in continuous time may have either a discrete or a continuous state space. For example, suppose that $X(t)$ gives the number of incoming calls at a switchboard in an interval $(0, t)$. Here the state space of $X(t)$ is discrete though $X(t)$ is defined for a continuous range of time. We have a process in continuous time having a discrete state space. Suppose that $X(t)$ represents the maximum temperature at a particular place in $(0, t)$, then the set of possible values of $X(t)$ is continuous. Here we have a system in continuous time having a continuous state space.

So far we have assumed that the values assumed by the r.v. X_n (or $X(t)$) are one-dimensional, but the process $\{X_n\}$ (or $\{X(t)\}$) may be multi-dimensional. Consider $X(t) = (X_1(t), X_2(t))$, where X_1 represents the maximum and X_2 the minimum temperature at a place in an interval of time $(0, t)$. We have here a two-dimensional stochastic process in continuous time having continuous state space. One can similarly have multi-dimensional processes. One-dimensional processes can be classified, in general, into the following four types of processes:

(i) Discrete time, discrete state space
(ii) Discrete time, continuous state space
(iii) Continuous time, discrete state space
(iv) Continuous time, continuous state space.

All the four types may be represented by $\{X(t), t \in T\}$. In case of discrete time, the parameter generally used is n, i.e. the family is represented by $\{X_n, n = 0, 1, 2, \ldots\}$. In case of continuous time both the symbols $\{X_t, t \in T\}$ and $\{X(t), t \in T\}$ (where T is a finite or infinite interval) are used. The parameter t is usually interpreted as time, though it may represent such characters as distance, length, thickness and so on. Some authors call the discrete parameter family a stochastic *sequence*, and the continuous parameter family a stochastic *process*.

We shall use here the notation $\{X(t), t \in T\}$ both in the cases of discrete and continuous parameters and shall specify T, whenever necessary.

Relationship

In some of the cases, the r.v. X_n, i.e. members of the family $\{X_n, n \geq 1\}$ are mutually independent (as in Example 1(a)(i)), but more often they are not. We generally come across processes whose members are mutually dependent. The relationship among them is often of great importance.

The nature of dependence could be infinitely varied. Here dependence of some special types, which occurs quite often and is of great importance, will be considered. We may broadly describe some stochastic processes according to the nature of dependence relationship existing among the members of the family.

Processes with independent increments

If for all $t_1, \ldots, t_n$, $t_1 < t_2 < \ldots < t_n$, the random variables

$$X(t_2) - X(t_1), X(t_3) - X(t_2), \ldots, X(t_n) - X(t_{n-1})$$

are independent, then $\{X(t), t \in T\}$ is said to be a process with independent increments.

Suppose that we wish to consider the discrete parameter case. Consider a process in discrete time with independent increments. Writing

$$T = \{0, 1, 2, \ldots\}, t_i = i - 1, X(t_i) = X_{i-1},$$

$$Z_i = X_i - X_{i-1}, i = 1, 2, \ldots \text{ and } Z_0 = X_0,$$

we have a sequence of independent random variables $\{Z_n, n \geq 0\}$.

Markov process. If $\{X(t), t \in T\}$ is a stochastic process such that, given the value $X(s)$, the values of $X(t), t > s$, do not depend on the values of $X(u), u < s$, then the process is said to be a Markov process.

A definition of such a process is given below.

If, for, $t_1 < t_2 < \ldots < t_n < t$

$$\Pr\{a \leq X(t) \leq b \mid X(t_1) = x_1, \ldots, X(t_n) = x_n\}$$

$$= \Pr\{a \leq X(t) \leq b \mid X(t_n) = x_n\}$$

the process $\{X(t), t \in T\}$ is a Markov process.

A discrete parameter Markov process is known as a Markov chain.

2.3 STATIONARY PROCESSES

2.3.1 Second-Order Processes

We consider here real-valued stochastic processes in continuous time. A stochastic process $\{X(t), t \in T\}$ is called a *second-order process* if $E\{X(t)\}^2 < \infty$. It is a collection of *second-order random variables* (i.e. r.v.'s with finite second-order moments).

The mean function is defined by

$$m(t) = E\{X(t)\}$$

and the covariance function is defined by

$$C(s,t) = \text{cov}\{X(s), X(t)\}$$
$$= E\{X(s)X(t)\} - E\{X(s)\}E\{X(t)\}.$$

The notation $C_{s,t}$ is also used instead of $C(s, t)$

A covariance function (also called autocovariance function) satisfies the following properties

(1) It is symmetric in t and s, i.e.

$$C(s, t) = C(t, s), s, t, \in T$$

(2) Application of Schwarz inequality yields

$$C(s, t) \leq \sqrt{\{C(s,s)C(t,t)\}}$$

(3) It is *non-negative definite*, that is

$$\sum_{j=1}^{n} \sum_{k=1}^{n} a_j a_k C(t_j, t_k) = E\left\{\sum_{j=1}^{n} a_j X_{t_j}\right\}^2 \geq 0$$

  where $a_1, \ldots, a_n$ is a set of real numbers and $t_i \in T$.
(4) Closure properties. The sum and the product of two covariance functions are covariance functions.

2.3.2 Stationarity

Many important questions relating to a stochastic process can be adequately, answered on the basis of the knowledge of the first two moments of the process. An important concept is stationarity.

A second order process is called *covariance stationary* or *weakly stationary* or *wide sense stationary*, if its mean function $m(t)$ is independent of t and its covariance function $C(s, t)$ is a function only of the time difference $|t - s|$, for all t, s, that is,

$$C(s, t) = f(t - s).$$

This also implies that, for any t_0,

$$C(s + t_0, t + t_0) = \text{cov}\{X(s + t_0), X(t + t_0)\}$$
$$= \text{cov}\{X(s), X(t)\}$$
$$= C(s, t).$$

If for arbitrary $t_1, \ldots t_n$, the joint distribution of the vector random variables

$$(X\,(t_1), X\,(t_2), \ldots, X\,(t_n)), \text{ and } (X\,(t_1 + h), \ldots, X\,(t_n + h))$$

are the same for all $h > 0$, then the stochastic process $\{X\,(t), t \in T\}$ is said to be *stationary* of order n. It is *strictly stationary* if it is stationary of order n for any integer n.

Stationarity of a process implies that the probabilistic structure of the process is invariant under translation of the time axis. Many processes encountered in practice exhibit such a characteristic. Now for such a strictly stationary process, the statistical properties of the associated joint distribution of $\{X\,(t_1 + h), \ldots, X\,(t_n + h)\}$ will obviously be independent of h. This will hold, in particular, in case of the moments of the distribution, when these exist.

If the mean of the process $\{X\,(t)\}$ exists, then $E\,\{X\,(t)\}$ must be equal to $E\,\{X\,(t + h)\}$ for any h, so that $E\{X\,(t)\}$ must be a constant m, independent of t. Without loss of generality, we shall assume that m is zero. If, moreover, the covariance function $C\,(s\ t)$ (or $C_{s,t}$) exists, then

$$C\,(s, t) = \text{cov}\,\{X(t), X(s)\} = E\{X(t)X(s)\}$$

$$= E\{X(t + h)X(s + h)\}, \text{ for any } h$$

$$= E\{X(t - s)X(0)\}.$$

This shows that $C\,(s, t)$ is a function of the time difference $|t - s|$.

Higher moments, when these exist, can also be considered. Many of the important questions relating to a stochastic process can be adequately answered on the basis of the knowledge of the first two moments of the process.

Note that a strictly stationary process will not necessarily be a weakly stationary process, nor will a weakly stationary process be necessarily strictly stationary. A process which is not stationary (in any sense) is said to be *evolutionary*. The processes considered here will be assumed to be real throughout. We now define a class of stochastic processes in terms of an important specific distribution – normal distribution.

2.3.3 Gaussian processes

If the distribution of $(X\,(t_1), \ldots, X\,(t_n))$ for all $t_1, \ldots, t_n$ is multivariate normal, then $\{X\,(t), t \in T\}$ is said to be a *Gaussian* process.

Such processes derive their importance from the fact that there are many physical processes which are approximately normal, e.g. the process describing atmospheric turbulence. Further, such processes have some very useful and simple properties. We state below such a property.

If a Gaussian process X (t) is covariance stationary, then it is strictly stationary.

The explanation is simple. The multivariate normal distribution of $(X\,(t_1), \ldots, X\,(t_n))$ is completely determined by its mean vector $(\mu_1, \ldots, \mu_n)$ (where

$\mu_i = E\{X(t_i)\}$ and the variance-covariance matrix $(C(i,j))$, whose elements are
$C(i,j) = \text{cov }\{X(t_i), X(t_j)\}, i, j = 1, 2, \ldots, n$.
If the Gaussian process is covariance stationary, then $E\{X^2(t_i)\}, E\{X^2(t_j)\}$ are finite
and the covariance function $C(i,j)$ is a function only of the difference $i - j$.

Example 3(a). Let $X_n, n \geq 1$ be uncorrelated r.v.'s with mean 0 and variance 1. Then

$$C(n,m) = \text{cov}\{X_n, X_m\} = E\{X_n X_m\} = 0, \quad m \neq n$$

$$= 1, \quad m \neq n$$

and so $\{X_n, n \geq 1\}$ is covariance stationary.
If X_n are also identically distributed, then $\{X_n, n \geq 1\}$ is strictly stationary.

Example 3(b). Poisson process. Consider the process $\{X(t), t \in T\}$ with

$$\Pr\{X(t) = n\} = \exp(-at)(at)^n/n!, \quad a > 0, n = 0, 1, 2, \ldots$$

We see that

$$m(t) = at, \text{ var }\{X(t)\} = at$$

are functions of t. The process is evolutionary. The distribution of the process is
functionally dependent on t.

Example 3(c). Consider the process

$$X(t) = A_1 + A_2 t,$$

where A_1, A_2 are independent r.v.'s with $E(A_i) = a_i$, var $(A_i) = \sigma_i^2, i = 1, 2$. We have

$$m(t) = a_1 + a_2 t.$$

$$E\{X(t)X(s)\} = E\{(A_1 + A_2 t)(A_1 + A_2 s)\}$$

$$= E(A_1^2) + (s + t)E(A_1 A_2) + ts\, E(A_2^2)$$

$$= \sigma_1^2 + a_1^2 + (s + t)a_1 a_2 + ts(\sigma_2^2 + a_2^2).$$

Thus

$$E\{X(t)\}^2 = \sigma_1^2 + a_1^2 + 2t\, a_1 a_2 + t^2(\sigma_2^2 + a_2^2),$$

$$\text{var}\{X(t)\} = \sigma_1^2 + t^2\sigma_2^2,$$

and

$$C(s,t) = \text{cov }\{X(t), X(s)\} = \sigma_1^2 + ts\sigma_2^2.$$

The process is evolutionary.

Example 3(d). Consider the process

$$X(t) = A \cos \omega t + B \sin \omega t$$

where A, B are uncorrelated r.v.'s each with mean 0 and variance 1 and ω is a positive constant.
We have

$$m(t) = 0,$$

$$E\{X(t)X(s)\} = E\{(A \cos \omega t + B \sin \omega t)(A \cos \omega s + B \sin \omega s)\}$$

$$= \cos \omega t \cos \omega s + \sin \omega t \sin \omega s$$

$$= \cos (s - t)\omega.$$

Thus

$$E\{(X(t))^2\} = \text{var } \{X(t)\} = 1,$$

$$C(s,t) = \text{cov } \{X(t), X(s)\} = \cos (s - t)\omega.$$

Here the first two moments are finite and the covariance function is a function of $(s - t)$. Thus the process is covariance stationary.

Example 3(e). Consider the process $\{X(t), t \in T\}$ whose probability distribution, under a certain condition, is given by

$$\Pr\{X(t) = n\} = \frac{(at)^{n-1}}{(1 + at)^{n+1}}, \quad n = 1, 2, \ldots$$

$$= \frac{at}{1 + at}, \quad n = 0.$$

We have

$$E\{X(t)\} = \sum_{n=0}^{\infty} n \Pr\{X(t) = n\}$$

$$= \sum_{n=1}^{\infty} \frac{n(at)^{n-1}}{(1 + at)^{n+1}}$$

$$= \frac{1}{(1 + at)^2} \sum_{n=1}^{\infty} n \left(\frac{at}{1 + at}\right)^{n-1} = 1$$

and

$$E\{X^2(t)\} = \sum_{n=2}^{\infty} \frac{n^2(at)^{n-1}}{(1 + at)^{n+1}}$$

$$= \frac{at}{(1+at)^3} \left\{ \sum_{n=2}^{\infty} n(n-1) \left(\frac{at}{1+at} \right)^{n-2} \right\}$$

$$+ \sum_{n=1}^{\infty} \frac{n(at)^{n-1}}{(1+at)^{n+1}}$$

$$= 2at + 1.$$

Thus, var $\{X(t)\} = 2\,at$.

Here the first moment is constant but the second moment (and the variance) increases with t. The process $\{X(t), t \in T\}$ is not stationary.

2.4 MARTINGALES

Another way of characterizing dependence relationship between members of a stochastic process is through martingale property which involves conditional expectation. This property, which is an important basic tool has been used in several contexts both in theoretical and applied probability. The development of martingale theory is due to Doob (1952); Neveu (1965, 1972) deals extensively with this theory. An exhaustive account is given by Karlin and Taylor (1975).

2.4.1 Definitions and Examples

Definition 1. A discrete parameter stochastic process $\{X_n, n \geq 0\}$ is called a *martingale* (or a *martingale process*) if,

(i) $E\{|X_n|\} < \infty$

and

(ii) $E\{X_{n+1} \mid X_n, X_{n-1}, \ldots, X_0\} = X_n$ \hfill (4.1)

Example 4(a). Let $\{Z_i\}$, $i = 1, 2, \ldots$ be a sequence of i.i.d. random variables with mean 0 and let $X_n = \sum_{i=1}^{n} Z_i$. Then $\{X_n, n \geq 1\}$ is a martingale.

We have

$$X_{n+1} = X_n + Z_{n+1}$$

so that

$$E\{X_{n+1} \mid X_n, X_{n-1}, \ldots, X_1\}$$
$$= E\{X_n + Z_{n+1} \mid X_n, \ldots, X_1\}$$
$$= E\{X_n \mid X_n, \ldots, X_1\} + E\{Z_{n+1} \mid X_n, \ldots, X_1\}$$
$$= X_n + E\{Z_{n+1}\}, \text{ since } Z_i\text{'s are independent}$$
$$= X_n, \text{ since } E\{Z_{n+1}\} = 0$$

Suppose that a player plays against an infinitely rich adversary: he stands to gain 1 unit of money with probability p and to lose 1 unit of money (or gain -1 unit) with probability q ($= 1 - p$). Denote by Z_n his gain in nth game and by $X_n = Z_1 + \ldots + Z_n$ his cumulative gain in the first n games. We have $E\{Z_n\} = p - q$ which is equal to 0 *iff* $p = q$; then $\{X_n, n \geq 1\}$ is a martingale. A game is said to be fair when $p = q$.

Example 4(b). Let $\{Z_i, i = 1, 2, \ldots\}$ be a sequence of i.i.d. random variables with $E\{Z_i\} = 1$ and let $X_n = \prod_{i=1}^{n} Z_i$. Then $\{X_n, n \geq 1\}$ is a martingale.

We have, $X_{n+1} = Z_{n+1} X_n$ and thus

$$
\begin{aligned}
E\{X_{n+1} &| X_n, \ldots, X_1\} \\
&= E\{Z_{n+1} X_n | X_n, \ldots, X_1\} \\
&= [E\{X_n | X_n, \ldots, X_1\}]\,[E\{Z_{n+1} | X_n, \ldots, X_1\}]
\end{aligned}
$$

because of indepdendence of Z_i's

$$
\begin{aligned}
&= X_n E\{Z_{n+1}\} \\
&= X_n, \quad \text{since } E\{Z_i\} = 1.
\end{aligned}
$$

Note :

(1) A martingale has a constant mean.
    Taking expectations of both sides of (4.1) we get

$$E\{X_{n+1}\} = E\{X_n\}$$

$$= E\{X_{n-1}\}$$

$$\cdots \qquad \cdots$$

$$= E\{X_1\} = E(X_0).$$

(2) The concept of martingale has been extended to continuous parameter stochastic processes as well.

(3) While the concept of martingale involves expectations, the Markovian concept involves distributions. It is to be noted that a Markov process need not necessarily be a martingale.

Another definition, a more general one, is given below.

Definition 2. Let $\{X_n, n \geq 0\}$ and $\{Y_n, n \geq 0\}$ be two discrete parameter stochastic processes. The process $\{X_n, n \geq 0\}$ is said to be a martingale w.r.t. the process $\{Y_n, n \geq 0\}$, if, for $n = 0, 1, 2, \ldots$

(i) $E\{|X_n|\} < \infty$

and   (ii) $E\{X_{n+1} | Y_n, Y_{n-1}, \ldots, Y_0\} = X_n$. $\qquad\qquad$ (4.2)

Consider Example 4(a). We have

$E\{X_{n+1}|Z_n, \ldots, Z_1\} = X_n$ (because of independence of Z_i's), so that $\{X_n, n \geq 1\}$ is a martingale w.r.t. $\{Z_n, n \geq 0\}$.

Here X_n is assumed to be a function of $(Y_0, Y_1, \ldots, Y_n)$ so that knowledge of $Y_0, Y_1, \ldots, Y_n$ determines X_n.

A more general concept can be obtained by relaxing the equality condition.

Definition 3. A stochastic process $\{X_n, n \geq 0\}$ with $E\{|X_n|\} < \infty$ is said to be a *sub-martingale*, if

$$E\{X_{n+1} \mid X_n, X_{n-1}, \ldots, X_0\} \geq X_n \tag{4.3}$$

and is said to be a *super-martingale*, if

$$E\{X_{n+1} \mid X_{n-1}, \ldots, X_0\} \leq X_n \tag{4.4}$$

It follows from (4.3) that, in case of a sub-martingale,

$$E\{X_{n+1}\} \geq E\{X_n\} \geq E\{X_0\}$$

and in case of a super-martingale

$$E\{X_{n+1}\} \leq E\{X_n\} \leq \ldots \leq E\{X_0\}$$

The concept of sub-martingale and super-martingale can be extended (on the lines of Definition 2) by considering another stochastic process $\{Y_n, n \geq 0\}$ such that X_n is a function of $(Y_0, Y_1, \ldots, Y_n)$.

2.4.2 Martingale Convergence Theorem

The martingale concept is used in several contexts. There are two important results that are useful in several applications: these are the *optional sampling theorem* and the *martingale convergence theorem*. We state below the second theorem without proof. This theorem deals with the convergence of a martingale X_n to a limit random variable as $n \to \infty$.

Theorem 2.1. If $\{X_n, n \geq 1\}$ is a martingale such that, for some finite M,

$$E\{|X_n|\} \leq M$$

then X_n converges to a random variable (denoted by X_∞) with probability 1, that is,

$$\Pr\{\lim_{n \to \infty} X_n = X_\infty\} = 1. \tag{4.5}$$

The above is one form of the martingale convergence theorem. Several other forms of the theorem are given.

For an exhaustive discussion including applications, refer to Karlin and Taylor (1975). We shall discuss applications of martingale theory in Markov chains and branching process at appropriate places.

EXERCISES

2.1. Let X_n, for n even, take values $+1$ and -1 each with probability $\frac{1}{2}$, and for

n odd, take values $\sqrt{a}$, $-1/\sqrt{a}$, with probability $1/(a+1)$, $a/(a+1)$ respectively (a is a real number > -1 and $\neq 0$, 1). Further let X_n's be independent. Show that $\{X_n, n \geq 1\}$ is covariance stationary but not strictly stationary. (Compare this with Example 3 (a).)

2.2. Let $Y_n = a_0 X_n + a_1 X_{n-1}$ ($n = 1, 2, \ldots$), where a_0, a_1 are constants and X_n, $n = 0, 1, 2, \ldots$ are i.i.d. random variables wtih mean 0 and variance σ^2. Is $\{Y_n, n \geq 1\}$ covariance stationary? Is it a Markov process?

2.3. Let $X(t) = \sum_{r=1}^{k} (A_r \cos \theta_r t + B_r \sin \theta_r t)$

where A_r, B_r are uncorrelated random variables with mean 0, variance σ^2 and θ_r are constants. Show that $\{X(t), t \geq 0\}$ is covariance stationary.

2.4. Let $X(t) = A_0 + A_1 t + A_2 t^2$, where A_i, $i = 0, 1, 2$ are uncorrelated random variables with mean 0 and variance 1. Find the mean value function and the covariance function of $\{X(t), t \in T\}$.

2.5. Find the covariance function of $\{Y_n, n \geq 1\}$ given by

$$Y_n = a_0 X_n + a_1 X_{n-1} + \ldots + a_k X_{n-k}, \quad n = 1, 2, \ldots,$$

where a's are constants and X_n's are uncorrelated random variables.

2.6 Show that every stochastic process $\{X, t = 0, 1, 2, \ldots\}$ with independent increments is a Markov process. Is the converse true?

2.7 If $\{X(t), t \geq 0\}$ with $X(0) = 0$ is a process having independent increments, then

$$\text{cov}\{X(s), X(t)\} = \text{var}\{X(s)\}, \text{ for } 0 \leq s \leq t$$

$$= \text{var}\{X(\min(s, t))\}, \text{ for any } s, t (t > 0, s > 0).$$

(Such a process is not covariance stationary.)

Deduce that, for a process $\{X(t), t \geq 0\}$ having independent increments,

$$\text{var}\{X(t + \Delta t) - X(t)\} = \text{var}\{X(t + \Delta t)\} - \text{var}\{X(t)\}.$$

In Exercises 8–9, the distribution of $N(t)$ (the number of events that occur in $(0, t]$ is assumed to be Poisson with mean at, i.e. $\{N(t), t \geq 0\}$ is assumed to be a Poisson Process.

2.8. Poisson increment process. The process $\{X(t), t \geq 0\}$ defined by

$$X(t) = \frac{N(t+\epsilon) - N(t)}{\epsilon},$$

where $\epsilon > 0$ is a given constant, is known as Poisson increment process. Show that the process is covariance stationary having

$$E\{X(t)\} = a$$

and

$$C(s,t) = 0, \quad |t-s| > \epsilon$$

$$= \frac{a}{\epsilon} - \frac{a|t-s|}{\epsilon^2}, |t-s| < \epsilon.$$

2.9.(a) Semi-random telegraph signal: The process $\{X(t), t \geq 0\}$ defined by

$$X(t) = (-1)^{N(t)}$$

is called a semi-random telegraph signal. Show that the process is evolutionary having

$$E\{X(t)\} = \exp(-2at)$$

and

$$C(s,t) = \exp\{-2a|t-s|\} - \exp\{-2a(t+s)\}.$$

The process $\{X(t)\}$ is a two-valued process with two possible values 1 and -1. It is a *one-minus-one* process.

(b) Random telegraph signal : The process $\{Y(t), t \geq 0\}$ defined by

$$Y(t) = \alpha X(t),$$

where α is a random variable which assumes the values -1 and 1 with equal probability and which is independent of $X(t)$ (defined in (a)), is called a random telegraph signal. Show that the process is covariance stationary having

$$E\{Y(t)\} = 0$$

and

$$C(s,t) = \exp\{-2a|t-s|\}.$$

(Random telegraph signals are useful in the construction of random signal generators).

2.10. Semi-binary random transmission: Assume that a fair coin is tossed every T units of time. The process $\{Z(t), t \geq 0\}$ such that $Z(t)$ takes the values 1 or -1 according as n th toss results in a head or tail respectively, where $(n-1)T < t < nT$, is called semi-random binary transmission. Show that the process is covariance stationary having

$$E\{Z(t)\} = 0$$

and

$$C(s,t) = 1, \text{ if } s,t \text{ are in the same tossing interval}$$

$$(n-1)T < s,t < nT$$

$$= 0, \text{ otherwise.}$$

REFERENCES

M.S. Bartlett. *An Introduction to Stochastic Processes*, 3rd Ed., Cambridge University Press, London (1978).

U.N. Bhat. *Elements of Applied Stochastic Processes*, 2nd ed., Wiley, New York (1984).

S. Chandrasekhar. Stochastic problems in physics and astronomy, *Review of Modern Physics*, **15**, 1–89 (1943). [Reprinted in *Selected Papers on Noise and Stochastic Processes*, (Ed. Wax, N.) 1–91, Dover Publications, New York (1954)].

J.L. Doob. *Stochastic Processes*, Wiley, New York (1953).

S. Karlin and H.M. Taylor. *A First Course in Stochastic Processes*, 2nd ed. Academic Press, New York (1975).

J. Neveu. *Mathematical Foundations of the Calculus of Probabilities*, Holden-Day, San Francisco (1965).

———— *Martingales A Temps Discret*, Masson, Paris (1972).

J. Neyman. Indeterminism in Science and new demands for statisticians, *J. Am. Stat. Ass.* **55**, 625–639 (1960).

N.U. Prabhu. *Stochastic Processes*, Macmillan, New York (1965).

S.M. Ross. *Stochastic Processes*, Wiley, New York (1983).

E. Wong, and B. Hajek. *Stochastic Processes in Engineering Systems*, Springer-Verlag, New York (1985).

CHAPTER 3

Markov Chains

3.1 DEFINITION AND EXAMPLES

Consider a simple coin tossing experiment repeated for a number of times. The possible outcomes at each trial are two: head with probability, say, p and tail with probability $q, p + q = 1$. Let us denote head by 1 and tail by 0 and the random variable denoting the result of the nth toss by X_n. Then for $n = 1, 2, 3, \ldots$,

$$\Pr\{X_n = 1\} = p, \Pr\{X_n = 0\} = q.$$

Thus we have a sequence of random variables $X_1, X_2, \ldots$. The trials are independent and the result of the nth trial does not depend in any way on the previous trials numbered $1, 2, \ldots, (n-1)$. The random variables are independent.

Consider now the random variable given by the partial sum $S_n = X_1 + \ldots + X_n$. The sum S_n gives the accumulated number of heads in the first n trials and its possible values are $0, 1, \ldots, n$.

We have $S_{n+1} = S_n + X_{n+1}$. Given that $S_n = j$ $(j = 0, 1, \ldots n)$, the r.v. S_{n+1} can assume only two possible values : $S_{n+1} = j$ with probability q and $S_{n+1} = j + 1$ with probability p; these probabilities are not at all affected by the values of the variables $S_1, \ldots, S_{n-1}$. Thus

$$\Pr\{S_{n+1} = j + 1 \mid S_n = j\} = p$$

$$\Pr\{S_{n+1} = j \mid S_n = j\} = q.$$

We have here an example of a Markov* chain, a case of simple dependence that the outcome of $(n + 1)$st trial depends directly on that of nth trial and *only* on it. The conditional probability of S_{n+1} given S_n depends on the value of S_n and the manner in which the value of S_n was reached is of no consequence.

Example 1(a). *Polya's urn model*: an urn contains b black and r red balls. A ball is drawn at random and is replaced after the drawing, i.e. drawing is with replacement. The outcome at the nth drawing is either a black ball or a red ball. Let the random variable X_n be defined as

$$X_n = 1, \text{ if } n\text{th drawing results in a black ball, and}$$

---

*Andrei Andreivich Markov (1856-1922).

$X_n = 0$, if it results in a red ball.

There are two possible outcomes of X_n with

$$\Pr\{X_n = 1\} = b/(b+r), \Pr\{X_n = 0\} = r/(b+r), \quad \text{for all } n \geq 1.$$

We have

$$\Pr\{X_1 = j, \ldots, X_n = k\} = \Pr\{X_1 = j\} \ldots \Pr\{X_n = k\}, j, k = 0, 1,$$

because of independence of $X_1, \ldots, X_n$.

Polya's urn model is such that after each drawing not only that the ball drawn is replaced but $c\ (> 0)$ balls of the colour drawn (at that drawing) are added to the urn so that the number of balls of the colour drawn increases, while the number of balls of the other colour remains unchanged as at the drawing, i.e. if the nth drawing results in a black ball, then the number of black balls for $(n + 1)$st drawing will be c more than the number of black balls at the nth drawing, while the number of red balls will be the same for both the drawings. We have

$$\Pr\{X_2 = 1\} = \Pr\{X_2 = 1, X_1 = 0\} + \Pr\{X_2 = 1, X_1 = 1\}$$

$$= \Pr\{X_2 = 1 \mid X_1 = 0\} \Pr\{X_1 = 0\} + \Pr\{X_2 = 1 \mid X_1 = 1\} \Pr(X_1 = 1)$$

$$= \frac{b}{b+r+c} \cdot \frac{r}{b+r} + \frac{b+c}{b+r+c} \cdot \frac{b}{b+r}$$

$$= \frac{b}{b+r}.$$

It can be shown that $\Pr\{X_n = 1\} = b/(b+r)$, $\Pr\{X_n = 0\} = r/(b+r)$ for all n. Again,

$$\Pr\{X_3 = 1, X_2 = 1\}$$

$$= \Pr(X_3 = 1, X_2 = 1, X_1 = 0) + \Pr\{X_3 = 1, X_2 = 1, X_1 = 1\}$$

$$= \Pr\{X_3 = 1 \mid X_2 = 1, X_1 = 0\} \Pr\{X_2 = 1 \mid X_1 = 0\} \Pr\{X_1 = 0\}$$

$$+ \Pr\{X_3 = 1 \mid X_2 = 1, X_1 = 1\} \Pr\{X_2 = 1 \mid X_1 = 1\} \Pr\{X_1 = 1\}$$

$$= \frac{b+c}{b+r+2c} \cdot \frac{b}{b+r+c} \cdot \frac{r}{b+r} + \frac{b+2c}{b+r+2c} \cdot \frac{b+c}{b+r+c} \cdot \frac{b}{b+r}$$

$$= \frac{b(b+c)}{(b+r)(b+r+c)}.$$

Hence

$$\Pr\{X_3 = 1 \mid X_2 = 1\} = \frac{\Pr\{X_3 = 1, X_2 = 1\}}{\Pr\{X_2 = 1\}}$$

$$= \frac{b+c}{b+r+c};$$

but

$$\Pr\{X_3 = 1 \mid X_2 = 1, X_1 = 1\} = \frac{b+2c}{b+r+2c}$$

and

$$\Pr\{X_3 = 1 \mid X_2 = 1, X_1 = 0\} = \frac{b+c}{b+r+2c}.$$

Hence

$$\Pr\{X_3 = 1 \mid X_2 = 1\} \neq \Pr\{X_3 = 1 \mid X_2 = 1, X_1 = k\}, k = 0, 1.$$

Here the conditional probability of X_3 given X_2 is not equal to that of X_3 given X_2 and X_1.

Now, consider the random variable Y_n which gives the number of black balls at the time of nth drawing. Y_1 takes only one value b, Y_2 takes values b and $b + c$, Y_3 values b, $b + c$ and $b + 2c$ etc. In general, the r.v. Y_n can assume values $b, b + c, \ldots, b + (n - 1)c$, given that Y_{n-1} assumes one of the possible values $b, b + c, \ldots, b + (n - 2)c$ say, $i = b + kc, k = 0, 1, \ldots, n - 2$, then Y_n assumes the value $j = i$ with probability a_1 and the value $j = i + c$ with probability b_1, where

$$a_1 = \Pr\{X_{n-1} = 0\} = \frac{r}{b+r}$$

$$b_1 = \Pr\{X_{n-1} = 1\} = \frac{b}{b+r}.$$

Let $\{X_n, n = 0, 1, 2, \ldots\}$ be a sequence of random variables (or a discrete parameter stochastic process). Let the possible outcomes of X_n be j ($j = 0, 1, 2, \ldots$), where the number of outcomes may be finite (say, m) or denumerable. The possible values of X_n constitute a set $S = \{0, 1, 2 \ldots\}$, and that the process has the state space S. Unless otherwise stated, by state space of a Markov chain, we shall imply *discrete state space* (having a finite or a countably infinite number of elements); it could be $N = \{0, 1, 2, \ldots\}$ or some other subset of the set of integers $I = \{\ldots, -2, -1, 0, 1, 2, \ldots\}$.

Definition. The stochastic process $\{X_n, n = 0, 1, 2, \ldots\}$ is called a Markov chain, if, for $j, k, j_1, \ldots j_{n-1} \in N$ (or any subset of I),

$$\Pr\{X_n = k \mid X_{n-1} = j, X_{n-2} = j_1, \ldots, X_0 = j_{n-1}\}$$

$$= \Pr\{X_n = k \mid X_{n-1} = j\} = p_{jk} \text{ (say)}, \tag{1.1}$$

whenever the first member is defined.

The outcomes are called the states of the Markov chain; if X_n has the outcome j (i.e. $X_n = j$), the process is said to be at state j at nth trial. To a pair of states (j, k) at the two successive trials (say, nth and $(n+1)$st trials) there is an associated conditional probability p_{jk}. It is the probability of transition from the state j at nth trial to the state k at $(n+1)$st trial. The transition probabilities p_{jk} are basic to the study of the structure of the Markov chain.

The transition probability may or may not be independent of n. If the transition probability p_{jk} is independent of n, the Markov chain is said to be *homogeneous* (or to have *stationary transition probabilities*). If it is dependent on n, the chain is said to be non-homogeneous. *Here we shall confine to homogeneous chains.* The transition probability p_{jk} refers to the states (j, k) at two *successive* trials (say, nth and $(n+1)$st trial); the transition is one-step and p_{jk} is called one-step (or unit step) transition probability. In the more general case, we are concerned with the pair of states (j, k) at two non-successive trials, say, state j at the nth trial and state k at the $(n+m)$th trial. The corresponding transition probability is then called m-step transition probability and is denoted by $p_{jk}^{(m)}$, i.e.

$$p_{jk}^{(m)} = \Pr\{X_{n+m} = k \mid X_n = j\}.$$

This will be considered in Sec. 3.2 below.

3.1.1 Transition Matrix. The transition probabilities p_{jk} satisfy

$$p_{jk} \geq 0, \quad \sum_k p_{jk} = 1 \text{ for all } j. \tag{1.2}$$

These probabilities may be written in the matrix form

$$P = \begin{pmatrix} p_{11} & p_{12} & p_{13} & \cdots \\ p_{21} & p_{22} & p_{23} & \cdots \\ \cdots & \cdots & \cdots & \cdots \\ \cdots & \cdots & \cdots & \cdots \end{pmatrix} \tag{1.3}$$

This is called the *transition probability matrix* or *matrix of transition probabilities* (t.p.m.) of the Markov chain. P is a *stochastic* matrix i.e. a square matrix with non-negative elements and unit row sums.

Example 1(b). A particle performs a random walk with absorbing barriers, say, at 0 and 4. Whenever it is at any position r $(0 < r < 4)$, it moves to $r+1$ with probability p or to $(r-1)$ with probability q, $p + q = 1$. But as soon as it reaches 0 or 4 it remains there itself. Let X_n be the position of the particle after n moves. The different states of X_n are the different positions of the particle. $\{X_n\}$ is a Markov chain whose unit-step transition probabilities are given by

$$\Pr\{X_n = r + 1 \mid X_{n-1} = r\} = p$$

$$\Pr\{X_n = r - 1 \mid X_{n-1} = r\} = q \qquad 0 < r < 4$$

$$\Pr\{X_n = 0 \mid X_{n-1} = 0\} = 1,$$

and
$$\Pr\{X_n = 4 \mid X_{n-1} = 4\} = 1.$$

The transition matrix is given by

States of X_n

States of X_{n-1}		0	1	2	3	4
	0	1	0	0	0	0
	1	q	0	p	0	0
	2	0	q	0	p	0
	3	0	0	q	0	p
	4	0	0	0	0	1

Example 1(c). *Random walk between two barriers*: *General Case*

Consider that a particle may be at any position r, $r = 0, 1, \ldots, k \ (\geq 1)$ of the x – *axis*. From state r it moves to state $r + 1$, $1 \leq r \leq k - 1$ with probability p and to state $r - 1$ with probability q. As soon as it reaches state 0 it remains there with probability a and is reflected to state 1 with probability $1 - a$ $(0 < a < 1)$; if it reaches state k it remains there with probability b and is reflected to $k - 1$ with probability $1 - b$ $(0 < b < 1)$. Then $\{X_n\}$, where X_n is the position of the particle after n steps or moves, is a Markov chain with state space $S = \{0, 1, \ldots, k\}$. The transition matrix is

$$P = \quad\begin{array}{c} 0 \\ 1 \\ \cdot \\ k-1 \\ k \end{array}\begin{pmatrix} a & 1-a & 0 & \cdots & 0 & 0 & 0 \\ q & 0 & p & \cdots & 0 & 0 & 0 \\ \cdots & \cdots & \cdots & \cdots & \cdots & \cdots & \cdots \\ 0 & 0 & 0 & \cdots & q & 0 & p \\ 0 & 0 & 0 & \cdots & 0 & 1-b & b \end{pmatrix}$$

If $a = 1$, then 0 is an *absorbing barrier* and if $a = 0$, then 0 is a *reflecting barrier*; if $0 < a < 1$, 0 is an *elastic barrier*. Similar is the case with state k. The case when both 0 and k are absorbing barriers corresponds to the familiar Gambler's ruin problem (with total capital between the two gamblers amounting to k).

Example 1(d). Suppose that a coin with probability p for a head is tossed indefinitely. Let X_n, the outcome of the nth trial, be k, where $k \ (= 0, 1, \ldots, n)$ denotes that there is a run of k successes, i.e. the length of the uninterrupted block of heads is k. $\{X_n, n \geq 0\}$ constitutes a Markov chain, with unit-step transition probabilities

$$p_{jk} = \Pr\{X_n = k \mid X_{n-1} = j\} = p, \quad k = j+1$$

$$= q, \quad k = 0$$

$$= 0, \quad \text{otherwise.}$$

The transition matrix is given by

States of X_n

States of X_{n-1}	0	1	2	...	k	k+1	...
0	q	p	0	...	0	0	...
1	q	0	p	...	0	0	...
2	q	0	0	...	.	.	...
.	.	.	.	...	.	.	...
.	.	.	.	...	.	.	...
k	q	0	0	...	0	p	...
.	.	.	.	...	.	.	...
.	.	.	.	...	.	.	...
.	.	.	.	...	.	.	...

Example 1(e). Partial sum of independent random variables: Consider a series of coin tossing experiments, where the outcomes of nth trial are denoted by 1 (for a head) and 0 (for a tail). Let X_n be the random variable denoting the outcome of nth trial and $S_n = X_1 + \ldots + X_n$ be the nth partial sum. The possible values of S_n are 0, 1, . . ., n, i.e. the states of S_n are r, $r = 0, 1, \ldots n$, $\{S_n, n \geq 0\}$ is a Markov chain with transition matrix as given below.

Transition matrix

States of S_n

States of S_{n-1}	0	1	2	...	r-1	r	...
0	q	p	0	...	0	0	...
1	0	q	p	...	0	0	...
2	0	0	q	...	0	0	...
.	.	.	.	...	.	.	...
.	.	.	.	...	.	.	...
.	.	.	.	...	.	.	...
r-1	0	0	0	...	q	p	...
.	.	.	.	...	.	.	...
.	.	.	.	...	.	.	...

A Markov chain $\{X_n, n \geq 0\}$ with k states, where k is finite, is said to be a finite Markov chain. The transition matrix P is, in this case, a square matrix with k rows and k columns. Examples 1(b) and 1(c) deal with finite Markov chians.

The number of states could however be infinite. When the possible values of X_n form a denumerable set, then the Markov chain is said to be *denumerably infinite or denumerable* and the chain is said to have a countable state space. Examples 1(d) and 1(e) are of denumerable Markov chains.

Note: $\{X_n\}$ discussed in Example 1(a) is non-Markov unless $c = 0$ (in which case X_n, $n \geq 1$ are i.i.d. random variables).

Example 1(f) . *A simple queueing model*

Consider a counter to which customers arrive for service. There is one server who serves one customer (if any present) only at epochs of time $0, 1, 2, \ldots$. Assume that in the time interval $(n, n + 1)$, a number of customers Y_n arrive, where Y_n, $n = 0$, $1, 2, \ldots$ are i.i.d. r.v.'s with mass function $\Pr\{Y_n = k\} = p_k, k = 0, 1, 2, \ldots$ The waiting room can accommodate at most m customers including the one being served and customers who find the waiting room full leave without receiving service and do not return. Let X_n be the number of customers present at epoch n, including the one being served, if any. Then $\{X_n, n \geq 0\}$ is a Markov chain with state space $S = \{0, 1, \ldots, m\}$. We have

$$
\begin{aligned}
X_{n+1} &= Y_n, & \text{if } X_n = 0 \text{ and } 0 \leq Y_n \leq m - 1 \\
&= X_n - 1 + Y_n, & \text{if } 1 \leq X_n \leq m \text{ and } 0 \leq Y_n \leq m + 1 - X_n \\
&= m, & \text{otherwise.}
\end{aligned}
$$

Denoting

$$
P_m = p_m + p_{m+1} + \ldots, \quad P_0 = 1
$$

we can write the transition matrix as

$$
\begin{bmatrix}
p_0 & p_1 & p_2 & \cdots & p_{m-2} & p_{m-1} & P_m \\
p_0 & p_1 & p_2 & \cdots & p_{m-2} & p_{m-1} & P_m \\
0 & p_0 & p_1 & \cdots & p_{m-3} & p_{m-2} & P_{m-1} \\
\cdots & \cdots & \cdots & \cdots & \cdots & \cdots & \cdots \\
0 & 0 & 0 & \cdots & p_0 & p_1 & P_2 \\
0 & 0 & 0 & \cdots & 0 & p_0 & P_1
\end{bmatrix}
$$

3.1.1.1 Probability Distribution

It may be seen that the probability distribution of $X_r, X_{r+1}, \ldots, X_{r+n}$ can be computed in terms of the transition probabilities p_{jk} and the initial distribution of X_r. Suppose, for simplicity, that $r = 0$, then

$$
\Pr\{X_0 = a, X_1 = b, \ldots, X_{n-2} = i, X_{n-1} = j, X_n = k\}
$$

$$
= \Pr\{X_n = k \mid X_{n-1} = j, \ldots, X_0 = a\} \Pr\{X_{n-1} = j, \ldots, X_0 = a\}
$$

$$
= \Pr\{X_n = k \mid X_{n-1} = j\} \Pr\{X_{n-1} = j \mid X_{n-2} = i\} \Pr\{X_{n-2} = i, \ldots, X_0 = a\}
$$

$$
= \Pr\{X_n = k \mid X_{n-1} = j\} \Pr\{X_{n-1} = j \mid X_{n-2} = i\} \ldots
$$

$$
\ldots \Pr\{X_1 = b \mid X_0 = a\} \Pr\{X_0 = a\}
$$

$$
= \{\Pr(X_0 = a)\} p_{ab} \cdots p_{ij} p_{jk}. \tag{1.4}
$$

Thus,

$$\Pr\{X_r = a, X_{r+1} = b, ..., X_{r+n-2} = i, X_{r+n-1} = j, X_{r+n} = k\}$$

$$= \{\Pr(X_r = a)\}\, p_{ab}\cdots p_{ij}\, p_{jk}. \tag{1.4a}$$

Example 1(g). Let $\{X_n, n \geq 0\}$ be a Markov chain with three states 0, 1, 2 and with transition matrix

$$\begin{pmatrix} 3/4 & 1/4 & 0 \\ 1/4 & 1/2 & 1/4 \\ 0 & 3/4 & 1/4 \end{pmatrix}$$

and the initial distribution $\Pr\{X_0 = i\} = \frac{1}{3}, i = 0, 1, 2$.

We have

$$\Pr\{X_1 = 1 \mid X_0 = 2\} = \tfrac{3}{4}$$

$$\Pr\{X_2 = 2 \mid X_1 = 1\} = \tfrac{1}{4}$$

$$\Pr\{X_2 = 2, X_1 = 1 \mid X_0 = 2\}$$

$$= \Pr\{X_2 = 2 \mid X_1 = 1\}\Pr\{X_1 = 1 \mid X_0 = 2\} = \tfrac{1}{4}\cdot\tfrac{3}{4} = \tfrac{3}{16}$$

$$\Pr\{X_2 = 2, X_1 = 1, X_0 = 2\}$$

$$= \Pr\{X_2 = 2, X_1 = 1 \mid X_0 = 2\}\Pr\{X_0 = 2\} = \tfrac{3}{16}\cdot\tfrac{1}{3} = \tfrac{1}{16}$$

$$\Pr\{X_3 = 1, X_2 = 2, X_1 = 1, X_0 = 2\}$$

$$= \Pr\{X_3 = 1 \mid X_2 = 2, X_1 = 1, X_0 = 2\} \times$$
$$\Pr\{X_2 = 2, X_1 = 1, X_0 = 2\}$$

$$= \Pr\{X_3 = 1 \mid X_2 = 2\}\left(\tfrac{1}{16}\right) = \tfrac{3}{4}\cdot\tfrac{1}{16} = \tfrac{3}{64}.$$

Remark

   The matrix of transition probabilities together with the initial distribution completely specifies a Markov chain $\{X_n, n = 0, 1, 2, ...\}$.
We state (without proof) the general existence theorem of Markov chains.

   Given the set N and the sequence of stochastic matrices $(^{(n)}p_{jk}) = {}^{(n)}P$, there exists a Markov chain $\{X_n, n \geq 0\}$ with state space N and transition probability matrix, $^{(n)}P$. (For proof, see Iosifescu & Tautu, *Stochastic Processes–I*, Springer-Verlag (1973), Chung (1967)).

3.1.1.2 *Strong Markov Property*

Stopping time for a sequence of r.v.'s $\{X_n\}$ is a random variable (see Sec. 6.4.1).

Let N be a stopping time for a Markov chain $\{X_n, n > 0\}$ and let A and B be two events (relating to X_n and happening) *prior* and *posterior* respectively to N. Then

$$\Pr\{B \mid X_N = i, A\} = Pr\{B \mid X_N = i\}.$$

This is called the *strong Markov property*. It shows that if N is a stopping time for a Markov chain $\{X_n, n > 0\}$, then the evolution of the chain starts afresh from the state reached at time N.

Strong Markov property is implied by the Markov property; both the properties are equivalent when N is constant (a degenerate r.v.).

Every discrete time Markov chain $\{X_n, n \geq 0\}$ possesses the strong Markov property.

3.1.2 Order of a Markov chain

Definition. A Markov chain $\{X_n\}$ is said to be of order $s(s = 1, 2, 3, \ldots)$, if, for all n,

$$\Pr\{X_n = k \mid X_{n-1} = j, X_{n-2} = j_1, \ldots, X_{n-s} = j_{s-1}, \ldots\}$$

$$= \Pr\{X_n = k \mid X_{n-1} = j, \ldots, X_{n-s} = j_{s-1}\}. \tag{1.5}$$

whenever the l.h.s. is defined.

A Markov chain $\{X_n\}$ is said to be of order one (or simply a Markov chain) if

$$\Pr\{X_n = k \mid X_{n-1} = j, X_{n-2} = j_1, \ldots\} = \Pr\{X_n = k \mid X_{n-1} = j\}$$

$$= p_{jk} \tag{1.6}$$

whenver $\Pr\{X_{n-1} = j, X_{n-2} = j_1, \ldots\} > 0$.

Unless explicitly stated otherwise, we shall mean by Markov chain, a *chain of order one*, to which we shall mostly confine ourselves here. A chain is said to be of order zero if $p_{jk} = p_k$ for all j. This implies independence of X_n and X_{n-1}. For example, for the Bernoulli coin tossing experiment, the t.p.m. is $\begin{pmatrix} q & p \\ q & p \end{pmatrix} = \begin{pmatrix} 1 \\ 1 \end{pmatrix}(q, p) = \mathbf{e}\,(q, p)$.

Denote the state that a day is rainy by 1 and that a day is not rainy by 0.

Let (1.5) hold for $s = 2$ and let

$$p_{ijk} = \Pr\{\text{actual day is in state } k \mid \text{the preceding day was in state } j,$$
$$\text{the day before the preceding was in state } i\}, i, j, k = 0, 1.$$

We then have a Markov chain of order two. Note that the matrix (p_{ijk}) is not a stochastic matrix. It is a (4×2) matrix and not a square matrix.

Let (1.5) hold for $s = 1$ and let

$p_{jk} = \Pr$ {actual day is in state k I preceding day was in state j}. We then have a Markov chain (i.e. a chain of order one) with t.p.m. (p_{jk}), $j, k = 0, 1$.

3.1.3 Markov Chains as Graphs

Before we proceed to discuss classification of states we explain how Markov chains can be described as graphs. The states of a Markov chain may be represented by the vertices (nodes) of the graph and one step transitions between states by directed arcs; if $i \rightarrow j$, then vertices i and j are joined by a directed arc with arrow towards j, the value of p_{ij}, which corresponds to the arc weight, may be indicated in the directed arc. If $S = \{1, 2, \ldots m\}$ is the set of vertices corresponding to the state space of the chain and a is the set of directed arcs between these vertices, then the graph $G = \{S, a\}$ is the *directed graph* or *digraph* or *transition graph* of the chain. A digraph such that its arc weights are positive and sum of the arc weights of the arcs from each node is unity is a called a *stochastic graph*; the digraph (or transition graph) of a Markov chain is a stochastic graph.

A path in a digraph is any sequence of arcs where the final vertex of one arc is the initial vertex of the next arc.

The digraph of the chain of Example 1(g) is shown in Fig. 3.1.

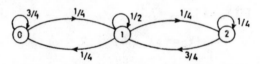

Fig. 3.1 Transition graph of the Markov chain of Example 1(g)

A transition graph is of great aid in visualizing a Markov chain; it is an useful tool in studying the properties (e.g. irreducibility) of the chain.

3.2 HIGHER TRANSITION PROBABILITIES

Chapman-Kolmogorov equation: We have so far considered unit-step or one-step transition probabilities, the probability of X_n given X_{n-1}, i.e. the probability of the outcome at the nth step or trial given the outcome at the previous step; p_{jk} gives the probability of unit-step transition from the state j at a trial to the state k at the next following trial. The m-step transition probability is denoted by

$$\Pr\{X_{m+n} = k \mid X_n = j\} = p_{jk}^{(m)}; \tag{2.1}$$

$p_{jk}^{(m)}$ gives the probability that from the state j at nth trial, the state k is reached at $(m + n)$th trial in m steps, i.e. the probability of transition from the state j to the state k in exactly m steps. The number n does not occur in the r.h.s. of the relation (2.1) and the chain is *homogeneous*. The one-step transition probabilities $p_{jk}^{(1)}$ are denoted by p_{jk} for simplicity. Consider

$$p_{jk}^{(2)} = \Pr \{X_{n+2} = k \mid X_n = j\}. \tag{2.2}$$

The state k can be reached from the state j in two steps through some intermediate state r. Consider a fixed value of r; we have

$$\Pr (X_{n+2} = k, X_{n+1} = r \mid X_n = j\}$$
$$= \Pr \{X_{n+2} = k \mid X_{n+1} = r, X_n = j\} \Pr \{X_{n+1} = r \mid X_n = j\}$$
$$= p_{rk}^{(1)} p_{jr}^{(1)} = p_{jr} \, p_{rk}.$$

Since these intermediate states r can assume values $r = 1, 2, \ldots$, we have

$$p_{jk}^{(2)} = \Pr \{X_{n+2} = k \mid X_n = j\} = \sum_r \Pr \{X_{n+2} = k, X_{n+1} = r \mid X_n = j\}$$

$$= \sum_r p_{jr} \, p_{rk} \tag{2.3}$$

(summing over for all the intermediate states).

By induction, we have

$$p_{jk}^{(m+1)} = \Pr \{X_{n+m+1} = k \mid X_n = j\}$$

$$= \sum_r \Pr \{X_{n+m+1} = k \mid X_{n+m} = r\} \Pr \{X_{n+m} = r \mid X_n = j\}$$

$$= \sum_r p_{rk} p_{jr}^{(m)}.$$

Similarly, we get

$$p_{jk}^{(m+1)} = \sum_r p_{jr} p_{rk}^{(m)}.$$

In general, we have

$$p_{jk}^{(m+n)} = \sum_r p_{rk}^{(n)} p_{jr}^{(m)} = \sum_r p_{jr}^{(n)} \, p_{rk}^{(m)}. \tag{2.4}$$

This equation is a special case of *Chapman-Kolmogorov* equation, which is satisfied by the transition probabilities of a Markov chain.

From (2.4) we get

$$p_{jk}^{(m+n)} \geq p_{jr}^{(m)} p_{rk}^{(n)}, \quad \text{for any } r.$$

We can put the results in terms of transition matrices as follows. Let $P = (p_{jk})$ denote the transition matrix of the unit-step transitions and $P^{(m)} = (p_{jk}^{(m)})$ denote the transition matrix of the m-step transitions. For $m = 2$, we have the matrix $P^{(2)}$ whose elements are given by (2.3). It follows that the elements of $P^{(2)}$ are the elements of the matrix obtained by multiplying the matrix P by itself, i.e.

$$P^{(2)} = P \cdot P = P^2.$$

Similarly,

$$P^{(m+1)} = P^m \cdot P = P \cdot P^m$$

and

$$P^{(m+n)} = P^m \cdot P^n = P^n \cdot P^m. \tag{2.5}$$

It should be noted that there exist non-Markovian chains whose transition probabilities satisfy Chapman-Kolmogorov equation (for example, see Feller I, p. 423, Parzen p. 203).

Example 2(a). Consider the Markov chain of Example 1(g). The two-step transition matrix is given by

$$\begin{pmatrix} 3/4 & 1/4 & 0 \\ 1/4 & 1/2 & 1/4 \\ 0 & 3/4 & 1/4 \end{pmatrix}\begin{pmatrix} 3/4 & 1/4 & 0 \\ 1/4 & 1/2 & 1/4 \\ 0 & 3/4 & 1/4 \end{pmatrix} = \begin{pmatrix} 5/8 & 5/16 & 1/16 \\ 5/16 & 1/2 & 3/16 \\ 3/16 & 9/16 & 1/4 \end{pmatrix}.$$

Hence

$$p_{01}^{(2)} = \Pr\{X_{n+2} = 1 \mid X_n = 0\} = 5/16 \text{ for } n \geq 0.$$

Thus

$$\Pr\{X_2 = 1 \mid X_0 = 0\} = 5/16,$$

and

$$\Pr\{X_2 = 1, X_0 = 0\} = \Pr\{X_2 = 1 \mid X_0 = 0\}\Pr\{X_0 = 0\}$$

$$= (5/16) \cdot (1/3) = 5/48.$$

Example 2(b). *Two-state Markov chain.* Suppose that the probability of a dry day (state 0) following a rainy day (state 1) is 1/3 and that the probability of a rainy day following a dry day is 1/2. We have a two-state Markov chain such that $p_{10} = 1/3$ and $p_{01} = 1/2$ and t.p.m.

$$\begin{array}{cc} & 0 \quad\ 1 \end{array}$$

$$P = \begin{array}{c} 0 \\ 1 \end{array}\begin{pmatrix} 1/2 & 1/2 \\ 1/3 & 2/3 \end{pmatrix}$$

We have

$$P^2 = \begin{pmatrix} 5/12 & 7/12 \\ 7/18 & 11/18 \end{pmatrix}, \quad P^4 = \begin{pmatrix} 173/432 & 259/432 \\ 259/648 & 389/648 \end{pmatrix}$$

Given that May 1 is a dry day, the probability that May 3 is a dry day is 5/12, and that May 5 is a dry day is 173/432. For calculation of powers of P, refer to Sec. 3.5.

Example 2(c). Consider a communication system which transmits the two digits 0 and 1 through several stages. Let X_n, $n \geq 1$ be the digit leaving the nth stage of system and X_0 be the digit entering the first stage (leaving the 0th stage). At each stage there is a constant probability q that the digit which enters will be transmitted unchanged (i.e. the digit will remain unchanged when it leaves), and probability p otherwise (i.e. the digit changes when it leaves), $p + q = 1$.

Here $\{X_n, n \geq 0\}$ is a homogeneous two-state Markov chain with unit-step transition matrix

$$P = \begin{pmatrix} q & p \\ p & q \end{pmatrix}.$$

It can be shown (by mathematical induction or otherwise) that

$$P^m = \begin{pmatrix} \dfrac{1}{2} + \dfrac{1}{2}(q-p)^m & \dfrac{1}{2} - \dfrac{1}{2}(q-p)^m \\ \dfrac{1}{2} - \dfrac{1}{2}(q-p)^m & \dfrac{1}{2} + \dfrac{1}{2}(q-p)^m \end{pmatrix}.$$

Here

$$p_{00}^{(m)} = p_{11}^{(m)} = \frac{1}{2} + \frac{1}{2}(q-p)^m$$

and

$$p_{01}^{(m)} = p_{10}^{(m)} = \frac{1}{2} - \frac{1}{2}(q-p)^m,$$

and as $m \to \infty$,

$$\lim p_{00}^{(m)} = \lim p_{01}^{(m)} = \lim p_{10}^{(m)} = \lim p_{11}^{(m)} \to \frac{1}{2}.$$

Suppose that the initial distribution is given by

$$\Pr\{X_0 = 0\} = a \text{ and } \Pr\{X_0 = 1\} = b = 1 - a.$$

Then we have

$$\Pr\{X_m = 0, X_0 = 0\} = \Pr\{X_m = 0 \mid X_0 = 0\} \Pr\{X_0 = 0\}$$
$$= a p_{00}^{(m)}$$

and

$$\Pr\{X_m = 0, X_0 = 1\} = b p_{10}^{(m)}.$$

The probability that the digit entering the first stage is 0 given that the digit leaving the mth stage is 0 can be evaluated by applying Bayes' rule. We have

$$\Pr\{X_0 = 0 \mid X_m = 0\}$$

$$= \frac{\Pr\{X_m = 0 \mid X_0 = 0\} \Pr\{X_0 = 0\}}{\Pr\{X_m = 0 \mid X_0 = 0\} \Pr\{X_0 = 0\} + \Pr\{X_m = 0 \mid X_0 = 1\} \Pr\{X_0 = 1\}}$$

$$= \frac{a p_{00}^{(m)}}{a p_{00}^{(m)} + b p_{10}^{(m)}}$$

$$= \frac{a\left\{\frac{1}{2} + \frac{1}{2}(q - p)^m\right\}}{a\left\{\frac{1}{2} + \frac{1}{2}(q - p)^m\right\} + b\left\{\frac{1}{2} - \frac{1}{2}(q - p)^m\right\}}$$

$$= \frac{a\{1 + (q - p)^m\}}{1 + (a - b)(q - p)^m}.$$

3.3 GENERALISATION OF INDEPENDENT BERNOULLI TRIALS: SEQUENCE OF CHAIN-DEPENDENT TRIALS

Consider a sequence or series of trials such that each trial has only two outcomes S and F (denoted by 1 and 0 respectively). Assume that the trials are *not independent* (and so are not Bernoulli trials) but the dependence is connected by a simple Markov chain having t.p.m.

$$\begin{array}{c} & X_n \\ & \begin{array}{cc} 0 & \quad 1 \end{array} \\ X_{n-1} \begin{array}{c} 0 \\ 1 \end{array} \left(\begin{array}{cc} 1 - a & a \\ b & 1 - b \end{array} \right) \end{array} \qquad \begin{array}{l} 0 < a, b < 1 \\ n \geq 1. \end{array}$$

Suppose that the initial distribution is given by

$$\Pr\{X_0 = 1\} = p_1 = 1 - \Pr\{X_0 = 0\}.$$

Denote $\qquad p_n = \Pr\{X_n = 1\}, n \geq 1$

and $\qquad q_n = 1 - p_n = \Pr\{X_n = 0\}.$

The probability that S occurs at the nth trial can happen in two mutually exclusive ways:
(i) S occurs at $(n-1)$th trial and again at the next trial, and (ii) S does not occur at $(n-1)$th trial but occurs at the nth trial. Thus

$$p_n = \Pr\{X_n = 1\} = \Pr\{X_n = 1, X_{n-1} = 1\}$$
$$+ \Pr\{X_n = 1, X_{n-1} = 0\}$$
$$= p_{n-1}(1-b) + q_{n-1}\,a$$
$$= (1-a-b)\,p_{n-1} + a, \ n \geq 1.$$

The solution of this difference equation yields

$$p_n = \frac{a}{a+b} + \left(p_1 - \frac{a}{a+b}\right)(1-a-b)^n, n \geq 1$$

The above result can also be obtained from (5.6) whence

$$\Pr\{X_n = 1 \mid X_0 = 1\} = p_{11}^{(n)}$$
$$= \frac{a}{a+b} + (1-a-b)^n\left(\frac{b}{a+b}\right)$$

$$\Pr\{X_n = 1 \mid X_0 = 0\} = p_{01}^{(n)}$$
$$= \frac{a}{a+b} + (1-a-b)^n\left(\frac{-a}{a+b}\right)$$

$$p_n = \Pr\{X_n = 1\}, \quad n \geq 1$$
$$= \Pr\{X_n = 1 \mid X_0 = 1\} \Pr\{X_0 = 1\}$$
$$+ \Pr\{X_n = 1 \mid X_0 = 0\} \Pr\{X_0 = 0\}.$$

We get p_n as given above.

Note: (1)   In particular, when $1 - a = b$, then the t.p.m. becomes

$$\begin{pmatrix} 1-a & a \\ 1-a & a \end{pmatrix}$$

so that the sequence reduces to an independent Bernoulli sequence.

(2) The above model can be used as a model to study the sequence of days (rainy and dry), where S denotes a rainy day, i.e. a day *with* some precipitation and F denotes a dry day *without* any precipitation.

(3) The above is a generalization of independent Bernoulli trials. Another generalization, considered by Wang (1981) is given below.

3.3.1 Markov-Bernoulli Chain

Wang (1981) considers a chain having t.p.m.

$$X_n$$

$$\begin{array}{cc} 0 & 1 \end{array}$$

$$X_{n-1} \quad \begin{array}{c} 0 \\ 1 \end{array} \begin{pmatrix} 1-(1-c)p & (1-c)p \\ (1-c)(1-p) & (1-c)p+c \end{pmatrix} \quad 0<p<1,\ 0\le c\le 1,$$

with initial distribution

$$p_1 = \Pr\{X_0 = 1\} = p = 1 - \Pr\{X_0 = 0\}.$$

When $c = 0$, we get an independent Bernoulli sequence. When $c = 1$, the t.p.m. is

$$\begin{pmatrix} 1 & 0 \\ 0 & 1 \end{pmatrix}$$

and the chain remains is its initial state for ever with probability 1.
Consider the case $0 < c < 1$

$$p_n = p_{n-1}[(1-c)p + c] + q_{n-1}[(1-c)p]$$

$$= c\, p_{n-1} + (1-c)p$$

so that

$$p_n = Ac^{n-1}p_1 + \frac{(1-c)p}{1-c},$$

(where A is a constant)

$$= Ac^{n-1}p + p\ ,\ n \ge 1,\ (p_1 = p).$$

Using $p_1 = p$, we get $p = A p + p$ so that $A = 0$.
Thus

$$p_n = p = \Pr \{X_0 = 1\} \text{ for all } n,$$

i.e. the probability that the event occurs is the same at all trials.
We get

$$E \{X_n\} = p$$

$$\text{var} \{X_n\} = p - p^2 = p(1-p)$$

$$E \{X_{n-1} X_n\} = [(1-c)p + c] \, p$$

so that

$$\text{cov} \{X_{n-1}, X_n\} = cp(1-p), n \geq 1$$

$$\text{and correlation} \{X_{n-1}, X_n\} = c.$$

It can be easily seen that

$$\text{cov} \{X_{n-2}, X_n\} = c^2 p(1-p)$$

and more generally,

$$\text{cov} \{X_{n-k}, X_n\} = c^k p(1-p),$$

$$\text{corr} \{X_{n-k}, X_n\} = c^k, k \geq 1.$$

Denote

$$S_n = X_1 + ... + X_n;$$

S_n gives the accumulated number of successes in n trials in the Markov-Bernoulli sequence (of dependent trials).
Then

$$E \{S_n\} = np$$

and

$$\text{var} \{S_n\} = \sum_{k=1}^{n} \text{var} \{X_k\}$$

$$+ 2 \sum_{\substack{j,k \\ j < k \\ j=1}}^{k=n} \text{cov} \{X_j, X_k\}.$$

Now

$$\sum \text{var}(X_k) = np(1-p)$$

and

$$\frac{\sum \text{cov}(X_j, X_k)}{p(1-p)} = (c + c^2 + \ldots + c^{n-1})$$

$$+ (c + c^2 + \ldots + c^{n-2})$$

$$\ldots$$

$$+ c$$

$$= \frac{c}{1-c}[(n-1) - c\frac{(1-c^{n-1})}{1-c}]$$

so that

$$\text{var}\{S_n\} = np(1-p) + 2p(1-p)[\frac{c(n-1)}{1-c} - c^2\frac{(1-c^{n-1})}{(1-c)^2}].$$

If as

$$n \to \infty, \quad \text{and} \quad p \to 0, np \to \lambda$$

then

$$E\{S_n\} \to \lambda$$

and

$$\text{var}\{S_n\} \to \lambda + \frac{2\lambda c}{1-c}.$$

When $c = 0$, the sequence becomes a sequence of independent Bernoulli trials and then in the limit we get Poisson distribution with $E\{S_n\} = \text{var}\{S_n\} = \lambda$, as should be evident.

In this section we assumed that X_n assumes two values 0 and 1.

3.3.2 Correlated Random Walk

Consider a sequence of random variable X_n, $n = 0, 1, 2, \ldots$ such that each of X_n assumes only two values -1 and 1 with conditional probabilities

$$
\begin{array}{c}
X_n \\
\begin{array}{cc} -1 & 1 \end{array} \\
X_{n-1} \quad \begin{array}{c} -1 \\ 1 \end{array} \begin{pmatrix} 1-a & a \\ b & 1-b \end{pmatrix} \quad 0 < a, b < 1.
\end{array}
$$

X_n denotes the direction of movement (to the left (right)) corresponding to the value -1 (1) at the nth step).

$$\{X_n, n \geq 0\} \text{ is a Markov chain.}$$

Let

$$\Pr\{X_0 = 1\} = p_1 = 1 - \Pr\{X_0 = -1\}$$

give the initial distribution. Let

$$p_n = \Pr\{X_n = 1\}$$

$$q_n = \Pr\{X_n = -1\} = 1 - p_n.$$

Then as before

$$p_n = (1 - a - b)\, p_{n-1} + a$$

$$= \frac{a}{a+b} + \left(p_1 - \frac{a}{a+b}\right)(1 - a - b)^n, \; n \geq 1$$

We have

$$E\{X_n\} = p_n - (1 - p_n) = 2p_n - 1,$$

$$\text{var}\{X_n\} = 1 - (2p_n - 1)^2,$$

$$E\{X_{n-1}X_n\} = 1 \cdot \{(1 - b)p_{n-1} + (1 - a)q_{n-1}\}$$

$$-1 \cdot \{b\, p_{n-1} + a\, q_{n-1}\}$$

$$= (1 - 2b)\, p_{n-1} + (1 - 2a)\{1 - p_{n-1}\}$$

$$= (1 - 2a) + 2(a - b)\, p_{n-1}$$

so that

$$\text{cov}\{X_{n-1}, X_n\} = (1 - 2a) + 2(a - b)\, p_{n-1}$$

$$- (2p_{n-1} - 1)(2p_n - 1).$$

In particular, when $a = b$, we get the case considered by Seth (*J. Roy Stat. Soc.* **B** (1963) 394-400).
Writing $1 - 2a = c$, we get

$$p_n = \frac{1}{2}\{1 + (2p_1 - 1)c^n\}$$

$$E\{X_n\} = (2p_1 - 1)c^n, \; \text{var}(X_n) = 1 - \{(2p_1 - 1)c^n\}^2$$

$$E\{X_{n-1}X_n\} = c$$

$$\text{cov}\{X_{n-1},X_n\} = c - (2p_1 - 1)^2 c^{2n-1}, \qquad p_1 \neq \frac{1}{2}$$

$$= c \qquad\qquad\qquad p_1 = \frac{1}{2}$$

and

$$\text{corr}\{X_{n-1},X_n\}$$

$$= \frac{c - c^{2n-1}(2p_1 - 1)^2}{\sqrt{\{\text{var}(X_n)\,\text{var}(X_{n-1})\}}} \qquad p_1 \neq \frac{1}{2}$$

$$= c \qquad\qquad\qquad p_1 = \frac{1}{2}$$

We have then a correlated random walk with correlation between X_{n-1} and X_n as given above.

Note :   The correlated random walks have been studied in great detail by several researchers. While the earlier authors, for example, Mohan, Seth, Jain, Proudfoot and Lampard considered mainly techniques of difference equations and p.g.f., a method based on transition probability matrices has been considered, for example, by Nain and Kanwar Sen (*J. Appl. Prob.* **17** (1980) 253-258 and *Metron*, **37** (1979) 150-163). Gore has considered a generalization of the correlated random walk and its application (in his Ph.D. dissertation, *Markov Random Walk*, University of Poona, India, 1987, which includes an extensive bibliography on the subject).

Taylor (1990) discusses random walks and martingales.

3.4 CLASSIFICATION OF STATES AND CHAINS

The states $j, j = 0, 1, 2, \ldots$ of a Markov chain $\{X_n, n \geq 0\}$ can often be classified in a distinctive manner according to some fundamental properties of the system. By means of such classification it is possible to identify certain types of chains.

3.4.1 Communication Relations

If $p_{ij}^{(n)} > 0$ for some $n \geq 1$, then we say that state j *can be reached* or state j is *accessible* from state i; the relation is denoted by $i \rightarrow j$. Conversely, if for all n, $p_{ij}^{(n)} = 0$, then j is not accessible from i; in notation $i \nrightarrow j$.

If two states i and j are such that each is accessible from the other then we say that the two states *communicate*: it is denoted by $i \leftrightarrow j$; then there exist integer m and n such that

$$p_{ij}^{(n)} > 0 \text{ and } p_{ji}^{(m)} > 0.$$

The relation $\rightarrow$ is transitive, i.e. if $i \rightarrow j$ and $j \rightarrow k$ then $i \rightarrow k$.
From Chapman-Kolmogorov equation

$$p_{ik}^{(m+n)} = \sum_r p_{ir}^{(m)} p_{rk}^{(n)}$$

we get

$$p_{ik}^{(m+n)} \geq p_{ij}^{(m)} p_{jk}^{(n)}$$

whence the transitivity property follows.
The relation $\leftrightarrow$ is also transitive; i.e. $i \leftrightarrow j, j \leftrightarrow k$ imply $i \leftrightarrow k$.
The relation is clearly symmetric, i.e. if $i \leftrightarrow j$, then $j \leftrightarrow i$.

The digraph of a chain helps in studying the communication relations.

From Fig. 3.1. we see that $0 \leftrightarrow 1$ and $1 \leftrightarrow 2$. Thus $0 \leftrightarrow 2$.

The states of this chain are such that every state can be reached from every other state.

3.4.2 Class Property

A class of states is a subset of the state space such that every state of the class communicates with every other and there is no other state outside the class which communicates with all other states in the class. A property defined for all states of a chain is a *class property* if its possession by one state in a class implies its possession by all states of the same class. One such property is the periodicity of a state.

Periodicity: State i is a *return state* if $p_{ii}^{(n)} > 0$ for some $n \geq 1$. The period d_i of a return to state i is defined as the greatest common divisor of all m such that $p_{ii}^{(m)} > 0$. Thus

$$d_i = \text{G.C.D.} \{m : p_{ii}^{(m)} > 0\} \ ;$$

state i is said to be *aperiodic* if $d_i = 1$ and *periodic* if $d_i > 1$. Clearly state i is aperiodic if $p_{ii} \neq 0$.

It can be shown that two distinctive states belonging to the same class have the same period.

3.4.3 Classification of Chains

If C is a set of states such that no state outside C can be reached from any state in C, then C is said to be *closed*. If C is closed and $j \in C$ while $k \notin C$, then $p_{jk}^{(n)} = 0$ for all, n. i.e. C is closed *iff* $\sum_{j \in C} p_{ij} = 1$ for every $i \in C$. Then the submatrix $P_1 = (p_{ij})$, $i, j \in C$, is also stochastic and P can be expressed in the canonical form as:

$$P = \begin{pmatrix} P_1 & 0 \\ R_1 & Q \end{pmatrix}$$

A closed set may contain one or more states. If a closed set contains only one state j then state j is said to be *absorbing* : j is absorbing *iff* $p_{jj} = 1$, $p_{jk} = 0$, $k \neq j$. In Example 1(b), states 0 and 4 are absorbing.

Every finite Markov chain contains at least one closed set, i.e. the set of all states or the state space. If the chain does not contain any other proper closed subset other than the state space, then the chain is called *irreducible*; the t.p.m. of irreducible chain is an irreducible matrix. In an irreducible Markov chain every state can be reached from every other state. The Markov chain of Example 1(g) is irreducible. Chains which are not irreducible are said to be *reducible* or *non-irreducible*; the t.p.m. is reducible. The irreducible matrices may be subdivided into two classes: *primitive* (aperiodic) and *imprimitive* (cyclic or periodic) (See Section A.4 Appendix). A Markov chain is primitive (aperiodic) *iff* the corresponding t.p.m. is primitive. In an irreducible chain all states belong to the same class.

3.4.4 Classification of States: Transient and Persistent (Recurrent) States

We now proceed to obtain a more sensitive classification of the states of a Markov chain.

Suppose that a system starts with the state j. Let $f_{jk}^{(n)}$ be the probability that it reaches the state k for the *first time* at the nth step (or after n transitions) and let $p_{jk}^{(n)}$ be the probability that it reaches state k (not necessarily for the first time) after n transitions. Let τ_k be the first passage time to state k, i.e. $\tau_k = \min \{n \geq 1, X_n = k\}$ and $\{f_{jk}^{(n)}\}$ be the distribution of τ_k given that the chain starts at state j. A relation can be established between $f_{jk}^{(n)}$ and $p_{jk}^{(n)}$ as follows. The relation allows $f_{jk}^{(n)}$ to be expressed in terms of $p_{jk}^{(n)}$.

Theorem 3.1. *First Entrance Theorem*

Whatever be the states j and k

$$p_{jk}^{(n)} = \sum_{r=0}^{n} f_{jk}^{(r)} p_{kk}^{(n-r)}, \quad n \geq 1 \tag{4.1}$$

with

$$p_{kk}^{(0)} = 1, f_{jk}^{(0)} = 0, f_{jk}^{(1)} = p_{jk}.$$

Intuitively, the probability that starting, with j, state k is reached for the first time at the rth step and again after that at $(n - r)$th step is given by $f_{jk}^{(r)} p_{kk}^{n-r}$ for all $r \leq n$. These cases are mutually exclusive. Hence the result. The recursive relation (4.1) can also be written as

$$p_{jk}^{(n)} = \sum_{r=1}^{n-1} f_{jk}^{(r)} p_{kk}^{(n-r)} + f_{jk}^{(n)}, \quad n > 1. \tag{4.1a}$$

▲

Note. (1) For a rigorous proof which uses the strong Markov property, see Iosifescu (1980).

(2) In practice, it is sometimes convenient to compute $f_{jk}^{(r)}$ from the diagraph of a chain.

3.4.4.1 First Passage Time Distribution

Let F_{jk} denote the probability that starting with state j the system will ever reach state k. Clearly

$$F_{jk} = \sum_{n=1}^{\infty} f_{jk}^{(n)}. \tag{4.2}$$

We have

$$\sup_{n \geq 1} p_{jk}^{(n)} \leq F_{jk} \leq \sum_{m \geq 1} p_{jk}^{(m)} \quad \text{for all } n \geq 1. \tag{4.3}$$

We have to consider two cases, $F_{jk} = 1$ and $F_{jk} < 1$.

When $F_{jk} = 1$, it is certain that the system starting with state j will reach state k; in this case $\{f_{jk}^{(n)}, n = 1, 2, \ldots\}$ is a proper probability distribution and this gives the *first passage time distribution* for k given that the system starts with j.

The mean (first passage) time from state j to state k is given by

$$\mu_{jk} = \sum_{n=1}^{\infty} n f_{jk}^{(n)}. \tag{4.4}$$

In particular, when $k = j$, $\{f_{jj}^{(n)}, n = 1, 2, \ldots\}$ will represent the distribution of the *recurrence times* of j; and $F_{jj} = 1$ will imply that the return to the state j is certain. In this case

$$\mu_{jj} = \sum_{n=1}^{\infty} n f_{jj}^{(n)} \tag{4.5}$$

is known as the *mean recurrence time* for the state j.

Thus, two questions arise concerning state j: first, whether the return to state j is certain and secondly, when this happens, whether the mean recurrence time μ_{jj} is finite.

Note. It can be shown that

$$d_i = \text{G.C.D.} \{m : p_{ii}^{(m)} > 0\} = \text{G.C.D.} \{m : f_{ii}^{(m)} > 0\}.$$

Definitions. A state *j* is said to be *persistent* (the word *recurrent* is also used by some authors; we shall however use the word persistent) if $F_{jj} = 1$ (i.e. return to state *j* is certain) and *transient* if $F_{jj} < 1$ (i.e. return to state *j* is uncertain). A persistent state *j* is said to be *null persistent* if $\mu_{jj} = \infty$, i.e. if the mean recurrence time is infinite, and is said to be *non-full* (or *positive*) *persistent* if $\mu_{jj} < \infty$.

Thus the states of a Markov chain can be classified as transient and persistent and persistent states can be subdivided as non-null and null persistent.

A persistent non-null and aperiodic state of a Markov chain is said to be *ergodic*. Consider the following example.

Example 4(a). Let $\{X_n, n \geq 0\}$ be a Markov chain having state space $S = \{1, 2, 3, 4\}$ and transition matrix

$$P = \begin{pmatrix} \frac{1}{3} & \frac{2}{3} & 0 & 0 \\ 1 & 0 & 0 & 0 \\ \frac{1}{2} & 0 & \frac{1}{2} & 0 \\ 0 & 0 & \frac{1}{2} & \frac{1}{2} \end{pmatrix}$$

Here $f_{33}^{(1)} = \frac{1}{2}, f_{33}^{(2)} = f_{33}^{(3)} = \ldots = 0$ so that $F_{33} = \frac{1}{2}$ and thus state 3 is transient. Again $f_{44}^{(1)} = \frac{1}{2}, f_{44}^{(n)} = 0, n \geq 2$, so that $F_{44} = \frac{1}{2}$ and thus state 4 is also transient. Now $f_{11}^{(1)} = \frac{1}{3}, f_{11}^{(2)} = \frac{2}{3}$ and $F_{11} = \frac{1}{3} + \frac{2}{3} = 1$, so that state 1 is persistent.

Further since $\mu_{11} = 1 \cdot \frac{1}{3} + 2 \cdot \frac{2}{3} = \frac{5}{3}$, state 1 is non-null persistent. Again $p_{11} = \frac{1}{3} > 0$, so that state 1 is aperiodic. State 1 is ergodic.

For State 2:

$$f_{22}^{(1)} = 0, \quad f_{22}^{(2)} = 1 \cdot \frac{2}{3}, \quad f_{22}^{(3)} = 1 \cdot \frac{1}{3} \cdot \frac{2}{3}, \quad f_{22}^{(4)} = 1 \cdot \left(\frac{1}{3}\right)^2 \cdot \frac{2}{3}$$

$$\cdots f_{22}^{(n)} = 1 \cdot \left(\frac{1}{3}\right)^{n-2} \cdot \frac{2}{3}, \quad n \geq 2$$

so that

$$F_{22} = \sum_{n=1}^{\infty} f_{22}^{(n)} = \sum_{k=2}^{\infty} \left(\frac{1}{3}\right)^{k-2} \cdot \frac{2}{3} = 1.$$

Thus state 2 is persistent. We have

$$\mu_{22} = \sum_{k=1}^{\infty} k f_{22}^{(k)} = \sum_{k=2}^{\infty} k \left(\frac{1}{3}\right)^{k-2} \cdot \frac{2}{3} = 2 \sum_{k=2}^{\infty} k \left(\frac{1}{3}\right)^{k-1} = \frac{5}{2}$$

so that state 2 is non-null persistent. It is also aperiodic, and hence ergodic.

In the above example, calculation of $f_{ii}^{(n)}$ and so also of $F_{ii} = \sum f_{ii}^{(n)}$ was easy. But sometimes it is not so easy to calculate $f_{ii}^{(n)}$ for $n \geq 2$. In view of this, another characterization of persistence is given in Theorem 3.2.

Example 4(b). Consider a Markov chain with transition matrix

$$
P = \begin{array}{c} 1 \\ 2 \\ 3 \\ 4 \end{array}
\begin{array}{cccc}
1 & 2 & 3 & 4 \\
\left(\begin{array}{cccc}
0 & 0 & 1 & 0 \\
0 & 0 & 0 & 1 \\
0 & 1 & 0 & 0 \\
1/4 & 1/8 & 1/8 & 1/2
\end{array}\right)
\end{array}
$$

It can be easily seen that the chain is irreducible. Consider state 4: we have $p_{44} = \frac{1}{2} > 0$; state is aperiodic and $f_{44}^{(1)} = \frac{1}{2}, f_{44}^{(2)} = \frac{1}{8}, f_{44}^{(3)} = \frac{1}{8}, f_{44}^{(4)} = \frac{1}{4}, f_{44}^{(n)} = 0, n > 4$ so that $F_{44} = 1$ and

$$\mu_{44} = 1 \cdot \frac{1}{2} + 2 \cdot \frac{1}{8} + 3 \cdot \frac{1}{8} + 4 \cdot \frac{1}{4} = \frac{17}{8} < \infty.$$

Thus state 4 is ergodic. Hence all the states are ergodic.

Theorem 3.2. State j is persistent *iff*

$$\sum_{n=0}^{\infty} p_{jj}^{(n)} = \infty \tag{4.6}$$

Proof: Let

$$P_{jj}(s) = \sum_{n=0}^{\infty} p_{jj}^{(n)} s^n = 1 + \sum_{n=1}^{\infty} p_{jj}^{(n)} s^n, \quad |s| < 1$$

and

$$F_{jj}(s) = \sum_{n=0}^{\infty} f_{jj}^{(n)} s^n = \sum_{n=1}^{\infty} f_{jj}^{(n)} s^n, \quad |s| < 1$$

be the generating functions of the sequences $\{p_{jj}^{(n)}\}$ and $\{f_{jj}^{(n)}\}$ respectively.

We have from (4.1)

$$p_{jj}^{(n)} = \sum_{r=0}^{n} f_{jj}^{(r)} p_{jj}^{(n-r)} \tag{4.7}$$

Multiplying both sides of (4.7) by s^n and adding for all $n \geq 1$ we get

$$P_{jj}(s) - 1 = F_{jj}(s) P_{jj}(s).$$

The r.h.s. of the above is immediately obtained by considering the fact that the r.h.s. of (4.7) is a convolution of $\{f_{jj}\}$ and $\{p_{jj}\}$ and that the generating function of the convolution is the product of the two generating functions. Thus we have

$$P_{jj}(s) = \frac{1}{1 - F_{jj}(s)}, \quad |s| < 1. \tag{4.8}$$

Assume that state j is persistent which implies that $F_{jj} = 1$. Using Abel's lemma we get

$$\lim_{s \to 1} F_{jj}(s) = 1.$$

Thus

$$\lim_{s \to 1} P_{jj}(s) \to \infty.$$

Since the co-efficients of $P_{jj}(s)$ are non-negative Abel's lemma applies and we get $\sum p_{jj}^{(n)} = \infty$.

Conversely, if state j is transient then by Abel's lemma we get

$$\lim_{s \to 1} F_{jj}(s) < 1$$

and from (4.8)

$$\lim_{s \to 1} P_{jj}(s) < \infty.$$

Since the co-efficients $p_{jj}^{(n)} \geq 0$, we get

$$\sum_{n} p_{jj}^{(n)} < \infty.$$

The result of Theorem 3.2 can also be deduced from the following result.

Doeblin's Formula: Whatever be the states j and k,

$$F_{jk} = \lim_{m \to \infty} \frac{\sum\limits_{n=1}^{m} p_{jk}^{(n)}}{1 + \sum\limits_{n=1}^{m} p_{kk}^{(n)}} \qquad (4.9)$$

and, in particular,

$$F_{jj} = 1 - \lim_{m \to \infty} \frac{1}{1 + \sum\limits_{n=1}^{m} p_{jj}^{(n)}}. \qquad (4.9a)$$

Remarks: (1) State j is transient if $\sum p_{jj}^{(n)} < \infty$; this implies that if j is transient then $p_{jj}^{(n)} \to 0$ as $n \to \infty$.

(2) The state space of a finite Markov chain must contain at least one persistent state.

(3) If k is a transient state and j is an arbitrary state then $\sum p_{jk}^{(n)}$ converges and $\lim\limits_{n \to \infty} p_{jk}^{(n)} \to 0$.

(4) If a Markov chain having a set of transient states T, starts in a transient state, then with probability 1, it stays at the transient set of states T only a finite number of times after which it enters a recurrent state where it remains forever.

Example 4(c). Consider the Markov chain with t.p.m.

$$
\begin{array}{c}
\quad\quad 0 \quad 1 \quad 2 \\
P = \begin{array}{c} 0 \\ 1 \\ 2 \end{array}\!\!\left(\begin{array}{ccc} 0 & 1 & 0 \\ \dfrac{1}{2} & 0 & \dfrac{1}{2} \\ 0 & 1 & 0 \end{array}\right)
\end{array}
$$

The chain is irreducible as the matrix is so. We have

$$P^2 = \begin{pmatrix} \dfrac{1}{2} & 0 & \dfrac{1}{2} \\ 0 & 1 & 0 \\ \dfrac{1}{2} & 0 & \dfrac{1}{2} \end{pmatrix}, \quad P^3 = P;$$

in general,

$$P^{2n} = P^2, \quad P^{2n+1} = P,$$

so that

$$p_{ii}^{(2n)} > 0, \ p_{ii}^{(2n+1)} = 0 \text{ for each } i.$$

The states are periodic with period 2.

We find that $f_{11} = 0$, $f_{11}^{(2)} = 1$ so that $F_{11} = \sum_n f_{11}^{(n)} = 1$, i.e. state 1 is persistent and hence the other states 0 and 2 are also persistent.

Now

$$\mu_{11} = \sum_n n f_{11}^{(n)} = 2,$$

i.e. state 1 is non-null. Thus the states of the chain are periodic (each with period 2) and persistent non-null.

Further,

$$p_{11}^{(2n)} \to t/\mu_{11} = 2/2 = 1 \text{ for all } n.$$

We now state a lemma without proof (for proof, see Feller, Vol. I).

Basic limit theorem of renewal theory

Lemma 3.1. Let $\{f_n\}$ be a sequence such that $f_n \geq 0$, $\Sigma f_n = 1$ and $t \ (\geq 1)$ be the greatest common divisor of those n for which $f_n > 0$.

Let $\{u_n\}$ be another sequence such that $u_0 = 1$ and $u_n = \sum_{r=1}^{n} f_r u_{n-r} \ (n \geq 1)$. Then

$$\lim_{n \to \infty} u_{nt} = t/\mu \tag{4.10}$$

where $\mu = \sum_{n=1}^{\infty} n f_n$, the limit being zero when $\mu = \infty$;

and $\lim_{N \to \infty} u_N = 0$ whenever N is not divisible by t.

The lemma will be used to obtain some important results.

Theorem 3.3. If state j is persistent non-null, then as $n \to \infty$

(i) $p_{jj}^{(nt)} \to t/\mu_{jj}$ (4.11)

   when state j is periodic with period t;

and (ii) $p_{jj}^{(n)} \to 1/\mu_{jj}$ (4.12)

   when state j is aperiodic.

In case state j is persistent null, (whether periodic or aperiodic), then

$$p_{jj}^{(n)} \to 0, \text{ as } n \to \infty \tag{4.13}$$

Proof: Let state j be persistent; then

$$\mu_{jj} = \sum_n n f_{jj}^{(n)} \qquad \text{is defined.}$$

Since (4.7) holds, we may put

$$f_{jj}^{(n)} \text{ for } f_n, \, p_{jj}^{(n)} \text{ for } u_n, \quad \text{and } \mu_{jj} \text{ for } \mu$$

in Lemma 3.1 above. Applying the lemma, we get

$$p_{jj}^{(nt)} \to t/\mu_{jj}, \quad \text{as } n \to \infty$$

when state j is periodic with period t.
When state j is aperiodic (i.e. $t = 1$), then

$$p_{jj}^{(n)} \to 1/\mu_{jj} \quad \text{as } n \to \infty.$$

In case state j is persistent null, $\mu_{jj} = \infty$, and

$$p_{jj}^{(n)} \to 0 \quad \text{as } n \to \infty. \qquad\qquad \blacktriangle$$

Note: (1) If j is persistent non-null, then

$$\lim_{n \to \infty} p_{jj}^{(n)} > 0$$

and (2) if j is persistent null or transient then

$$\lim_{n \to \infty} p_{jj}^{(n)} \to 0$$

Theorem 3.4. If state k is persistent null, then for every j

$$\lim_{n \to \infty} p_{jk}^{(n)} \to 0 \tag{4.14}$$

and if state k is aperiodic, persistent non-null then

$$\lim_{n \to \infty} p_{jk}^{(n)} \to \frac{F_{jk}}{\mu_{kk}}. \tag{4.15}$$

Proof: We have

$$p_{jk}^{(n)} = \sum_{r=1}^{n} f_{jk}^{(r)} p_{kk}^{(n-r)}.$$

Let $n > m$, then

$$p_{jk}^{(n)} = \sum_{r=1}^{m} f_{jk}^{(r)} p_{kk}^{(n-r)} + \sum_{r=m+1}^{n} f_{jk}^{(r)} p_{kk}^{(n-r)}$$

$$\leq \sum_{r=1}^{m} f_{jk}^{(r)} p_{kk}^{(n-r)} + \sum_{r=m+1}^{n} f_{jk}^{(r)}. \qquad (4.16)$$

Since state k is persistent null,

$$p_{kk}^{(n-r)} \to 0, \quad \text{as } n \to \infty.$$

Further, since

$$\sum_{m=1}^{\infty} f_{jk}^{(m)} < \infty, \quad \sum_{r=m+1}^{n} f_{jk}^{(r)} \to 0 \text{ as } n, m \to \infty.$$

Hence as $n \to \infty$

$$p_{jk}^{(n)} \to 0.$$

From (4.16)

$$p_{jk}^{(n)} - \sum_{r=1}^{m} f_{jk}^{(r)} p_{kk}^{(n-r)} \leq \sum_{r=m+1}^{n} f_{jk}^{(r)} \qquad (4.16a)$$

Since j is aperiodic, persistent and non-null then by Theorem 3.3.

$$p_{kk}^{(n-r)} \to 1/\mu_{kk} \quad \text{as } n \to \infty.$$

Hence from (4.16a) we get, as $n, m \to \infty$

$$p_{jk}^{(n)} \to F_{jk} / \mu_{kk}. \qquad \blacktriangle$$

Theorem 3.5. In an irreducible chain, all the states are of the same type. They are either all transient, all persistent null, or all persistent non-null. All the states are aperiodic and in the latter case they all have the same period.

Proof: Since the chain is irreducible, every state can be reached from every other state. If i, j are any two states, then i can be reached from j and j from i, i.e.

$$p_{ij}^{(N)} = a > 0 \text{ for some } N \geq 1,$$

and

$$p_{ji}^{(M)} = b > 0 \text{ for some } M \geq 1.$$

We have

$$p_{jk}^{(n+m)} = p_{jk}^{(m+n)} = \sum_r p_{jr}^{(m)} p_{rk}^{(n)}$$

$$\geq p_{jr}^{(m)} p_{rk}^{(n)} \text{ for each } r.$$

Hence

$$p_{ii}^{(n+N+M)} \geq p_{ij}^{(N)} p_{jj}^{(n)} p_{ji}^{(M)} = ab \, p_{jj}^{(n)} \tag{4.17}$$

and

$$p_{jj}^{(n+N+M)} = p_{ji}^{(M)} p_{ii}^{(n)} p_{ij}^{(N)} = ab \, p_{ii}^{(n)}. \tag{4.18}$$

From the above it is clear that the two series $\sum_n p_{ii}^{(n)}$, and $\sum_n p_{jj}^{(n)}$ converge or diverge together. Thus the two states i, j are either both transient or both persistent.

Suppose that i is persistent null, then $p_{ii}^{(n)} \to 0$ as $n \to \infty$; then from (4.17), $p_{jj}^{(n)} \to 0$ as $n \to \infty$, so that j is also persistent null, i.e. they are both persistent null.

Suppose that i is persistent non-null and has period t, then $p_{ii}^{(n)} > 0$ whenever n is a multiple of t. Now

$$p_{ii}^{(N+M)} \geq p_{ij}^{(N)} p_{ji}^{(M)} = ab > 0,$$

so that $(N + M)$ is a multiple of t. From (4.18)

$$p_{jj}^{(n+N+M)} \geq ab \, p_{ii}^{(n)} > 0.$$

Thus $(n + N + M)$ is a multiple of t and so t is the period of the state j also. ▲

Corollary: In a *finite irreducible* Markov chain all states are *non-null persistent*.

Proof : Let $S = \{1, 2, \ldots, k\}$ be the state space of the chain and P be its t.p.m. Suppose, if possible, that state 1 is null persistent. Then all other states are null persistent. This implies that

$$\lim_{n \to \infty} p_{ij}^{(n)} = 0 \text{ for all } j \in S$$

Now $\sum_{j \in S} p_{ij}^{(n)} = 1$ for all n.

Thus since S is finite we are lead to a contradiction. Hence all states must be non-null persistent.

Note :   Thus transience or null persistence of an irreducible chain need be examined only when it is *not* finite. There is no necessity to examine the characteristic of non-null persistence in case of a finite irreducible chain, because such a chain necessarily possesses this characteristic.

3.5 DETERMINATION OF HIGHER TRANSITION PROBABILITIES

Consider a Markov chain with m states (m finite) having transition probability matrix (t.p.m.) $P = (p_{ij})$. The n-step transition probabilities $p_{ij}^{(n)}$ (i.e. the elements of P^n) can be obtained by using

$$p_{ij}^{(n)} = \sum_k p_{ik} p_{kj}^{(n-1)}, \quad n = 1, 2, ...,$$

where $$p_{jj}^{(0)} = 1, p_{jk}^{(0)} = 0, \quad k \neq j, p_{ij}^{(1)} = p_{ij}.$$

We shall now show how some of the results of spectral theory of matrices could be used to determine $p_{ij}^{(n)}$ or P^n more expeditiously. We refer to the discussions on matrices given in Appendix A.4.

Assume that the eigenvalues $t_i, i = 1, 2, \ldots, m$, of the stochastic matrix P are all distinct and different from zero. Suppose that (apart from multiplicative constants)

$$\mathbf{x}_i = (x_{i1}, x_{i2}, ..., x_{im})'$$

and $$\mathbf{y}_i' = (y_{i1}, y_{i2}, ..., y_{im})$$

are, respectively, the right and left-eigenvectors of P corresponding to t_i and that $X = (\mathbf{x}_1, \ldots \mathbf{x}_m)$ is the matrix formed with the eigenvectors $\mathbf{x}_1, \ldots, \mathbf{x}_m$. Suppose that $D = (d_{ij})$ is a diagonal matrix such that

$$d_{ii} = t_i, d_{ij} = 0, j \neq i.$$

Then from the spectral theorem we have

$$P = XDX^{-1} \tag{5.1}$$

and $$P^n = XD^nX^{-1} \tag{5.2}$$

$$= \sum_{k=1}^{m} t_k^n c_k \mathbf{x}_k \mathbf{y}_k', \text{ where } c_k = 1/(\mathbf{y}_k' \mathbf{x}_k).$$

Thus $$p_{ij}^{(n)} = \sum_{k=1}^{m} t_k^n c_k x_{ki} y_{kj}, n = 0, 1, 2, \ldots . \tag{5.3}$$

Note: Replacement of $\mathbf{x}_i$ and y'_i by $A_i\mathbf{x}_i$ and $B_iy'_i$ (by inclusion of multiplicative constants) would lead to the same results [(5.2) & (5.3)].

3.5.1 Aperiodic Chain: Limiting Behaviour

Assume that P is primitive, i.e. the chain is aperiodic; then there will be no other eigenvalue with modulus equal to 1 except 1. Since all the eigenvalues are assumed to be distinct, we may put

$$t_1 = 1, |t_i| < 1, i \neq 1.$$

Then $x_{ii} = 1$ for all i and we have

$$p_{ij}^{(n)} = c_1 y_{1j} + \sum_{k=2}^{m} t_k^n c_k x_{ki} y_{kj}. \tag{5.4}$$

(see Section A 4.4 of Appendix A).
Further, in the limit, as $n \to \infty$,

$$p_{ij}^{(n)} \to c_1 y_{1j}, (c_1 = 1/(y_1' x_1) = 1/ \sum_{j=1}^{m} x_{1j} y_{1j}, \tag{5.5}$$

i.e. P^n tends to a matrix with all rows equal to

$$c_1 y_1' = (c_1 y_{11}, c_1 y_{12}, ..., c_1 y_{1m}).$$

The above shows that $\lim p_{ij}^{(n)}$ exists and is independent of the initial state i, if P is primitive (see also Theorem A.4.6 of Appendix A).

To calculate this limit, when it exists, one needs determine only the left-eigenvectors y_1' corresponding to $t_1 = 1$.

Remark: The above method used in the study of limiting behaviour of aperiodic chains is based on spectral theory of matrices. A simpler method for deriving $\lim p_{ij}^{(n)}$ will be discussed in the next section.

Example 5(a). Consider the two-state Markov chain

$$P = \begin{pmatrix} 1-a & a \\ b & 1-b \end{pmatrix}, \quad 0 < a, b < 1.$$

The eigenvalues of P are $t_1 = 1$ and $t_2 = 1 - a - b$, $|t_2| < 1$. The right eigenvectors corresponding to $t_1 = 1$ and $t_2 = 1 - a - b$ are given, respectively, by x_1 and x_2 where $x_1' = (1, 1)$ and $x_2' = (1, -a/b)$. The left eigenvectors corresponding to t_1 and t_2 are given by $y_1' = (1, a/b)$ and $y_2' = (1, -1)$ respectively. We have

$$c_1 = 1/(y_1' x_1) = b/(a+b), c_2 = 1/(y_2' x_2) = a/(a+b)$$

$$\mathbf{x}_1\mathbf{y}_1' = \begin{pmatrix} 1 & a/b \\ 1 & a/b \end{pmatrix}, \quad \mathbf{x}_2\mathbf{y}_2' = \begin{pmatrix} 1 & -1 \\ -b/a & b/a \end{pmatrix}.$$

Thus

$$P = \sum_{j=1}^{2} t_j c_j \mathbf{x}_j \mathbf{y}_j' = \frac{1}{a+b}\begin{pmatrix} b & a \\ b & a \end{pmatrix} + \frac{1-a-b}{a+b}\begin{pmatrix} a & -a \\ -b & b \end{pmatrix},$$

$$P^n = \sum t_j^n c_j \mathbf{x}_j \mathbf{y}_j' = \frac{1}{a+b}\begin{pmatrix} b & a \\ b & a \end{pmatrix}$$

$$+ \frac{(1-a-b)^n}{a+b}\begin{pmatrix} a & -a \\ -b & b \end{pmatrix}, \quad n = 0, 1, 2, \ldots \tag{5.6}$$

Further,

$$\lim_{n \to \infty} P^n = \frac{1}{a+b}\begin{pmatrix} b & a \\ b & a \end{pmatrix} = \begin{pmatrix} b/(a+b) & a/(a+b) \\ b/(a+b) & a/(a+b) \end{pmatrix}.$$

As $n \to \infty$ $\lim p_{i1} \to b/(a+b)$ and $\lim p_{i2} \to a/(a+b)$, $i = 1, 2$. \tag{5.6a}

Remarks:

(1) Suppose that the states of a chain are 0 and 1 and that the initial distribution is $(1 - p_1, p_1)$, i.e. $\Pr\{X_0 = 1\} = p_1 = 1 - \Pr\{X_0 = 0\}$.

From (5.6) we get

$$\Pr\{X_n = 1 \mid X_0 = 0\}$$

$$= \frac{a}{a+b} - \frac{a}{a+b}(1-a-b)^n$$

and

$$\Pr\{X_n = 1 \mid X_0 = 1\}$$

$$= \frac{a}{a+b} + \frac{b}{a+b}(1-a-b)^n.$$

Thus

$$p_n = \Pr\{X_n = 1\} = \Pr\{X_n = 1 \mid X_0 = 0\}\, \Pr = \{X_0 = 0\}$$

$$+ \Pr\{X_n = 1 \mid X_0 = 1\}\, \Pr\{X_0 = 1\}$$

$$= \frac{a}{a+b} + (1-a-b)^n\left[p_1 - \frac{a}{a+b}\right].$$

This can also be obtained from the difference equation

$$p_n = a + (1-a-b)p_{n-1}, \quad n \geq 2$$

(satisfied by p_n).

As $n \to \infty$, $\qquad\qquad\qquad\qquad p_n \to \dfrac{a}{a+b}$.

(2) When $1 - a = b$, i.e. in case of independence,

$$P^n = P \text{ for all } n,$$

as should be intuitively clear.

(3) The case $a = b = p$, $0 \leq p \leq 1$ is discussed in the Remarks at the end of section 3.6. p. 112.

Example 5(b). Consider the three-state Markov chain with t.p.m.

$$P = \begin{pmatrix} 0.5 & 0.3 & 0.2 \\ 0.2 & 0.4 & 0.4 \\ 0.1 & 0.5 & 0.4 \end{pmatrix}$$

The eigenvalues of P are $t_1 = 1$, $t_2 = 0.1$, $t_3 = 0.2$. The left eigenvector $\mathbf{y}_1' = (y_{11}, y_{12}, y_{13})$ corresponding to $t_1 = 1$ satisfies

$$-0.5y_{11} + 0.2y_{12} + 0.1y_{13} = 0$$

$$0.3y_{11} - 0.6y_{12} + 0.5y_{13} = 0$$

$$0.2y_{11} + 0.4y_{12} - 0.6y_{13} = 0.$$

We have $\mathbf{y}_1' = (0.16, 0.28, 0.24)$ and $c_1 = 1 / (\sum_{j=1}^{3} y_{1j}) = 1/0.68$. Hence

$$p_{ij}^{(n)} \to c_1 y_{1j}, \quad (j = 1, 2, 3),$$

i.e. $\qquad p_{i1}^{(n)} \to 0.2353, \quad p_{i2}^{(n)} \to 0.4118, \quad p_{i3}^{(n)} \to 0.3529.$

Example 5(c). Let the genotypes AA, Aa, aa be denoted by 1, 2, 3 respectively. Let X_n, $(n \geq 1)$ be the genotype of the offspring of a father of genotype X_{n-1}.

$\{X_n, n \geq 0\}$ is a Markov chain with state space $S = \{1, 2, 3\}$ and transition matrix

$$P = \begin{pmatrix} p & q & 0 \\ p/2 & 1/2 & q/2 \\ 0 & p & q \end{pmatrix}, \quad p + q = 1, 0 < p < 1.$$

Here

$$A = I - zP = \begin{pmatrix} 1-zp & -zq & 0 \\ -zp/2 & 1-z/2 & -zq/2 \\ 0 & -z/p & 1-zq \end{pmatrix}$$

The eigenvalues of P are 0, 1/2 and 1; the eigenvalues of the matrix $I - zP$ are 1, $(1 - z/2)$ and $1 - z$, so that

$$\det (I - zP) = (1 - z/2)(1 - z).$$

Thus

$$(I - zP)^{-1} = \text{Adj } (I - zP) \Big/ \left\{ \left(1 - \frac{z}{2}\right)(1 - z) \right\}.$$

Now the cofactor

$$A_{11} = (-1)^{1+1} \begin{vmatrix} 1-z/2 & -zq/2 \\ -zp & 1-zq \end{vmatrix}$$

$$= (1 - z/2)(1 - zq) - z^2 pq/2.$$

Using partial fractions and separating components, we get

$$\frac{A_{11}}{(1-z/2)(1-z)} = \frac{a_1}{1-z} + \frac{a_2}{1-z/2} = \frac{a_1(1-z/2)+a_2(1-z)}{(1-z/2)(1-z)}$$

where

$$\frac{1}{2}a_1 = \lim_{z \to 1} A_{11} = p^2/2$$

$$-a_2 = \lim_{z \to 2} A_{11} = -2pq.$$

That is,

$$\frac{A_{11}}{(1-z/2)(1-z)} = \frac{p^2}{1-z} + \frac{2pq}{1-z/2}.$$

Similarly the other cofactors of $A = I - zP$ can be found and elements of $(I - z P)^{-1}$ can be resolved into partial fractions.

We thus have

$$(I - zP)^{-1} = \frac{1}{1-z}\begin{pmatrix} p^2 & 2pq & q^2 \\ p^2 & 2pq & q^2 \\ p^2 & 2pq & q^2 \end{pmatrix} + \frac{2}{1-z/2}\begin{pmatrix} pq & q(q-p) & -q^2 \\ p(q-p)/2 & (1-4pq)/2 & q(p-q)/2 \\ -p^2 & p(p-q) & pq \end{pmatrix}$$

so that, for $n \geq 1$

$$P^n = \text{coefficient of } z^n \text{ in } (I - zP)^{-1}$$

$$= e\,(p^2, 2pq, q^2) + \frac{1}{2^{n-1}}\begin{pmatrix} pq & q(q-p) & -q^2 \\ p(q-p)/2 & (1-4pq)/2 & q(p-q)/2 \\ -p^2 & p(p-q) & pq \end{pmatrix}.$$

As $n \to \infty$

$$P^n \to e\,(p^2,\, 2pq,\, q^2)$$

which is a matrix with identical rows with elements $(p^2, 2pq, q^2)$.

This gives the limiting distribution of the genotypes.

Note . Since $p_{12} > 0$, $p_{21} > 0$, $p_{13}^{(2)} > 0$, $p_{31}^{(2)} > 0$, $p_{23} > 0$, $p_{32} > 0$ each pair of states communicates, hence the chain is irreducible. Since $p_{11} > 0$, state 1 is aperiodic; as $\sum p_{11}^{(n)} \to \infty$ state 1 is persistent; and as $\lim p_{11}^{(n)} \to p^2 (> 0)$ the persistent state 1 is non-null persistent. Thus the chain is an irreducible chain with aperiodic non-null persistent states.

Example 5(d). Consider a three-state Markov chain with t.p.m.

$$P = \begin{pmatrix} 0 & 1 & 0 \\ 0 & 0 & 1 \\ 1 & 0 & 0 \end{pmatrix}.$$

The eigenvalues are $t_1 = 1$, $t_2 = \omega$, $t_3 = \omega^2$, ω, ω^2 being the imaginary cube roots of unity. As $|\omega| = |\omega^2| = 1$, the matrix is not primitive; the chain is periodic with period 3, $(P^3 = I)$. Equation (5.5) cannot be applied in this case.

Remarks:

Let us now examine some of the restrictions imposed in deriving (5.1)–(5.5).

(a) The results hold even when the chain is denumerable.

(b) Suppose that 0 is an eigenvalue (say $t_m = 0$) and the other eigenvalues t_i, $i = 1, \ldots, m - 1$ are distinct and non-zero. Then (5.5) holds for $n = 1, 2, \ldots$ with $t_m = 0$, and so (5.5) also holds.

(c) In the discussion of the spectral representation of a matrix (Sec. Appendix A.4.4.) it is assumed that the eigenvalues of the matrix are all distinct. Suppose now that the characteristic equation has some multiple roots. In this case there will be some modification in the spectral representations of P and P^n given in (5.1) and (5.2).

Before the spectral representation in the general case (with multiple root) is given, the following categories of multiple roots may be noted:

(i) $t_1 = 1$ is a multiple root irrespective of whether some of the other roots t_i, $i \neq 1$ are multiple. This case which is of special importance occurs whenever the chain contains two or more closed subchains.

(ii) $t_1 = 1$ is not a multiple root but some t_i ($i \neq 1$) are multiple roots.

Consider the general case when t_i, $i = 1, \ldots, k$ are multiple roots of order r_i (≥ 1) respectively, i.e. $\sum_{i=1}^{k} r_i = m$, the order of the chain. In this case there exists a non-singular matrix H such that P has the canonical representation,

$$P = HZH^{-1}, \tag{5.7}$$

where

$$Z = \begin{pmatrix} A_{r_1}(t_1) & & & 0 \\ A_{r_2}(t_2) & & \cdots & \\ & & \cdots & \\ 0 & & & A_{r_k}(t_k) \end{pmatrix} \tag{5.7a}$$

and

$A_r(t) = (a_{i,j}(t))$ is a $(r \times r)$ matrix such that
$a_{i,j}(t) = t$ when $i = j$, $a_{i,j}(t) = 1$, when $j = i + 1$

and

$a_{i,j}(t) = 0$ for $j \neq i, i, +1$. Clearly, $A_1(t) = a_{1,1} = t$.

Now it can be seen that

$$(A_r(t))^n = \begin{bmatrix} t^n & \binom{n}{1}t^{n-1} & \cdots & \binom{n}{r-1}t^{n-r+1} \\ 0 & t^n & \cdots & \binom{n}{r-2}t^{n-r+2} \\ \cdots & \cdots & \cdots & \cdots \\ 0 & 0 & \cdots & t^n \end{bmatrix}, \tag{5.8}$$

where $\binom{n}{k} = 0$ for $n < k$.

We also get z^n by replacing $A_r(t)$ in the r.h.s. of (5.7) by $(A_r(t))^n$. Thus

$$P^n = H Z^n H^{-1}, \quad n = 0, 1, 2, \ldots \tag{5.9}$$

In the general case with multiple roots we thus have (5.7) and (5.9) in place of (5.1) and (5.2) respectively.

3.6 STABILITY OF A MARKOV SYSTEM

Limting Behaviour: Finite Irreducible Chains

Definition: *Stationary distribution*: Consider a Markov Chain with transition probabilities p_{jk} and t.p.m. $P = (p_{jk})$. A probability distribution $\{v_j\}$ is called *stationary* (or *invariant*) for the given chain if

$$v_k = \sum_j v_j \, p_{jk}$$

such that $v_j \geq 0, \sum_j v_j = 1$.

Again

$$v_k = \sum_j v_j \, p_{jk} = \sum_j \left\{ \sum_i v_i \, p_{ij} \right\} p_{jk} = \sum_i v_i \, p_{ik}^{(2)}$$

and, in general,

$$v_k = \sum_i v_i \, p_{ik}^{(n)}, \quad n \geq 1.$$

We now study the question whether a Markov system, regardless of the initial state j, reaches a steady or stable state after a large number of transitions or moves. In other words, under what conditions, if any, as n tends to ∞, $p_{jk}^{(n)}$ tends to a limit v_k independent of the initial state j (i.e. P^n tends to a stochastic matrix whose rows are identical). This property of limiting distribution of $p_{jk}^{(n)}$ being independent of the initial state j (or P^n tending to a matrix with identical rows) is known as *ergodicity*; the Markov chain is called *ergodic*. When such limits exist, the probabilities settle down and become stable. The system then shows some long run regularity properties.

Theorem 3.7 considered below gives a sufficient condition for the existence of $\{v_k\}$. It also shows how the distribution can be obtained, in case where this exists.

We first prove Theorem 3.6 which will be needed to prove Theorem 3.7.

Theorem 3.6.   If state j is persistent, then for every state k that can be reached from state j, $F_{kj} = 1$.

Proof : Let a_k be the probability that starting from state j, the state k is reached without previously returning to state j. The probability of never returning to state j once state k is reached is $(1 - F_{kj})$.

The probability of the compound event that starting from state j, the system reaches state k (without returning to state j) and never returns to state j is $a_k (1 - F_{kj})$. If there are some other states, say, $r, s, \ldots$, then we get similar terms $a_r (1 - F_{rj}), a_s (1 - F_{sj})$. Thus the probability Q that starting from state j the system never returns to state j is given by

$$Q = a_k (1 - F_{kj}) + a_r (1 - F_{rj}) + a_s (1 - F_{sj}) + \ldots \ .$$

But since the state j is persistent, $F_{jj} = 1$ and the probability of never returning to state j is $1 - F_{jj} = 0$. Thus $Q = 0$. This implies that each term is zero, so that $F_{kj} = 1$.   ▲

For a rigorous proof which uses strong Markov property, see Iosifescu (1980).

Theorem 3.7   *(Ergodic Theorem)*. For a finite irreducible, aperiodic chain with t.p.m. $P = (p_{jk})$, the limits

$$v_k = \lim_{n \to \infty} p_{jk}^{(n)} \tag{6.1}$$

exist and are independent of the initial state j. The limits v_k are such that $v_k \geq 0$, $\sum v_k = 1$, i.e. the limits v_k define a probability distribution.

Furthermore, the limiting probability distribution $\{v_k\}$ is identical with the stationary distribution for the given chain, so that

$$v_k = \sum_j v_j \, p_{jk}, \quad \sum v_k = 1; \tag{6.2}$$

writing $V' = (v_1, \ldots, v_k, \ldots)$, $\sum v_k = 1$ the relation (6.2) may also be written as

$$V' = V'P$$

or
$$V'(P - I) = 0. \tag{6.3}$$

Proof :   Since the states are aperiodic, persistent non-null, for each pair of j, k, $\lim_{n \to \infty} p_{jk}^{(n)}$ exists and is equal to F_{jk}/μ_{kk} (Theorem 3.4). Again by Theorem 3.6, since k is persistent, $F_{jk} = 1$, so that

$$v_k = \lim_{n \to \infty} p_{jk}^{(n)} = \frac{1}{\mu_{kk}} > 0$$

and is independent of j.

Since
$$\sum_k p_{ik}^{(n)} = 1, \ \sum_{k=1}^{N} p_{ik}^{(n)} \leq 1 \ \text{ for all } N.$$

Thus,
$$\sum_{k=1}^{N} \lim_{n \to \infty} p_{ik}^{(n)} \leq 1, \quad \text{i.e. } \sum_{k=1}^{N} v_k \leq 1.$$

Since it holds for all N, $\sum\limits_{k=1}^{\infty} v_k \leq 1$.

We have
$$p_{jk}^{(n+m)} = \sum_i p_{ji}^{(n)} p_{ik}^{(m)}$$

and
$$v_k = \lim_{n \to \infty} p_{jk}^{(n+m)} = \lim_{n \to \infty} \sum_i p_{ji}^{(n)} p_{ik}^{(m)}$$

$$\geq \sum_i \left\{ \lim_{n \to \infty} p_{ji}^{(n)} \right\} p_{ik}^{(m)}$$

(by Fatou's lemma)

$$= \sum v_i p_{ik}^{(m)}, \text{ for all } m$$

i.e. $v_k \geq \sum\limits_i v_i \, p_{ik}^{(m)}$.

Suppose, if possible, that $v_k > \sum\limits_i v_i \, p_{ik}^{(m)}$ then summing over all k, we get

$$\sum_k v_k > \sum_k \sum_i v_i p_{ik}^{(m)} = \sum_i v_i$$

which is impossible. Hence the sign of equality holds, i.e.

$$v_k = \sum_i v_i p_{ik}^{(m)} \qquad \text{for all } m \geq 1.$$

In particular, when $m = 1$,

$$v_k = \sum_i v_i p_{ik}.$$

For large m, we have

$$v_k = \sum_i v_i v_k = \left(\sum_i v_i \right) v_k.$$

Hence $\sum v_i = 1$. This shows that $\{v_k\}$ is a probability distribution; the distribution is unique. ▲

We state, without proof, the *converse* which also holds.

If a chain is irreducible and aperiodic and if there exist a unique stationary distribution $\{v_k\}$ for the chain, then the chain is ergodic and $v_k = 1/\mu_{kk}$.

Thus ergodicity is a necessary and sufficient condition for the existence of $\{v_k\}$ satisfying (6.2) in case of an irreducible and aperiodic chain.

Remarks:

(1) The probability vector V' is itself an eigenvector of P corresponding to the eigenvalue 1 of P.

(2) The matrix $(P - I)$ in (6.3) is such that its off-diagonal elements are non-negative and diagonal elements are strictly negative. From these properties and the irreducibility condition it follows that the matrix $(P - I)$ is of rank $(m - 1)$, where m is the number of states of the chain. The matrix $(P - I)$ (as well as $(I - P)$) is singular. If the number of states is finite, then $\{v_k\}$ can be obtained from the homogeneous equations $V'(P - I) = 0$ (which are finite in number), and the normalising condition $\Sigma v_k = 1$. Some examples are considered below.

(3) The row vector V' satisfying $V'P = V', V'e = 1$, is known as the *invariant* measure of P.

(4) The ergodic theorem may also be stated as follows:
As $n \to \infty$, $P^n \to e V'$, where e is the column vector with all entries equal to 1.

(5) Further, it can be shown that the rate of approach to the limit is geometric (fast). The rate of approach to the limit is geometrically fast, when, there exist $a > 0$ and b $(0 < b < 1)$ such that

$$|p_{jk}^{(n)} - v_k| \leq ab^n.$$

Example 6(a). Consider the two state Markov chain of Example 5(a). The chain is irreducible and consists of two aperiodic persistent non-null states. The equations (6.2) giving v_1, v_2 become

$$av_1 - bv_2 = 0 \text{ and } v_1 + v_2 = 1$$

whence we get $v_1 = b/(a + b)$ and $v_2 = a/(a + b)$. These are directly obtained as limits in Example 5(a).

Example 6(b). Consider the three-state Markov Chain having t.p.m.

$$\begin{pmatrix} 0 & 2/3 & 1/3 \\ 1/2 & 0 & 1/2 \\ 1/2 & 1/2 & 0 \end{pmatrix}.$$

The chain is irreducible and the states are aperiodic persistent non-null. The equations (6.2) giving v_k $(k = 1, 2, 3)$ become

$$0 = -v_1 + \frac{1}{2}v_2 + \frac{1}{2}v_3$$

$$0 = \frac{2}{3}v_1 - v_2 + \frac{1}{2}v_3$$

$$1 = v_1 + v_2 + v_3.$$

We get

$$v_1 = 1/3, v_2 = 10/27, v_3 = 8/27.$$

Example 6(c). For the ergodic chain of Example 4(b), $V'P = V'$ leads to
$V' = (2/17, 4/17, 3/17, 8/17)$.

Verify that (4.12) holds.

Example 6(d). For the ergodic chain considered in Example 5(c), $V'P = V'$ leads to
$V' = (p^2, 2pq, q^2)$, as shown there as lim of P^n as $n \to \infty$.

Example 6(e). *One-dimensional random walk with two elastic barriers.* Let the states
be $1, 2, \ldots, k$. The particle moves from the state i to state $(i + 1)$ with probability p
and to state $(i - 1)$ with probability $q = 1 - p$ for $i = 2, \ldots, k - 1$. From states 1 and
k (elastic barriers) the particle is reflected, i.e. from state 1 it moves to state 2 with
probability p and remains at state 1 with probability q and similarly from state k it
moves to state $(k - 1)$ with probability q and remains at state k with probability p
$(0 < p < 1)$.

The transition probabilities are:

$$p_{ij} = p, \quad j = i + 1$$

$$= q, \quad j = i - 1 \text{ for } 1 < i < k$$

$$p_{ii} = 0, \quad i \neq 1, k$$

$$p_{12} = p, \quad p_{11} = q$$

$$p_{k,k} = p, \quad p_{k,k-1} = q.$$

The chain is irreducible and the states are aperiodic persistent non-null. The limiting
probabilities v_k are solutions of the equation (6.2). The solution can be obtained
recursively. From the first $(k - 1)$ equations we have

$$v_2 = (p/q)v_1, v_3 = (p/q)^2 v_1, \ldots, v_k = (p/q)^{k-1}v_1.$$

Using $\Sigma v_j = 1$, we get

$$v_1 = (1 - (p/q)) / (1 - (p/q)^k).$$

Hence

$$v_r = (1 - (p/q)) (p/q)^{r-1} [1 - (p/q)^k]^{-1}, r = 1, \ldots, k.$$

Remarks : The Inverse Problem

Consider a Markov chain with state space $S = \{1, 2, \ldots k\}$, transition matrix P and unique stationary distribution V'. To understand the limitations of equilibrium (stationary) distribution one has to examine the inverse problem: what is the set of all transition matrices having a given equilibrium distribution?

To give an idea of the problem, Whitt (1983) considers the example of simple two state chain having t.p.m.

$$P = \begin{pmatrix} 1-p & p \\ p & 1-p \end{pmatrix}, \quad 0 \le p \le 1$$

Now for $p = 0$, $P = I$ and for $p = 1$, P is periodic so that for p near 0, P is nearly decomposable and for p near 1, P is nearly periodic.

We have, for all p in $0 < p < 1$

$$P^n = \begin{pmatrix} \dfrac{1}{2} & \dfrac{1}{2} \\ \dfrac{1}{2} & \dfrac{1}{2} \end{pmatrix} + (1-2p)^n \begin{pmatrix} \dfrac{1}{2} & -\dfrac{1}{2} \\ -\dfrac{1}{2} & \dfrac{1}{2} \end{pmatrix}$$

so that, as $n \to \infty$

$$P^n \to e\left(\frac{1}{2}, \frac{1}{2}\right)$$

the equilibrium distribution is $\left(\dfrac{1}{2}, \dfrac{1}{2}\right)$.

The example gives an idea as to how much can the transient behaviour vary over the class of transient matrices for a given equilibrium distribution. For further details, see Karr (1978), Karr and Pittenger, *St. Pro. & Appl.* 7 (1978) 165 and Whitt (1983).

3.7 GRAPH THEORETIC APPROACH

A simple graph theoretic formula for finding v_k has been advanced by Solberg (1975). The formula, which is similar in spirit to the well-known flow-graph formulas, is conceptually simple and is claimed to have computational advantages. Further, the formula is applicable to the continuous parameter case also (see Sec. 4.5.4). This approach is considered below.

We recall the digraph representation of a Markov chain (Sec. 3.1.3).

Definition. An *intree* T_j to a specific point j in a directed graph G, is a spanning subgraph of G in which every point (node or vertex) but j has exactly one arc emanating from it and j has none. Thus an intree to point j is just a tree with the arcs all oriented towards the point j.

The *weight w* (T_j) of an intree j is defined to be the product of the weights of all arcs appearing in T_j.

We state the theorem due to Solberg (for proof refer to the paper).

Theorem 3.8. The limiting distribution $\{v_k\}$ of an aperiodic, irreducible Markov chain (discrete or continuous parameter) with a finite number of states is given by

$$v_k = \frac{c_k}{\sum_j c_j}$$

where $c_j = \sum_j w_i\,(T_j)$ is the sum over all intrees to the point j in the transition diagram of the process. ▲

The above theorem is illustrated with the help of an example. Consider the Markov chain of Example 6(b).

The intrees to the point 1 are shown below:

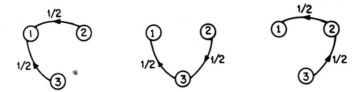

Fig. 3.2 Intrees to point 1

The weights of the intrees to the point 1 are:

$$w_1(T_1) = p_{21}p_{31} = 1/4, \quad w_2(T_1) = p_{23}p_{31} = 1/4, \quad w_3(T_1) = p_{32}p_{21} = 1/4$$

so that

$$c_1 = \sum_i w_i(T_1) = 3/4.$$

The weights of the intrees to the point 2 are:

$$w_1(T_2) = p_{32}p_{12} = 1/3, \quad w_2(T_2) = p_{31}p_{12} = 1/3, \quad w_2(T_3) = p_{13}p_{32} = 1/6$$

so that

$$c_2 = \sum_i w_i(T_2) = 5/6.$$

The weights of the intrees to the point 3 are:

$$w_1(T_3) = p_{13}p_{23} = 1/6, \quad w_2(T_3) = p_{21}p_{13} = 1/6, \quad w_3(T_3) = p_{12}p_{23} = 1/3$$

so that

$$c_3 = \sum_i w_i(T_3) = 2/3.$$

Thus

$$\sum c_j = 3/4 + 5/6 + 2/3 = 9/4.$$

We have

$$v_1 = c_1/\sum c_j = 1/3$$

$$v_2 = c_2/\sum c_j = 10/27$$

and

$$v_3 = c_3/\sum c_j = 8/27.$$

3.8   MARKOV CHAIN WITH DENUMERABLE NUMBER OF STATES (OR COUNTABLE STATE SPACE)

So far we discussed Markov chains with a finite number of states. The results can be generalised to chains with a denumerable number of states (or with a countable state space). Let $P = (p_{ij})$ be the t.p.m. of the chain $\{X_n, n \geq 1\}$ with countable states space $S = \{0, 1, 2, \ldots\}$. Then $P^k = (p_{ij}^{(k)})$ is well defined. The states of the chain may not constitute even a single closed set. For example when

$$p_{ij} = 1, j = i + 1$$

$$= 0, \quad \text{otherwise,}$$

the states do not belong to any closed set, including S.

For dealing with a chain with a countable state space, we need a more sensitive classification of states – transient, persistent null and persistent non-null. Besides irreducibility and aperiodicity, non-null persistence is required for erodicity for such a chain (a chain with countable state space) while aperiodicity and irreducibility (or some type of reducibility) were enough for ergodicity for a finite chain. We shall state the theorem without proof. For proof see Seneta (1981).

Theorem 3.9. General Erogodic Theorem

Let $\{X_n\}$ be an irreducible, aperiodic Markov chain with state space $S = \{\ldots, k, \ldots\}$ and having t.p.m. $P = (p_{ij})$.

If the chain is transient or persistent-null then $\lim\limits_{n \to \infty} p_{jk}^{(n)} = 0$.

If the chain is persistent non-null then the limits $\lim\limits_{n \to \infty} p_{jk}^{(n)} = v_k$ exist and are independent of the initial state j. The limits are such that $v_k \geq 0$, $\sum\limits_k v_k = 1$, i.e. $V' = \{v_1, v_2, \ldots, v_k, \ldots\}$ is a probability vector. The vector V' is given by the solution of

$$V' = V' P$$

that is

$$v_k = \sum_j v_j \, p_{jk}.$$

In other words $P^n \to e \, V'$ where V' is given by $V' = V' P$.

The above theorem is a generalization of Theorem 3.7 concerning ergodicity of finite, irreducible, aperiodic chain.

In this case, the number of unknowns and so also the number of equations are infinite. The limiting distribution can be obtained more conveniently in terms of its p.g.f. We consider an example below.

Example 8(a). A Markov chain occurring in queueing theory is a chain with a countable state space $S = \{0, 1, 2, \ldots\}$ and transition probability matrix

$$P = \begin{pmatrix} p_0 & p_1 & p_2 & p_3 & \cdots \\ p_0 & p_1 & p_2 & p_3 & \cdots \\ 0 & p_0 & p_1 & p_2 & \cdots \\ 0 & p_0 & p_1 & p_2 & \cdots \\ 0 & 0 & p_0 & p_1 & \cdots \\ \cdots & \cdots & \cdots & \cdots & \cdots \end{pmatrix}$$

where $\sum p_k = 1$.

Let

$$P(s) = \sum_k p_k s^k \text{ and } V(s) = \sum_k v_k s^k$$

be generating functions of $\{p_k\}$ and $\{v_k\}$ respectively. Assume that $p_i > 0$ for all i. The chain is irreducible and aperiodic.

It can be shown that the states are transient, persistent null or persistent non-null according as $P'(1) > 1, = 1$ or < 1. Thus when $P'(1) < 1$, v_k's are the unique solutions of the equations (6.2). The equations (6.2) become:

$$v_0 = p_0 v_0 + p_0 v_1$$

$$v_1 = p_1 v_0 + p_1 v_1 + p_0 v_2$$

$$v_2 = p_2 v_0 + p_2 v_1 + p_1 v_2 + p_0 v_3$$

$$\cdots \quad \cdots \quad \cdots \quad \cdots \quad \cdots$$

$$v_k = p_k v_0 + p_k v_1 + p_{k-1} v_2 + \cdots + p_0 v_{k+1}$$

$$\cdots \quad \cdots \quad \cdots \quad \cdots \quad \cdots$$

Multiplying both sides of the $(k + 1)$ st equation by s^k $(k = 0, 1, 2, \ldots)$ and adding over k, we get

$$V(s) = v_0 P(s) + v_1 P(s) + v_2 s P(s) + v_3 s^2 P(s) + \cdots$$

$$= P(s)\{v_0 + (V(s) - v_0)/s\}.$$

This gives

$$V(s) = v_0(1 - s)P(s) \big/ (P(s) - s),$$

in terms of v_0 which can be evaluated from $\sum v_k = 1$. We have

$$\lim_{s \to 1} \frac{V(s)}{v_0} = \lim_{s \to 1} \frac{\{1 - s\}P(s)}{P(s) - s}$$

whence

$$v_0 = 1 - P'(1)\,(> 0).$$

Thus

$$V(s) = \frac{\{1 - P'(1)\}\,(1 - s)P(s)}{P(s) - s}.$$

3.9 REDUCIBLE CHAINS

So far we have considered limiting properties of irreducible Markov chains. Now we shall examine such behaviour of certain types of reducible chains.

3.9.1 Finite Reducible Chains with a Single Closed Class

Consider a finite Markov chain with state space S such that it has a *single closed class* C of states which communicate with one another; the other states form one or more sets such that members of a set communicate with one another and lead to a state of C.

Further assume that the states of C are *aperiodic*.

The ergodicity of finite irreducible Markov chain was considered in Theorem 3.7. Another ergocity theorem for reducible chains having a single closed class of aperiodic states is given below.

Theorem 3.10. *Ergodic theorem for reducible chains*
Chain with a single closed class
Let $\{X_n, n \geq 0\}$ be a finite Markov chain with aperiodic states.

Let P be the transition matrix of the m-state chain with state space S, and P_1 the transition (submatrix) of transitions among the k ($\leq m$) members of the closed class C. Let $V_1' = \{\ldots, v_j, \ldots\}$ be the stationary distribution corresponding to the stochastic submatrix P_1, i.e. $P_1^n \to e\, V_1'$

If $V' = (V_1', \mathbf{0}')$, then, as $n \to \infty$, $P^n \to e\, V'$.

In other words, elementwise V' is the stationary distribution corresponding to the matrix P.

An outline of proof is given below:

The transition matrix of the chain can be put in cononical form

$$P = \begin{pmatrix} P_1 & \mathbf{0} \\ R_1 & Q \end{pmatrix} \tag{9.1}$$

where the stochastic (sub) matrix corresponds to transitions among the members of class C and Q corresponds to transitions among the other states (of $S - C$).

We have

$$P^n = \begin{pmatrix} P_1^n & \mathbf{0} \\ R_n & Q^n \end{pmatrix} \tag{9.2}$$

where $R_n = R_{n-1} P_1 + Q^{n-1} R_1$.

Writing $R_1 = R$, we get

$$R_{n+1} = \sum_{i=0}^{n} Q^i R P_1^{n-i} = \sum_{i=0}^{n} Q^{n-i} R P_1^i.$$

As $n \to \infty$ $\qquad P_1^n \to e V_1'$

and $\qquad\qquad Q^n \to \mathbf{0}$.

Again it can be shown that, as $n \to \infty$

$$R_{n+1} \to e\, V_1'$$

so that, writing $V' = (V_1', \mathbf{0}')$ we have

$$P^n \to e V'. \tag{9.3}$$

3.9.2 Chain with One Single class of Persistent non-null Aperiodic States

Now suppose that the states of the closed class C are non-null persistent and aperiodic, the remaining states of S being transient; the transient states constitute a set T.

Then we have, for each pair of i, j,

$$\lim_{n \to \infty} p_{ij}^{(n)} = v_j$$

is independent of i, when i, j are persistent, and also when j is persistent and i is transient; again

$$\lim_{n \to \infty} p_{ij}^{(n)} = 0 \qquad \text{when } j \text{ is transient.}$$

In this case we shall write the transition matrix as

$$P = \begin{pmatrix} P_1 & 0 \\ R_1 & M \end{pmatrix} \tag{9.4}$$

where M gives the matrix of transitions among the transient states.

Example 9(a). Consider a reducible chain with $S = \{1, 2, 3, 4\}$ and t.p.m.

$$P = \begin{pmatrix} P_1 & 0 \\ R_1 & M \end{pmatrix}$$

where

$$P_1 = \begin{matrix} 1 \\ 3 \end{matrix}\begin{pmatrix} 1/2 & 1/2 \\ 2/3 & 1/3 \end{pmatrix}, R_1 = \begin{pmatrix} 0 & 1/4 \\ 0 & 1/4 \end{pmatrix}, M = \begin{pmatrix} 3/4 & 0 \\ 1/2 & 1/4 \end{pmatrix}.$$

We have

$$P_1^n \sim e\, V'_1 \quad \text{where} \quad V'_1 = (4/7, 3/7).$$

Thus

$$P^n \sim e\, V', \quad \text{where} \quad V' = (V'_1, 0')$$

$$= (4/7, 0, 3/7, 0).$$

In other words, for all i, as $n \to \infty$

$$p_{i1}^{(n)} \to 4/7$$

$$p_{i3}^{(n)} \to 3/7$$

$$p_{i2}^{(n)} \to 0$$

$$p_{i4}^{(n)} \to 0.$$

Example 9(b). *Stochastic Inventory Model* (Seneta, 1981)

Consider that a store stocks a certain item, the demand for which is given by

$$p_k = \Pr\{k \text{ demands of the item in a week}\},$$
$$p_k > 0, k = 0, 1, 2 \text{ and } p_k = 0, k \geq 3.$$

Stocks are replenished at weekends according to the policy: not to replenish if there is any stock in store and to obtain 2 new items if there is no stock. Let X_n be the number of items at the end of nth week, just before week's replenishment, if any, and $\Pr\{X_0 = 3\} = 1$.

Then $\{X_n, n \geq 0\}$ is a Markov chain with state space $S = \{0, 1, 2, 3\}$ and t.p.m.

$$P = \begin{array}{c} \\ 0 \\ 1 \\ 2 \\ 3 \end{array}\begin{array}{cccc} 0 & 1 & 2 & 3 \\ \left(\begin{array}{cccc} p_2 & p_1 & p_0 & 0 \\ p_1 + p_2 & p_0 & 0 & 0 \\ p_2 & p_1 & p_0 & 0 \\ 0 & p_2 & p_1 & p_0 \end{array} \right) \end{array}, \quad p_0 + p_1 + p_2 = 1.$$

The Markov chain is reducible, with a single closed class C with states 0, 1 and 2, the states being persistent non-null and aperiodic. The t.p.m. (submatrix) is

$$P_1 = \begin{array}{c} \\ 0 \\ 1 \\ 2 \end{array}\begin{array}{ccc} 0 & 1 & 2 \\ \left(\begin{array}{ccc} p_2 & p_1 & p_0 \\ p_1 + p_2 & p_0 & 0 \\ p_2 & p_1 & p_0 \end{array} \right) \end{array}$$

The state 3 is transient.

Thus

$$P = \begin{pmatrix} P_1 & \mathbf{0} \\ R_1 & M \end{pmatrix}$$

where $R_1 = (0, p_2, p_1), M = (p_0)$ and

$$\mathbf{0} = \begin{pmatrix} 0 \\ 0 \\ 0 \end{pmatrix}.$$

Using Theorem 3.10 we get, $V_1' = (v_0, v_1, v_2)$

$$P_1^n \to \mathbf{e} V_1'$$

where $\quad V_1' P_1 = V_1', \quad V_1' \mathbf{e} = 1,$

$$p_{ij}^{(n)} \to v_j, j = 0, 1, 2.$$

We get

$$v_0 = (1 - p_0)^2/c$$

$$v_1 = p_1/c$$

$$v_2 = p_0(1 - p_0)/c$$

where $c = (1 - p_0 + p_1).$

Further $P^n \to e\, V'$ where $V' = (v_0, v_1, v_2, 0)$.

3.9.3 Absorbing Markov Chains

Assume that all the persistent states of a Markov chain with finite state space S are absorbing and that the chain has a set T of transient states. We can then rearrange the states so that all the absorbing states are taken first and then the transient states. The transition matrix can be put in the canonical form

$$P = \begin{pmatrix} I & 0 \\ R & Q \end{pmatrix} \tag{9.5}$$

where Q corresponds to transitions among the members of the set T of transent states; and I, the unit matrix corresponds to the transitions among the members of the set of absorbing states.
We have

$$P^n = \begin{pmatrix} I & 0 \\ R_n & Q^n \end{pmatrix}.$$

As $n \to \infty$, $Q^n \to 0$ as $\lim_{n \to \infty} p_{ij}^{(n)} = 0, \quad i,j \in T.$

If the chain starts with $i \in T$, then it stays in T for a finite number of steps after which it enters the set of absorbing states where it remains forever. Our interest lies in the probability of absorption.

Theorem 3.11. Let a_{ik} denote the probability that the chain starting with a transient state i eventually gets absorbed in an absorbing state k.
If we denote the absorption probability matrix by

$$A = (a_{ik}), i \in T, k \in S - T$$

then

$$A = (I - Q)^{-1} R = NR, N = (I - Q)^{-1}. \tag{9.6}$$

Proof : We have

$$a_{ik} = F_{ik} = \sum_n f_{ik}^{(n)}$$

since transitions between absorbing states are impossible. Now

$$F_{ik} = \Pr\left[\bigcup_{n \geq 1} \{X_n = k \mid X_0 = i\}\right].$$

Since state k is absorbing, once the chain reaches an absorbing state k after n steps, it remains there after steps $(n + 1), (n + 2) \ldots$. Thus

$$\{X_n = k\} \subset \{X_{n+1} = k\} \subset \{X_{n+2} = k\} \cdots$$

Using the result for the ascending sequence $A_1 \subset A_2 \subset \ldots$.

$$\Pr\left\{\bigcup_{i \geq 1} A_i\right\} = \lim_{n \to \infty} \Pr(A_n)$$

we get

$$a_{ik} = F_{ik} = \Pr\left\{\bigcup_{n \geq 1} p_{ik}^{(n)}\right\} = \lim_{n \to \infty} p_{ik}^{(n)}. \tag{9.7}$$

Chapman-Kolmogrov equation can be written as

$$p_{ik}^{(n+1)} = \sum_{j \in S} p_{ij} p_{jk}^{(n)}.$$

Using

$$p_{kk}^{(n)} = 1, \text{ and } p_{jk} \neq 0 \text{ only when } j \in T, k \in S - T$$

we get

$$p_{ik}^{(n+1)} = p_{ik} + \sum_{j \in T} p_{ij} p_{jk}^{(n)}.$$

Taking limits of both sides as $n \to \infty$ and using (9.7), we get

$$a_{ik} = p_{ik} + \sum_{j \in T} p_{ij} a_{jk}. \tag{9.8}$$

Thus in matrix notation,

$$A = (a_{ik}) = R + Q A.$$

Thus

$$A = (I - Q)^{-1} R$$

$$= NR. \tag{9.9}$$

Note:

(1) The matrix $N = (I - Q)^{-1}$ is known as the *fundamental matrix*.

(2) Suppose that the state space S has m elements, the set T has m_1 ($< m$) elements, then the set $S - T$ of absorbing states has $m - m_1$ elements. Then I and Q in (9.5) are square matrices of order $(m - m_1)$ and m_1 respectively and R is a rectangular matrix of order $m_1 \times (m - m_1)$.

A is a rectangular matrix of order $m_1 \times (m - m_1)$ and I in (9.9) is a unit matrix of order $m_1 \times m_1$.

(3) As $n \to \infty$, the limits $\lim p_{ik}^{(n)}$ exist but are *not* independent of the initial state i.

3.9.3.1 *Time upto Absorption from a transient state to an absorbing state*

Let γ_i be the number of steps, including the starting position at i, in which the Markov chain remains in a transient state, that is, γ_i is the number of steps that a chain starting with $i \in T$ stays in T before entering an absorbing state. The r.v. γ_i is known as the *time until absorption* for the chain starting at state i. We proceed to find the distribution of γ_i and its mean.

The random events $\{\gamma_i > n\}$ and $\{X_n \in T\}$ are equivalent. Again, the event

$$\{\gamma_i > n - 1\} = \{\gamma_i > n\} \ U \ \{\gamma_i = n\}, n \geq 1$$

so that the distribution is given by

$$\Pr\{\gamma_i = n\} = \Pr\{\gamma_i \geq n - 1\} - \Pr\{\gamma_i \geq n\}$$

$$= \Pr\{X_{n-1} \in T\} - \Pr\{X_n \in T\}$$

$$= \sum_{j \in T} \{p_{ij}^{(n-1)} - p_{ij}^{(n)}\}. \tag{9.10}$$

Now, for $i, j \in T$

$$(p_{ij}^{(k)}) = Q^k$$

so that the column vector

$$\left(\dots, \sum_{j \in T} p_{ij}^{(k)}, \dots \right)_{i \in T} = Q^k \, \mathbf{e}$$

where **e** is the column vector with all elements equal to unity. Denote the mean time upto absorption starting at state i by $\mu_i = E(\gamma_i) = \sum_n n \Pr\{\gamma_i = n\}$; let

$$\gamma(n) = (\cdots, \Pr\{\gamma_i = n\}, \cdots)_{i \in T}$$

be the vector whose elements (probability mass functions) correspond to the transient states and $\mu = (\cdots, \mu_i, \cdots)_{i \in T}$ be the vector whose elements (expectations) correspond to the transient states.

Then the vector

$$\gamma(n) = \left(\cdots, \sum_{j \in T}\{p_{ij}^{(n-1)} - p_{ij}^{(n)}\}, \cdots\right)_{i \in T}$$

$$= (Q^{n-1} - Q^n)\mathbf{e}$$

$$= Q^{n-1}(I - Q)\mathbf{e}, \; n \geq 1 \qquad (9.11)$$

and $\qquad\qquad \gamma(0) = \mathbf{0};$

further

$$\mu = \sum_{n=0}^{\infty} n\, \gamma(n) = \sum_{n=1}^{\infty} nQ^{n-1}(I - Q)\mathbf{e}$$

$$= (I - Q)^{-2}(I - Q)\mathbf{e}$$

$$= (I - Q)^{-1}\mathbf{e}$$

$$= N\,\mathbf{e}. \qquad (9.12)$$

Thus $\mu = (\cdots, \mu_i, \cdots)$ can be found out from the fundamental matrix $N = I - Q$.

Example 9(c). Gambler's ruin problem (Random walk with two absorbing states 0 and k). We can write the transition matrix as

$$
P = \begin{array}{c}
\\ 0 \\ k \\ 1 \\ 2 \\ \cdot \\ \cdot \\ \cdot \\ k-1
\end{array}
\begin{array}{c}
\begin{array}{ccccccc}
0 & k & 1 & 2 & \cdots & k-1 \\
\end{array} \\
\left(
\begin{array}{cccccc}
1 & 0 & 0 & 0 & & 0 \\
0 & 1 & 0 & 0 & & 0 \\
q & 0 & 0 & p & & 0 \\
0 & 0 & q & 0 & & \\
& & \cdots & & & \cdots \\
& & \cdots & & & \cdots \\
0 & p & 0 & 0 & & 0 \\
\end{array}
\right)
\end{array}
$$

so that

$$Q = \begin{array}{c} 1 \\ 2 \\ \cdots \\ k-1 \end{array} \begin{array}{ccccc} 1 & 2 & \cdots & k-1 & \\ 0 & p & 0 & \cdots & 0 & 0 \\ q & 0 & p & \cdots & 0 & 0 \\ & & \cdots & & & \\ 0 & 0 & 0 & & q & 0 \end{array}$$

is a $(k-1) \times (k-1)$ matrix

and

$$R = \begin{array}{c} 1 \\ 2 \\ \cdot \\ \cdot \\ \cdot \\ k-1 \end{array} \begin{array}{cc} 0 & k \\ q & 0 \\ 0 & 0 \\ & \\ \cdots & \\ 0 & p \end{array}$$

is a $(k-1) \times 2$ matrix.

Using (9.8) we get the absorption probability from a transient state i to the absorbing state 0 as

$$a_{i0} = p_{i0} + \sum_{j=1}^{k-1} p_{ij} \, a_{j0}$$

and putting $i = 1, 2, \ldots, l, \ldots, k-1$ we get

$$a_{1,0} = q + p \, a_{2,0}$$

$$a_{2,0} = q a_{1,0} + p a_{3,0}$$

$$\cdots \qquad \cdots$$

$$a_{k-1,0} = q a_{k-2,0} \, .$$

This can be written as

$$\left. \begin{array}{l} a_{i,0} = q a_{i-1,0} + p a_{i+1,0}, \, 1 \le i \le k-1 \\ \\ a_{0,0} = 1, \, a_{k,0} = 0. \end{array} \right\} \qquad (9.13)$$

with

The difference equation (9.13) admits of the solution
(for $1 \le i \le k-1$),

$$a_{i,0} = \frac{\left(\frac{p}{q}\right)^{k-i} - 1}{\left(\frac{p}{q}\right)^{k} - 1} \ , \ p \neq q$$

$$= 1 - \frac{i}{k} \ , \qquad p = q \ .$$

(9.14)

We get the absorption probabilities from a transient state i, $1 \leq i \leq k - 1$ to the absorbing state k as

$$a_{i,k} = 1 - a_{i,0}, \ 1 \leq i \leq k - 1.$$

(9.15)

Note: We can also get $a_{i,0}$ by computing $(I - Q)^{-1} R$.

Special Case:
Martingale: Suppose that the sequence of r.v.'s. $\{Y_n, n \geq 0\}$ is a martingale. Then

$$E\{Y_n \mid Y_{n-1} = y_{n-1}, \cdots, Y_0 = y_0\} = y_{n-1}, \text{ for every } y\text{'s}.$$

(9.16)

In particular, if $\{X_n, n \geq 0\}$ is a martingale as well as a Markov chain with t.p.m. $P = (p_{ij})$ then the above implies that

$$E\{X_n \mid X_{n-1} = i\} = i$$

(9.17)

for every i. Thus

$$\sum_j j \, \Pr\{X_n = j \mid X_{n-1} = i\} = i$$

$$\text{or} \quad \sum_j j \, p_{ij} = i \ .$$

(9.18)

Writing (9.17) as

$$E\{X_{m+1} \mid X_m = i\} = i$$

we get

$$E\{X_{m+2} \mid X_m = i\} = i$$

and by induction

$$E\{X_{m+n} \mid X_m = i\} = i, n = 1, 2, \cdots$$

(9.17b)

This implies that for all i

$$\sum_j j\, p_{ij}^{(n)} = i.\qquad\qquad(9.18b)$$

We shall obtain the following result for a certain type of martingale Markov Chain.

Theorem 3.12. Let a finite Markov chain with state space $S = \{0, 1, \ldots, l\}$ be also a martingale. Then, as $n \to \infty$

$$p_{ij}^{(n)} = 0,\; j = 1, 2, \ldots, l - 1$$

$$\left.\begin{array}{l} p_{il}^{(n)} = \dfrac{i}{l} \\[2mm] p_{i0}^{(n)} = 1 - \dfrac{i}{l} \end{array}\right\} \quad i = 1, 2, \ldots, l - 1.$$

and

Proof: The relation (9.18) is satisfied for $i = 0$ *iff* $p_{00} = 1$, and for $i = l$ *iff* $p_{ll} = 1$. Thus if a finite Markov chain is also a martingale then the terminal states are absorbing.

Assuming that there are no further closed sets we get that the interior states $1, 2, \ldots (l - 1)$ are transient. Hence

$$\lim_{n \to \infty} p_{ij}^{(n)} = 0,\; j = 1, 2, \ldots, l - 1\,;$$

from (9.18b) we get, as $n \to \infty$,

$$l\, p_{il}^{(n)} = i$$

so that

$$p_{il}^{(n)} \to \frac{i}{l},\; i = 1, 2, \ldots, l - 1$$

and

$$p_{i0}^{(n)} \to 1 - \frac{i}{l}.$$

Thus the probabilities of ultimate absorption from a transient state $i, i = 1, 2, \ldots, l - 1$ to the absorbing states 0 and l are $(1 - i/l)$ and i/l respectively; these are dependent on i. ▲

Example 9(d). *Random genetic drift model* (Wright-Fisher model)

Each individual produces infinitely many A-and/or a gametes from which the $2m$ gametes are picked up at random to form the next generation. The population is said to be in state $i, i = 0, 1, \ldots, 2m$ *iff* there are i A-genes and $(2m - i)$ a-genes. The state i of the population in the next generation corresponds to the number of successes in

$2m$ Bernoulli trials, the probability of success (number of A-genes) being $i/2m$. If X_n denotes the state of the population in nth generation then $\{X_n, n \geq 0\}$ is a Markov chain with state space $S = \{0, 1, \ldots, 2m\}$ and transition matrix $P = (p_{ij})$, where

$$p_{ij} = \binom{2m}{j}\left(\frac{i}{2m}\right)^j \left(1 - \frac{i}{2m}\right)^{2m-j} \tag{9.19}$$

$$0 \leq i, j \leq 2m$$

(with the convention $0^0 = 1$).

We have for this Markov chain with state space $S = \{0, 1, \ldots, 2m\}$

$$\sum_{j=0}^{2m} j \, p_{ij} = \frac{i}{2m} \times 2m = i \qquad \text{for all } i$$

since, for fixed i, (9.19) corresponds to binomial distribution with parameters $p\left(= \frac{i}{2m}\right)$ and $n \,(= 2m)$. The chain is a martingale with 0 and $2m$ as absorbing states and 1, 2, $\ldots, 2m - 1$ as transient states. The probabilities of absorption from state i to state $2m$ and state 0 are respectively

$$a_{i,2m} = \frac{i}{2m}$$

$$a_{i,0} = 1 - \frac{i}{2m} \tag{9.20}$$

$$i = 1, 2, \cdots, 2m - 1.$$

Note. Straight forward calculation of these probabilities by using (9.6) is rather tedious.

3.9.4 Extension: Reducible Chain with one Closed Class of Persistent Aperiodic States

A set of states H of a finite Markov chain with state space S is called *open iff* from every state of H leads to a state of $S - H$.

Now the set of transient states is open. Consider a finite Markov chain with a set T of transient states and a set $S - T$ of persistent aperiodic states. If $P = (p_{ij})$ is its transition matrix then we can have a new Markov chain with transition matrix

$$\overline{P} = (\overline{p}_{ij})$$

such that

$$\overline{p}_{ij} = p_{ij}, i \in T, j \in S$$

$$= \delta_{ij}, i \in S - T, j \in S.$$

Then all the states of $S - T$ are absorbing and we can write

$$\overline{P} = \begin{pmatrix} I & 0 \\ R & M \end{pmatrix}$$

where M corresponds to transitions among states of T. Then we shall have

$$A = (a_{ik}) = NR, \text{ where } N = (I - M)^{-1}.$$

Example 9(e). Consider the chain with transition matrix

$$P = \begin{array}{c} 0 \\ 1 \\ 2 \\ 3 \end{array} \begin{array}{cccc} 0 & 1 & 2 & 3 \\ \left(\begin{array}{cccc} 1/6 & 1/3 & 1/2 & 0 \\ 1/2 & 1/2 & 0 & 0 \\ 1/6 & 1/3 & 1/2 & 0 \\ 0 & 1/6 & 1/3 & 1/2 \end{array} \right) \end{array}.$$

States 0, 1, 2 communicate with one another and form a closed class. The chain has one single closed class. The t.p.m. can be written as

$$P = \begin{pmatrix} P_1 & 0 \\ R_1 & Q \end{pmatrix}$$

where
$$P_1 = \begin{pmatrix} 1/6 & 1/3 & 1/2 \\ 1/2 & 1/2 & 0 \\ 1/6 & 1/3 & 1/2 \end{pmatrix}.$$

The equation

$$V' P_1 = V'$$

leads to
$$V' = (3/10, 2/5, 3/10), \text{ so that}$$

$$P^n \rightarrow e V_1 = e(3/10, 2/5, 3/10, 0).$$

Again state 3 is transient, since it is open (also $f_{33}^{(1)} = 1/2, f_{33}^{(n)} = 0, n \geq 0, F_{33} = 1/2 < 1$.

There must be at least one persistent state and since other states communicate with one another, the states 0, 1, 2 are all persistent.

Suppose that we are interested in finding mean first passage times from states 1, 2, 3 to state 0. Then we have to consider the new chain with t.p.m. $\overline{P}$ where

$$\overline{p}_{0j} = \delta_{0j}, \quad \overline{p}_{ij} = p_{ij}, i \neq 0.$$

The chain is then an absorbing chain with 0 as absorbing state. The matrix can be written as

$$\bar{P} = \begin{pmatrix} I & 0 \\ R & M \end{pmatrix}$$

where

$$M = \begin{pmatrix} 1/2 & 0 & 0 \\ 1/3 & 1/2 & 0 \\ 1/6 & 1/3 & 1/2 \end{pmatrix}.$$

Then

$$N = (I - M)^{-1} = \begin{pmatrix} 2 & 0 & 0 \\ 4/3 & 2 & 0 \\ 14/9 & 4/3 & 2 \end{pmatrix}$$

and

$$\mu' = N\mathbf{e} = \left(2, \frac{10}{3}, \frac{44}{9}\right), \text{ so that}$$

$$\mu_{10} = 2, \ \mu_{20} = \frac{10}{3}, \ \mu_{30} = \frac{44}{9}.$$

3.9.5 Further Extension : Reducible Chains with more than one Closed Class

Consider a finite reducible chain with *more than one* irreducible closed class of persistent states and some transient states and t.p.m. P. To fix our ideas suppose that there are two irreducible closed classes C_1 and C_2 of non-null persistent states with transition (sub) matrix P_1 and P_2 respectively. Then we can write

$$P = \begin{pmatrix} P_1 & 0 & 0 \\ 0 & P_2 & 0 \\ R_1 & R_2 & Q \end{pmatrix} \qquad (9.21)$$

where Q is the (sub) matrix of transitions among the m_1 transient states (which form a set T). Let $C = C_1 \cup C_2$ then $S = C \cup T$. Consider a new Markov chain with transition probabilities

$$\tilde{p}_{ij} = \begin{cases} p_{ij}, i \in T, j \in S \\ \delta_{ij}, i \in C, j \in S, C = C_1 \cup C_2. \end{cases}$$

The new chain then becomes an absorbing chain with the same transient states. The transition matrix becomes

$$\tilde{P} = \begin{pmatrix} I & 0 & 0 \\ 0 & I & 0 \\ R_1 & R_2 & Q \end{pmatrix}$$

$$(9.22)$$

$$\equiv \begin{pmatrix} I & 0 \\ R & Q \end{pmatrix}$$

where R is the augmented matrix having the same number of rows as Q and with columns of R_1 augmented by those of R_2.

The absorption probabilities can be computed in the same manner from

$$A = (a_{ik}) = NR \tag{9.23}$$

where $N = I - Q$ is the fundamental matrix.

Suppose that $i \in T$ and $k_j, l_j, \ldots, r_j \in C_j$, then the probability of absorption into the closed set equals

$$a_{ik_1} + a_{il_1} + \cdots a_{ir_1} = \alpha_i \text{ (say)} \tag{9.24}$$

and

$$a_{ik_2} + a_{il_2} + \cdots + a_{ir_2} = \beta \text{ (say)}. \tag{9.25}$$

Limit of $p_{ij}^{(n)}$ as $n \to \infty$, $i \in T$, $j \in S - T \equiv C$

Suppose that the states of the closed class C_1 are also aperiodic. Let

$$V' = (v_1, v_2 \cdots)$$

be the stationary distribution of the ergodic (sub) chain C_1, i.e., V' is a solution of the equation

$$V' P_1 = V'.$$

Then

$$\lim_{n \to \infty} p_{ij}^{(n)} = \alpha_i v_j, \quad j \in C_1, i \in T. \tag{9.26}$$

Similarly, if C_2 is aperiodic and if

$$U' = (u_1, u_2 \cdots)$$

is the stationary distribution of the ergodic (sub) chain C_2, then

$$\lim_{n \to \infty} p_{ik}^{(n)} = \beta_i u_k, \quad k \in C_2, i \in T. \tag{9.27}$$

The limits (9.26) and (9.27) are not independent of the initial state $i (\in T)$; the Markov chain is not ergodic.

In case C_1 is periodic, then

$$\lim_{n \to \infty} p_{ij}^{(n)} \text{ does not exist, } i \in T, j \in T.$$

Note that

$$\sum_j \alpha_i v_j = \alpha_i \le 1, \quad \sum_k \beta_i u_k = \beta_i \le 1, \quad \alpha_i + \beta_i = 1.$$

Example 9(f) Consider the Markov chain having t.p.m.

$$
\begin{array}{c}
1 \\ 2 \\ 3 \\ 4 \\ 5 \\ 6
\end{array}
\left(
\begin{array}{cccccc}
1/3 & 0 & 2/3 & 0 & 0 & 0 \\
0 & 1/2 & 1/4 & 0 & 1/4 & 0 \\
2/5 & 0 & 3/5 & 0 & 0 & 0 \\
0 & 1/4 & 1/4 & 1/4 & 0 & 1/4 \\
0 & 0 & 0 & 0 & 1/2 & 1/2 \\
0 & 0 & 0 & 0 & 1/4 & 3/4
\end{array}
\right)
$$

The digraph is given in Fig. 3.3

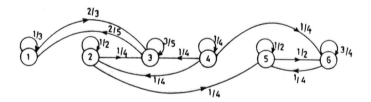

Fig. 3.3 Transition graph of the Markov chain (Example 9(f))

The chain consists of two classes of persistent non-null aperiodic states $C_1 = \{1, 3\}$, $C_2 = \{5, 6\}$ and transient states 2 and 4.
We can rewrite as

$$
P =
\begin{array}{c}
\\ \\ \\ \\ \\ \\
\end{array}
\begin{array}{c}
1 \\ 3 \\ 5 \\ 6 \\ 2 \\ 4
\end{array}
\begin{array}{cccccc}
1 & 3 & 5 & 6 & 2 & 4 \\
\end{array}
\left(
\begin{array}{cccccc}
1/3 & 2/3 & 0 & 0 & 0 & 0 \\
2/5 & 3/5 & 0 & 0 & 0 & 0 \\
0 & 0 & 1/2 & 1/2 & 0 & 0 \\
0 & 0 & 1/4 & 3/4 & 0 & 0 \\
0 & 1/4 & 1/4 & 0 & 1/2 & 0 \\
0 & 1/4 & 0 & 1/4 & 1/4 & 1/4
\end{array}
\right)
$$

$$R = \begin{pmatrix} 0 & 1/4 & 1/4 & 0 \\ 0 & 1/4 & 0 & 1/4 \end{pmatrix}, Q = \begin{pmatrix} 1/2 & 0 \\ 1/4 & 1/4 \end{pmatrix}$$

$$I - Q = \begin{pmatrix} 1/2 & 0 \\ -1/4 & 3/4 \end{pmatrix}, \quad N = (I - Q)^{-1} = \begin{pmatrix} 2 & 0 \\ 2/3 & 4/3 \end{pmatrix}$$

$$
A = (a_{ik}) = NR =
\begin{array}{c}
\\ 2 \\ 4
\end{array}
\begin{array}{cccc}
1 & 3 & 5 & 6 \\
\end{array}
\left(
\begin{array}{cccc}
0 & 1/2 & 1/2 & 0 \\
0 & 1/2 & 1/6 & 1/3
\end{array}
\right) \quad i, k = 2, 4
$$

The probability of absorption to class $C_1 = \{1, 3\}$ equals

$$\alpha_2 = a_{21} + a_{23} = 1/2, \quad \alpha_4 = a_{41} + a_{43} = 1/2$$

$$\beta_2 = a_{25} + a_{26} = 1/2, \quad \beta_4 = a_{45} + a_{46} = 1/2$$

$$\lim_{n \to \infty} p_{21}^{(n)} = \alpha_2 v_1 = \frac{1}{2} \cdot \frac{3}{8} = \frac{3}{16}$$

$$\lim_{n \to \infty} p_{23}^{(n)} = \frac{1}{2} \cdot \frac{5}{8} = \frac{5}{16}, \quad \lim_{n \to \infty} p_{41}^{(n)} = \frac{3}{16},$$

$$\lim_{n \to \infty} p_{43}^{(n)} = \frac{5}{16}, \quad \lim_{n \to \infty} p_{25}^{(n)} = \frac{1}{6},$$

$$\lim_{n \to \infty} p_{26}^{(n)} = \frac{1}{3}, \quad \lim_{n \to \infty} p_{45}^{(n)} = \frac{1}{6},$$

$$\lim_{n \to \infty} p_{46}^{(n)} = \frac{1}{3}$$

$$\lim_{n \to \infty} p_{jk}^{(n)} = \lim_{n \to \infty} p_{kj}^{(n)} = 0, j = 1, 3, k = 5, 6.$$

We have

$$\mu = N\mathbf{e} = \begin{pmatrix} 2 & 0 \\ \dfrac{2}{3} & \dfrac{4}{3} \end{pmatrix} \begin{pmatrix} 1 \\ 1 \end{pmatrix} = \begin{pmatrix} 2 \\ 2 \end{pmatrix}.$$

Thus the mean time upto absorption from transient states 2 and 4 are both equal to 2.

3.10 STATISTICAL INFERENCE FOR MARKOV CHAINS

3.10.1 M.L. Estimation and Hypothesis Testing

Inference problems, such as estimation and hypothesis testing, involving Markov chains have been considered by several authors (e.g., Bartlett, Whittle, Anderson and Goodman, Billingsley, and so on) not only because of their theoretical interest but also for their applications in diverse areas. Methods put forward for estimation of transition probabilities under different situations include one involving linear and quadratic programming procedures to produce least squares estimates (Lee *et al.*, 1970). We shall discuss here the maximum-likelihood method of estimation of transition probabilities from individual or micro-unit data. Some tests based on these estimates will also be discussed.

Basawa and Prakasa Rao (1980) discuss statistical inference for stochastic processes at great length.

Maximum-likelihood estimation: Consider a time-homogeneous Markov chain with a finite number, m, of states $(1, 2, \ldots, m)$ and having transition probability matrix $P = (p_{jk})$, $j, k = 1, 2, \ldots, m$. Suppose that the number of observed direct transitions from the state j to the state k is n_{jk}, and that the total number of observations is $(N + 1)$. Put

$$\sum_{k=1}^{m} n_{jk} = n_{j\cdot}. \quad \text{and} \quad \sum_{j=1}^{m} n_{jk} = n_{\cdot k}, \quad j, k = 1, 2, \ldots, m.$$

That there is a striking similarity between a sample from a Markov chain and one from a set of independent multinomial trials has been observed, among others, by Whittle (1955) who obtained the exact probability of the observed n_{jk} in the form:

$$f(n_{jk}) = T(n_{jk}) \frac{\prod_j (n_{j\cdot})!}{\prod_j \prod_k (n_{jk})!} \prod_j \prod_k p_{jk}^{n_{jk}}; \tag{10.1}$$

the factor $T(n_{jk})$ which involves the joint distribution of the $n_{j\cdot}$'s is independent of the p_{jk}'s.

The logarithm of the likelihood function can be put as

$$L(p_{jk}) = C + \sum_{j=1}^{m} \sum_{k=1}^{m} n_{jk} \log p_{jk} \tag{10.2}$$

where C contains all terms independent of p_{jk}.

Since $\sum_k p_{jk} = 1$, (10.2) can be written as

$$L(p_{jk}) = C + \sum_{j=1}^{m} \sum_{k=1}^{m-1} n_{jk} \log p_{jk}$$

$$+ \sum_{j=1}^{m} n_{jm} \log\left(1 - \sum_{k=1}^{m-1} p_{jk}\right). \tag{10.3}$$

Let r be a specific value of j. The maximum-likelihood estimates $\hat{p}_{rk}$ are given by the solutions of the equations

$$\frac{\partial L(p_{rk})}{\partial p_{rk}} = 0, \quad k = 1, 2, \ldots, (m-1).$$

These equations give

$$\frac{n_{rk}}{p_{rk}} - \frac{n_{rm}}{1 - \sum\limits_{k=1}^{m-1} p_{rk}} = 0, k = 1, 2, ..., (m-1). \tag{10.4}$$

To fix our ideas, let us take a specified value, s, of k. Then

$$\frac{n_{rs}}{p_{rs}} = \frac{n_{rk}}{p_{rk}} = \frac{n_{rm}}{1 - \sum\limits_{k=1}^{m-1} p_{rk}}, \quad k = 1, 2, ..., s, ...(m-1).$$

Thus

$$1 - \sum_{k=1}^{m-1} p_{rk} = \frac{n_{rm}}{n_{rs}} p_{rs} \tag{10.5}$$

and

$$p_{rk} = \frac{n_{rk}}{n_{rs}} p_{rs}, \quad k = 1, 2, ..., s, ...(m-1). \tag{10.6}$$

Summing (10.6) over all k and adding (10.5), we get

$$1 = \frac{\sum\limits_{k=1}^{m-1} n_{rk}}{n_{rs}} p_{rs}$$

and hence the estimate $\hat{p}_{rs}$ is given by

$$\hat{p}_{rs} = \frac{n_{rs}}{\sum\limits_{k=1}^{m} n_{jk}} \tag{10.7}$$

Now r, s are two arbitrary values of j, k respectively. Hence, for

$$j, k = 1, 2, \ldots, (m-1)$$

$$\hat{p}_{jk} = \frac{n_{jk}}{\sum\limits_{k=1}^{m} n_{jk}} = \frac{n_{jk}}{n_{j.}} \tag{10.8}$$

are the maximum-likelihood estimates of p_{jk}.

Hypothesis testing: Some of the tests developed (Anderson and Goodman) on the above estimates are given below.

(i) Suppose that one wishes to test the null hypothesis that the observed realisation comes from a Markov chain with a given transition, i.e. matrix $P^0 = (p_{jk}^0)$; suppose that the null hypothesis is

$$H_0 : P = P^0.$$

Then, for large N, and for $\hat{p}_{jk}$ given by (10.8), the statistic

$$\sum_{k=1}^{m} \frac{n_j.(\hat{p}_{jk} - p_{jk}^0)^2}{p_{jk}^0}, \quad j = 1, 2, \ldots m,$$

is distributed as χ^2 with $m - 1$ d.f. (degrees of freedom). Here p_{jk}^0's which are equal to zero are excluded and the d.f. is reduced by the number of p_{jk}^0's equal to 0. Alternatively, a test for all $\hat{p}_{jk}$ can be obtained by adding over all j, and the statistic

$$\sum_{j=1}^{m} \sum_{k=1}^{m} \frac{n_j.(\hat{p}_{jk} - p_{jk}^0)^2}{p_{jk}^0} \tag{10.9}$$

has an asymptotic χ^2-distribution with $m(m-1)$ d.f. (the number of d.f. being reduced by the number of p_{jk}^0's equal to zero, if any, for $j, k = 1, 2, \ldots, m$).

The likelihood ratio criterion for H_0 is given by

$$\lambda = \prod_{j} \prod_{k} \left(\frac{\hat{p}_{jk}}{p_{jk}^0} \right)^{n_{jk}}.$$

Thus, under the null hypothesis, the statistic

$$-2 \log \lambda = 2 \sum_{j} \sum_{k} n_{jk} \log \frac{n_{jk}}{(n_j.)p_{jk}^0} \tag{10.10}$$

has an asymptotic χ^2-distribution with $m(m-1)$ d.f.

(ii) The maximum-likelihood estimates could also be used to test the order of a Markov chain. For testing the null hypothesis that the chain is of order zero, i.e. $H_0 : p_{jk} = p_k$ for all j, against the alternative that the chain is of order 1, the test criterion is

$$\lambda = \prod_{j} \prod_{k} \left(\frac{\hat{p}_k}{\hat{p}_{jk}} \right)^{n_{jk}}$$

where

$$\hat{p}_k = n_{\cdot k} \Big/ \Big(\sum_j \sum_k n_{jk}\Big), \quad \hat{p}_{jk} = n_{jk}/n_j .$$

Under the null hypothesis, the statistic

$$-2\log\lambda = 2\sum_j\sum_k n_{jk} \log\frac{n_{kj}(\sum\sum n_{jk})}{(n_{\cdot k})(n_{j\cdot})}$$

has an asymptotic χ^2-distribution with $(m-1)^2$ d.f. Similar test can be constructed for testing the null hypothesis that the chain is of order one against the alternative that it is of order two.

These tests do not prove that the observed chain is of order one. They provide, however, a rationale for using a Markov model. Similar likelihood ratio tests based on maximum-likelihood estimates are given (iii) for testing the stationarity (or time-homogeneity) of a Markov chain, i.e. $H_0: p_{jk}(t) = p_{jk}$ and also (iv) for testing that several samples are from the same Markov chain of a given order.

3.10.2 Determination of the Order of a Markov Chain by MAICE

A procedure for the determination of the order of a Markov chain by Akaike's Information Criterion (AIC) has been developed by Tong (1975). The AIC is defined as

AIC = (− 2) log (Maximum-likelihood) + 2 (Number of independent
parameters in the model).

This statistic is introduced as a measure of deviation of the fitting model from the true structure. Given several models, the procedure envisages adoption of the model that minimises the AIC and is called *Minimum AIC Estimation* (MAICE). (It is argued that the MAICE procedure represents an attempt to strike a balance between overfitting which needs more parameters, and underfitting which incurs an increased residual variance).

Denote the transition probability for a r-order chain by $p_{ij\ldots kl}$, $i = 1, 2, \ldots s$, s being the (finite) number of states of the chain.

Denote the ML estimates by $\hat{p}_{ij\ldots kl} = \dfrac{n_{ij\ldots kl}}{n_{ij\ldots k\cdot}}$.

where $n_{ij\ldots k\cdot} = \sum_l n_{ij\ldots kl}$. The hypothesis tested is $H_{r-1}: p_{ij\ldots kl} = p_{j\ldots kl}$, $i = 1, \ldots, s$

(that the chain is $(r-1)$-dependent against H_r: that the chain is r-dependent). The statistic constructed is

$$_{r-1}A_r \equiv -2\log\lambda_{r-1,r} = 2\sum_{i\ldots l}(n_{ij\ldots kl})\log\frac{(n_{ij\ldots kl})\Big(\sum_{i\ldots l}(n_{ij\ldots kl})\Big)}{(n_{ij\ldots k\cdot})(n_{\cdot j\ldots kl})}$$

which is a χ^2-variate with $s^{r-1}(s-1)^2$ d.f.

The hypothesis H_k that the chain is k-dependent implies the hypothesis H_r, that the chain is r-dependent for $k < r$. Denote by $\lambda_{k,r}$ the ratio of the maximum-likelihood given H_k (that the chain is of order k) to that given H_r (that the chain is of order r); then we get

$$\lambda_{k,r} = \lambda_{k,k+1} + \ldots + \lambda_{r-1,r}$$

and so

$${}_kA_r = -2\log\lambda_{k,k+1} - \ldots - 2\log\lambda_{r-1,r}, k < r.$$

Assume the variables $-2\log\lambda_{r-1,r}$ ($r = 0, 1, \ldots$) to be asymptotically independent, given H_k; then it follows that ${}_kA_r$ has χ^2-distribution with d.f.

$$\nabla s^{r+1} - \nabla s^{k+1}, \qquad k \geq 0$$

and

$$\nabla s^{r+1}, \qquad k = -1$$

where

$$\nabla s^r = s^r - s^{r-1}, \qquad r \geq 1.$$

Now the question of choice of an appropriate loss function arises once this identification procedure is considered as a decision procedure. The loss functions considered in classical theory of hypothesis testing are defined by the probabilities of accepting the incorrect hypothesis or rejecting the correct hypothesis.

Tong proposes the choice of the loss function, based on the AIC approach as

$$R(k) = {}_kA_m - 2(\nabla s^{m+1} - \nabla s^{k+1})$$

where m is the highest order model to be considered and k is the order of the fitting model. The MAICE of the best approximating order of the Markov chain is that value of k which gives the minimum of $R(k)$ over all orders considered.
Note that $R(m) = 0$.

Gabriel and Neumann described the occurrences and non-occurrences of rainfall (of Tel Aviv) by a two-state Markov-chain. A dry date is denoted by state 0 and a wet date by state 1. In the following example we consider a similar model for weather data (Medhi, 1976).

Example 10(a). The following data are obtained from observations of the occurrence and non-occurrence of rainfall in Gauhati, India. A dry day is denoted by state 0 and a wet day by state 1. The period covered is from the last day of February to the last day of May (i.e. before the onset of monsoon in early June) for four consecutive years 1967-70 and N, the total number of transitions in the 4 years is $4 \times 92 = 368$. The transitions are as noted below.

	0	1	Total
0	175	49	224
1	48	96	144
Total	223	145	368

The estimated transition matrix $P = (p_{ij})$ is given by

$$\begin{array}{cc} & \begin{array}{cc} 0 & \ 1 \end{array} \\ \begin{array}{c} 0 \\ 1 \end{array} & \left(\begin{array}{cc} 25/32 & 7/32 \\ 1/3 & 2/3 \end{array} \right) \end{array}.$$

We have

$$P^5 = \begin{pmatrix} 0.61091 & 0.38909 \\ 0.59289 & 0.40711 \end{pmatrix}$$

$$P^{10} = \begin{pmatrix} 0.60390 & 0.39610 \\ 0.60358 & 0.39642 \end{pmatrix}, \quad P^{20} = \begin{pmatrix} 0.60377 & 0.39623 \\ 0.60377 & 0.39623 \end{pmatrix}.$$

Using (5.6a), we get $\lim_{n \to \infty} P^n \to (0.60377, 0.39623)$.

We wish to test the null hypothesis that the chain is of order zero against the alternative that it is of order one:

$$H_0 : p_{ij}^0 = p_j, \, i, j = 1, 2.$$

We have

$$-2 \log \lambda = 2 \sum_{i=1}^{2} \sum_{j=1}^{2} n_{ij} \log \frac{(N) n_{ij}}{(n_j.)(n_{.j})}$$

$$= 2 \{ 175 \log \frac{(368)(175)}{(223)(224)} + 49 \log \frac{(368)(49)}{(145)(224)} + 48 \log \frac{(368)(48)}{(223)(144)}$$

$$+ 96 \log \frac{(368)(96)}{(145)(144)} \}$$

$$= 53.28.$$

For 1 d.f., $\Pr \{ \chi^2 \geq 53.28 \}$ being very very small the null hypothesis that the chain is of order zero is rejected.

Again, we wish to test the null hypothesis that the transition probability matrix is P against the alternative $P^0 = (p_{ij}^0)$, where

$$H_0 : P^0 = \begin{pmatrix} 7/15 & 8/15 \\ 1/5 & 4/5 \end{pmatrix}$$

is the transition probability matrix of dry and wet days during the monsoon months (June-August).

We have

$$\sum_{i=1}^{2} \sum_{j=1}^{2} \frac{n_i . (p_{ij} - p_{ij}^0)^2}{p_{ij}^0} = 105.07.$$

For 2 d.f., $\Pr\{\chi^2 \geq 105.07\}$ being extremely small, the null hypothesis $H_0 : P = P^0$ is rejected.

3.11 MARKOV CHAINS WITH CONTINUOUS STATE SPACE

So far we have discussed Markov chains $\{X_n, n = 0, 1, 2, \ldots\}$ with discrete state space, i.e. with $0, \pm 1, \pm 2, \ldots$ as possible values of X_n. Here we consider chains $\{X_n\}$ with continuous state space, i.e. with $(-\infty, \infty)$ as possible range of values of X_n. We shall have to use either probability distribution function (d.f.) or probability density function (p.d.f.), when this exists, in place of probability mass function.

Definition. If, for all m, and for all possible values of X_m in $(-\infty, \infty)$

$$\Pr\{X_{m+1} \leq x \mid X_m = y, X_{m-1} = y_1, \ldots, X_0 = y_m\}$$

$$= \Pr\{X_{m+1} \leq x \mid X_m = y\} \tag{11.1}$$

then $\{X_m, m \geq 0\}$ is said to constitute a Markov chain with continuous state space. If the conditional d.f. as given by (11.1) is independent of m, then the chain is homogeneous; (11.1) gives (one-step) transition probability d.f. More generally, the n-step transition probability d.f. is defined by

$$\Pr\{X_{m+n} \leq x \mid X_m = y, \ldots, X_0 = y_m\}$$

$$= \Pr\{X_{m+n} \leq x \mid X_m = y\} \tag{11.2}$$

$$= P_n(y;x).$$

Denote $P(y; x) = P_1(y; x) = \Pr\{X_{m+1} \leq x \mid X_m = y\} = \Pr\{X_{n+1} \leq x \mid X_n = y\}$.
Let $P_n(x) = \Pr(X_n \leq x)$. The transition d.f. $P_n(y; x)$ and the initial distribution $\Pr\{X_0 \leq x\} = P_0(x)$ can unquiely determine $P_n(x)$. The Chapman-Kolmogorov equation takes the form

$$P_{n+m}(y;x) = \int_{\infty}^{\infty} d_z P_n(y;z) P_m(z;x), \quad m, n \geq 0, \tag{11.3}$$

which corresponds to $p_{jk}^{(n+m)} = \sum_s p_{js}^{(n)} p_{sk}^{(m)}$ for Markov chains with discrete state space.

For $m = 1$, we can write (11.3) as

$$P_{n+1}(y;x) = \int d_z P_n(y; z) P_1(z; x)$$

$$= \int d_z P_n(y; z) P(z; x). \tag{11.4}$$

Suppose that as $n \to \infty$, $P_n(y; x)$ tends to a limit $P(x)$ independent of the initial value. Then the limiting distribution $P(x)$ satisfies the integral equation

$$P(x) = \int d_z P(z) P(z; x)$$

$$= \int P(z; x) dP(z). \tag{11.5}$$

The relation (11.5) corresponds to (6.2) for chains with discrete state space. In the above relations, distribution functions can be replaced by p.d.f.'s when these exist. It is to be noted that the theory of Markov chains with continuous state space is not so fully developed as that of chains with discrete state space. Special methods are however developed to study particular types of chains such as random walks. We consider here two examples.

Example 11(a). Random walk: Suppose that the steps of the unrestricted random walk are i.i.d. continuous random variables $Z_1, Z_2 \ldots$ with a p.d.f. $f(z)$, $-\infty < z < \infty$. We have $X_n = Z_1 + \ldots + Z_n$, so that $X_{n+1} = X_n + Z_{n+1}$. Clearly, $\{X_n, n \geq 0\}$ is a Markov chain with a continuous state space. The (unit-step) transition d.f. can be written as

$$P(y; x) = \Pr\{X_{n+1} \leq x \mid X_n = y\}$$

$$= \Pr\{X_n + Z_{n+1} \leq x \mid X_n \leq y\}$$

whence we get

$$f(x) = \int f^{n*}(z) f(x - z) dz$$

where $f^{n*}(z)$ is the n-fold convolution of $f(z)$. Here $f(x - z)$ which is the p.d.f. of $X_{n+1} = X_n + Z_{n+1}$ given that $X_n = z$, is the transition p.d.f.

Example 11(b). Queueing process*: Suppose that customers arrive at a facility at instants $t_0 (= 0), t_1, t_2, \ldots$, i.e. $a_n = t_{n+1} - t_n$ is the interarrival time between the nth and $(n + 1)$st customers. Suppose that $b_0, b_1, b_2, \ldots$ are the durations of the successive service times. Assume that $\{a_n\}, \{b_n\}$ are sequences of independent, positive random

---

*For a description see Chapter 10.

variables with d.f. $A(x) = \Pr\{a_n \leq x\}$ and $B(x) = \Pr\{b_n \leq x\}$ respectively. Let the random variable $W_n = W(t_n - 0)$, denote the waiting time of the nth arrival. Let $X_n = b_n - a_n$ have d.f. $F(x) = \Pr\{X_n \leq x\}$.

We have, for $n = 0, 1, 2, \ldots$

$$
\begin{aligned}
W_{n+1} &= W_n + X_n, & &\text{if } W_n + X_n > 0 \\
&= 0, & &\text{if } W_n + X_n \leq 0.
\end{aligned}
\tag{11.6}
$$

It is clear that $\{W_n, n \geq 1\}$ constitutes a Markov chain with $(0, \infty)$ as possible range of its value. We have

$$P_0(y;x) = 0, \text{ if } x < y$$

$$= 1, \text{ if } x \geq y$$

and $\qquad P_n(y;x) = \Pr\{W_{m+n} \leq x \mid W_n = y\}$ for $n = 1, 2, \ldots$.

In particular, the unit-step transition d.f. is

$$
\begin{aligned}
P(y;x) = P_1(y;x) &= \Pr\{W_{n+1} \leq x \mid W_n = y\} \\
&= \Pr\{W_{n+1} = 0 \mid W_n = y\} \\
&\qquad + \Pr\{0 < W_{n+1} \leq x \mid W_n = y\} \\
&= \Pr\{y + b_n - a_n \leq 0\} \\
&\qquad + \Pr\{0 < y + b_n - a_n \leq x\} \\
&= \Pr\{y + b_n - a_n \leq x\} \\
&= \Pr\{b_n - a_n \leq x - y\} \\
&= F(x - y).
\end{aligned}
\tag{11.7}
$$

From (11.4), we get, for $n = 0, 1, 2, \ldots$

$$P_{n+1}(y;x) = \int_0^\infty F(x-z)\, d_z P_n(y;z) \tag{11.8}$$

$$= F(x-z) P_n(y;z) \big|_{z=0}^\infty$$

$$- \int_0^\infty dF(x-z) P_n(y;z)$$

$$= - \int_0^\infty P_n(y; z)dF(x - z).$$ (11.9)

Suppose that $\lim_{n \to \infty} P_n(y; z) = P(z)$ exists, then we get

$$P(x) = \int_0^\infty x^{\cdot}(x - z)dP(z)$$

$$= - \int_0^{\cdot} P(z)dF(x - z).$$ (11.10)

3.12 NON-HOMOGENEOUS CHAINS

The Markov chains considered so far are time-homogeneous or stationary. In other words, we have assumed that for a discrete space Markov chain $\{X_n\}$, the transition probabilities $p_{ij} = \Pr \{X_n = j \mid X_{n-1} = i\}$ (and therefore the corresponding t.p.m. $P = (p_{ij})$) are independent of n. This assumption leads to some elegant results, as already discussed. However, the model of a homogeneous (or stationary) chain is not appropriate in many real situations arising, for example, in epidemiology, learning theory, etc. This points to the necessity of considering also non-homogeneous chains, which have so long received little attention in literature, at least, as much as the general case is concerned.

Example 12(a). Consider Polya's urn scheme discussed in Example 1(a). Let S_n be the accumulated number of black balls drawn in the first n drawings. Then

$S_n = i$,   if the accumulated number of black balls drawn in the first n drawings is $i, i = 0, 1, \ldots, n$.

$\{S_n, n \geq 0\}$ is clearly a Markov chain. We have

$$\Pr \{S_n = j \mid S_{n-1} = i, S_{n-2} = ., \ldots, S_0 = .\}$$

$$= \Pr \{S_n = j \mid S_{n-1} = i\}$$

$$= p_{ij}(n), \text{ say}$$

where

$$p_{ij}(n) = \frac{b + ic}{(b + r) + (n - 1)c}, \quad j = i + 1$$

$$= \frac{r+(n-1-i)c}{(b+r)+(n-1)c}, \quad j=i$$

$$= 0, \text{ otherwise.}$$

The transition probability $p_{ij}(n)$ is a function of n. The chain is non-homogeneous. Further, the matrix

$$P_n = (p_{ij}(n))$$

is not a square matrix but is a rectangular $n \times (n+1)$ matrix.

We shall restrict here to non-homogeneous Markov-chains with square transition matrix.

While homogeneous chains have the advantage of being amenable to a complete analysis in most cases, it is not so in case of non-homogeneous chains. Even the problem of classification of non-homogeneous chains has not yet been sorted out. One approach of study in the ergodicity approach.

Definition. Denote

$$p(m,i;m+n,j) = \Pr\{X_{m+n}=j \mid X_m=i\}.$$

The Markov chain $\{X_n, n \geq 0\}$ is said to be *strongly ergodic* if the limits

$$\lim_{n \to \infty} p(m,i;m+n,j)$$

exist and are independent of i for all $m \geq 0$. It is said to be *weakly ergodic* if

$$\lim_{n \to \infty} [p(m,i;m+n,j) - p(m,k;m+n,j)] = 0$$

for all $m \geq 0$.

For an account of ergodicity approach, reference may be made to Iosifescu (1980) and Isaacson and Madsen (1976).

We consider Matrix Approach (Harary et al (1970)) below.

3.12.1 Matrix Approach for Finite Non-homogeneous Chain

This makes use of the notion of a causative matrix that, where multiplies of a t.p.m. yields a subsequent one.

Suppose that the transition probabilities p_{ij} are functions of n, i.e.

$$\Pr\{X_n=j \mid X_{n-1}=i\} = p_{ij}(n).$$

Then the probabilistic situation is described by a sequence of t.p.m.'s

$$P_1, P_2, \ldots, P_n, \ldots$$

In the case of homogeneous chains, the situation is described by a single t.p.m. P, i.e. $P_n = P$ for all n. A sequence of causative matrices $C_1, C_2 \ldots, C_n, \ldots$ are introduced to describe the change in the transition probability matrices. These causative matrices will be defined by

$$P_1 C_1 = P_2, \quad P_2 C_2 = P_3, \ldots, \quad P_n C_n = P_{n+1} \ldots \tag{12.1}$$

From (12.1) we get $C_n = P_n^{-1} P_{n+1}$, provided the matrix P_n is nonsingular. The matrices C_n multiply P_n from the right to yield P_{n+}. Likewise one may consider causative matrices which involve multiplication from the left.

The homogeneous chain can be considered as a special case obtained by taking

$$C_n = I \quad \text{for all } n.$$

We shall assume that all the causative matrices C_n are equal, i.e. $C_n = C$ for all n. We then have

$$P_{n+1} = P_1 C^n \tag{12.2}$$

and, in general,

$$P_{n+m} = P_m C^n, \quad n = 0, 1, 2, \ldots.$$

Further, if the matrices are non-singular, then

$$P_{n+1} = P_n P_{n-1}^{-1} P_n$$

and
$$P_n = (P_2 P_1^{-1})^{n-2} P_2 = P_2 (P^{-1} P_2)^{n-2}, \quad n = 2, 3, \ldots$$

Denote the t.p.m. from period u to v by $P(u, v)$ and $P_1 = P(0, 1)$ by Q. Then

$$P(n, n+1) = P_{n+1}$$

$$= Q C^n, \quad n = 0, 1, 2, \ldots,$$

and
$$P(0, n+1) = P(0, 1) P(1, 2) \ldots P(n, n+1)$$

$$= Q.QC \ldots. QC^n$$

$$= \prod_{r=0}^{n}(QC^r) = T_n \text{ (say).} \tag{12.3}$$

A good deal of information about a chain can be had from the matrix C itself. The limiting properties of a constant chain (with $C_1 = C_2 = \ldots = C$) depend on the convergence of C^n and T_n as $n \to \infty$.

First we consider the following result:

Theorem 3.13. If Q and R are two stochastic matrices of the same order and if Q is non-singular, then the causative matrix $C = Q^{-1}R$ has unit row sums.

Proof: Let $\mathbf{e} = (1, 1 \ldots, 1)'$ be a column vector with unity as each of its elements. Then $R\mathbf{e} = \mathbf{e}$, $Q\mathbf{e} = \mathbf{e}$ and hence $Q^{-1}\mathbf{e} = \mathbf{e}$. Thus

$$C\mathbf{e} = Q^{-1}R\mathbf{e} = Q^{-1}\mathbf{e} = \mathbf{e}.$$

Note that C can have negative elements and so C may not be a stochastic matrix.

Some of the relevant questions are the existence of the limit and stochasticity of C_n and T_n. We examine them for the two-state non-stationary chains (for details, see Harary *et al*).

Two-state non-stationary (non-homogeneous) chains

Let $P_1 = Q$ and $P_2 = R$ be given by

$$P_1 = \begin{pmatrix} a & 1-a \\ 1-b & b \end{pmatrix}, \quad P_2 = \begin{pmatrix} c & 1-c \\ 1-d & d \end{pmatrix}$$

and the causative matrix be

$$C = \begin{pmatrix} u & 1-u \\ u-m & 1-u+m \end{pmatrix}. \tag{12.4}$$

C is a stochastic matrix *iff*

$$0 \leq u \leq 1, 0 \leq u - m \leq 1, \tag{12.5}$$

i.e. *iff* $\qquad \max(0, m) \leq u \leq \min(1, 1+m).$

Since

$$C = P_1^{-1}P_2,$$

we have

$$u = [(bc - (1-a)(1-d)] / (a+b-1)$$

$$= (a+d-ad+bc-1)/(a+b-1),$$

and $$m = (c+d-1)/(a+b-1).$$

The characteristic roots of C are 1 and m, so that the characteristic roots, of C^n are 1 and m^n. Since $t_r(C^n)$ = sum of the characteristic roots, and C^n has unit row sums, we may write

$$C^n = \begin{pmatrix} u_n & 1-u_n \\ u_n - m^n & 1-u_n+m^n \end{pmatrix}, \quad n = 1,2,\ldots,$$

taking $u_1 = u$. Since $C^{n+1} = C \cdot C^n = C^n \cdot C$, we have

$$u_{n+1} = u_n - m^n + um^n = u - m + mu_n, \quad n = 1,2,\ldots.$$

Solving this difference equation of order 1, we get u_n.
     For $m = 1$,

$$u_{n+1} = u_n + u - 1$$

$$= u_{n-1} + 2(u-1)$$

$$\ldots \quad \ldots$$

$$= u_1 + n(u-1)$$

or $$u_n = 1 + n(u-1) \text{ (since } u_1 = u).$$

For $m \neq 1$, we get

$$u_n = \{u - m + m^n(1-u)\}/(1-m).$$

Thus, we get

$$\lim_{n \to \infty} u_n = 1, \quad m = u = 1$$

$$= (u-m)/(1-m), \quad -1 < m < 1.$$

The limit does not exist otherwise. Thus, for $m = u = 1$

$$\lim_{n \to \infty} C^n = \begin{pmatrix} 1 & 0 \\ 0 & 1 \end{pmatrix}.$$

When $m = u = 1$, the causative matrix C reduces to the identity matrix and the chain becomes homogeneous.

For $-1 < m < 1$,

$$\lim_{n \to \infty} C^n = \begin{pmatrix} (u-m)/(1-m) & (1-u)/(1-m) \\ (u-m)/(1-m) & (1-u)/(1-m) \end{pmatrix}$$

$$= ((u-m)/(1-m), (1-u)/(1-m)). \tag{12.6}$$

The limiting matrix is stochastic *iff*

$$0 \le (u-m)/(1-m) \le 1$$

or
$$m \le u \le 1; \tag{12.7}$$

when $0 \le m < 1$, then (12.5) and (12.7) are equivalent, i.e. $\lim C^n$ is stochastic *iff* C is stochastic.

More important are the questions like existence and stochasticity of

$$\lim_{n \to \infty} T_n = \lim_{n \to \infty} \prod_{r=0}^{n} (QC^r).$$

These are considered in the following theorem.

Theorem 3.14. Let m be the non-unity characteristic root of the causative matrix C. If $-1 < m < 1$, $\lim_{n \to \infty} T_n$ exists and is equal to

$$((u-m)/(1-m), (1-u)/(1-m)).$$

If $0 \le m < 1$, the matrix $\lim_{n \to \infty} T_n$ is stochastic *iff* C is stochastic.

If $-1 < m \le 0$, QC^r is stochastic for all $r = 0, 1, 2, 3, \ldots$ and so are T_n and $\lim_{n \to \infty} T_n$,

but then C is a stochastic matrix provided only $0 \le u \le 1 + m$.

For the proof which involves algebra of stochastic matrices refer to Harary *et al.*

Note. 1 The analysis of non-homogeneous chains provides interesting insight into problems otherwise intractable and also furnishes basis for useful models, for many situations in real life.

2. The elegant result concerning the asymptotic behaviour of the n-step transition probabilities of homogeneous Markov chains does not hold good in case of non-homogeneous chains. For some results, on such behaviour of non-homogeneous chains, reference may be made to a paper by Cohn (1976), who presents a general treatment of the asymptotic behaviour of the transition probabilities between various groups of states. For an account of the historical development of the theory of non-homogeneous chains, see Seneta (1973).

For further contributions refer to Kingman, (1975), Cohn (1976), Isaacson and Madsen (1976), Paz (1971), Iosifescu (1980), Seneta (1981), Adke and Manjunath (1984).

HISTORICAL NOTE

The concept of Markov dependence is attributed to the Russian mathematician Andrei Andreivich Markov (1856-1922). It is said that the idea first occurred to him as he was watching an opera of the famous Russian writer Pushkin's (1799-1837) verse novel *"Eugene (Evgeni) Onegin"* adapted by the famous composer Tchaikovsky (1840-93). As an application of the concept, Markov made a study of the succession of consonants and vowels of Pushkin's *Onegin*. Starting with a paper in 1906 in which the concept was given in an explicit form, Markov wrote a number of papers on this topic. Almost about the same time, the French mathematician Henri Poincaré (1854-1912), known as the last great universalist, also came across sequences of random variables which are in fact Markov chains; but he did not undertake an indepth study of the topic as did Markov and thus the recognition given to Markov is quite justified. The literature of the subject has grown enormously over the last 87 years and is still growing.

Some of the early researchers in this area include beside Markov, mathematicians Fréchet, Hadamard, Kolmogorov, Bernstein, Doeblin, Feller, to name a few. The study of non-homogeneous chain was initiated by Markov himself; important early contributions to continuous parameter case (Ch IV) were made by Kolmogorov, Feller, Doeblin. Apart from the classical methods developed by the early workers, an entirely different approach based on Perron (1907) and Frobenius (1912) theory of non-negative matrices was developed by Von Mises in 1931. A systematic study of Markov chain based on the theory of non-negative matrices, particularly on Perron-Frobenius theorem, is given in the book by Seneta (1981).

Concluding Remarks

Apart from its rich theoretical contents, the Markov chain has a wealth of applications in several areas such as demography, biology, genetics and medicine, chemistry, physics, geophysics, meteorology, geography, education, linguistics, economics, marketing, finance, and other diverse aspects of management, social and labour mobility, manpower systems, statistical quality control, metric theory of numbers etc. etc. Markov chains serve as one of the most important methods in application of applied probability theory to real-world models involving uncertainty.

EXERCISES*

3.1   A communication system transmits the two digits 0 and 1, each of them passing through several stages. Suppose that the probability that the digit that enters remains unchanged when it leaves, is p and that it changes is $q = 1 - p$. Suppose

---

*P denotes the transition probability matrix of the Markov chain referred to in the exercise.

further that X_0 is the digit which enters the first stage of the system and $X_n (n \geq 1)$ is the digit 0 leaving the nth stage of the system. Show that $\{X_n, n \geq 1\}$ forms a Markov chain. Find P, P^2, P^3 and calculate

$$\Pr(X_2 = 0 \mid X_0 = 1) \text{ and } \Pr(X_3 = 1 \mid X_0 = 0).$$

3.2   The t.p.m. of a Markov chain $\{X_n, n = 1, 2, \ldots\}$ having three states 1, 2 and 3 is

$$P = \begin{pmatrix} 0.1 & 0.5 & 0.4 \\ 0.6 & 0.2 & 0.2 \\ 0.3 & 0.4 & 0.3 \end{pmatrix}$$

and the initial distribution is

$$\Pi_0 = (0.7, 0.2, 0.1).$$

Find

(i) $\Pr\{X_2 = 3\}$,

(ii) $\Pr\{X_3 = 2, X_2 = 3, X_1 = 3, X_0 = 2\}$.

3.3   Three children (denoted by 1, 2, 3) arranged in a circle play a game of throwing a ball to one another. At each stage the child having the ball is equally likely to throw it into any one of the other two children. Suppose that X_0 denotes the child who had the ball initially and X_n ($n \geq 1$) denotes the child who had the ball after n throws. Show that $\{X_n, n \geq 1\}$ forms a Markov chain. Find P. Calculate

$$\Pr(X_2 = 1 \mid X_0 = 1), \quad \Pr(X_2 = 2 \mid X_0 = 3), \quad \Pr(X_2 = 3 \mid X_0 = 2),$$

and also the probability that the child who had originally the ball will have it after 2 throws.
Find P if the number of children is m (≥ 3).

3.4   Consider a sequence $\{X_n, n \geq 2\}$ of independent coin-tossing trials with probability p for head (H) in a trial. Denote the states of X_n by states 1, 2, 3, 4, according as the trial numbers ($n - 1$) and n results in HH, HT, TH, TT respectively. Show that $\{X_n, n \geq 2\}$ is a Markov chain. Find P; show that P^m ($m \geq 2$) is a matrix with all the four rows having the same elements.

3.5   Suppose that a fair die is tossed. Let the states of X_n be k ($= 1, 2, \ldots, 6$), where k is the maximum number shown in the first n tosses. Find P, P^2 and P^n. Calculate $\Pr(X_2 = 6)$.

3.6   Suppose that a ball is thrown at random to one of the 3 cells. Let X_n ($n \geq 1$) be said to be in state k (= 1, 2, 3) if after n throws k cells are occupied. Find P and P^n.

3.7   *Occupancy problem* (Generalisation of Exercise 3.6): Suppose that a ball is thrown at random to one of the r cells. Let X_n ($n \geq 1$) be said to be in state k (= 1, 2, ... r) if after n throws k cells are occupied. Find P and P^2.

3.8   Show that in a Markov chain with a *finite* state space S, there is at least one non-null persistent state and therefore at least one invariant distribution.

3.9   A coin is tossed, p being the probability of head in a toss. Let $\{X_n, n \geq 1\}$ have the two states 0 or 1 according as the accumulated number of heads and tails in n tosses are equal or unequal. Show that the states are transient when $p \neq \frac{1}{2}$, and persistent null when $p = \frac{1}{2}$.

3.10   A particle starting from the origin moves from position j to position $(j + 1)$ with probability a_j and returns to origin with probability $(1 - a_j)$. Suppose that the states, after n moves are, 0, 1, 2, ... Show that the state 0 is persistent *iff*

$$\lim \prod_{i=1}^{n} a_i \to 0 \text{ as } n \to \infty.$$

3.11   Suppose that for a Markov chain $p_{j,j+2} = v_j$, $p_{j,0} = 1 - v_j$ and $p_{j,k} = 0, k \neq j+2, 0$. Show that the state 0 (that X_n is at the origin after n moves) is persistent *iff* $(v_0 \ldots v_{2n-2}) \to 0$ and non-null *iff* $\sum_n (v_0 \ldots v_{2n-2})$ is finite. In particular, when $v_j = e^{-aj}$ ($a > 0$), state 0 is persistent non-null.
(In Exercises 10 and 11, the random walk is non-homogeneous.)

3.12   *A random walk on the set of integers.* Show that the Markov chain with countable state space $S = \{\ldots, -2, -1, 0, 1, 2, \ldots\}$ and transition probabilities

$$p_{i,i+1} = p = 1 - p_{i,i-1}, \quad i = 0, \pm 1, \pm, 2, \ldots$$

is persistent when $p = \frac{1}{2}$ and transient when $p \neq \frac{1}{2}$.

3.13   Find P^n and the limiting probability vector V for the chain having t.p.m.

$$P = \begin{pmatrix} 1 & 0 & 0 \\ 0 & 1 & 0 \\ p_1 & p_2 & p_3 \end{pmatrix}, \quad \Sigma p_i = 1.$$

3.14   Consider Example 9(b) with

$$p_0 = 0.5, \ p_1 = 0.3, \ p_2 = 0.2.$$

Find the mean time upto the first replenishment.

3.15 Let $\{Y_n, n \geq 1\}$ be a sequence of independent random variables with

$$\Pr(Y_n = 1) = p = 1 - \Pr(Y_n = -1).$$

Let X_n be defined by

$$X_0 = 0, X_{n+1} = X_n + Y_{n+1}.$$

Examine whether $\{X_n, n \geq 1\}$ is a Markov chain.
Find $\Pr\{X_n = k\}, k = 0, 1, 2, \ldots$

3.16 Consider Gambler's ruin problem of Example 9(c). Show that the Markov chain $\{X_n, n \geq 1\}$ is a martingale *if and only if* $p = q$.
Hence find the absorption probabilities.

3.17 Show that, for a chain with a finite number, m, of states and having a doubly stochastic matrix (p_{jk}) for its transition matrix,

$$p_{jk}^{(n)} \to v_k = 1/m \text{ for all } j, k = 1, 2, \ldots, m.$$

(A non-negative square matrix is said to be *doubly stochastic* if all the row and column sums are unity.)
A *necessary* and *sufficient* condition that all the states of an irreducible and aperiodic Markov chain possessing a finite number of states are equally likely in the limit is that the transition probability matrix is doubly stochastic.

3.18 A Markov chain with states $0, 1, \ldots$ has transition probabilities

$$p_{jk} = e^{-a} \sum_{r=0}^{k} \binom{j}{r} p^r (1-p)^{j-r} a^{k-r} \Big/ (k-r)!,$$

$$j, k = 0, 1, 2, \ldots, 0 < p < 1.$$

Show that the limiting distribution $\{v_k\}$ is Poisson with parameter $a/(1-p)$.

3.19 *Grouping of states of a Markov chain*: Consider that three states of the Markov chain $\{X_n\}$ described in Exercise 3.4 above are grouped together as follows: Y_n is said to be in state 0 if both the $(n-1)$th and nth tosses result in Heads and 1 otherwise, (i.e. state 0 for HH and state 1 for HT, TH, TT). Show that the chain $\{Y_n\}$ is non-Markovian.
If the states of a Markov chain are grouped or lumped together, the new chain does not, in general, have the Markov property.

3.20 *Urn Process*: Consider an urn which initially contains m balls, out of which a certain number are red and the rest black. Suppose that a red ball is added to the urn with probability $f(x)$ and a black ball with probability $1 - f(x)$ in each operation. Let X_n be the proportion of red balls after the nth operation. Suppose that $X_0 = x$. Then show that $\{X_n, n \geq 0\}$ is a discrete parameter Markov chain (called urn process) having *urn function f* and *initial urn composition* (x, m), and that it is non-homogeneous with transition probability depending on n.

What is the nature of the state space? Examine the special case $f(p) = p_0$ for all p in $[0, 1]$ (*Bernoulli urn process*).

The limiting distribution of X_n exists and is independent of the initial composition (x, m).

[Refer to Hill et al., *Ann. Prob.* **8** (1980) 214-226 for generalised urn processes ; to Johnson & Kotz (1977) for urn models and Markov chains.]

3.21   Suppose that there are m urns numbered (1), (2), . . ., (m) and that each urn contains balls numbered 1, 2, . . ., m but with different proportions. Let the proportion of balls numbered j in urn (i) be p_{ij}, $0 \le p_{ij} \le 1$

$$\sum_{j=1}^{m} p_{ij} = 1 \text{ for } i = 1, 2, ...m.$$

From an arbitrarily chosen urn, say, urn (a_1) a ball is drawn at random, its number, say, A_2 is noted and the ball is returned to the urn (a_1). Next drawing is made from urn (A_2), the number on the ball drawn A_3 is noted and the ball is returned to the urn (A_2); the next drawing is made from urn (A_3) and the process is repeated. Let X_n be the number noted in the nth drawing ($X_0 = a_1$, $X_1 = A_2, \ldots, X_2 = A_3, \ldots$). Then $\{X_n, n \ge 0\}$ forms a Markov chain with t.p.m. (p_{ij}), $i, j = 1, \ldots, m$.

3.22   *Last Exit Theorem*

Let $l_{jk}^{(m)}$ be the probability of going from state j to k in m steps without revisiting j in the meantime, that is, for $m \ge 2$,

$$l_{jk}^{(m)} = \Pr \{X_1 \ne j, 1 \le l \le m - 1, X_m = k \mid X_0 = j\}$$

and $$l_{jk}^{(1)} = p_{jk}, l_{jk}^{(0)} = 0.$$

Let $$L_{jk}^{(s)} = \sum_{m=0}^{\infty} l_{jk}^{(m)} s^m, |s| < 1,$$

be the generating function, and let for all j, k,

$$L_{jk} = \sum_{m=0}^{\infty} l_{jk}^{(m)} s^m$$

$$= \lim_{s \to 1-0} L_{jk}^{(s)}.$$

Show that

$$p_{jk}^{(n)} = \sum_{m=0}^{n} p_{jj}^{(m)} l_{jk}^{(n-m)}, \quad n \ge 1.$$

Deduce that

$$l_{ij}^{(n+1)} = \sum_{r \neq i} l_{ir}^{(n)} p_{rj}, \quad n \geq 1$$

$$P_{jk}(s) = P_{jj}^{(s)} L_{jk}^{(s)}, \quad j \neq k$$

$$P_{jj}(s) = 1 + P_{jj}(s) L_{jj}(s)$$

and hence that

$$F_{jj} = L_{jj}$$

and

$$f_{jj}^{(k)} = l_{jj}^{(k)}.$$

Further, show that

$$\sum_{m=0}^{\infty} p_{jj}^{(m)} = \infty \Leftrightarrow L_{jj} = 1$$

$$\sum_{m=0}^{\infty} p_{jj}^{(m)} < \infty \Leftrightarrow L_{jj} < 1$$

that is, j is persistent *iff* $L_{jj} = 1$ and j is transient *iff* $L_{jj} < 1$.

A persistent state j is non-null if $L_{jj}'(1) < \infty$. The sequence $\{l_{jk}^{(n)}\}$ provides a dual approach to the classification of states as persistent and transient.

3.23 *Inventory model with (s, S) policy:* Suppose that the stock position of an item at an establishment is examined at epochs of time $t_1, t_2, \ldots$ and the following policy is adopted. If at epoch t_n, the stock level is less than or equal to s, procurement is immediately done to bring it to the level S and if the level is greater than s, no replenishment is done. It is supposed that the demand a_n during (t_{n-1}, t_n) obeys the probability law

$$\text{Pr}\,(a_n = k) = v_k, \quad \sum_{k=0}^{\infty} v_k = 1,$$

and that $\{a_n\}$ are i.i.d. random variables. Suppose that the states of $\{X_n, n \geq 1\}$ consist of possible values of stock size just before examination at epoch t_n, i.e. states are $S, S-1, \ldots, 0, -1, -2, \ldots$, negative values being interpreted as unfulfilled demand for stock to be fulfilled as soon as replenishment is done. Show that $\{X_n, n \geq 1\}$ is a Markov chain.

3.24 *Branching process* (with one type of organism or particle): Suppose that at the end of a lifetime an organism produces similar offspring (or a particle at the end of a transformation gives rise to similar new particles) with probability $\text{Pr}\,(Z = i) = p_i, i = 0, 1, 2, \ldots, \sum p_i = 1$. Suppose that the production of offspring does not depend on the evolution of the process and that an offspring produces new offspring, independently of others and in the same way as the parents. Let X_n denote the population size at the nth generation ($n \geq 1$). Assume that $X_0 = 1$, i.e. the process starts with one ancestor. Show that $\{X_n, n \geq 1\}$ is a Markov chain.

A detailed account is given in Chapter 9. (See also Athreya and Ney, 1972).)

3.25   *Consumer Brand switching model* (Fourt, Whitaker): Consider that there are N brands of a consumer product competing in a market. Define

$d_i \equiv$ brand loyalty for the ith brand (equal to the proportion of consumers repurchasing the same brand i without persuation next time),

$w_j \equiv$ purchasing pressure for the jth brand (equal to the proportion of consumers who are persuaded to purchase the jth brand next time).

Then the brand switching transition matrix is given by $A = (a_{ij})$, where

$$a_{ij} = d_i + (1 - d_i)w_j, \quad j = i, j = 1, ..., N$$

$$= (1 - d_i)w_j, \qquad j \neq i, j = 1, ..., N$$

$$0 \leq d_i \leq 1, 0 \leq w_j \leq 1, \sum_{j=1}^{N} w_j = 1.$$

Thus

a_{ii} gives the proportion of customers d_i, remaining with the ith brand plus the proportion of disloyal customers $(1 - d_i)$ who remain with the same brand after being persuaded by the purchasing pressure w_j;

a_{ij} ($i \neq j$) gives the proportion of disloyal customers who switch from ith to jth brand being subjected to purchasing pressure w_j. The brand share of ith brand in period t is given by the elements of

$$Y_t = (y_{jt}),$$

where $$Y_t = Y_{t-1}A.$$

Show that the equilibrium brand shares v_j (equilibrium probabilities v_j) are given by

$$v_j = \frac{w_j/(1 - d_j)}{\sum_{i=1}^{N} \{w_i/(1 - d_i)\}}, \quad j = 1, ...N.$$

Examine the particular cases (i) $d_i = d$; (ii) $d_i = 1$.

Whitaker (1978) develops a technique of estimating brand loyalty by using least squares method to fit the above Markov model to aggregate data.

REFERENCES

S.R. Adke and S.M. Manjunath. *An Introduction to Finite Markov Processes*, Wiley Eastern, New Delhi (1984).

T.W. Anderson and L.A. Goodman. Statistical inference about Markov chains, *Ann. Math. Stat.*, **28**, 89-110 (1957).

S. Asmussen. *Applied Probability and Queues*, Wiley, New York (1987).

K.B. Athreya and P. Ney. *Branching Processes*, Springer-Verlag, Berlin, (1972).

—— A new approach to the limit theory of recurrent Markov chains, *Trans. Am. Math. Soc.*, **245**, 493-501 (1978).

I.V. Basawa and B.L.S. Prakasa Rao. *Statistical Inference for Stochastic Processes*, Academic Press, New York (1980).

A.T. Bharucha-Reid. *Elements of the Theory of Markov Processes and their Applications*, McGraw-Hill, New York, (1960).

P. Billingsley. Statistical methods in Markov chains, *Ann. Math. Stat.*, **32**, 12-40 (1961).

—— *Statistical Inference for Markov Processes*, University of Chicago Press, Chicago (1961).

H. Cohn. Finite non-homogeneous Markov chains: Asymptotic behaviour, *Adv. Appl. Prob.*, **8**, 502-516 (1976).

K.L. Chung, *Markov Chains with Stationary Transition Probabilities*, 2nd ed., Springer-Verlag, New York (1967).

L. Collin. Estimating Markov transition probabilities from micro-unit data, *Appl. Stat.*, **23**, 355–371 (1974).

D.R. Cox and H.D. Miller. *The Theory of Stochastic Processes*, Methuen, London (1965).

C. Derman. On sequential decisions and Markov chains, *Mgmt. Sci.* **29**, 16-24 (1982).

W. Feller. *An Introduction to Probability Theory and its Applications*, Vol. **I**, 3rd ed. Wiley, New York, (1968).

A.F. Forbes. Markov chain models in manpower systems: in *Manpower and Management Science*, (Eds. D.J. Bartholomew and A.R. Smith), English Univ. Press, 93–113 (1971).

D. Freedman. *Markov Chains*, Holden-Day, San Francisco (1971).

W.K. Grassman, M.I. Taksar and D.P. Heyman. Regenerative analysis and steady state distribution for Markov chains, *Opns. Res.* **33**, 1107-1116 (1985).

G. Guardabassi and S. Rinaldi. Two problems in Markov chains: A topological approach, *Opns. Res.* **18**, 324–333 (1970).

F. Harary, B. Lipstein and G.P.H. Styan. A matrix approach to non-stationary chains, *Opns. Res.*, **18**, 1168-1181 (1970).

B. Hill, D. Lane and W. Sudderth. A strong law for generalized urn processes, *Ann. Prob.* **8**, 214-226 (1980).

M. Iosifescu. *Finite Markov Processes and Their Applications*, Wiley, New York (1980).

D.L. Isaacson and R.W. Madsen. *Markov Chains: Theory and Applications*, Wiley, New York (1976).

N.L. Johnson and S. Kotz. *Urn Models and Their Applications*, Wiley, New York (1977).

M. Kaplan. Sufficient conditions for non-ergodicity of a Markov chain, *I.E.E.E. Trans. Inf. Th.* **IT 25**, 470-471 (1975).

S. Karlin and H.M. Taylor. *A First Course in Stochastic Processes*, 2nd ed., Academic Press, New York (1975).

A.F. Karr. Markov chains and processes with a prescribed invariant measure, *Stoch. Proc. & Appl.* **7**, 272-290 (1978).

J. Keilson. *Markov Chain Models : Rarity and Exponentiality*, Springer-Verlag, New York (1979).

F.P. Kelly. Stochastic models for computer communication systems, *J. Roy. Stat. Soc.* **B 47**, 379-395 (1985).

J.G. Kemeny and J.L. Snell. *Finite Markov Chains*, Van Nostrand, Princeton, NJ (1960).

—— and A.W. Knapp. *Denumerable Markov Chains*, 2nd. ed., Springer-Verlag, New York (1976).

J.F.C. Kingman. Geometrical aspects of the theory of non-homogeneous Markov chains, *Math. Proc. Camb. Phil. Soc.* **77**, 171-185 (1975).

T.C. Lee, G.G. Judge and A. Zellner. *Estimating the Parameters of the Markov Probability Model for Aggregate Time Series Data*, North Holland, Amsterdam (1970).

J. Medhi. A Markov chain model for the occurrence of dry and wet days, *Ind. J. Met. Hydro. & Geophys.* **27**, 431-435 (1976).

—— *Stochastic Models in Queueing Theory* (Chapter 1), Academic Press, San Diago (1991).

M.F. Neuts. *Structured Stochastic Matrices of M/G/1 Type and Their Applications*, Prob. : Pure and Appl. V. **5**, Marcel Dekker, New York (1989).

E. Nummelin. *General Irreducible Markov Chains and Non-Negative Operators*, Camb. Univ. Press, London (1984).

A.G. Pakes. Some conditions of ergodicity and recurrence of Markov chains, *Opns. Res.* **17**, 1058-1081 (1969).

E. Parzen. *Stochastic Processes*, Holden-Day, San Francisco, (1962).

A. Paz. *Introduction to Probabilistic Automata*, Academic Press, New York (1971).

B.L.S. Prakasa Rao. *Asymptotic Theory of Statistical Inference*, Wiley, New York (1987).

S.M. Ross. *Stochastic Processes*, Wiley, New York (1983).

I. Sahin. *Regenerative Inventory Systems*, Springer-Verlag, New York (1990).

E. Seneta. On the historical development of the finite non-homogeneous Markov chains, *Proc. Camb. Phil. Soc.* **74**, 507-513 (1973).

—— *Non-Negative Matrices and Markov Chains*, 2nd. ed., Springer-Verlag, New York (1981).

L.I. Sennott. Conditions for non-ergodicity of Markov chains with applications to multiclass communication systems, *J. Appl. Prob.* **24**, 338-346 (1987).

J.J. Solberg. A graph theoretic formula for the steady state distribution of finite Markov processes, *Mgmt. Sci.* **21**, 1040-1048 (1975).

E.J. Subelman. On the class of Markov chains with finite convergence time. *Stoch. Proc. Appl.* **4**, 253-259 (1976).

W. Szpankowski. Some sufficient conditions for non-ergodicity of Markov chains, *J. Appl. Prob.* **22**, 138-147 (1985).

—— and V. Rego. Some theorems on instability with applications to multiaccess protocols, *Opns. Res.* **36**, 958-966 (1988).

H.M. Taylor. Martingales and Random walks: in *Stochastic Models* Handbooks, Vol. 2, (Eds. D.P. Heyman and M.J. Sobel), North Holland, Amsterdam (1990).

H. Tong. Determination of the order of a Markov chain by Akaike's information croterion, *J. Appl. Prob.* **21**, 488-497 (1975).

R.L. Tweedie. Criteria for classifying Markov chains, *Adv. Appl. Prob.* **8**, 737-771 (1976).

Y.H. Wang. On the limit of Markov-Bernoulli distribution, *J. Appl. Prob.* **18**, 937-942 (1981).

W. Whitt. Untold horrors of the waiting room; what the equilibrium distribution will never tell about the queue length process, *Mgmt. Sci.* **29**, 395-408 (1983).

D. Whitaker. The derivative of a measure of brand loyalty using a Markov brand switching model, *J. Op. Res. Soc.* **29**, 959-970 (1978).

P. Whittle. Some distributions and moment formulae for Markov chains, *J. Roy. Stat. Soc.* **B 17**, 235-242 (1955).

P.W. Wolff. *Stochastic Modeling and the Theory of Queues*, Prentice-Hall, Engelwood Cliffs, NJ (1989).

CHAPTER 4

Markov Processes With Discrete State Space : Poisson Process and its Extensions

4.1 POISSON PROCESS

4.1.1 Introduction

Here we shall deal with some stochastic processes in continuous time and with discrete state space, which, apart from their theoretical importance, play an important role in the study of a large number of phenomena. One such process is Poisson process to which we shall confine ourselves here. To fix our ideas, consider a random event E such as (i) incoming telephone calls (at a swtichboard), (ii) arrival of customers for service (at a counter), (iii) occurrence of accidents (at a certain place). Let us consider the total number $N(t)$ of occurrences of the event E in an interval of duration t, i.e. if we start from an initial epoch (or instant) $t = 0$, $N(t)$ will denote the number of occurrences upto the epoch (or instant) t (more precisely to $t + 0$). For example, if an event actually occurs at instants of time $t_1, t_2, t_3, \ldots$ then $N(t)$ jumps abruptly from 0 to 1 at $t = t_1$, from 1 to 2 at $t = t_2$ and so on; the situation can be represented graphically as in Fig. 4.1.

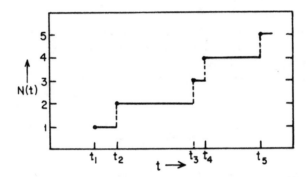

Fig. 4.1 Number $N(t)$ of occurrences of an event in an interval of duration t

The values of $N(t)$ given here are observed values of the random variable $N(t)$. Let $p_n(t)$ be the probability that the random variable $N(t)$ assumes the value n, i.e.

$$p_n(t) = \Pr\{N(t) = n\}. \tag{1.1}$$

This probability is a function of the time t. Since the only possible values of n are $n = 0, 1, 2, 3, \ldots,$

$$\sum_{n=0}^{\infty} p_n(t) = 1 \; ; \qquad (1.2)$$

thus, $\{p_n(t)\}$ represents the probability distribution of the random variable $N(t)$ for *every value of t*. The family of random variables $\{N(t), t \geq 0\}$ is a stochastic process. Here the time t is continuous, the state space of $N(t)$ is discrete and integral-valued and the process is integral-valued. One of the most important integral-valued processes is Poisson process, which serves as a mathematical model for a wide range of empirical phenomena with remarkable accuracy. The justification for this is based on the concept of rare events.

We proceed to show that, under certain conditions, $N(t)$ follows Poisson distribution with mean λt (λ being a constant). In case of many empirical phenomena, these conditions are approximately true and the corresponding stochastic process $\{N(t), t \geq 0\}$ follows the Poisson law.

A stochastic process $\{X(t)\}$ with integral-valued state space [associated with counting (one by one) of an event] such that as t increases, the cumulative count can only increase is called a *counting* or *point process*. Cox and Isham (1980) deal this subject exhaustively. Poisson process is such a process.

4.1.2 Postulates for Poisson Process

I. *Independence*: $N(t)$ is independent of the number of occurrences (of the event E) in an interval prior to the interval $(0, t)$, i.e. future changes in $N(t)$ are independent of the past changes.

II. *Homogeneity in time*: $p_n(t)$ depends only on the length t of the interval and is independent of where this interval is situated, i.e. $p_n(t)$ gives the probability of the number of occurrences (of E) in the interval $(t_1, t + t_1)$ (which is of length t) for every t_1.

III. *Regularity*: In an interval of infinitesimal length h, the probability of exactly one occurrence is $\lambda h + o(h)$ and that of more than one occurrence is of $o(h)$. (Occurrences are regular or orderly).

$o(h)$ is used as a symbol to denote a function of h which tends to 0 more rapidly than h, i.e.

$$\text{as } h \to 0, \ \frac{o(h)}{h} \to 0. \qquad (1.3)$$

In other words, if the interval between t and $t + h$ is of short duration h, then

$$p_1(h) = \lambda h + o(h) \qquad (1.4)$$

$$\sum_{k=2}^{\infty} p_k(h) = o(h). \qquad (1.5)$$

Since

$$\sum_{n=0}^{\infty} p_n(h) = 1,$$

it follows that

$$p_0(h) = 1 - \lambda h + o(h) \tag{1.6}$$

Theorem 4.1. Under the postulates I, II, III, $N(t)$ follows Poisson distribution with mean λt, i.e. $p_n(t)$ is given by the Poisson law:

$$p_n(t) = \frac{e^{-\lambda t}(\lambda t)^n}{n!}, \quad n = 0, 1, 2, 3, ... \tag{1.7}$$

Proof: Consider $p_n(t+h)$ for $n \geq 0$:

0	t	$t+h$

n events by epoch $t + h$ can happen in the following mutually exclusive ways

$$A_1, A_2, A_3 \ldots, A_{n+1}.$$

For $n \geq 1$

A_1: n events by epoch t and no event between t and $t + h$;
   we have

$$\Pr(A_1) = \Pr\{N(t) = n\} \Pr\{N(h) = 0 \mid N(t) = n\} \tag{1.8}$$

$$= p_n(t)p_0(h)$$

$$= p_n(t)(1 - \lambda h) + o(h)$$

A_2: $(n - 1)$ events by epoch t and 1 event between t and $t + h$;
   we have

$$\Pr(A_2) = \Pr\{N(t) = n - 1\} \Pr\{N(h) = 1 \mid N(t) = n - 1\} \tag{1.9}$$

$$= p_{n-1}(t)p_1(h)$$

$$= p_{n-1}(t)(\lambda h) + o(h)$$

For $n \geq 2$

A_3: $(n - 2)$ events by epoch t and 2 events between t and $t + h$;

we have $\qquad \Pr(A_3) = p_{n-2}(t)\{p_2(h)\} \leq p_2(h),$

and so on for $\Pr(A_4) \Pr(A_5), \ldots$ We thus have

$$\sum_{k=2}^{n} \Pr\{A_{k+1}\} \leq \sum_{k=2}^{n} p_k(h) = o(h)$$

and so

$$p_n(t+h) = p_n(t)(1 - \lambda h) + p_{n-1}(t)(\lambda h) + o(h), \quad n \geq 1 \tag{1.10}$$

or,
$$\frac{p_n(t+h) - p_n(t)}{h} = -\lambda p_n(t) + \lambda p_{n-1}(t) + \frac{o(h)}{h}$$

In the limit, as $h \to 0$,

$$p_n'(t) = -\lambda[p_n(t) - p_{n-1}(t)], n \geq 1. \tag{1.11}$$

For $n = 0$, we get

$$p_0(t+h) = p_0(t)p_0(h) = p_0(t)(1 - \lambda h) + o(h)$$

or
$$\frac{p_0(t+h) - p_0(t)}{h} = -\lambda p_0(t) + \frac{o(h)}{h}$$

whence, as $h \to 0$,

$$p_0'(t) = -\lambda p_0(t). \tag{1.12}$$

Initial condition: Suppose that the process starts from scratch at time 0 (or at the origin of epoch of measurement), so that $N(0) = 0$, i.e.

$$p_0(0) = 1; \quad p_n(0) = 0 \quad \text{for } n \neq 0. \tag{1.13}$$

The differential-difference equations (1.11) and the differential equation (1.12) together with (1.13) completely specify the system. Their solutions give the probability distribution $\{p_n(t)\}$ of $N(t)$. The solutions are given by

$$p_n(t) = \frac{e^{-\lambda t}(\lambda t)^n}{n!}, n = 0, 1, 2, \dots \tag{1.14}$$

(see Example 3(b), Section A.3, where the equations are considered to illustrate the use of Laplace transform in solving differential-difference equations).

We indicate here two other methods of solving these equations.

(1) The method of induction: The solution of (1.12) is given by $p_0(t) = Ce^{-\lambda t}$. Since $p_0(0) = 1$, we have $C = 1$ so that $p_0(t) = e^{-\lambda t}$. Consider (1.11) for $n = 1$. Substituting the value of p_0 and solving the equation and using (1.13), we find $p_1(t) = \lambda t \cdot e^{-\lambda t}$. Thus (1.14) is seen to hold for $n = 0$ and 1. Assuming that it holds for $(n-1)$ it can be shown likewise that it holds for any n. Hence, by induction, we get (1.14) for all n.

(2) The method of generating function : Define the probability generating function

$$P(s,t) = \sum_{n=0}^{\infty} p_n(t)s^n = \sum_{n=0}^{\infty} \Pr\{N(t) = n\}s^n = E\{s^{N(t)}\}. \tag{1.15}$$

Now $P(s,0) = \sum\limits_{n=0}^{\infty} p_n(0)s^n = p_0(0) + p_1(0)s + \ldots = 1.$ (1.16)

We have

$$\frac{\partial}{\partial t}P(s,t) = \sum\limits_{n=0}^{\infty} \frac{\partial}{\partial t}p_n(t)s^n = \sum\limits_{n=1}^{\infty} p_n'(t)s^n + p_0'(t);$$

$$\sum\limits_{n=1}^{\infty} p_n(t)s^n = P(s,t) - P_0(t);$$

and

$$\sum\limits_{n=1}^{\infty} p_{n-1}(t)s^n = sP(s,t).$$ (1.17)

Multiplying (1.11) by s^n and adding over for $n = 1, 2, 3, \ldots$ and using (1.17), we get

$$\frac{\partial}{\partial t}P(s,t) - p_0'(t) = -\lambda[\{P(s,t) - p_0(t)\} - sP(s,t)]$$

or $\quad \dfrac{\partial}{\partial t}P(s,t) + \lambda p_0(t) = P(s,t)\{\lambda(s-1)\} + \lambda p_0(t).$

Thus

$$\frac{\partial}{\partial t}P(s,t) = P(s,t)\{\lambda(s-1)\}.$$ (1.18)

Solving (1.18), we get

$$P(s,t) = Ae^{\lambda(s-1)t}.$$ (1.19)

Now $P(s, 0) = 1$ from (1.16), so that $A = 1$.
Hence the p.g.f. of Poisson process is given by

$$P(s,t) = e^{\lambda t(s-1)}$$ (1.20)

$$= e^{-\lambda t}\left\{\sum\limits_{n=0}^{\infty} \frac{(\lambda s t)^n}{n!}\right\},$$ (1.21)

so that

$$p_n(t) \equiv \text{coefficient of } s^n \text{ in } P(s,t)$$

$$= \frac{e^{-\lambda t}(\lambda t)^n}{n!}, \quad n \geq 0.$$ (1.22)

▲

Corollary 1. We have

$$E\{N(t)\} = \lambda t \qquad\qquad (1.23)$$

and $$\mathrm{var}\,\{N(t)\} = \lambda t. \qquad\qquad (1.24)$$

The mean number of occurrences in an interval of length t is λt, so that the mean number of occurrences per unit time ($t = 1$), i.e. in an interval of unit length is λ. The mean rate λ per unit time is known as the *parameter* of the Poisson process.

   The mean and the variance of $N(t)$ are functions of t; in fact, its distribution is functionally dependent on t. As such the process $\{N(t), t \geq 0\}$ is not stationary—it is evolutionary (Sec. 2.3).

Note: While $\{N(t), t \geq 0\}$ is a continuous parameter stochastic process with discrete state space, $E\{N(t)\}$ is a non-random continuous function of t.

Corollary 2. If E occurred r times upto the initial instant 0 from which t is measured, then the initial condition will be

$$p_r(0) = 1, p_n(0) = 0, n \neq r. \qquad\qquad (1.25)$$

Then $p_n(t) = \mathrm{Pr}\,\{\text{Number } N(t) \text{ of occurrences by epoch } t \text{ is } n - r, n \geq r\}$

$$= \frac{e^{-\lambda t}(\lambda t)^{n-r}}{(n-r)!}, \qquad n \geq r \qquad\qquad (1.26)$$

$$= 0 \qquad\qquad n < r.$$

Remarks:

   1. Postulate I implies that Poisson process is Markovian; postulate II that Poisson process is time-homogeneous; postulate III that in an infinitesimal interval of length h, the probability of exactly one occurrence is approximately proportional to the length h of that interval and that of the simultaneous occurrence of two (or more) events is extremely small.

   2. Poisson process has independent as well as stationary (time-homogeneous) increments.

   Again for every t, future increments of a Poisson process are independent of the process generated. This is what is termed by Wolff (1982) as *lack of anticipation assumption*; with this assumption, he proves an extremely useful result: PASTA (Poisson Arrivals See Time Averages).

   3. Poisson process is a Markov process such that the conditional probabilities are constant (independent of t) and are given by (1.4)-(1.6). We shall consider time-dependent Poisson process in Sec. 4.3.5. It can be easily verified that the Poisson process satisfies the Chapman-Kolmogorov equation (5.11) (see Sec. 4.5 and also Example 5(a)).

Example 1(a). Suppose that customers arrive at a Bank according to a Poisson process with a mean rate of a per minute. Then the number of customers $N(t)$ arriving in an interval of duration t minutes follows Poisson distribution with mean at. If the rate of arrival is 3 per minute, then in an arrival of 2 minutes, the probability that the number of customers arriving is:

(i) exactly 4 is

$$\frac{e^{-6}(6)^4}{4!} = 0.133,$$

(ii) greater than 4 is

$$\sum_{k=5}^{\infty} \frac{e^{-6}(6)^k}{k!} = 0.714,$$

(iii) less than 4 is

$$\sum_{k=0}^{3} \frac{e^{-6}(6)^k}{k!} = 0.152,$$

(using tables of Poisson distribution).

Example 1(b). A machine goes out of order whenever a component part fails. The failure of this part is in accordance with a Poisson process with mean rate of 1 per week.

Then the probability that two weeks have elapsed since the last failure is $e^{-2} = 0.135$, being the probability that in time $t = 2$ weeks, the number of occurrences (or failures) is 0.

Suppose that there are 5 spare parts of the component in an inventory and that the next supply is not due in 10 weeks. The probability that the machine will not be out of order in the next 10 weeks is given by

$$\sum_{k=0}^{5} \frac{e^{-10}(10)^k}{k!} = 0.068,$$

being the probability that the number of failures in $t = 10$ weeks will be less than or equal to 5.

Example 1(c). Estimation of the parameter of Poisson process. For a Poisson process $\{N(t)\}$, as $t \to \infty$

$$\Pr\left\{ \left| \frac{N(t)}{t} - \lambda \right| \ge \varepsilon \right\} \to 0, \tag{1.27}$$

where $\varepsilon > 0$ is a preassigned number.

This can be proved by applying Tshebyshev's lemma (for a r.v. X)

$$\Pr\{ |X - E(X)| \ge a \} \le \frac{\text{var}(X)}{a^2}, \quad \text{for } a > 0.$$

From the above, we have, for $X = N(t)$,

$$\Pr\{| N(t) - \lambda t | \geq a\} \leq \frac{\lambda t}{a^2}$$

or

$$\Pr\left\{\left|\frac{N(t)}{t} - \lambda\right| \geq \frac{a}{t}\right\} \leq \frac{\lambda t}{a^2}$$

or

$$\Pr\left\{\left|\frac{N(t)}{t} - \lambda\right| \geq \varepsilon\right\} \leq \frac{\lambda}{t\varepsilon^2}.$$

Hence

$$\Pr\left\{\left|\frac{N(t)}{t} - \lambda\right| \geq \varepsilon\right\} \to 0 \text{ as } t \to \infty.$$

This implies that for large t, the observation $N(t)/t$ may be used as a reasonable estimate of the mean rate λ of the process $\{N(t)\}$.

4.1.3 Properties of Poisson Process

1. *Additive property*: Sum of two independent Poisson processes is a Poisson process
     Let $N_1(t)$ and $N_2(t)$ be two Poisson processes with parameters λ_1, λ_2 respectively and let

$$N(t) = N_1(t) + N_2(t).$$

The p.g.f. of $N_i(t)$ ($i = 1, 2$) is

$$E\{s^{N_i(t)}\} = e^{\lambda_i(s-1)t}$$

The p.g.f. of $N(t)$ is

$$E\{s^{N(t)}\} = E\left\{s^{N_1(t) + N_2(t)}\right\}$$

and because of independence of $N_1(t)$ and $N_2(t)$, we have

$$E\{s^{N(t)}\} = E\left\{s^{N_1 t}\right\} E\left\{s^{N_2(t)}\right\}$$

$$= \left\{e^{\lambda_1(s-1)t}\right\}\left\{e^{\lambda_2(s-1)t}\right\}$$

$$= e^{(\lambda_1 + \lambda_2)(s-1)t}$$

Thus $N(t)$ is a Poisson process with parameter $\lambda_1 + \lambda_2$.

The result can also be proved as follows:

$$\Pr\{N(t)=n\} = \sum_{r=0}^{n} \Pr\{N_1(t)=r\} \Pr\{N_2(t)=n-r\}$$

$$= \sum_{r=0}^{n} \frac{e^{-\lambda_1 t}(\lambda_1 t)^r}{r!} \cdot \frac{e^{-\lambda_2 t}(\lambda_2 t)^{n-r}}{(n-r)!}$$

$$= \frac{e^{-(\lambda_1+\lambda_2)t}\{(\lambda_1+\lambda_2)t\}^n}{n!}, \quad n \ge 0. \tag{1.28}$$

Hence $N(t)$ is a Poisson process with parameter $(\lambda_1 + \lambda_2)$.

2. *Difference of two independent Poisson processes:* The probability distribution of $N(t) = N_1(t) - N_2(t)$ is given by,

$$\Pr\{N(t)=n\} = e^{-(\lambda_1+\lambda_2)t}\left(\frac{\lambda_1}{\lambda_2}\right)^{n/2} I_{|n|}(2t\sqrt{\lambda_1\lambda_2}), \quad n=0,\pm 1,\pm 2...,$$

$$\tag{1.29}$$

where

$$I_n(x) = \sum_{r=0}^{\infty} \frac{(x/2)^{2r+n}}{r!\Gamma(r+n+1)} \tag{1.30}$$

is the modified Bessel function of order $n \,(\ge -1)$.

Proof: (i) The p.g.f. of $N(t)$ is

$$E\{s^{N(t)}\} = E\left\{s^{N_1(t)-N_2(t)}\right\}$$

$$= E\left\{s^{N_1(t)}\right\} E\left\{s^{-N_2(t)}\right\},$$

because of the independence of $N_1(t)$ and $N_2(t)$. Thus

$$E\{s^{N(t)}\} = E\left\{s^{N_1(t)}\right\} E\left\{(1/s)^{N_2(t)}\right\}$$

$$= \exp\{\lambda_1 t(s-1)\} \exp\{\lambda_2 t(s^{-1}-1)\}$$

$$= \exp\{-(\lambda_1+\lambda_2)t\} \exp\{\lambda_1 ts + \lambda_2 t/s\}; \tag{1.31}$$

$\Pr\{N(t)=n\}$ is given by the coefficient of s^n in the expansion of the right hand side of (1.31) as a series in positive and negative powers of s.

(ii) $\Pr\{N(t) = n\}$ can also be obtained directly as follows:

$$\Pr\{N(t) = n\} = \sum_{r=0}^{\infty} \Pr\{N_1(t) = n + r\}\ \Pr\{N_2(t) = r\}$$

$$= \sum_{r=0}^{\infty} \frac{e^{-\lambda_1 t}(\lambda_1 t)^{n+r}}{(n+r)!}\frac{e^{-\lambda_2 t}(\lambda_2 t)^r}{r!}$$

$$= e^{-(\lambda_1 + \lambda_2)t}\left(\frac{\lambda_1}{\lambda_2}\right)^{n/2}\sum_{r=0}^{\infty}\frac{(t\sqrt{\lambda_1\lambda_2})^{2r+n}}{r!(r+n)!}$$

$$= e^{-(\lambda_1 + \lambda_2)t}\left(\frac{\lambda_1}{\lambda_2}\right)^{n/2}I_{|n|}(2t\sqrt{\lambda_1\lambda_2}). \qquad \blacktriangle$$

It may be noted that:
(1) the difference of two independent Poisson processes is *not* a Poisson process;
(2) $I_{-n}(t) = I_n(t) = I_{|n|}(t),\ = 1, 2, 3, \ldots;$
(3) the first two moments of $N(t)$ are given by

$$E\{N(t)\} = (\lambda_1 - \lambda_2)t \quad \text{and} \quad E\{N^2(t)\} = (\lambda_1 + \lambda_2)t + (\lambda_1 - \lambda_2)^2 t^2.$$

Example 1(d). If passengers arrive (singly) at a taxi stand in accordance with a Poisson process with parameter λ_1 and taxis arrive in accordance with a Poisson process with parameter λ_2 then $N(t) = N_1(t) - N_2(t)$ gives the excess of passengers over taxis in an interval t. The distribution of $N(t)$, i.e., $\Pr\{N(t) = n\}$, $n = 0, \pm 1, \pm, 2 \ldots$ is given by (1.29). The mean of $N(t)$ is $(\lambda_1 - \lambda_2)t$, which is $>$ = or < 0 according as

$$\lambda_1 > = \text{or} < \lambda_2; \quad \text{and} \quad \text{var}\{N(t)\} = (\lambda_1 + \lambda_2)t.$$

3. *Decomposition of a Poisson process:* A random selection from a Poisson process yields a Poisson process. Suppose that $N(t)$, the number of occurrences of an even E in an interval of length t is a Poisson process with parameter λ. Suppose also that each occurrence of E has a constant probability p of being recorded, and that the recording of an occurrence is independent of that of other occurrences and also of $N(t)$.

If $M(t)$ is the number of occurrences recorded in an interval of length t, then $M(t)$ is also a Poisson process with parameter λp.

Proof: The event $\{M(t) = n\}$ can happen in the following mutually exclusive ways:

A_r : E occurs $(n + r)$ times by epoch t and exactly n out of $(n + r)$ occurrences are recorded, probability of each occurrence recorded being p, $(r = 0, 1, 2, \ldots)$.

We have

$$\Pr(A_r) = \Pr\{E \text{ occurs } (n + r) \text{ times by epoch } t\}.\ \Pr\{n \text{ occurrences are recorded given that the number of occurrences is } n + r\}$$

$$= \frac{e^{-\lambda t}(\lambda t)^{n+r}}{(n+r)!}\binom{n+r}{n}p^n q^r.$$

Hence

$$\Pr\{M(t)=n\} = \sum_{r=0}^{\infty} \Pr(A_r)$$

$$= \sum_{r=0}^{\infty} \frac{e^{-\lambda t}(\lambda t)^{n+r}}{(n+r)!}\binom{n+r}{n}p^n q^r$$

$$= e^{-\lambda t} \sum_{r=0}^{\infty} \frac{(\lambda p t)^n (\lambda q t)^r}{n!\, r!}$$

$$= e^{-\lambda t} \frac{(\lambda p t)^n}{n!} \sum_{r=0}^{\infty} \frac{(\lambda q t)^r}{r!}$$

$$= e^{-\lambda t} \frac{(\lambda p t)^n}{n!} e^{\lambda q t} = \frac{e^{-\lambda p t}(\lambda p t)^n}{n!}. \tag{1.32}$$

▲

We can interpret the above as follows

For a Poisson process $\{N(t)\}$, the probability of an occurrence in an infinitesimal interval h is (approximately) proportional to the length of the interval h, the constant of proportionality being λ.

Now for $\{M(t)\}$, the probability of a recording in the interval h is proportional to the length h, the constant of proportionality being λp. Thus $\{M(t), t \geq 0\}$ is a Poisson process with parameter λp.

4. (*Continuation of property 3*): The number $M_1(t)$ of occurrences not recorded is also a Poisson process with parameter $\lambda q = \lambda(1-p)$ and $M(t)$ and $M_1(t)$ are independent.

Thus by random selection a Poisson process $\{N(t), t \geq 0\}$ of parameter λ is decomposed into two independent Poisson processes $\{M(t), t \geq 0\}$ and $\{M_1(t), t \geq 0\}$ with parameters λp and $\lambda(1-p)$ respectively.

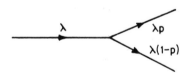

Fig. 4.2 Decomposition rates

As an example, suppose that the births (in a certain town hospital) occur in accordance with a Poisson process with parameter λ. If the probability that an individual born is male is p, then the male births form a Poisson process with parameter λp and the female births form an independent Poisson process with parameter $\lambda (1 - p)$.

More generally, a Poisson process $\{N(t)\}$ with parameter λ may be decomposed into r streams of Poisson processes. If $p_1, p_2, \ldots p_r$, are the probability of the process decomposing into r independent streams such that $p_1 + \ldots p_r = 1$ then the Poisson process is decomposed into r independent Poisson processes with parameters λp_1, $\lambda p_2, \ldots, \lambda p_r$.

5. *Poisson process and binomial distribution*: If $\{N(t)\}$ is a Poisson process and $s < t$, then

$$\Pr\{N(s) = k \mid N(t) = n\} = \binom{n}{k}(s/t)^k (1 - (s/t))^{n-k}.$$

Proof:

$$\Pr\{N(s) = k \mid N(t) = n\} = \frac{\Pr\{N(s) = k \text{ and } N(t) = n\}}{\Pr\{N(t) = n\}}$$

$$= \frac{\Pr\{N(s) = k \text{ and } N(t - s) = n - k\}}{\Pr\{N(t) = n\}}$$

$$= \frac{\Pr\{N(s) = k\} \Pr\{N(t - s) = n - k\}}{\Pr\{N(t) = n\}}$$

$$= \frac{e^{-\lambda s}(\lambda s)^k}{k!} \cdot \frac{e^{-\lambda(t-s)}[\lambda(t-s)]^{n-k}}{(n-k)!} \Big/ \frac{e^{-\lambda t}(\lambda t)^n}{n!}$$

$$= \frac{n!}{k!(n-k)!} \frac{s^k(t-s)^{n-k}}{t^n}$$

$$= \binom{n}{k}(s/t)^k (1 - (s/t))^{n-k}. \qquad (1.33)$$

▲

6. If $\{N(t)\}$ is a Poisson process then the (correlation) auto-correlation coefficient between $N(t)$ and $N(t + s)$ is $\{t/(t + s)\}^{1/2}$.

Proof: Let λ be the parameter of the process; then

$$E\{N(T)\} = \lambda T, \text{ var }\{N(T)\} = \lambda T,$$

and
$$E\{N^2(T)\} = \lambda T + (\lambda T)^2 \text{ for } T = t \text{ and } t + s.$$

Since $N(t)$ and $\{N(t + s) - N(t)\}$ are independent, $\{N(t), t \geq 0\}$ being a Poisson process,

$$E\{N(t) N(t + s)\} = E[N(t)\{N(t + s) - N(t) + N(t)\}]$$
$$= E[N(t)N(t)] + E[N(t)\{N(t + s) - N(t)\}]$$

$$= E\{N^2(t)\} + E\{N(t)\}\, E\{N(t+s) - N(t)\}.$$

Hence $\qquad E\{N(t)\, N(t+s)\} = (\lambda t + \lambda^2\, t^2) + \lambda t \cdot \lambda s.$

Thus the autocovariance between $N(t)$ and $N(t+s)$ is given by

$$C(t,t+s) = E\{N(t)N(t+s)\} - E\{N(t)\}E\{N(t+s)\}$$

$$= (\lambda t + \lambda^2 t^2 + \lambda^2 ts) - \lambda t(\lambda t + \lambda s) = \lambda t.$$

Hence the autocorrelation function

$$\rho\,(t,t+s) = \frac{C(t,t+s)}{\{\mathrm{var}\,N(t)\,\mathrm{var}\,N(t+s)\}^{1/2}}$$

$$= \{(t\,/\,(t+s))\}^{1/2}. \tag{1.34}$$

It can be shown that $\qquad \rho\,(t,t') = \dfrac{\min\,(t,t')}{(tt')^{1/2}}. \tag{1.35}$

This is the autocorrelation function of the process.

Example 1(e). A radioactive source emits particles at a rate of 5 per minute in accordance with a Poisson process. Each particle emitted has a probability 0.6 of being recorded. The number $N(t)$ of particles recorded in an interval of length t is a Poisson process with rate $5 \times 0.6 = 3$ per minute, i.e. with mean $3t$ and variance $3t$. In a 4-minute interval the probability that the number of particles recorded is 10 is equal to $e^{-12}(12)^{10}/10! = 0.104$.

Example 1(f). A person enlists subscriptions to a magazine, the number enlisted being given by a Poisson process with mean rate 6 per day. Subscribers may subscribe for 1 or 2 years independently of one another with respective probabilities $\frac{2}{3}$ and $\frac{1}{3}$.

The number of subscribers $N(t)$ enrolled by the person in time t days is a Poisson process with mean rate $6t$: the number $N_1(t)$ enrolled for 1-year period is a Poisson process with mean $(6 \times \frac{2}{3})t = 4t$ and the number $N_2(t)$ enrolled for 2-year period is a Poisson process with mean $(6 \times \frac{1}{3})t = 2t$.

If he received commission a for a 1-year subscription and b for a 2-year one, then the total commission earned in period t is given by

$$X(t) = aN_1(t) + bN_2(t)$$

We have

$$E\{X(t)\} = aE\{N_1(t)\} + bE\{N_2(t)\}$$
$$= 4\,at + 2\,bt$$

and

$$\mathrm{var}\,X(t) = 4\,a^2 t + 2\,b^2\,t.$$

Note: The process $\{X(t), t > 0\}$ is a compound Poisson process (see Sec. 4.3.2).

4.2 POISSON PROCESS AND RELATED DISTRIBUTIONS

4.2.1 Interarrival Time

With a Poisson process, $\{N(t), t \geq 0\}$, where $N(t)$ denotes the number of occurrences of an event E by epoch t, there is associated a random variable--the interval X between two successive occurrences of E. We proceed to show that X has a negative exponential distribution.

Theorem 4.2. The interval between two successive occurrences of a Poisson process $\{N(t), t \geq 0\}$ having parameter λ has a negative exponential distribution with mean $1/\lambda$.

Proof: Let X be the random variable representing the interval between two successive occurrences of $\{N(t), t \geq 0\}$ and let $\Pr(X \leq x) = F(x)$ be its distribution function.

Let us denote two successive events by E_i and E_{i+1} and suppose that E_i occurred at the instant t_i. Then

$$
\begin{aligned}
\Pr\{X > x\} &= \Pr\{E_{i+1} \text{ did not occur in } (t_i, t_i + x) \text{ given that } E_i \text{ occurred at the} \\
&\quad \text{instant } t_i\} \\
&= \Pr\{E_{i+1} \text{ did not occur in } (t_i, t_i + x) \mid N(t_i) = i\} \\
&\quad \text{(because of the postulate of independence)} \\
&= \Pr\{\text{no occurrence takes place in an interval } (t_i, t_i + x) \text{ of length } x \mid \\
&\qquad\qquad\qquad\qquad\qquad\qquad\qquad\qquad\qquad\qquad N(t_i) = i\} \\
&= \Pr\{N(x) = 0 \mid N(t_i) = i\} = p_0(x) = e^{-\lambda x}, x > 0.
\end{aligned}
\tag{2.1}
$$

Since i is arbitrary, we have for the interval X between *any two* successive occurrences,

$$
F(x) = \Pr\{X \leq x\} = 1 - \Pr\{X > x\} = 1 - e^{-\lambda x}, x > 0.
$$

The density function is

$$
f(x) = F'(x) = \lambda e^{-\lambda x}, (x > 0).
\tag{2.2}
$$

▲

It can be further be proved that if X_i denotes the interval between E_i and E_{i+1}, $i = 1, 2, \ldots.$, then $X_1, X_2 \ldots$ are also *independent*. We omit the proof which is outside the scope of this book. We state the result as follows:

Theorem 4.3. The intervals between successive occurrences (called interarrival times) of a Poisson process (with mean λt) are identically and independently distributed random variables which follow the negative exponential law with mean $1/\lambda$.

The converse also holds; this is given in Theorem 4.4 below. These two theorems give a characterisation of the Poisson process.

Theorem 4.4. If the intervals between successive occurrences of an event E are independently distributed with a common exponential distribution with mean $1/\lambda$, then the events E form a Poisson process with mean λt.

Proof: Let Z_n denote the interval between $(n-1)$th and nth occurrence of a process $\{N(t), t \geq 0\}$ and let the sequence $Z_1, Z_2, \ldots$ be independently and identically distributed random variables having negative exponential distribution with mean $1/\lambda$. The sum $W_n = Z_1 + \ldots + Z_n$ is the waiting time upto the nth occurrence, i.e. the time from the origin to the nth subsequent occurrence. W_n has a gamma distribution with parameters λ, n (Theorem 1.10). The p.d.f. $g(x)$ and the distribution function F_{W_n} are given respectively by

$$g(x) = \frac{\lambda^n x^{n-1} e^{-\lambda x}}{\Gamma(n)}, x > 0$$

and
$$F_{W_n}(t) = \Pr\{W_n \leq t\} = \int_0^t g(x)\, dx.$$

The events $\{N(t) < n\}$ and $\{W_n = Z_1 + \ldots + Z_n > t\}$ are equivalent. Hence the distribution functions $F_{N(t)}$ and F_{W_n} satisfy the relation

$$F_{W_n}(t) = \Pr\{W_n \leq t\} = 1 - \Pr\{W_n > t\}$$

$$= 1 - \Pr\{N(t) < n\} = 1 - \Pr\{N(t) \leq (n-1)\}$$

$$= 1 - F_{N(t)}(n-1).$$

Hence the distribution function of $N(t)$ is given by

$$F_{N(t)}(n-1) = 1 - F_{W_n}(t)$$

$$= 1 - \int_0^t \frac{\lambda^n x^{n-1} e^{-\lambda x}}{\Gamma(n)}\, dx$$

$$= 1 - \frac{1}{\Gamma(n)} \int_0^{\lambda t} y^{n-1} e^{-y}\, dy$$

$$= \frac{1}{\Gamma(n)} \int_{\lambda t}^{\infty} y^{n-1} e^{-y}\, dy$$

$$= \sum_{j=0}^{n-1} \frac{e^{-\lambda t}(\lambda t)^j}{j!} \quad \text{(integrating by parts)}.$$

Thus the probability law of $N(t)$ is

$$p_n(t) = \Pr \{N(t) = n\} = F_{N(t)}(n) - F_{N(t)}(n-1)$$

$$= \sum_{j=0}^{n} \frac{e^{-\lambda t}(\lambda t)^j}{j!} - \sum_{j=0}^{n-1} \frac{e^{-\lambda t}(\lambda t)^j}{j!}$$

$$= \frac{e^{-\lambda t}(\lambda t)^n}{n!}, \; n = 0, 1, 2, \ldots$$

Thus the process $\{N(t), t \ge 0\}$ is a Poisson process with mean λt.   ▲

Note that Poisson process has independent exponentially distributed interarrival times and gamma distributed waiting times.

Example 2(a). Suppose that customers arrive at a counter in accordance with a Poisson process with mean rate of 2 per minute ($\lambda = 2$/minute). Then the interval between any two successive arrivals follows exponential distribution with mean $1/\lambda = \frac{1}{2}$ minute. The probability that the interval between two successive arrivals is

(i)   more than 1 minute is

$$\Pr (X > 1) = e^{-2} = 0.135$$

(ii)   4 minutes or less

$$\Pr (X \le 4) = 1 - e^{-4 \times 2} = 1 - e^{-8} = 0.99967$$

(iii)   between 1 and 2 minutes is

$$\Pr (1 \le X \le 2) = \int_1^2 2e^{-2x} \, dx = e^{-2} - e^{-4} = 0.0179.$$

Example 2(b). Suppose that customers arrive at a counter independently from two different sources. Arrivals occur in accordance with a Poisson process with mean rate of λ per hour from the first source and μ per hour from the second source. Since arrivals at the counter (from either source) constitute a Poisson process with mean $(\lambda + \mu)$ per hour, the interval between any two successive arrivals has a negative exponential distribution with mean $\{1/(\lambda + \mu)\}$ hours.

For example, it taxis arrive at a spot from the north at the rate of 1 per minute and from the south at the rate of 2 per minute in accordance with two independent Poisson processes, the interval between arrival of two taxis has a (negative) exponential distribution with mean $\frac{1}{3}$ minute; the probability that a lone person will have to wait more than a given time t can be found.

Poisson type of occurrences are also called purely random events and the Poisson process is called a *purely* (or *completely*) *random process*. The reason for this is that the occurrence is equally likely to happen anywhere in $[0, T]$ given that only one occurrence has taken place in that interval. We state this by the following theorem.

Theorem 4.5. Given that only one occurrence of a Poisson process $N(t)$ has occurred by epoch T, then the distribution of the time interval γ in $[0, T]$ in which it occurred is uniform in $[0, T]$, i.e.

$$\Pr\{t < \gamma \le t + dt \mid N(T) = 1\} = \frac{dt}{T}, \quad 0 < t < T.$$

Proof: We have

$$\Pr\{t < \gamma \le t + dt\} = \lambda e^{-\lambda t}\, dt,$$
$$\Pr\{N(T) = 1\} = e^{-\lambda T}(\lambda T),$$

and

$$\Pr\{N(T) = 1 \mid \gamma = t\} = e^{-\lambda(T-t)}$$

the last one being the probability that there was no occurrence in the interval of length $(T - t)$. Hence

$$\Pr\{t < \gamma < t + dt \mid N(T) = 1\}$$

$$= \Pr\{t < \gamma < t + dt \text{ and } N(T) = 1\} \;/\; \Pr\{N(T) = 1\}$$

$$= \Pr\{t < \gamma < t + dt\}\Pr\{N(T) = 1 \mid \gamma = t\} \;/\; \Pr\{N(T) = 1\}$$

$$= \lambda e^{-\lambda t}dt\, e^{-\lambda(T-t)} \;/\; e^{-\lambda T}(\lambda T) = dt/T. \qquad \blacktriangle$$

It may be said that a Poisson process distributes points at random over the infinite interval $[0, \infty]$ in the same way as the uniform distribution distributes points at random over a finite interval $[a, b]$.

For a more general form of Theorem 4.5, see Exercise 4.13 (proof is similar).

4.2.2 Further Interesting Properties of Poisson Process

We consider here some interesting properties which are of great use in the study of various phenomena.

We have shown that the interval $X_i (= t_{i+1} - t_i)$ between two successive occurrences E_i, E_{i+1} ($i \ge 1$) of a Poisson process with parameter λ has an exponential distribution with mean $1/\lambda$. Further, the following result holds.

I. For a Poisson process with parameter λ, the interval of time X upto the *first* occurrence also follows an exponential distribution with mean $1/\lambda$. For,

$$\Pr\{X_0 > x\} = \Pr\{N(x) = 0\} = e^{-\lambda x}, \quad (x > 0).$$

In other words, the relation (2.1) does not depend on i nor on t_i.

The same argument also leads to another very important property.

Suppose that the interval X is measured from an *arbitrary* instant of time $t_i + \gamma$ (γ arbitrary) in the interval (t_i, t_{i+1}) and not just the instant t_i of the occurrence of E_i, and Y is the interval upto the occurrence of E_{i+1} measured from $t_i + \gamma$, i.e

$$Y = t_{i+1} - (t_i + \gamma);$$

Y is called *random modification of X or residual time of X*. It follows that:

II. If X is exponential then its random modification Y has also exponential distribution with the same mean.

We may state the result also as follows.

III. For a Poisson process with parameter λ, the interval upto the occurrence of the next event measured from any start of time (not necessarily from the instant of the previous occurrence) is independent of the elapsed time (since the previous instant of occurrence) and is a random variable having exponential distribution with mean $1/\lambda$.

The implication of the above would be clear when viewed in the light of an illustrative example (see, Karlin & Taylor (1975) p. 118) of fishing (incidentally, *poisson* is the French word for fish) as a Poisson process (named after the celebrated French mathematician Siméon Denis Poisson (1781-1840)). Suppose that the random variable $N(t)$ denotes the number of fish caught by an angler in $[0, t]$. Under certain "ideal conditions", such as (i) the number of fish available is very large, (ii) the angler stands in no better chance of catching fish than others, and (iii) the number of fish likely to nibble at one particular instant is the same as at another instant, the process $\{N(t), t \geq 0\}$ may be considered as a Poisson process. The interval upto the first catch, as also the interval between two successive catches has the same exponential distribution. So also is the time interval upto the next catch (from an arbitrary instant γ) which is independent of the elapsed time since the last catch to that instant γ. The long time spent since the last catch gives "no premium for waiting" so far as the next catch is concerned.

Example 2(c). Poisson process and geometric distribution. Consider two independent series of events E and F occurring in accordance with Poisson processes with mean at and bt respectively. The number N of occurrences of E between two successive occurrences of F has a geometric distribution.

Proof: The interval between two successive occurrences of F has the density $f(x) = be^{-bx}$ and therefore Pr $(N = k)$, the probability that k occurrences of E take place during an arbitrary interval (between two successive occurrences of F) is given by

$$\Pr\{N = k\} = \int_0^\infty \frac{e^{-at}(at)^k}{k!} f(t) dt$$

$$= \int_0^\infty \frac{e^{-at}(at)^k}{k!} be^{-bt} dt$$

$$= \frac{ba^k}{k!} \int_0^\infty t^k e^{-(a+b)t} dt$$

$$= \frac{ba^k}{(a+b)^{k+1}}$$

$$= \frac{b}{a+b} \left\{ \frac{a}{a+b} \right\}^k, \quad k = 0, 1, 2, 3, \dots .$$ ▲   (2.3)

Poisson Count Process: Let E and E' be two random sequences of events occurring at instants $(t_1, t_2, \dots)$ and $(t_1', t_2', \dots)$ respectively. The number N_n of occurrences of E' during the interval (t_{n-1}, t_n) of length $t_n - t_{n-1}$ is called the count process of E' in E. If E is Poisson then such a process is called *Poisson count process*. We have proved above that if E' is also Poisson, then the count process, N_n has geometric distribution. It can also be shown that N_n $(n = 1, 2, \dots)$ are *independently* and *identically* distributed with a common geometric distribution. For more general results and for applications of Poisson count process in Queueing theory, see Kingman (1963).

Example 2(d). Suppose that E and F occur independently and in accordance with Poisson processes with parameters a and b respectively.

The interval between two consecutive occurrences of F is the sum of two independent exponential distributions and thus has the density

$$f(x) = b^2 x e^{-bx}, \quad x > 0.$$

The probability that k occurrences of E take place between every second occurrence of F is given by

$$\int_0^\infty \frac{e^{-at}(at)^k}{k!} b^2 t e^{-bt} dt = \frac{a^k b^2}{k!} \int_0^\infty e^{-(a+b)t} t^{k+1} dt$$

$$= \frac{a^k b^2}{k!} \cdot \frac{\Gamma(k+2)}{(a+b)^{k+2}}$$

$$= (k+1)(b/(a+b))^2 (a/(a+b))^k$$

$$= \binom{k+1}{1} \left(\frac{b}{a+b} \right)^2 \left(\frac{a}{a+b} \right)^k, \quad k = 0, 1, 2, \dots$$

The distribution is negative binomial. This also follows from the fact that we have here a sum of two independent geometric distributions; we have two successive intervals (of occurrences of F) in which the number of occurrences of E follows a geometric distribution (see Example 1(h), Ch. 1).

4.3 GENERALISATIONS OF POISSON PROCESS

There are several directions in which the Poisson process (which may be referred to also as the *classical* Poisson process) discussed in the previous section can be generalised. We consider some of them here.

4.3.1 Poisson Process in Higher Dimensions

We have considered so far the one-dimensional case: the occurrences take place at random instants of time t (say, $t_1, t_2, \ldots$) and thus we were concerned with distribution of points on a line. Instead, we may have the two-dimensional case.

Consider the two-dimensional case, such that for the number $N(\Delta a)$ of occurrences in an element of area Δa, we have, for infinitesimal Δa,

$$\Pr\{N(\Delta a) = 1\} = \lambda \Delta a + o(\Delta a),$$
$$\Pr\{N(\Delta a) = k\} = o(\Delta a), k \geq 2$$

and
$$\Pr\{N(\Delta a) = 0\} = 1 - \lambda \Delta a + o(\Delta a). \tag{3.1}$$

Thus, if the number of occurrences in non-overlapping areas are mutually independent, the number $N(a)$ of occurrences in an area a will be a Poisson process with mean λa. Here in place of one-dimensional t, we consider two-dimensional a. Similarly, we can describe Poisson process in higher dimensions.

4.3.2 Poisson Cluster Process (Compound or Cumulative Poisson Process)

We considered in Sec. 4.1 that only one event can occur at an instant of occurrence. Now let us suppose that several events can happen simultaneously at such an instant, i.e. we have a cluster (of occurrences) at a point. We assume that:

    (i) The number $N(t)$ of clusters in time t, i.e. the points at which clusters occur constitute a Poisson process with mean rate λ.

    (ii) Each cluster has a random number of occurrences, i.e. the number X_i of occurrences in ith cluster is a.r.v. The various numbers of occurrences in the different clusters are mutually independent and follow the same probability distribution:

$$\Pr\{X_i = k\} = p_k, k = 1, 2, 3, \ldots$$

$$i = 1, 2, 3, \ldots \tag{3.2}$$

having p.g.f.

$$P(s) = \sum_{k=1}^{\infty} p_k s^k. \tag{3.3}$$

Theorem 4.6. If $M(t)$ denote the total number of occurrences in an interval of length t under the conditions (i) and (ii) stated above, then the generating function of $M(t)$ is given by

$$G(P(s)) = \exp[\lambda t \{P(s) - 1\}]. \qquad (3.4)$$

Proof: $M(t)$ is the sum of a random number of terms, i.e.

$$M(t) = \sum_{i=1}^{N(t)} X_i,$$

where $N(t)$ is a Poisson process with mean λt.

Now $P(s)$ is the p.g.f. of X_i and $G(s)$ is the p.g.f. of $N(t)$. Thus,

$$G(s) = \exp\{\lambda t (s - 1)\}.$$

Hence by Theorem 1.3. the p.g.f. of $M(t)$ is given by

$$G(P s)) = \exp\{\lambda t (P(s) - 1)\}. \qquad \blacktriangle$$

Note: $M(t)$ is called a *compound Poisson process* (or *cumulative Poisson process*). It is to be noted that $M(t)$ is not necessarily Poisson. Poisson cluster processes (compound Poisson processes) arise in bulk queues, where customers arrive or are served in groups. In fact, a large variety of practical problems can be reduced to compound Poisson process (Feller II, 1966, p. 179).

A compound Poisson process is *not* a point process; it is what is called a *jump process*.

An important application arises in '*Collective risk theory*'. Suppose that claims against a company (such as an insurance company) occur in accordance with a Poisson process with mean λt, and that individual claims X_i are i.i.d. with distribution $\{p_k\}$, then $M(t)$ represents the total claims at epoch t. If A represents initial reserve and c the rate of increase of the reserves in the absence of claims, then the total reserve at epoch t is $A + ct - M(t)$, and negative reserve implies 'ruin'.

Corollary: We have

$$E\{M(t)\} = \lambda t\, E(X_i)$$

and
$$\text{var}\{M(t)\} = \lambda t\, E(X_i^2). \qquad (3.5)$$

It can be shown that

$$\text{cov}\{M(t), M(s)\} = \{\min(s, t)\}\,\{\lambda E(X_i^2)\}.$$

Compound Poisson process and linear combination of independent Poisson processes:

Consider Example 1(f).

The process $X(t) = aN_1(t) + bN_2(t)$ is a linear combination of two independent Poisson processes. The process can also be expressed as the compound Poisson process

$$X(t) = \sum_{i=1}^{N(t)} X_i$$

where X_i, the amount of commission received from a subscription, is a random variable such that $\Pr(X_i = a) = \frac{2}{3}$ and $\Pr\{X_i = b\} = \frac{1}{3}$.

If follows that

$$E\{X(t)\} = E\{N(t)\}\, E(X_i) = (6t) \cdot \frac{1}{3} \cdot (2a + b) = 4at + 2bt$$

and

$$\mathrm{var}\{X(t)\} = 6t\, E\{X_i^2\} = 6t \cdot \frac{1}{3} \cdot (2a^2 + b^2) = 4a^2 t + 2b^2 t.$$

The two approaches are equivalent. The result may be stated in a more general form as follows:

Let $a_k > 0$, $k = 1, 2, \ldots, r\ (\geq 2)$ and $\Pr(X_i = a_k) = p_k$ for each i, $\sum p_k = 1$.

Then $X(t) = \sum_{i=1}^{N(t)} X_i$, where $\{N(t)\}$ is a Poisson process with parameter λ, is a compound Poisson process. For $t > 0$, let $N_i(t)$ be the number of jumps of value a_i for the process $\{X(t)\}$ which occur prior to t. Then we have

$$X(t) = a_1 N_1(t) + \ldots + a_r N_r(t)$$

where $\{N_k(t), t \geq 0\}$ is a Poisson process with parameter λp_k and $N_1(t), \ldots, N_r(t)$ are mutually independent. $N(t)$ is decomposed into r independent Poisson processes $N_k(t)$, $k = 1, 2, \ldots, r$.

Example 3(a). Customers arrive at a store in groups consisting of 1 or 2 individuals with equal probability and the arrival of groups is in accordance with a Poisson process with mean rate λ.

Here

$$p_k = \Pr\{X_i = k\} = \frac{1}{2} \text{ (for } k = 1, 2)$$

$$= 0 \text{ (otherwise)};$$

hence

$$P(s) = \sum_k p_k s^k = \frac{1}{2}s + \frac{1}{2}s^2$$

and so the generating function of $M(t)$, the total number of customers arriving in time t (by 3.4) is

$$G(s) = \exp\left\{\lambda t\left[\frac{1}{2}(s + s^2) - 1\right]\right\}.$$

The mean number of customers arriving in time t equals $E\{X_i\}\{\lambda t\} = \frac{3}{2}\lambda t$. For $\lambda = \frac{1}{2}$ per minute and $t = 4$ minutes, the generating function will be

$$\exp\left[2\left\{\frac{1}{2}(s + s^2) - 1\right\}\right] = \{\exp(-2)\}\{\exp(s + s^2)\},$$

and the probability that the total number of arrivals is exactly 4 is

$$e^{-2}\left(\frac{1}{4!} + \frac{1}{2!} + \frac{1}{2!}\right) = 0.141.$$

Example 3(b). Suppose that the number of arrival epochs in an interval of length t is given by a Poisson process $\{N(t), t \geq 0\}$ with mean a and that the number of units arriving at an arrival epoch is given by a zero-truncated Poisson variate X_i, $i = 1, 2,$ ... with parameter λ. Then the total number $M(t)$ of units which arrive in an interval of length t is a Poisson cluster process with p.g.f.

$$G(P(s)) = \exp[at\{P(s) - 1\}]$$

where $P(s)$ is the p.g.f. of zero-truncated Poisson process (Exercise 1.5(a)), namely,

$$P(s) = (\exp \lambda - 1)^{-1}(\exp(\lambda s) - 1).$$

Hence

$$G(P(s)) = \exp\left[at\left\{\frac{(\exp(\lambda s) - 1)}{(\exp \lambda - 1)} - 1\right\}\right].$$

Example 3(c). *An application in inventory theory*
Suppose that X_i are i.i.d. decapitated geometric r.v.'s such that

$$\Pr\{X_i = k\} = q^{k-1}p, \quad k = 1, 2, 3, \ldots, p + q = 1$$

then

$$G(s) = ps/(1 - qs)$$

and

$$G(P(s)) = \exp[\lambda t\{ps/(1 - qs) - 1\}].$$

We get

$$p_0 \equiv \Pr\{M(t) = 0\} = e^{-\lambda t}$$

and

$$p_k \equiv \Pr\{M(t) = k\} \text{ is given by}$$

$$p_k = \frac{p\lambda t}{k} \sum_{j=1}^{k} q^{j-1} j \, p_{k-j}, \quad k \geq 1.$$

Several authors have used this as a model for lead time demand of a commodity. It has been shown that this distribution fits actual data for demand of units of an EOQ or consumable type inventory item during stock replenishment or lead time. (See Mitchell *et al* (1983) & Sherbrooke (1968)).

4.3.3 Pure Birth Process : Yule-Furry Process

In the classical Poisson process we assume that the conditional probabilities are constant. Here, the probability that k events occur between t and $t + h$ given that n events occurred by epoch t is given by

$$
\begin{aligned}
p_k(h) = \Pr\{N(h) = k \mid N(t) = n\} &= \lambda h + o(h), & k &= 1 \\
&= o(h), & k &\geq 2 \\
&= 1 - \lambda h + o(h), & k &= 0;
\end{aligned}
$$

$p_k(h)$ is independent of n as well as t. We can generalise the process by considering that λ is not constant but is a function of n or t or both; the resulting processes will still be Markovian in character.

Here we consider that λ is a function of n, the population size at the instant. We assume that

$$
\begin{aligned}
p_k(h) = \Pr\{N(h) = k \mid N(t) = n\} &= \lambda_n h + o(h), & k &= 1 \\
&= o(h), & k &\geq 2 \qquad (3.6) \\
&= 1 - \lambda_n h + o(h), & k &= 0.
\end{aligned}
$$

Then proceeding as in Sec. 4.1.2 we shall have the following equation corresponding to (1.10)

$$p_n(t + h) = p_n(t)(1 - \lambda_n h) + p_{n-1}(t)\lambda_{n-1} h + o(h), \ n \geq 1.$$

Proceeding as before, we get

$$p_n'(t) = -\lambda_n p_n(t) + \lambda_{n-1} p_{n-1}(t), \ n \geq 1 \qquad (3.7)$$

$$p_0'(t) = -\lambda_0 p_0(t) \qquad (3.8)$$

For given initial conditions, explicit expressions for $p_n(t)$ can be obtained from the above equations.

We shall consider here only a particular case of interest, the case $\lambda_n = n\lambda$ and describe a situation where this can happen.

While the above process is called *pure birth process* the process corresponding to $\lambda_n = n\lambda$ is called *Yule-Furry process*.

Yule-Furry Process:

Consider a population whose members are either physical or biological entities. Suppose that members can give birth (by splitting or otherwise) to new members (who are exact replicas of themselves) but cannot die. We assume that in an interval of length h each member has a probability $\lambda_h + o(h)$ of giving birth to a new member. Then, if n individuals are present at time t, the probability that there will be one birth between t and $t + h$ is $n\,\lambda h + o(h)$. If $N(t)$ denotes the total number of members by epoch t and $p_n(t) = \Pr\{N(t) = n\}$, then by putting $\lambda_n = n\lambda$ in (3.7) and (3.8), we obtain the following equations for $p_n(t)$:

$$p_n'(t) = -n\lambda p_n(t) + (n-1)\lambda p_{n-1}(t), \; n \geq 1 \tag{3.9}$$

$$p_0'(t) = 0 \tag{3.10}$$

If the initial conditions are given, explicit expressions for $p_n(t)$ can be obtained.

Suppose that the initial condition is $p_1(0) = 1$, $p_i(0) = 0$ for $i \neq 1$, i.e. the process started with only one member at time $t = 0$. The solution can be obtained by the method of induction as follows:

For $n = 1$, we have

$$p_1'(t) = -\lambda\, p_1(t)$$

whose solution is

$$p_1(t) = c_1 e^{-\lambda t};$$

and putting $p_1(0) = 1$, we have $c_1 = 1$, so that, $p_1(t) = e^{-\lambda t}$.

For $n = 2$, we have

$$p_2'(t) = -2\lambda p_2(t) + \lambda p_1(t)$$

or

$$p_2'(t) + 2\lambda p_2(t) = \lambda p_1(t) = \lambda e^{-\lambda t}.$$

This linear equation has the integrating factor $e^{2\lambda t}$ and therefore

$$e^{2\lambda t} p_2(t) = \int \lambda e^{2\lambda t - \lambda t} dt = e^{\lambda t} + c_2 ;$$

since

$$p_2(0) = 0, \text{ we have } c_2 = -1$$

and

$$p_2(t) = e^{-2\lambda t}(e^{-\lambda t} - 1) = e^{-\lambda t}(1 - e^{-\lambda t}).$$

Proceeding in this way it can be shown that

$$p_n(t) = e^{-\lambda t}(1 - e^{-\lambda t})^{n-1}, n \geq 1. \tag{3.11}$$

Solving (3.10) and noting that $p_0(0) = 0$, we get

$$p_0(t) = 0.$$

The distribution is of the geometric form (see Exercise 1.2). Its p.g.f. is given by

$$P(s, t) = \sum_{n=1}^{\infty} \{e^{-\lambda t}(1 - e^{-\lambda t})^{n-1}\} s^n$$

$$= \frac{s e^{-\lambda t}}{1 - s(1 - e^{-\lambda t})}. \tag{3.12}$$

The mean of the process is given by

$$E\{N(t)\} = e^{\lambda t};$$

Further, var $\{N(t)\} = e^{\lambda t}(e^{\lambda t} - 1)$.

4.3.4 Birth-Immigration Process

Suppose that there is also immigration in addition to birth, such that in an interval of infinitesimal length h, the probability of a new member being added (by immigration) to the population is $\nu h + o(h)$. Then

$$p_n(t+h) = p_n(t)\{1 - n\lambda h\}\{1 - \nu h\}$$

$$+ p_{n-1}(t)\{(n-1)\lambda + \nu\} + o(h), n \geq 1$$

so that

$$p_n'(t) = -(n\lambda + \nu)p_n(t) + \{(n-1)\lambda + \nu\}p_{n-1}(t).$$

Let

$$P(s,t) = \sum_{n=0}^{\infty} p_n(t)s^n$$

then we have

$$\frac{\partial P}{\partial t} = \lambda s(s-1)\frac{\partial P}{\partial s} + \nu(s-1)P.$$

With $p_n(0) = 1$, $n = i$, the above differential equation admits of the solution

$$P(s,t) = \frac{s^i e^{-\lambda t} e^{-v(1-s)t}}{[1 - s(1 - e^{-\lambda t})]^i}$$

(see also Exercise 4.6).

Particular case: $i = 1$, $v = 0$: this gives (3.12).

4.3.5 Time-dependent Poisson Processes

Here we assume that λ is a non-random function of time t, and that

$$\begin{aligned}
\Pr \{N(h) = k \mid N(t) = n\} &= \lambda(t)h + o(h), & k &\geq 1 \\
&= o(h), & k &\geq 2 \qquad (3.13) \\
&= 1 - \lambda(t)h + o(h), & k &\geq 0;
\end{aligned}$$

then we get a time-dependent Poisson process. Here the postulate II of the classical Poisson process regarding homogeneity does not hold good and this process is called a non-homogeneous Poisson process. The equations of the process, which can be obtained from those of Poisson process by replacing λ by $\lambda(t)$, are

$$p_n'(t) = -\lambda(t)p_n(t) + \lambda(t)p_{n-1}(t), n \geq 1 \qquad (3.14)$$

$$p_0'(t) = \lambda(t)p_0(t). \qquad (3.15)$$

Theorem 4.7. The p.g.f. of a non-homogeneous process $\{N(t), t \geq 0\}$ is given by

$$Q(s,t) = \exp \{m(t)(s - 1)\} \qquad (3.16)$$

where $\qquad m(t) = \int_0^t \lambda(x)dx$ is the expectation of $N(t)$. $\qquad (3.17)$

Proof: Now $Q(s, t) = E\{s^{N(t)}\}$ with $Q(s, 0) = E(s^{N(0)}) = 1$.

Hence $\qquad\qquad Q(s,t+h) = E\{s^{N(t+h)}\}$

$$= E\{s^{N(t+h)-N(t)} s^{N(t)}\}$$

$$= E\{s^{N(t+h)-N(t)}\} E\{s^{N(t)}\}$$

$$= E\{s^{N(t+h)-N(t)}\} Q(s,t),$$

since $N(t)$ and $\{N(t + h) - N(t)\}$ are independent (by postulate I of Sec. 4.1 which still holds good; only postulate II does not).

Now $\qquad E\{s^{N(t+h)-N(t)}\} = E\{s^{N(h)}\}$

$$= \sum_{k=0}^{\infty} \Pr\{N(h)=k\}s^k$$

$$= \{1-\lambda(t)h+o(h)\} + \{\lambda(t)h+o(h)\}s+o(h)$$

$$= 1+(s-1)\lambda(t)h+o(h).$$

Hence we have

$$Q(s,t+h) = \{1+(s-1)\lambda(t)h+o(h)\}\,Q(s,t)$$

or $\qquad \dfrac{Q(s,t+h)-Q(s,t)}{h} = (s-1)\lambda(t)Q(s,t)+\dfrac{o(h)}{h}.$

We have, taking limits of both sides, as $h \to 0$

$$\frac{\partial}{\partial t}Q(s,t) = (s-1)\lambda(t)Q(s,t).$$

Integrating we get

$$Q(s,t) = C\,\exp\left\{(s-1)\int_0^t \lambda(x)dx\right\}.$$

As $\qquad Q(s,0)=0$, we have $C = 1$, so that
$$Q(s,t) = \exp\{m(t)(s-1)\},$$

where $\qquad m(t) = \int_0^t \lambda(x)dx.$

Now
$$E\{N(t)\} = \frac{\partial Q(s,t)}{\partial s}\Big|_{s=1} = m(t),$$

so that $m(t)$ is the expectation of $N(t)$. $\qquad\blacktriangle$

Corollary. The probability of no occurrence in an interval of length t is given by

$$p_0(t) = Q(s,t)|_{s=0} = \exp\{-m(t)\} = \exp\{-\int_0^t \lambda(x)\,dx\}. \qquad (3.18)$$

The probability of exactly k occurrences in an interval $(0, t)$ is given by

$$p_k(t) = \frac{1}{k!}\frac{\partial^k}{\partial s^k}Q(s,k)\Big|_{s=0} = \frac{1}{k!}\{m(t)\}^k e^{-m(t)}$$

$$= \frac{1}{k!}\left\{\int_0^t \lambda(x)\,dx\right\}^k \exp\left\{-\int_0^t \lambda(x)\,dx\right\}, \quad k \geq 0. \qquad (3.19)$$

That is, if we introduce $m(t)$ as *operational time* instead of absolute time t, the probability p_k of exactly k occurrences in the *operational time interval* $[0, m(t)]$ is given by the Poisson expression $\dfrac{[m(t)]^k \exp\{-m(t)\}}{k!}$. It can easily be verified that when $\lambda(t) = \lambda$, these lead to the corresponding probabilities of a homogeneous Poisson process.

Further generalisation can be obtained by considering λ as a function of both n and t, i.e. $\lambda = \lambda_n(t)$.

Polya process, which is such a generalisation, corresponds to

$$\lambda_n(t) = \frac{1 + an}{1 + at}.$$

4.3.6 Random Variation of the Parameter λ

In order to allow a for lack of homogeneity and the influence of random factors on the development of the process, some random variation is introduced in the parameter λ of a Poisson process $N(t)$. These are considered below (see Cramer, 1969).

Case I: Suppose that λ is a random variable with probability density function $f(\lambda)$, $0 \le \lambda \le \infty$.

The probability of n occurrences in an interval of length t is then given by,

$$p_n(t) = \Pr\{N(t) = n\} = \int_0^\infty \frac{(\lambda t)^n}{n!} e^{-\lambda t} f(\lambda) d\lambda.$$

The particular case when λ follows a gamma distribution is rather well known. Suppose that

$$f(\lambda) = \frac{h^k \lambda^{k-1} e^{-h\lambda}}{\Gamma(k)}, \quad \lambda > 0$$

where k is a constant (parameter). Then

$$p_n(t) = \binom{k+n-1}{n} p^k q^n, \, p = h/(t+h) = 1 - q, \, n = 0, 1, \dots$$

and $N(t)$ follows a negative binomial distribution with index k and mean kt/h.

Case II: Suppose that λ varies in a random way as time proceeds. Here, we consider λ to be a *random* function of time t, i.e. $\lambda = \lambda(t)$ is a stochastic process. A special case of Case II arises where the process $\lambda(t)$ is stationary. This has been studied as *conditional Poisson process* or '*doubly stochastic Poisson process.*'

A conditional Poisson process is essentially a non-homogeneous Poisson process whose intensity function $\lambda(t)$ (or mean value function) is a stationary stochastic process. These processes with strictly stationary intensity processes, have been studied by several authors (Bartlett, Kingman, McFadden, Serfozo etc.; see Grandell (1976) for a detailed account).

4.3.7 Renewal Process

In the classical Poisson process, the intervals between successive occurrences are independently and identically distributed with a negative exponential distribution. Suppose that there is a sequence of events E such that the intervals between successive occurrences of E are distributed independently and identically but have a distribution not necessarily negative exponential; we have then a certain generalisation of the classical Poisson process: the corresponding process is called a *renewal process* (for details, see Ch. 6).

4.4 BIRTH AND DEATH PROCESS

In Sec. 4.3 3., we considered a pure birth process, where
Pr {Number of births between t and $t + h$ is k, given that the number of individuals
at epoch t is n}
is given by

$$p(k, h \mid n, t) = \begin{cases} \lambda_n h + o(h), & k = 1 \\ o(h), & k \geq 2 \\ 1 - \lambda_n h + o(h), & k = 0. \end{cases} \tag{4.1}$$

The above holds for all $n \geq 0$; λ_0 may or may not be equal to zero. Here k is a non-negative integer which implies that there can only be an increase by k, i.e. only births are considered possible. Now we suppose that there could also be a decrease by k, i.e. death(s) is also considered possible. In this case we shall further assume that
Pr {Number of deaths between t and $t + h$ is k,
given that the number of individuals at epoch t is n}
is given by

$$q\{k, h \mid n, t\} = \begin{cases} \mu_n + o(h), & k = 1 \\ o(h), & k \geq 2 \\ 1 - \mu_n + o(h), & k = 0. \end{cases} \tag{4.2}$$

The above holds for $n \geq k$; further $\mu_0 = 0$. With (4.1) and (4.2) we have, what is known as a *birth and death process*. Through a birth there is an increase by one and through a death, there is a decrease by one in the number of "individuals". The probability of more than one birth or more than one death in an interval of length h is $o(h)$. Let $N(t)$ denote the total number of individuals at epoch t starting from $t = 0$ and let $p_n(t) = \text{Pr}\{N(t) = n\}$. Consider the interval between 0 and $t + h$; suppose that it is split into two periods $(0, t)$ and $[t, t + h)$. The event $\{N(t + h) = n, n \geq 1\}$, (having probability $p_n(t + h)$) can occur in a number of mutually exclusive ways.

These would include events involving more than one birth and/or more than one death between t and $t + h$. By our assumption, the probability of such an event is $o(h)$. There will remain four other events to be considered:

A_{ij} : $(n - i + j)$ individuals by epoch t, i birth and j death between t and $t + h$, $i, j = 0, 1$.

We have

$$\Pr(A_{00}) = p_n(t)\{1 - \lambda_n h + o(h)\}\{1 - \mu_n h + o(h)\}$$

$$= p_n(t)\{1 - (\lambda_n + \mu_n)h + o(h)\};$$

$$\Pr(A_{10}) = p_{n-1}(t)\{\lambda_{n-1} h + o(h)\}\{1 - \mu_{n-1} h + o(h)\}$$

$$= p_{n-1}(t)\lambda_{n-1} h + o(h);$$

$$\Pr(A_{01}) = p_{n+1}(t)\{1 - \lambda_{n+1} h + o(h)\}\{\mu_{n+1} h + o(h)\}$$

$$= p_{n+1}(t)\mu_{n+1} h + o(h);$$

and

$$\Pr(A_{11}) = p_n(t)\{\lambda_n h + o(h)\}\{\mu_n h + o(h)\}$$

$$= o(h).$$

Hence we have, for $n \geq 1$

$$p_n(t + h) = p_n(t)\{1 - (\lambda_n + \mu_n)h\} + p_{n-1}(t)\lambda_{n-1} h$$

$$+ p_{n+1}(t)\mu_{n+1} h + o(h) \qquad (4.3)$$

or

$$\frac{p_n(t + h) - p_n(t)}{h} = -(\lambda_n + \mu_n)p_n(t) + \lambda_{n-1}p_{n-1}(t)$$

$$+ \mu_{n+1}p_{n+1}(t) + \frac{o(h)}{h},$$

and taking limits, as $h \to 0$, we have

$$p_n'(t) = -(\lambda_n + \mu_n)p_n(t) + \lambda_{n-1}p_{n-1}(t) + \mu_{n+1}p_{n+1}(t), n \geq 1. \qquad (4.4)$$

For $n = 0$, we have

$$p_0(t + h) = p_0(t)\{1 - \lambda_0 h + o(h)\} + p_1(t)\{1 - \lambda_0 h + o(h)\}\{\mu_1 h + o(h)\}$$

$$\qquad (4.5)$$

$$= p_0(t) - \lambda_0 h p_0(t) + \mu_1 h p_1(t)$$

or

$$\frac{p_0(t + h) - p_0(t)}{h} = -\lambda_0 p_0(t) + \mu_1 p_1(t) + \frac{o(h)}{h}$$

whence we get

$$p_0'(t) = -\lambda_0 p_0(t) + \mu_1 p_1(t). \tag{4.6}$$

If at epoch $t = 0$, there were i (≥ 0) individuals, then the initial condition is

$$p_n(0) = 0, n \neq i; \, p_i(0) = 1. \tag{4.7}$$

The equations (4.4) and (4.6) are the equations of the birth and death process. The birth and death processes play an important role in queueing theory. They also have interesting applications in diverse other fields such as economics, biology, ecology, reliability theory etc.

Note: The result about existence of solutions of (4.4) and (4.6) is stated below without proof.

For arbitrary $\lambda_n \geq 0$, $\mu_n \geq 0$, there always exists a solution $p_n(t)$ (≥ 0) such that $\sum p_n(t) \leq 1$. If λ_n, μ_n are bounded, the solution is unique and satisfies $\sum p_n(t) = 1$.

Birth and Death Rates

Some particular values of λ_n and μ_n are of special interst.

When $\lambda_n = \lambda$, i.e. λ_n is independent of the population size n, then the increase may be thought of as due to an external source such as *immigration*.

When $\lambda_n = n\lambda$, we have the case of (linear) birth; $\lambda_n h = n\lambda h$ may be considered as the probability of one birth in an interval of length h given that n individuals are present (at the instant from which the interval commences), the probability of one individual giving a birth being λh, (i.e. rate of birth in unit internal is λ per individual). Here $\lambda_0 = 0$.

When $\mu_n = \mu$, the decrease may be thought of as due to a factor such as *emigration*.

When $\mu_n = n\mu$, we have the case of (linear) death, the rate of death in unit interval being μ per individual.

4.4.1 Particular Cases

I. Immigration-Emigration Process

For $\lambda_n = \lambda$ and $\mu_n = \mu$ we have what is known as *immigration-emigration* process. The process associated with the simple queueing model $M/M/1$ (which we discuss in Sec. 10.2) is such a process.

II. Linear Growth Process

(a) *Generating function*: In the Yule-Furry process (Sec. 4.3.3) one is concerned with a population whose members can give birth only (to new members) but cannot die. Let us consider the case where both births and deaths can occur. Suppose that the probability that a member gives birth to a new member in a small interval of length h is $\lambda h + o(h)$ and the probability that a member dies is $\mu h + o(h)$. Then, if n members are present at the instant t, the probability of one birth between t and $t + h$ is $n \lambda h + o(h)$ and that of one death is $n \mu h + o(h)$, $n \geq 1$.

We have thus a birth and death process with

$$\lambda_n = n\lambda, \; \mu_n = n\mu \; (n \geq 1), \; \lambda_0 = \mu_0 = 0.$$

If $X(t)$ denotes the total number of members at time t, then from (4.4) and (4.6) we have the following differential-difference equations for $p_n(t) = \Pr\{X(t) = n\}$:

$$p_n'(t) = -n(\lambda + \mu)p_n(t) + \lambda(n-1)p_{n-1}(t) + \mu(n+1)p_{n+1}(t), n \geq 1 \qquad (4.8)$$

$$p_0'(t) = \mu p_1(t). \qquad (4.9)$$

If the initial population size is i, i.e. $X(0) = i$, then we have the initial condition $p_i(0) = 1$ and $p_n(0) = 0$, $n \neq i$.

Let

$$P(s,t) = \sum_{n=1}^{\infty} p_n(t)s^n \qquad \text{be the p.g.f. of } \{p_n(t)\}.$$

Then

$$\frac{\partial P}{\partial s} = \sum_{n=1}^{\infty} np_n(t)s^{n-1} \quad \text{and} \quad \frac{\partial P}{\partial t} = \sum_{n=0}^{\infty} p_n'(t)s^n.$$

Multiplying (4.8) by s^n and adding over $n = 1, 2, 3, \ldots$ and adding (4.9) thereto, we get

$$\frac{\partial P}{\partial t} = -(\lambda + \mu) \sum_{n=1}^{\infty} np_n(t)s^n + \lambda \sum_{n=1}^{\infty} (n-1)p_{n-1}(t)s^n$$

$$+ \mu \left\{ \sum_{n=1}^{\infty} (n+1)p_{n+1}(t)s^n + p_1(t) \right\}$$

$$= -(\lambda + \mu)s \frac{\partial P}{\partial s} + \lambda s^2 \frac{\partial P}{\partial s} + \mu \frac{\partial P}{\partial s}$$

$$= \{\mu - (\lambda + \mu)s + \lambda s^2\} \frac{\partial P}{\partial s}. \qquad (4.10)$$

$P(s, t)$ thus satisfies a partial differential equation of Lagrangian type. We shall not discuss here the method of solution; the solution with the initial condition $X(0) = i$, is given by

$$P(s,t) = \left[\frac{\mu(1-s) - (\mu - \lambda s)e^{-(\lambda-\mu)t}}{\lambda(1-s) - (\mu - \lambda s)e^{-(\lambda-\mu)t}} \right]^i$$

$$= \left[\frac{\mu\{1 - e^{-(\lambda-\mu)t}\} - \{\mu - \lambda e^{-(\lambda-\mu)t}\}s}{\{\lambda - \mu e^{-(\lambda-\mu)t}\} - \lambda\{1 - e^{-(\lambda-\mu)t}\}s} \right]^i. \qquad (4.11)$$

Explicit expression for $p_n(t)$ can be obtained from the above by expanding $P(s, t)$ as a power series in s.

(b) *Mean population size*: We can obtain the mean population size by differentiating $P(s, t)$ partially with respect to s and putting $s = 1$. It can however be obtained directly from (4.8) and (4.9) without obtaining $P(s, t)$ as follows:

Let
$$E\{X(t)\} = M(t) = \sum_{n=1}^{\infty} n p_n(t)$$

and
$$E\{X^2(t)\} = M_2(t) = \sum_{n=1}^{\infty} n^2 p_n(t).$$

Multiplying both sides of (4.8) by n and adding over for $n = 1, 2, 3, \ldots$, we have

$$\sum_{n=1}^{\infty} n p'_n(t) = -(\lambda + \mu) \sum_{n=1}^{\infty} n^2 p_n(t) + \lambda \sum_{n=1}^{\infty} n(n-1) p_{n-1}(t)$$

$$+ \mu \sum_{n=1}^{\infty} n(n+1) p_{n+1}(t). \tag{4.12}$$

Now

$$\sum_{n=1}^{\infty} n(n-1) p_{n-1}(t) = \sum_{n=1}^{\infty} (n-1)^2 p_{n-1}(t) + \sum_{n=1}^{\infty} (n-1) p_{n-1}(t)$$

$$= M_2(t) + M(t);$$

$$\sum_{n=1}^{\infty} n(n+1) p_{n+1}(t) = \sum_{n=1}^{\infty} (n+1)^2 p_{n+1}(t) - \sum_{n=1}^{\infty} (n+1) p_{n+1}(t)$$

$$= \{M_2(t) - p_1(t)\} - \{M(t) - p_1(t)\}$$

$$= M_2(t) - M(t);$$

and
$$\sum_{n=1}^{\infty} n p'(t) = M'(t).$$

Hence from (4.12) we get

$$M'(t) = -(\lambda + \mu) M_2(t) + \lambda\{M_2(t) + M(t)\} + \mu\{M_2(t) - M(t)\}$$

$$= (\lambda - \mu) M(t).$$

The solution of the above differential equation (that $M(t)$ satisfies) is easily found to be

$$M(t) = C e^{(\lambda - \mu)t}$$

The initial condition gives $M(0) = \sum_{n=1}^{\infty} np_n(0) = i$, whence $C = M(0) = i$.

We have therefore,

$$M(t) = ie^{(\lambda-\mu)t}. \tag{4.13}$$

The second moment $M_2(t)$ of $X(t)$ can also be calculated in the same way.

Limiting case: As $t \to \infty$, the mean population size $M(t)$ tends to 0 for $\lambda < \mu$ (birth rate smaller than death rate) or to ∞ for $\lambda > \mu$ (birth rate greater than death rate) and to the constant value i when $\lambda = \mu$.

(c) *Extinction probability*: Since $\lambda_0 = 0$, 0 is an absorbing state, i.e. once the population size reaches 0, it remains at 0 thereafter. This is the interesting case of extinction of the population. We can determine the probability of extinction as follows:

Suppose, for simplicity, that $X(0) = 1$, i.e. the process starts with only one member at time 0. Then from (4.11) we can write $P(s, t)$ as

$$P(s,t) = \frac{a-bs}{c-ds} = \frac{a}{c} \cdot \frac{1-bs/a}{1-ds/c}$$

where

$$a = \mu\{1 - e^{-(\lambda-\mu)t}\}$$

and

$$c = \lambda - \mu e^{-(\lambda-\mu)t}$$

$\Pr\{X(t) = 0\} = p_0(t)$, the constant term in the expression of $P(s, t)$ as a power series in s, is given by

$$\frac{a}{c} = \frac{\mu\{1 - e^{-(\lambda-\mu)t}\}}{\lambda - \mu e^{-(\lambda-\mu)t}}. \tag{4.14}$$

The probability that the population will eventually die out is given by $\lim p_0(t)$ as $t \to \infty$ and can be obtained from the above by letting $t \to \infty$.

If $\lambda > \mu$, then

$$\lim_{t \to \infty} p_0(t) = \lim_{t \to \infty} \frac{\mu\{1 - e^{-(\lambda-\mu)t}\}}{\lambda - \mu e^{-(\lambda-\mu)t}}$$

$$= \frac{\mu}{\lambda} < 1.$$

If $\lambda < \mu$, then

$$\lim_{t \to \infty} p_0(t) = \lim_{t \to \infty} \frac{\mu\{1 - e^{-(\mu-\lambda)t}\}}{\lambda - \mu e^{-(\mu-\lambda)t}} = 1. \tag{4.15}$$

and
$$\lim_{t \to \infty} p_n(t) = 0 \text{ for } n \neq 0.$$

In other words, the probability of ultimate extinction is 1 when $\mu > \lambda$ (i.e. when the death rate is greater than the birth rate) and is $\mu/\lambda < 1$ when $\mu < \lambda$.

III. Linear Growth with Immigration

In II, we have $\lambda_0 = 0$ and, as a result, if the population size reaches zero at any time, it remains at zero thereafter. Here 0 is an absorbing state. If we consider $\lambda_n = n\lambda + \alpha$ ($\alpha > 0$), $\mu_n = n\mu$ ($n \geq 0$) we get what is known as a linear growth process with immigration, where 0 is not an absorbing state.

IV. Immigration-Death Process

If $\lambda_n = \lambda$ and $\mu_n = n\mu$, we get what is known as an immigration-death process. This corresponds to the Markovian queue with infinite number of channels, i.e. the queue $M/M/\infty$.

V. Pure Death Process

Here $\lambda_n = 0$ for all n, i.e. an individual cannot give birth to a new individual and the probability of death of an individual in $(t, t + h)$ is $\mu\, h + o\,(h)$. Then, if n individuals are present at time t, the probability of one death in $(t, t + h)$ is $n\mu h + o(h)$.

The birth and death process is a special type of continuous time Markov process with discrete state space $0, 1, 2, \ldots$ such that the probability of transition from state i to state j in (Δt) is $o(\Delta t)$ whenever $|i - j| \geq 2$. In other words, changes take place through transitions only from a state to its immediate neighbouring state.

In the next section we consider certain aspects of more general continuous time Markov processes with discrete space. Such processes are also known as *continuous time Markov chains*.

4.5   MARKOV PROCESSES WITH DISCRETE STATE SPACE (CONTINUOUS TIME MARKOV CHAINS)

4.5.1 Introduction

Let $X(t)$ be a continuous parameter Markov Process with state space $N = \{0, 1, 2, \ldots, i, \ldots, j, \ldots\}$.

Further let $X(t)$ be time-homogeneous then the (transition) probability of a transition from state i to state j during the time interval from epoch T to epoch $T + t$ does not depend on the initial time T but depends only on the elapsed time t and on the initial and terminal states i and j. We can thus write

$$\Pr\{X(T + t) = j \mid X(T) = i\} = p_{ij}(t), \tag{5.1}$$

$$i, j = 0, 1, 2, \ldots, \quad t \geq 0.$$

In particular,

$$\Pr\{X(t) = j \mid X(0) = i\} = p_{ij}(t).$$

We have

$$0 \le p_{ij}(t) \le 1 \text{ for each } i,j,t$$

and
$$\sum_j p_{ij}(t) = 1.$$

Let $p_j(t) = \Pr\{X(t) = j\}$ be the state probability at epoch t; then

$$p_j(t) = \Pr\{X(t) = j\}$$

$$= \sum_i \Pr\{X(t) = j \text{ and } X(0) = i\}$$

$$= \sum_i \Pr\{X(0) = i\} \Pr\{X(t) = j \mid X(0) = i\}$$

$$= \sum_i \Pr\{X(0) = i\} p_{ij}(t). \tag{5.2}$$

We have $\sum_j p_j(t) = 1$ for each $t \ge 0$.

Let us denote the transition probability matrix by

$$P(t) = (p_{ij}(t)). \tag{5.3}$$

Setting $p_{ij}(0) = \delta_{ij}$, we get

$$P(0) = I. \tag{5.4}$$

We shall assume here that the functions $p_{ij}(t)$ are continuous and differentiable for $t \ge 0$.

The waiting time for a change of state: Suppose that $X(t)$ is a homogeneous Markov process and that at time $t_0 = 0$ the state of the process $X(t_0) = X(0) = i$ is known. The time taken for a change of state from state i is a random variable, say τ, which is called the waiting time for a change of state from state i.

We have

$$\Pr\{\tau > s + t \mid X(0) = i\}$$
$$= \Pr\{\tau > s + t \mid X(0) = i, \tau > s\} \Pr\{\tau > s \mid X(0) = i\}$$
$$= \Pr\{\tau > s + t \mid X(s) = i\} \Pr\{\tau > s \mid X(0) = i\}.$$

If we denote $\overline{F}(t) = \Pr\{\tau > t \mid X(0) = i\}$, $t > 0$ then the above can be written as

$$\overline{F}(s+t) = \overline{F}(s)\overline{F}(t), \quad \text{for } s,t > 0.$$

which is satisfied *iff* $\overline{F}(t)$ is of the form

$$\overline{F}(t) = e^{-\lambda t}, \quad t > 0, \lambda > 0.$$

Thus the waiting time τ has exponential distribution with parameter λ, which is called the transition desnsity from state i. The distribution is the same for all i.

4.5.2 Chapman-Kolmogorov Equations

The transition probability $p_{ij}(t + T)$ is the probability that given that the state was i at epoch 0, it is state j at epoch $t + T$; but in passing from state i to state j in time $(t + T)$ the process passes through some state k in time t. Thus

$$p_{ij}(t+T) = \sum_k \Pr\{X(t+T) = j, X(t) = k \mid X(0) = i\}$$

$$= \sum_k \Pr\{X(0) = i, X(t) = k, X(t+T) = j\} \big/ \Pr\{X(0) = i\}$$

$$= \sum_k \frac{\Pr\{X(0) = i, X(t) = k\}}{\Pr\{X(0) = i\}} \times \frac{\Pr\{X(0) = i, X(t) = k, X(t+T) = j\}}{\Pr\{X(0) = i, X(t) = k\}}$$

$$= \sum_k \Pr\{X(t) = k \mid X(0) = i\} \Pr\{X(t+T) = j \mid X(0) = i, X(t) = k\}.$$

Since $\{X(t)\}$ is a Markov process,

$$\Pr\{X(t+T) = j \mid X(0) = i, X(t) = k\}$$

$$= \Pr\{X(t+T) = j \mid X(t) = k\}$$

$$= p_{kj}(T).$$

Thus we get the relation

$$p_{ij}(t+T) = \sum_k p_{ik}(t)p_{kj}(T), \tag{5.5}$$

which holds for all states i, j and $t \geq 0, T \geq 0$; it is called the Chapman-Kolmogorov equation. In matrix notation it can be written as

$$P(t+T) = P(t)P(T). \tag{5.6}$$

The equations (5.5) and (5.6) correspond to (2.4) and (2.5) (of Ch. 3) respectively which hold for discrete space Markov chain.

Denote the right-hand derivative at zero by

$$a_{ij} = \frac{d}{dt}p_{ij}(t)\big|_{t=0}; \quad i \neq j \tag{5.7}$$

$$a_{ii} = \frac{d}{dt}p_{ii}(t)\big|_{t=0}.$$

Then

$$a_{ij} = \lim_{\Delta t \to 0} \frac{p_{ij}(\Delta t) - p_{ij}(0)}{\Delta t} = \lim_{\Delta t \to 0} \frac{p_{ij}(\Delta t)}{\Delta t}$$

or
$$p_{ij}(\Delta t) = a_{ij}\Delta t + o(\Delta t), \, i \neq j \qquad (5.8)$$

and
$$a_{ii} = \lim_{\Delta t \to 0} \frac{p_{ii}(\Delta t) - p_{ii}(0)}{\Delta t} = \lim_{\Delta t \to 0} \frac{p_{ii}(\Delta t) - 1}{\Delta t}$$

or
$$p_{ii}(\Delta t) = 1 + a_{ii}\Delta t + o(\Delta t). \qquad (5.9)$$

It can be seen from the above relations that $a_{ij} \geq 0$, $i \neq j$ and $a_{ii} < 0$.

From
$$\sum_j p_{ij}(t) = 1, \text{ using (5.7) we get}$$

$$\sum_j a_{ij} = 0$$

or
$$\sum_{j \neq i} a_{ij} = -a_{ii}. \qquad (5.10)$$

The quantities a_{ij} are called *transition densities* and the matrix

$$A = (a_{ij})$$

is called the *transition density matrix* or *rate matrix* of the process. The matrix is such that

(1) its off-diagonal elements are non-negative and the diagonal elements are negative;
(2) the sum of the elements of each row is zero, the sum of the off-diagonal elements being equal in magnitude but opposite in sign to the diagonal elements.

Differentiating (5.5) with respects to T, we get

$$p'_{ij}(t+T) = \frac{\partial}{\partial T} p_{ij}(t+T) = \sum_k p_{ik}(t) \frac{d}{dT} p_{kj}(T).$$

Putting $T = 0$, we get

$$p'_{ij}(t) = \sum_k p_{ik}(t) a_{kj}. \qquad (5.11)$$

Or, in matrix notation

$$P'(t) = P(t)A. \qquad (5.11a)$$

Similarly we can get

$$\frac{d}{dT} p_{ij}(T) = \sum_k a_{ik} p_{kj}(T).$$

Replacing T by t, we can write this as

$$p_{ij}'(t) = \sum_k a_{ik} p_{kj}(t) \tag{5.12}$$

or $\qquad\qquad\qquad P'(t) = AP(t). \tag{5.12a}$

Equations (5.11) and (5.12) which give Chapman-Kolmogorov equations as differential equations are called respectively Forward and Backward Kolmogorov equations.

Solution of the Equations for a Finite State Process

When the rate matrix is given, the equations (5.11) or (5.11a) together with the initial conditions $p_{ij} = \delta_{ij}$ (or $P(0) = I$) yield as solution the unknown probabilities $p_{ij}(t)$. We consider below a method of solution for a process with finite number of states. From (5.11a) we see at once that the solutions can be written in the form

$$P(t) = P(0)e^{At} = e^{At} \tag{5.13}$$

where the matrix

$$e^{At} = I + \sum_{n=1}^{\infty} \frac{A^n t^n}{n!}. \tag{5.13a}$$

Assume that the eigenvalues of A are all distinct. Then from the spectral theorem of matrices (see Sec. 3.5 (Remarks)), we have

$$A = H D H^{-1}$$

where H is a non-singular matrix (formed with the right eigenvectors of A) and D is the diagonal matrix having for its diagonal elements the eigenvalues of A. Now, 0 is an eigenvalue of A and if $d_i \neq 0$, $i = 1, \ldots, m$ are the other distinct eigenvalues, then

$$D = \begin{pmatrix} 0 & 0 & \ldots & 0 \\ 0 & d_1 & \ldots & 0 \\ \ldots & & \ldots & \ldots \\ \ldots & & \ldots & \ldots \\ 0 & 0 & \ldots & d_m \end{pmatrix}.$$

We then have

$$D^n = \begin{pmatrix} 0 & 0 & \cdots & 0 \\ 0 & d_1^n & \cdots & 0 \\ \cdots & & \cdots & \cdots \\ \cdots & & \cdots & \cdots \\ 0 & 0 & \cdots & d_m^n \end{pmatrix}.$$

and

$$A^n = H D^n H^{-1}.$$

Substituting in (5.13), we get

$$P(t) = I + \sum_{n=1}^{\infty} \frac{(H D^n H^{-1})t^n}{n!}$$

$$= H \left\{ I + \sum_{n=1}^{\infty} \frac{D^n t^n}{n!} \right\} H^{-1}$$

$$= H e^{Dt} H^{-1} \qquad (5.14)$$

where

$$e^{Dt} = \begin{pmatrix} 1 & 0 & \cdots & 0 \\ 0 & e^{d_1 t} & \cdots & 0 \\ . & . & \cdots & . \\ . & . & \cdots & . \\ 0 & 0 & \cdots & e^{d_m t} \end{pmatrix}.$$

The right-hand side of (5.14) gives explicit solution of the matrix $P(t)$. Note that even in the general case when the eigenvalues of A are not necessarily distinct, a canonical representation of $A = SZS^{-1}$ exists (Sec. 3.5). Using this, $P(t)$ can be obtained in a modified form.

Example 5(a). Poisson process: If events occur in accordance with a Poisson process $N(t)$ with mean λt, then

$$\begin{aligned} p_{i,i+1}(\Delta t) &= \text{Pr \{the process goes to state } i+1 \text{ from state } i \text{ in time} \\ &\quad \Delta t\} \\ &= \text{Pr \{one event occurs in time } \Delta t\} \\ &= \text{Pr } \{N(\Delta t) = 1\} \\ &= \lambda \Delta t + o(\Delta t), \\ p_{i,i}(\Delta t) &= 1 - \lambda \Delta t + o(\Delta t) \\ p_{i,j}(\Delta t) &= o(\Delta t), j \neq i, i+1. \end{aligned}$$

and

By comparing with (5.8) and (5.9), we have

$$a_{i,i+1} = \lambda, \, a_{i,i} = -\lambda, \, a_{i,j} = 0 \text{ for } j \neq i, i+1.$$

The rate matrix is $A = (a_{ij}) = \begin{pmatrix} -\lambda & \lambda & 0_* & \dots & 0 \\ 0 & -\lambda & \lambda & \dots & 0 \\ \dots & & & \dots & \end{pmatrix}$

The Kolmogorov forward equations are

$$p'_{i,i}(t) = -\lambda p_{i,i}(t)$$

$$(5.15)$$

$$p'_{i,j}(t) = -\lambda p_{i,j}(t) + \lambda p_{i,i-1}(t), j = i+1, i+2, \dots.$$

Let $p_j(t) = \Pr\{N(t) = j\}$ and $p_0(0) = 1, p_n(0) = 0, n \neq 0$. Using (5.2) we get $p_j(t) \equiv p_{0j}(t), j = 0, 1, 2 \dots$ Thus (5.15) become identical with (1.11) and (1.12) so that $p_j(t) = e^{-\lambda t}(\lambda t)^j/j!$ Similarly, with $p_{ij}(0) = 1, j = i, p_{ij}(0) = 0, i \neq j$, we can get

$$p_{ij}(t) = \frac{e^{-\lambda t}(\lambda t)^{j-i}}{(j-i)!}.$$

Example 5(b). *Two-state process*: Suppose that a certain system can be considered to be in two states: "Operating" and "Under repair" (denoted by 1 and 0 respectively). Suppose that the lengths of operating period and the period under repair are independent random variables having negative exponential distributions with means $1/b$ and $1/a$ respectively ($a, b > 0$). The evolution of the system can be described by a Markov process with two states 0 and 1.

Now

$$
\begin{aligned}
p_{01}(\Delta t) &= \Pr\{\text{change of state from 0 to 1 in time } \Delta t\} \\
&= \Pr\{\text{repair being completed in time } \Delta t \\
&= a\Delta t + o(\Delta t)
\end{aligned}
$$

and

$$
\begin{aligned}
p_{10}(\Delta t) &= \Pr\{\text{change of state from 1 to 0 in time } \Delta t\} \\
&= b\,\Delta t + o(\Delta t).
\end{aligned}
$$

Thus the transition densities are

$$
\begin{array}{ll}
a_{01} = a, & a_{10} = b \\
a_{00} = -a, & a_{11} = -b
\end{array}
$$

and

so that

$$A = \begin{pmatrix} -a & a \\ b & -b \end{pmatrix}.$$

The Kolmogorov forward equations, for $i = 0, 1$, are

$$p'_{i0}(t) = -a\,p_{i0}(t) + b p_{i1}(t)$$

$$p'_{i1}(t) = a\,p_{i0}(t) - b p_{i1}(t).$$

Now we proceed to find the transition probabilities $p_{ij}(t)$.
Using

$$p_{00}(t) + p_{01}(t) = 1, p_{10}(t) + p_{11}(t) = 1,$$

we get

$$p'_{00}(t) - (a + b) p_{00}(t) = b$$

and

$$p'_{11}(t) + (a + b) p_{11}(t) = a.$$

The solution of the first of these differential equation is

$$p_{00}(t) = \frac{b}{a+b} + Ce^{-(a+b)t}.$$

With

$$p_{00}(0) = 1, \text{ we get } C = \frac{a}{a+b}, \text{ so that}$$

$$p_{00}(t) = \frac{b}{a+b} + \frac{a}{a+b} e^{-(a+b)t}.$$

Hence

$$p_{01}(t) = 1 - p_{00}(t) = \frac{a}{a+b} - \frac{a}{a+b} e^{-(a+b)t}.$$

Similarly, the solution of the second differential equation with the initial condition $p_{11}(0) = 1$, gives

$$p_{11}(t) = \frac{a}{a+b} + \frac{b}{a+b} e^{-(a+b)t}.$$

and hence

$$p_{10}(t) = 1 - p_{11}(t)$$

$$= \frac{b}{a+b} - \frac{b}{a+b} e^{-(a+b)t}.$$

Let $p_j(t)$ be the probability that the system is in state j at time t, $j = 0, 1$, and let $p_0(0) = 1, p_n(0) = 0, n \neq 0$. Then $p_j(t) \equiv p_{0j}(t), j = 0, 1$.

Alternative Method: We consider the above example to show how to proceed with matrix method of solution: this method is useful when Kolomogorov differential equations are not easily solvable.

Here $A = \begin{pmatrix} -a & a \\ b & -b \end{pmatrix}$ has eigenvalues 0 and $-(a + b)$, corresponding right

eigenvectors being $(1, 1)'$ and $(a, -b)'$ respectively. The Kolmogorov forward equation

$$P'(t) = (p'_{ij}(t)) = P(t)A$$

has as solution (as given in (5.14))

$$P(t) = He^{D(t)}H^{-1} \tag{5.16}$$

where

$$H = \begin{pmatrix} 1 & a \\ 1 & -b \end{pmatrix}$$

and

$$H^{-1} = \frac{1}{a+b}\begin{pmatrix} b & a \\ 1 & -1 \end{pmatrix}.$$

The diagonal matrix D whose elements are the eigenvalues of A is

$$D = \begin{pmatrix} 0 & 0 \\ 0 & -(a+b) \end{pmatrix}$$

so that

$$e^{Dt} = \begin{pmatrix} 1 & 0 \\ 0 & e^{-(a+b)t} \end{pmatrix}.$$

Thus from (5.16)

$$P(t) = \begin{pmatrix} 1 & a \\ 1 & -b \end{pmatrix}\begin{pmatrix} 1 & 0 \\ 0 & e^{-(a+b)t} \end{pmatrix}\begin{pmatrix} b & a \\ 1 & -1 \end{pmatrix} \Big/ [(a+b)]$$

$$= \frac{1}{a+b}\begin{pmatrix} b+ae^{-(a+b)t} & a-ae^{-(a+b)t} \\ b-be^{-(a+b)t} & a+be^{-(a+b)t} \end{pmatrix}.$$

We have

$$p_{00}(t) = \frac{b}{a+b} + \frac{a}{a+b}e^{-(a+b)t}, \quad p_{01}(t) = 1 - p_{00}(t)$$

$$p_{11}(t) = \frac{a}{a+b} + \frac{b}{a+b}e^{-(a+b)t}, \quad p_{10}(t) = 1 - p_{11}(t).$$

4.5.3 Limiting Distribution (Ergodicity of Homogeneous Markov Process)

We recall the result on limiting distribution of certain types of Markov chains as given in Theorems 3.7 & 3.11. We recall:

$$V(P - I) = 0, \ V = (v_0, v_1...), \ V\mathbf{e} = 1 \tag{5.17}$$

A similar elegant result holds for continuous parameter Markov processes as well. We shall state the result without proof. Here similar definitions for the classification of the states (which constitute a discrete state-space) will be used.

Theorem 4.8. Suppose that the time-homogeneous Markov process $\{X(t)\}$ is irreducible having aperiodic non-null persistent states; also that its t.p.m. is $P(t) = (p_{ij}(t))$, $i, j = 0, 1, 2, \ldots$ and the matrix of transition densities (or rate matrix) is

$$A = (a_{ij}),$$

where

$$a_{ij} = p'_{ij}(t)|_{t=0}.$$

Then given any state j,

$$\lim_{t \to \infty} p_{ij}(t) = v_j \tag{5.18}$$

exists and is the same for all initial states $i = 0, 1, 2, \ldots$. The asymptotic values v_j represent a probability distribution, i.e.

$$0 \le v_j \le 1, \sum_j v_j = 1.$$

The values v_j can then be determined as solutions of the system of linear equations

$$\sum_j v_i a_{ij} = 0, j = 0, 1, 2\ldots \tag{5.19}$$

or in matrix notation,

$$V A = \mathbf{0}, V = (v_0, v_1, ..) \tag{5.19a}$$

by using the normalising condition

$$V e = 1, \text{ that is, } \sum_j v_j = 1.$$

Note 1: The eq. (5.19) can be obtained from the forward Kolmogorov eq. (5.11) by putting

$$\lim_{t \to \infty} p_{ij}(t) = v_j, \text{ and } \lim_{t \to \infty} p'_{ij}(t) = 0.$$

Note 2: The eqs. (5.17) and (5.19a) for discrete and continuous parameter processes respectively are similar in structure. The matrices $(P - I)$ and A both have non-negative off-diagonal elements, strictly negative diagonal elements and zero row sums. If the number of states are finite, say, m, then both $(P - I)$ and A are of rank $(m - 1)$. Then V can be easily determined from any of the $(m - 1)$ equations (out of m equations contained in the relations (5.17) or (5.19a) and the normalising condition $\sum_j v_j = 1$.

Example 5(c). Consider the two-state process given Example 5(b). Here

$$v_0 = \lim_{t \to \infty} p_{00}(t) = \lim_{t \to \infty} p_{10}(t) = b/(a+b)$$

and

$$v_1 = \lim_{t \to \infty} p_{01}(t) = \lim_{t \to \infty} p_{11}(t) = a/(a+b).$$

These limiting probabilities can also be obtained from the equations (5.19) which become

$$-a\, v_0 + b\, v_1 = 0$$

$$a\, v_0 - b\, v_1 = 0.$$

Here there are two states and so the matrix A is of rank 1 and the two equations are equivalent. Each of them yields $v_1 = (a/b)\, v_0$. Using the normalising condition $v_0 + v_1 = 1$, we get

$$v_0 = b/(a+b),\ v_1 = a/(a+b).$$

When the number of states is finite and when only the limiting probabilities v_j are needed it is easier and more convenient to determine them from (5.19) or (5.19a).

Example 5(d). Suppose that in a queueing system which has m service channels, the demand for service arises in accordance with a Poisson process with parameter a. Suppose that the servicing time in each channel is exponential with parameter b. Further suppose that there is no storage facility, in the sense that a demand which arrives at the moment when any channel is free is received and is processed, whereas a demand which is received when all the m channels are busy is rejected and leaves the system. This is called a *loss system*.

Suppose that state k implies that k channels are busy. Then we get a Markov process $\{X(t), t \geq 0\}$, where $X(t)$ denotes the number of busy channels at time t; it has $(m+1)$ states $0, 1, \ldots, m$. The transition probabilities are given by

$$p_{i,i+1}(\Delta t) = \Pr\{\text{one demand is received for processing in time } \Delta t\}$$
$$= a\,\Delta t + o(\Delta t),\ 0 \leq i < m$$
$$p_{i,j}(\Delta t) = o(\Delta t),\ j > i+1$$

$$\Pr\{\text{one service demand is met in time } \Delta t\}$$
$$= b\,\Delta t + o(\Delta t)$$

and so if i channels are working

$$\Pr\{\text{one service demand is met in time } \Delta t\}$$
$$= i\,b\,\Delta t + o(\Delta t),$$

i.e.

$$p_{i,i-1}(\Delta t) = i\,b\,\Delta t + o(\Delta t)$$

and

$$p_{i,j}(\Delta t) = o(\Delta t),\ j < i-1.$$

Thus

$$a_{01} = a, a_{00} = -a$$

$$\begin{array}{rl} a_{ij} &= ib, \qquad\qquad j = i-1 \\ &= a, \qquad\qquad j = i+1 \\ &= -(a+ib), \qquad j = i \end{array} \Bigg\} \; 1 \le i < m$$

$$a_{m,m-1} = mb, \; a_{m,m} = -mb$$

$$A = \begin{pmatrix} -a & a & 0 & 0 & \cdots & 0 & 0 \\ b & -(a+b) & a & 0 & \cdots & 0 & 0 \\ 0 & 2b & -(a+2b) & a & \cdots & 0 & 0 \\ \cdots & \cdots & \cdots & \cdots & \cdots & \cdots & \cdots \\ 0 & 0 & 0 & 0 & \cdots & mb & -mb \end{pmatrix}.$$

The equations (5.19) become

$$-a\,v_0 + b\,v_1 = 0$$

$$a v_{j-1} - (a+jb)v_j + (j+1)b v_{j+1} = 0, \quad j = 1,2,...,m-1$$

$$a v_{m-1} - bm v_m = 0.$$

The solution of these equations can be obtained recursively. From the first equation

$$v_1 = (a/b)\,v_0.$$

Writing the second equation with $j = 1$ and putting there this value of v_1, we get

$$v_2 = \frac{1}{2}(a/b)^2 v_0.$$

Proceeding in this way, we get

$$v_j = (1/j!)(a/b)^j v_0, \quad j = 0,1,2,...,m.$$

Using $\sum_{j=0}^{m} v_j = 1$, one gets $v_0 = 1/(\sum_{i=0}^{m} (1/i!)(a/b)^i)$.

Formulas giving v_j are called *Erlang's formulas* (see Sec. 10.4.4).

Example 5(e). *Machine Interference Problem.* Consider that there are m identical machines. Each of the machines operates independently and is serviced by a single servicing unit in case of break down. The operating time and servicing time of each machine are independently distributed as exponential distribution with parameters b and a respectively. Then the number of machines in operating condition at time t constitutes a Markov process $\{X(t), t \ge 0\}$ with state space $\{0, 1, . . ., m\}$.

(Incidentally, the rate matrix here is exactly the same as that of the preceding example; the limiting probabilities v_j are given by the Erlang's formula.)

Example 5(f). Two-channel service system. Consider Example 5(d) with $m = 2$. Suppose that the service channels are numbered I and II and suppose that we are interested in whether particular channels are busy (B) or free (F). Let the ordered pair (i, j) denote the state of the system, where i refers to that of the first channel and j to that of the second. The four states of the system (F, F), (F, B), (B, F) and (B, B) may be denoted by 0, 1, 2, 3 respectively. The process $\{X(t)\}$ denoting that the states of the system in terms of the two channels may be described by a Markov process with state space 0, 1, 2, 3.

Assume that when both the channels are free a demand may join either of the channels for service with equal probability $(\frac{1}{2})$. Thus, when both the channels are free, demands to each of the channels flow in accordance with a Poisson process with parameter $a/2$. We have

$$p_{0,j}(\Delta t) = (a/2)\Delta t + o(\Delta t), j = 1, 2$$

$$p_{1,j}(\Delta t) = b\Delta t + o(\Delta t), \qquad j = 0$$

$$= a\Delta t + o(\Delta t), \qquad j = 3$$

$$p_{2,j}(\Delta t) = b\Delta t + o(\Delta t), \qquad j = 0$$

$$= a\Delta t + o(\Delta t), \qquad j = 3$$

$$p_{3,j}(\Delta t) = b\Delta t + o(\Delta t), \qquad j = 1, 2$$

and for all other combinations of $i \neq j$, $p_{i,j}(\Delta t) = o(\Delta t)$.
Thus the rate matrix A will be

$$\begin{pmatrix} -a & a/2 & a/2 & 0 \\ b & -(a+b) & 0 & a \\ b & 0 & -(a+b) & a \\ 0 & b & b & -2b \end{pmatrix}.$$

The normal equations (5.19) become

$$\begin{aligned}
-av_0 &+ bv_1 && + bv_2 && &&= 0 \\
(a/2)v_0 &- (a+b)v_1 && && + bv_3 &&= 0 \\
(a/2)v_0 && && - (a+b)v_2 && + bv_3 &&= 0 \\
& av_1 && + av_2 && - 2bv_3 &&= 0.
\end{aligned}$$

From the second and third equations we get $v_1 = v_2$ and then from the first we get $v_0 = (2b/a)v_1$ and from the last $v_3 = (a/b)v_1$. Utilising $\sum_{i=0}^{3} v_i = 1$, we at once get

$$v_1 = v_2 = \frac{ab}{a^2 + 2ab + 2b^2}$$

$$v_0 = \frac{2b^2}{a^2 + 2ab + 2b^2}$$

$$v_3 = \frac{a^2}{a^2 + 2ab + 2b^2}.$$

4.5.4 Graph Theoretic Approach for Determining V

The graph theoretic approach for determining V in case of discrete parameter Markov chain (see Sec. 3.7) is also applicable to finite, irreducible Markov processes with continuous parameter. In case of discrete Markov chains, arc "weights" correspond to the transition probabilities p_{ij}; in case of continuous Markov processes, the arc weights correspond to the transition densities $a_{ij} = (d/dt)\, p_{ij}\,(t)|_{t=0}$. The same result (Theorem 3.8) leading to steady-state solutions can be applied to finite-state discrete parameter Markov chains as well as to finite-state continuous parameter Markov processes, provided these are irreducible (Solberg). This is a definite advantage of this graph theoretic approach.

The procedure for continuous parameter process is illustrated below.

Example 5(g). *Erlang's formula*: Consider the machine interference (or m-channel loss system) problem discussed above.
Here for $i = 0, 1, 2, \ldots, m - 1, j \neq i$

$$
\begin{aligned}
a_{i,j} &= a, & j &= i + 1 \\
\text{and for } i = 1, 2, \ldots, m, & & j &\neq 1 \\
a_{i,j} &= ib, & j &= i - 1 \\
\text{and} \quad a_{i,i} &= -a, & i &= 0 \\
&= -mb, & i &= m \\
&= -(a + ib), & i &\neq 0, m.
\end{aligned}
$$

Also $\quad a_{ij} = 0, j \neq i - 1, i \text{ or } i + 1$

The transition diagram may be shown as indicated below:

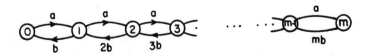

Fig. 4.3 Transition diagram

Denote c_r = sum of weights of all the intrees to the point r, $r = 0, 1, 2, ..., m$.

Here there is only one intree to each point. We have

$$c_0 = (mb)((m-1)b) \quad ... \quad (2b)(b) \quad = m!\, b^m$$

$$c_1 = (mb)((m-1)b) \quad ... \quad (2b)\, a \quad = m!\, ab^{m-1}/1!$$

$$...\qquad\qquad\qquad ...\qquad\qquad\qquad ...$$

$$c_r = (mb)((m-1)b) \quad ... \quad ((r+1)b)\, a^r = m!\, a^r b^{m-r}/r!,$$

so that $\displaystyle\sum_{r=0}^{m} c_r = m! \sum_{r=0}^{m} (a^r b^{m-r})/r!$. Thus

$$v_k = \frac{\frac{1}{k!}\left(\frac{a}{b}\right)^k}{\displaystyle\sum_{r=0}^{m}\frac{1}{r!}\left(\frac{a}{b}\right)^r}, \quad k = 0, 1, 2, ..., m.$$

4.6 RANDOMIZATION (UNIFORMIZATION) : DERIVED MARKOV CHAINS

Let $\{Y_n, n \geq 0\}$ be a countable Markov chain having state space $S = \{0, 1, 2 ...\}$ and t.p.m. P. Let $\{N(t), t \geq 0\}$ be a Poisson process with rate λ. Assume that $\{Y_n\}$ and $N(t)$ are independent.
Define a stochastic process

$$\{X(t) = Y_{N(t)}, t \geq 0\}.$$

It is a continuous time Markov chain having state space S. Changes of state of the process $\{X(t), t \geq 0\}$ take place at the epochs of occurrence of the events of the Poisson process $\{N(t)\}$. Thus $N(t)$ denotes the number of transitions by time t. The mean time spent in a state is the same as the mean inter-occurrence time between two Poisson events, i.e. mean time spent in a state is λ and is the same for all the states. Here rendomization of operational time is done through Poisson events: the process is therefore called *randomization* (*uniformization* by Ross).

The transition probabilities $p_{ij}(t)$ of $\{X(t), t \geq 0\}$ can be computed by conditioning on $N(t)$. Thus

$$p_{ij}(t) = \Pr\{X(t) = j \mid X(0) = i\}$$

$$= \sum_{n=0}^{\infty} \Pr\{X(t) = j \mid X(0) = i, N(t) = n\} \times$$

$$\Pr\{N(t) = n \mid X(0) = i\} \qquad (6.1)$$

Now $\Pr\{N(t) = n \mid X(0) = i\} = \Pr\{N(t) = n\} = e^{-\lambda t}\dfrac{(\lambda t)^n}{n!}$

and $\Pr\{X(t) = j \mid X(0) = i, N(t) = n\}$

gives the probability that the system goes from state i and state j in n transitions; so that

$$\Pr\{X(t) = j \mid X(0) = i, N(t) = n\} = p_{ij}^{(n)}$$

in terms of n-step transition probability of the chain $\{Y_n, n \geq 0\}$. Thus the matrix of transition function of $\{X(t), t \geq 0\}$ is

$$P(t) = (p_{ij}(t))$$

where

$$p_{ij}(t) = e^{-\lambda t} \sum_{n=0}^{\infty} p_{ij}^{(n)} \frac{(\lambda t)^n}{n!}. \tag{6.2}$$

Let us find the generator of the Q-matrix of $\{X(t), t \geq 0\}$. Suppose that h denotes the length of an infinitesimal interval. Then

$$\Pr\{N(h) = 1\} = \lambda h + o(h)$$

$$\Pr\{N(h) = 0\} = 1 - \lambda h + o(h)$$

$$\Pr\{N(h) \geq i\} = o(h), i \geq 2$$

From (6.2), we get, for $i \neq j$

$$p_{ij}(h) = \sum_{n=0}^{1} p_{ij}^{(n)} e^{-\lambda h} \frac{(\lambda h)^n}{n!} + o(h)$$

so that

$$q_{ij} = \lim_{h \to 0} \frac{p_{ij}(h)}{h} = \lambda p_{ij} \text{ exists and is finite; then}$$

$$q_i = -\sum_{j \neq i} q_{ij} = -\lambda(1 - p_{ii}) = \lambda(p_{ii} - 1).$$

We thus have

$$Q = \lambda(P - I).$$

Thus $\{X(t), t \geq 0\}$ is a Markov process (continuous time Markov chain) with the same state process S and the same initial distribution as $\{Y_n, n \geq 0\}$ and having as its infinitesimal generator $Q = \lambda (P - I)$.

The *converse* also holds as can be seen now. Let $\{X(t), t \geq 0\}$ be a Markov process with countable state space S with generator Q. Assume that $X(t)$ is uniformizable, that is, the diagonal elements of Q are uniformly bounded.

Let $\alpha = \sup_i q_i$ (q_i being the ith diagonal element of Q).

Then there exists a Markov chain $\{Y_n, n \geq 0\}$ with state space S and t.p.m. $P = Q/\lambda + I$, such that the relation (6.2) holds.

The relation (6.2) can be written in the matrix form as follows:

$$P(t) = (p_{ij}(t)) = \sum_{n=0}^{\infty} P^n e^{-\lambda t} \frac{(\lambda t)^n}{n!} \tag{6.3}$$

The R.H.S.
$$= e^{-\lambda t} \sum_{n=0}^{\infty} \left(\frac{Q}{\lambda} + I \right)^n \frac{(\lambda t)^n}{n!}$$

$$= e^{-\lambda t} \sum_{n=0}^{\infty} \frac{(\lambda t)^n}{n!} \sum_{k=0}^{n} \binom{n}{k} \left(\frac{Q}{\lambda} \right)^k \tag{6.4}$$

$$= \sum_{k=0}^{\infty} \frac{t^k}{k!} Q^k$$

so that $P(t)$ is independent of λ. Thus λ may be any real number not less than $\alpha = \sup_i q_i$.

Denote

$$\Pi_j(t) = \Pr \{X(t) = j\}$$

$$\Pi(t) = \{\Pi_1(t), \Pi_2(t), ...\}$$

then

$$\Pi(t) = \Pi(0) P(t)$$

$$= \Pi(0) e^{-\lambda t} \sum_{n=0}^{\infty} \frac{\lambda^n t^n}{n!} P^n. \tag{6.5}$$

Now let us see how Poisson process comes in. For a uniformizable Markov process $\{X(t), t \geq 0\}$, there exists a Poisson process $\{N(t), t \geq 0\}$ with parameter λ such that Y_n and $N(t)$ are independent of each other and that

$$\{X(t), t \geq 0\} \text{ and } \{Y_{N(t)}, t \geq 0\}$$

are probabilistically identical, that is

$$X(t) = Y_{N(t)}$$

This is a very interesting result.

The method of construction by using rendomization techniques is useful in some general areas of applied probability. Its usefulness lies in developing stochastic inequalities between processes and monotonicity between processes (refer to Keilson (1979)). It is also useful for establishing equivalence relations between discrete time and continuous time Markov decision processes (see, for example, Howard (1960) and Lippman (1975)).

We shall discuss later how the technique has been employed for numerical computation.

Equivalence of two limiting forms: Let $\{Y_n, n \geq 0\}$ be an irreducible, aperiodic Markov chain having countable state space S and t.p.m. P.

Let $\{X(t), t \geq 0\}$ be the Markov process constructed from the given Markov chain by randomization of operational time through a Poisson process with rate λ

$$Q = \lambda(P - I)$$

where Q is the generator of $\{X(t), t \geq 0\}$.

Now for $\{Y_n, n \geq 0\}$

$$\lim_{n \to \infty} p_{ij}^{(n)} = v_j.$$

exists and is independent of i.

The probability vector V is given by the unique solution of $VP = P$. Again for $\{X(t), t \geq 0\}$, the limiting probabilities

$$\lim_{t \to \infty} p_{ij}(t) = u_j, \quad i, j \in S$$

exist and are independent of i. From Theorem 4.8 we find that the probability vector $U = \{\cdots, u_j, \cdots\}$, $U\mathbf{e} = 1$ is given by the unique solution of

$$UQ = \mathbf{0}$$

or
$$U[\lambda(P - I)] = \mathbf{0}$$

or
$$UP = U$$

This implies that $U = V$ elementwise. In other words,

$$\lim_{t \to \infty} p_{ij}(t) = \lim_{n \to \infty} p_{ij}^{(n)}, \quad i, j, \in S. \tag{6.6}$$

Numerical Method. Numerical method is in itself a subject. Of importance, in applications are the time dependent or transient state probabilities. Computations of transient probabilities are useful for many stochastic systems, such as queueing system, inventory system, reliability system, maintenance system etc. Transient behaviour is specially important when convergence to steady state is slow and as such equilibrium behaviour does not properly indicate system behaviour.

We indicate here briefly a method based on randomization, technique for computation of $p_{ij}(t)$ or $\Pi_j(t)$, that is transient probabilities of a uniformizable Markov process.

First, $P = Q/\lambda + I$ is chosen where $\lambda \geq \sup_j q_j$. A truncation point N is chosen such

that $e^{-\lambda t} \sum\limits_{n=N+1}^{\infty} \frac{(\lambda t)^n}{n!} P^n$ is less than desired control error ϵ. That is, N is chosen such that

$$\left| p_{ij}(t) - \sum_{n=0}^{N} e^{-\lambda t} \frac{(\lambda t)^n}{n!} P^n \right| \leq \epsilon \qquad (6.7)$$

The same procedure is adopted for calculation of $\Pi_j(t)$. Efficient algorithms for computation have been put forward by Grassman (1977) and Gross and Miller (1984) who develop a SERT algorithm and show its application in a reliability problem.

4.7 ERLANG PROCESS

4.7.1 Introduction

Consider a series of events E' which occur in accordance with a Poisson process $N(t)$ with mean $\lambda' t$. The random variable X_i which represents the interval between two successive occurrences has density $h(x) = \lambda' e^{-\lambda' x}$ having mean $1/\lambda'$. The interval separating k occurrences, i.e. the interval between the events numbered r and $r + k$ (for $k \geq 1$) is given by the sum $S_k = X_1 + \ldots + X_k$ of k independent exponential distributions each with mean $1/\lambda'$. S_k is a gamma variate (Theorem 1.10) with density function

$$f(t) = \frac{(\lambda')^k t^{k-1} e^{-\lambda' t}}{\Gamma(k)}, t > 0$$

$$= 0, t \leq 0. \qquad (7.1)$$

We have $E\{S_k\} = k/\lambda'$ and var $\{S_k\} = k/(\lambda')^2$.

Suppose that every kth occurrence of E' is represented by a new series of events E, i.e. E occurs for the first time at the kth occurrence of E', E occurs for the second time at the $(2k)$th occurrence of E' and so on. If $X(t)$ is the number of occurrences of E in an interval of duration t, then we have

$$\Pr\{X(t) = n\} = \Pr\{nk \leq N(t) \leq (n+1)k - 1\}, n \geq 0$$

$$= \sum_{r=nk}^{(n+1)k-1} \Pr\{N(t)=r\}$$

$$= \sum_{r=nk}^{(n+1)k-1} \frac{e^{-\lambda't}(\lambda't)^r}{r!} , \qquad (7.2)$$

since $N(t)$ is a Poisson process.

Evidently, for positive integral k,

$$\sum_{n=0}^{\infty} \Pr\{X(t)=n\} = \sum_{n=0}^{\infty} \sum_{r=nk}^{(n+1)k-1} \frac{e^{-\lambda't}(\lambda't)^r}{r!}$$

$$= \sum_{s=0}^{\infty} \frac{e^{-\lambda't}(\lambda't)^s}{s!} = 1.$$

$\{X(t), t \geq 0\}$ is a renewal process with probability distribution given by (7.2). The interval between two successive occurrences of $X(t)$ which is the same as the interval between k occurrences of $N(t)$ is a gamma variate with density given by (7.1). For $k = 1$, $X(t)$ is identical with $N(t)$ and is a Poisson process.

(For the p.g.f. of $X(t)$, see Parzen, 1962, p. 175.)

4.7.2 Erlangian Distribution

Putting $\lambda' = k\lambda$, we find that the interval separating k events of E' (which is the sum of k exponential variables each with mean $1/k\lambda$) has the density

$$f(t) = \frac{(\lambda k)^k t^{k-1} e^{-k\lambda t}}{\Gamma(k)}, t > 0 \qquad (7.3)$$

$$= 0, t \leq 0$$

(obtained by putting $\lambda' = \lambda k$ in (7.1).)

The distribution having the above as its density function is called k-Erlangian or *Erlang-k* distribution (denoted by E_k), (see Sec. 1.3.4.6, Ch. 1).

If the occurrences of F are such that the intervals between two successive occurrences of F are identically and independently distributed with a common distribution E_k, then the number of occurrences $X(t)$, $(t \geq 0)$ of F in an interval of length t constitutes what is known as an *Erlang-k-process*. $X(t)$ is a renewal process (Sec. 4.3.7). We have

$$\Pr\{X(t)=n\} = \sum_{r=nk}^{(n+1)k-1} \frac{e^{-k\lambda t}(k\lambda t)^r}{r!}, n \geq 0. \qquad (7.4)$$

Note. It may be said that a *random* selection preserves a Poisson process (property 3 of Poisson process Sec. 4.1) whereas a *deterministic* (systematic) selection destroys it. A systematic selection of Poisson process results in an Erlang process.

Example 7(a). Suppose that a system fails at the cumulative effect of n shocks. If shocks occur in accordance with a Poisson process with parameter λ', then the lifetime T of the system is a random variable being the sum of n independent exponential variates each with mean $1/\lambda'$. Hence T is a gamma variate with its density given by (7.1).

Example 7(b). Suppose that customers arrive at the desk of a receptionist in accordance with a Poisson process with a mean rate 2 (per minute). The receptionist immediately directs them alternately to two counters A and B so that each counter gets every second customer.

The interval T between two successive arrivals at A (or at B) is the sum of two independent exponential variates each with mean $\frac{1}{2}$ (minute). The density function of T is obtained by putting $k = 2$ and $\lambda' = 2$ in (7.1). We have

$$f(t) = \frac{(2)^2 t e^{-2t}}{\Gamma(2)}, t > 0$$

$$= 0, \qquad t \le 0,$$

$$E(T) = 2/2 = 1 \text{ and var } (T) = 2/2^2 = 0.5$$

and

$$\Pr \{T > 1 \text{ minute}\} = \int_1^\infty f(t) \, dt = 3e^{-2} = 0.406.$$

The probability $\Pr \{X(1) = 0\}$ that there is no arrival at B during the first minute is obtained by putting $t = 1$, $\lambda' = 2$, $k = 2$ and $n = 0$ in (7.2), i.e.

$$\Pr \{X(1) = 0\} = \sum_{r=0}^{1} \frac{e^{-2}(2)^r}{r!} = 0.406.$$

In fact,

$$\Pr (T > 1) = \Pr \{X(1) = 0\}.$$

The probability $\Pr \{X(1) = 1\}$ of 1 arrival is obtained by putting $n = 1$, i.e.

$$\Pr \{X(1) = 1\} = \sum_{r=2}^{3} \frac{e^{-2}(2)^r}{r!} = 0.450.$$

Example 7(c). Suppose that particles reach a nuclear particle counter in accordance with a Poisson process with a mean rate of 4 (per minute), but that the counter records only every *third* particle actually arriving.

The probability that the number $X(t)$ of particles recorded in time t is n is given by (putting $k = 3$, $\lambda' = 4$ in (7.2)). We get

$$\Pr \{X(t) = n\} = \sum_{r=3n}^{3(n+1)-1} \frac{e^{-4t}(4t)^r}{r!}, n \ge 0.$$

If $t = 1, n = 1$, then

$$\Pr\{X(1) = 1\} = \sum_{r=3}^{5} \frac{e^{-4}4^r}{r!} = 0.547.$$

The interval T between two successive records (i.e. recorded particles) has the density function (putting $\lambda' = 4$, $k = 3$ in (7.1)),

$$f(t) = \frac{4^3 t^2 e^{-4t}}{\Gamma(3)}, t > 0$$

$$= 0, \qquad t \le 0,$$

with $E(T) = 3/4$ and var $(T) = 3/16$.

The probability that the interval T between two records is less than or equal to $\frac{1}{2}$ (minute) is given by

$$\Pr\left(T \le \frac{1}{2}\right) = 1 - \Pr\left(T > \frac{1}{2}\right)$$

$$= 1 - \Pr\left\{X\left(\frac{1}{2}\right) = 0\right\},$$

$$= 1 - \sum_{r=0}^{2} \frac{e^{-2}2^r}{r!} = 0.323.$$

EXERCISES

4.1 Divide the interval $[0, t]$ into a large number n of small intervals of length h and suppose that in each small interval, Bernoulli trials with probability of success λh are held (i.e. trials with only two outcomes, success with probability λh and failure with probability $(1 - \lambda h)$. Show that the number of successes in an interval of length t is a Poisson process with mean λt. State the assumptions you make.

4.2 If $N_1(t)$, $N_2(t)$ are two independent Poisson processes with parameters λ_1, λ_2 respectively, then show that

$$\Pr\{N_1(t) = k \mid N_1(t) + N_2(t) = n\} = \binom{n}{k} p^k q^{n-k},$$

where $p = \lambda_1/(\lambda_1 + \lambda_2), q = \lambda_2/(\lambda_1 + \lambda_2)$.

4.3 In a Poisson cluster process, suppose that each cluster X_i has the independent distribution of the geometric type:

$$p_k = \Pr\{X_i = k\} = q^{k-1}p, k = 1, 2, \dots.$$

Find the p.g.f., the mean and the variance of $M(t)$, (the total number of events). Show that the arrival stream of units is a renewal process with inter-arrival time distribution

$$A(t) = 1 - pe^{-\lambda t}(t \geq 0, 0 < p \leq 1).$$

4.4   The number of accidents in a town follows a Poisson process with a mean of 2 per day and the number X_i of people involved in the ith accident has the distribution (independent)

$$\Pr\{X_i = k\} = \frac{1}{2^k}(k \geq 1).$$

Find the mean and the variance of the number of people involved in accidents per week.

4.5   For the pure birth process show that the interval T_k between the kth and $(k + 1)$st birth has an exponential distribution with parameter λ_k.

4.6   *Birth-immigration process with multiple births*
Suppose that in an interval of infinitesimal length h, the probability of an individual giving birth or creating simultaneously m (≥ 1) new members of the same type is $\lambda h + o(h)$, so that with each birth a change of state from j to $j + m$ occurs; and that the probability of a new individual being added from outside is $vh + o(h)$. If $N(t)$ denotes the number of individuals at epoch t and

$$p_{i,n}(t) = \Pr\{N(t) = n \mid N(0) = i\}$$

then show that

$$p'_{i,n}(t) = -(n\lambda + v)p_{i,n}(t) + \lambda(n - m)p_{i,n-m}(t) + vp_{i,n-1}(t), n \geq 1.$$

Denote

$$F(s,t) = \sum_{n=0}^{\infty} p_{i,n}(t)s^n, |s| \leq 1;$$

show that F satisfies the differential equation

$$\frac{\partial F}{\partial t} = \lambda s(s^m - 1)\frac{\partial F}{\partial s} + v(s - 1)F.$$

Show that its solution is given by

$$F(s,t) = s^m(1 - b)^\lambda e^{-a+sa}(1 - bs^m)^{-\lambda}$$

where $\qquad \lambda = i/m (> 0), a = -\nu t, b = 1 - e^{-m\lambda}.$

When $i = 1$, $m = 1$, $\nu = 0$ we get Yule-Furry process.

$F(s, t)$ can also be expressed in terms of the generating function $e^{sa} (1 - bs^m)^{-\lambda}$ of generalized Charlier polynomials.

(See Jain & Gupta (1975) and Medhi & Borah (1986)).

4.7 *Zipf's rank size law*: Let Pr {a settlement S has n members at time t} $= p_n(t)$ and let Pr {a new member arrives in the settlement in $(t, t + h)$ given that there are n at time t} $= nah$, $a > 0$. Assume that the settlement starts with 1 individual at time 0, i.e. $p_1(0) = 1$ and that different types of individual have different rates of arrival a, so that a is a random variable with density $g(a) = be^{-ab}$, $a > 0$, $b > 0$. Show that the probability distribution of S is given by

$$\text{Pr} \{S = n \text{ at time } t\} = \int_0^\infty p_n(t) g(a) da = rB(n, r + 1), n = 1, 2, ...,$$

where B is a Beta function and $r = b/t$. Show that for large n,

$$\text{Pr} \{S \geq n \mid r\} \propto n^{-r}$$

(J.K. Ord (1975)).

4.8 Find the differential equation of pure death process. If the process starts with i individuals, find the mean and variance of the number $N(t)$ present at time t.

4.9(a) Show that for the linear growth process, the second moment $M_2(t)$ satisfies the differential equation:

$$M'_2(t) = 2(\lambda - \mu)M_2(t) + (\lambda + \mu)M(t).$$

Show that

$$\text{var} \{X(t)\} = \{i(\lambda + \mu)/(\lambda - \mu)\} e^{(\lambda - \mu)t} \{e^{(\lambda - \mu)t} - 1\}, \lambda \neq \mu$$

$$= 2i\mu t, \lambda = \mu,$$

i being the population size at $t = 0$.

(b) Find the probability of ultimate extinction in the case of the linear growth process starting with i individuals at time 0.

4.10 Write down the differential-difference equations for the linear growth process with immigration having

$$\lambda_n = n\lambda + a, \mu_n = n\mu.$$

Show that

$$M(t) = E\{X(t)\}$$

satisfies the differential equation

$$M'(t) = (\lambda - \mu) M(t) + a.$$

Hence show that with the initial condition $M(0) = i$, when $X(0) = i$,

$$M(t) = \frac{a}{\lambda - \mu}\{e^{(\lambda - \mu)t} - 1\} + ie^{(\lambda - \mu)t}, \lambda \neq \mu$$

and taking limit as $\lambda \to \mu$, show that

$$M(t) = at + i \text{ for } \lambda = \mu.$$

What is the limit of $M(t)$ as $t \to \infty$ for $\lambda < \mu$?

4.11   Continuation of 4.10: Linear growth process with immigration. Given the initial conditions $p_0(0) = 1$, $p_n(0) = 0$, $n \neq 0$, show that the population size, after the population has been developing for time T, has negative binomial distribution with index a/λ, and mean

$$\frac{a\{\exp(\lambda - \mu)T - 1\}}{\lambda - \mu}.$$

If $\lambda < \mu$ the population size will have a steady state distribution. When a is small compared to λ, show that, the population size, given that a population exists, follows a logarithmic series distribution.
[Williamson and Bretherton, *Ann. Math. Stat.* **35** (1964) 284–297.]

4.12   For the Yule-Furry process starting with only one individual at $t = 0$, show that the density function of T_n (the time at which the population *first* reaches the value n (> 1)) is given by

$$f(t) = \lambda(n-1)e^{-\lambda t}(1 - e^{-\lambda t})^{n-2}, t \geq 0.$$

Find $E(T_n)$. [Note that $T_n \leq t \Leftrightarrow N(t) \geq n$].

4.13   *General form of Theorem 4.5:* For a Poisson process with parameter λ it is given that n occurrences take place in time t; show that the density function of the time of occurrence Z_k upto the kth event ($k < n$) is given by

$$f(x) = \frac{n!}{(k-1)!(n-k)!} \frac{x^{k-1}!}{t^k}\left(1 - \frac{x}{t}\right)^{n-k}, 0 < x < t.$$

$$= 0, x \geq t.$$

Find the mean and variance of Z_k.
Show that (Z_k/t) has Beta distribution with parameters $(k-1)$ and $(n-k)$.

4.14 In Example 7(a), let $n = 10$, $\lambda' = 1$ per hour; find the probability that the lifetime T (i) exceeds 10 hours, (ii) is less than or equal to 16 hours.

4.15 If $\{N(t), t \geq 0\}$ is a Poisson process, show that the process

$$X(t) = N(t+L) - N(t)$$

(where L is a positive constant) is covariance stationary.

4.16 Consider a Yule-Furry process starting with a single member at time $t = 0$, and having birth rate λ. Suppose that this first member (parent) is also subject to death, his lifetime being distributed as an exponential variable with parameter μ. Find the distribution of the number N of offspring due to this parent as well as his descendants at the time of death of the parent.

4.17 For a count process N of E' in E, where E and E' are independent Poisson processes with parameters μ and λ respectively, show that the joint distribution of N and T, T being the interval between two successive occurrences of E, is given by

$$f(n,t) = \frac{\mu e^{-(\lambda+\mu)t}(\lambda t)^n}{n!}, \quad n = 0, 1, 2, \ldots, \quad t \geq 0.$$

Hence find the (marginal) distribution of N as

$$\Pr\{N = n\} = \int_0^\infty f(n,t)dt,$$

and the distribution of T as

$$f(t) = \sum_{n=0}^\infty f(n,t).$$

4.18 Suppose that $N_i(t)$, $i = 1, 2$ are two independent Poisson processes with parameters λ_i and $X(t) = N_1(t) + N_2(t)$ is their sum. Suppose that T_i and Z are the intervals *upto the first occurrences* of $N_i(t)$ and $X(t)$ respectively. Show that the random variable Z can be expressed in terms of T_i as $Z = min$ (T_1, T_2). What is the distribution of Z?

How do you generalise the above for $i = 1, 2, \ldots, n$? Does the resuld hold if T_i, Z are the intervals between two successive occurrences of $N_i(t)$ and $X(t)$ respectively?

4.19 Write down the rate matrix for the birth and death process (Sec. 4.4). Assume that the limits $\lim_{t \to \infty} p_{ij}(t) = v_j$ exist.

Obtain the difference equations involving v_j from the Forward Kolmogorov equations. Using the method of induction, show that the solutions of these equations can be written as

$$v_{j+1} = \alpha_{j+1} v_0$$

where $\qquad \alpha_0 = 1$ and $\alpha_j = \prod_{r=1}^{j} \dfrac{\lambda_{r-1}}{\mu_r}, \quad j = 1, 2, ...$

When does $\{v_j\}$ define a probability distribution?

4.20   *Two-sex birth and death process with sex-specific birth and death rates* (Keyfitz).

Let the probability of birth (to a female member of the population) of a male and a female individual be b_m and b_f respectively and let the probability of death of males and females be d_m and d_f respectively (in an interval of infinitesimal length h).

Let

$$p_{m,f}(t) = \text{Pr} \{ \text{of } m \text{ males and } f \text{ females at epoch } t\}$$

and $\qquad p_{m_0, f_0}(0) = 1.$

Then

$$\frac{d}{dt} p_{m,f}(t) = -[m d_m + f(d_f + b_m + b_f)] p_{m,f}(t)$$

$$+ b_m f p_{m-1,f}(t)$$

$$+ b_f(f-1) p_{m,f-1}(t)$$

$$+ d_m(m+1) p_{m+1,f}(t)$$

$$+ d_f(f+1) p_{m,f+1}(t).$$

Show that the expected numbers of females and males are given by

$$E(F) = f_0 \exp \{(b_f - d_f)t\}, \text{ and}$$

$$E(M) = m_0 \exp(-d_m t) + \frac{f_0 b_m}{d_m + b_f - d_f} \{\exp(b_f - d_f)t - \exp(-d_m t)\}.$$

4.21   *Three-state process.* Consider Example 5 (*d*) with $m = 2$. Find $p_{ij}(t)$. Obtain V from normal equation and also by using the graph theoretic technique.

4.22   Obtain V for Example 5 (*f*) by using the graph theoretic technique .

4.23   Suppose that two machines working independently are looked after by one repairman. Suppose further that the break-downs of each of the machines occur in accordance with a Poisson process with parameter a, while the repair times of the two machines have independent exponential distributions with parameter b_1, and b_2 respectively. Suppose that W (R) represents the working (under repair) state of a machine and the ordered pair (i, j) $(i, j = W, R)$ represents the state of the whole system so that the four states of the system are 0: (W, W), 1: (W, R), 2: (R, W) and 3: (R, R).

Determine the rate matrix. Evaluate V
   (i) from normal equations,
   (ii) by applying the graph theoretic technique.

4.24   *A three-state process: Application in Reliability.* The working of an automatic machine has been cited as an example of a three-state model by Proctor and Singh (1976). The states of an automatic machine on an assembly line could be good (state 0), (failed in mode 1 (state 1) and failed in mode 2 (state 2), (failure in mode 1 and 2 occuring due to carrying out some unwanted operations and due to ceasing to operate, respectively). If a_i is the constant failure rate to failure mode i and b_i the constant repair rate of failure mode i, then show that the rate matrix is

$$A = \begin{pmatrix} -(a_1 + a_2) & a_1 & a_2 \\ b_1 & -b_1 & 0 \\ b_2 & 0 & -b_2 \end{pmatrix}.$$

If $p_i(t)$ is the probability that the devices is in state i at time t, $i = 0, 1, 2$, then show that the differential-difference equations of the system are

$$p_0'(t) = -(a_1 + a_2)p_0(t) + b_1(t)p_1(t) + b_2(t)p_2(t)$$

$$p_i'(t) = a_i p_0(t) - b_i p_i(t), i = 1, 2.$$

If, at $t = 0$, the machine is in good state, i.e. $p_0(0) = 1$, $p_i(0) = 0$, $i = 1, 2$, then show that the solutions of the equations are (by the classical integration method or by applying Fourier transforms)

$$p_0(t) = 1 - C_1 - C_2 - (A_1 + A_2)e^{-rt} + (B_1 + B_2)e^{-st}$$

$$p_i(t) = C_i + A_i e^{-rt} - B_i e^{-st}, i = 1, 2,$$

where $C_i = \dfrac{a_i b_{3-i}}{rs}, A_i = \dfrac{a_i(b_{3-i} - r)}{r(r-s)}, B_i = \dfrac{a_i(b_{3-i} - s)}{r(r-s)}$

and r, s are the roots of

$$x^2 - x(a_1 + a_2 + b_1 + b_2) + (b_1 b_2 + a_1 b_2 + a_2 b_1) = 0.$$

Show that, in steady state

$$p_0 = \frac{b_1 b_2}{rs}, p_1 = \frac{a_1 b_2}{rs}, p_2 = \frac{a_2 b_1}{rs}.$$

REFERENCES

N.T.J. Bailey. *Elements of Stochastic Processes with Applications to the Natural Sciences*, Wiley, New York (1964).

M.S. Bartlett. *Introduction to Stochastic Processes*, Cambridge University Press, New York (1951), 3rd Edition (1978).

A.T. Bharucha-Reid. *Elements of the Theory of Markov Processes and Their Applications*, McGraw-Hill, New York (1960).

U.N. Bhat. *Elements of Applied Stochastic Processes*, 2nd ed., Wiley, New York (1984).

R.N. Bhattacharya and E.C. Waymire. *Stochastic Processes With Applications*, Wiley, New York (1990).

A. Blanc-Lapiérre and R. Fortet. *Théorie des Fonctions Aléatoires*, Masson et Cie. Paris (1953).

E. Cinlar. *Introduction to Stochastic Processes*, Prentice-Hall, Englewood Cliffs (1975).

H. Cramer. On streams of random events, *Skand. Akt. Supp.* **52**, 13–23 (1969).

D.R. Cox and H.D. Miller. *The Theory of Stochastic Processes*, Methuen, London (1965).

D.R. Cox and V. Isham. *Point Processes*, Chapman & Hall, London (1980).

S.N. Ethier and T.G. Kurtz. *Markov Processes : Characterization and Convergence*, Wiley, New York (1986).

W. Feller. *An Introduction to Probability Theory and Its Applications*, Vol. I, 3rd ed., (1968), Vol. II (1966), Wiley, New York

M. Girault. *Stochastic Processes*, Springer-Verlag, Berlin (1966).

J. Grandell. *Doubly Stochastic Poisson Processes*, Lecture notes in Maths, No. 529, Springer-Verlag, Berlin (1976).

W.K. Grassman. Transient solutions of Markovian queueing systems, *Comp. & Opns. Res.* **4**, 47–53 (1977).

D. Gross and D.R. Miller. The randomization technique as a modeling tool and solution procedure for transient Markov process, *Opns. Res.* **32**, 343–361 (1984).

R.A. Howard. *Dynamic Probabilistic Systems*, **Vol. I**: *Markov Models*, Wiley, New York (1960).

G.C. Jain and R.P. Gupta. On a class of polynomials and associated probabilities, *Utilitas Mathematica*, **7**, 363–381 (1975).

S. Karlin and H.M. Taylor. *A First Course in Stochastic Processes*, 2nd ed., Academic Press, New York (1975).

A.F. Karr. Markov Processes: in *Stochastic Models*, Handbooks in O.R. & Mgmt. Sc., **2** (Eds. D.P. Heyman & M.J. Sobel), North Holland, Amsterdam (1990).

J. Keilson. *Markov Chain Models: Rarity and Exponentiality*, Springer-Verlag, New York (1979).

N. Keyfitz. *The Mathematics of Population*, Addison-Wesley, Reading, MA (1968) (with revision (1977)).

J.F.C. Kingman. Poisson counts for random sequence of events, *Ann. Math. Statist.* **34**, 1217–1232 (1963).

—— Markov population process, *J. Appl. Prob.* **6**, 1–18 (1969).

S. Lippman. Applying a new device in the optimization of exponential queueing systems, *Opns. Res.* **23**, 687–710 (1975).

J. Medhi and M. Borah. On generalized four parameter Charlier distribution, *J. Statist. Planning & Inference* **14**, 69–77 (1986).

C.R. Mitchell, R.A. Rappold and W.B. Faulkner. An analysis of airforce EOQ data with an application to recorder point calculation, *Mgmt. Sci.* **29**, 440–446 (1983).

J.K. Ord. The size of human settlements: in *A Modern Course on Statistical Distributions in Scientific Work*, Vol. 2, Reidel, Dordrecht, Holland, 141–150 (1975).

E. Parzen. *Stochastic Processes*, Holden-Day, San Francisco (1962).

N.U. Prabhu. *Stochastic Processes*, Macmillan, New York (1965).

C.L. Proctor and B. Singh. A repairable 3-state device, *IEEE Transactions on Reliability*, Vol. R-25, No. 3, 210–211 (1976).

S.M. Ross. *An Introduction to Probability Models*, 2nd ed., Academic Press. New York (1980).

C.C. Sherbrooke. Discrete compound Poisson processes and tables of geometric Poisson distribution, *Nav. Res. Log. Qrly.* **15**, 189–204 (1968).

R.W. Wolff. Poisson Arrivals See Time Averages, *Opns. Res.* **30**, 223–231 (1982).

CHAPTER 5

Markov Processes With Continuous State Space

5.1 INTRODUCTION : BROWNIAN MOTION

Poisson process is a process in continuous time with a discrete state space. Here in a small interval of time Δt, there is either no change of state or there is only one change, the probability of more than one change being of the order of Δt. In this chapter we shall consider Markov processes such that in an infinitesimal interval, there is a small change of state or displacement. In such a process, changes of state occur continually all the time and the state space is continuous. An example of a continuous parameter continuous state space Markov process is the *Brownian motion process*. The process derives its name from the English botanist Robert Brown (1773-1858) who first noticed, in 1827, the erractic movement of particles suspended in a fluid under the random impact of neighbouring particles. Einstein (1879-1955) was the first to put forward a satisfactory mathematical model for the physical phenomenon in 1905, while Wiener, in 1923, gave it a rigorous mathematical form. Considerable attention has since been given which has resulted in remarkable progress in the subject. It has now wide applications in such fields as economics, communication theory, management science etc. Because of the connection with the theory of diffusion, Markov processes with continuous state space are also known as *diffusion processes*. A particle under diffusion or undergoing Brownian motion is also known as a Brownian particle. We shall confine ourselves to one-dimentional Brownian motion, that is, to motion along a fixed axis.

At epoch t, let $X(t)$ be the displacement along a fixed axis of a particle undergoing Brownian motion and let $X(0) = x_0$. Consider an interval (s, t) of time; let us regard this interval as the sum of a large number of small intervals. The total displacement $\{X(t) - X(s)\}$ in this interval can be regarded as the limit of the sum of random displacements over the small intervals of time. Suppose that the random displacements are independently distributed. Then it can be seen that the central-limit theorem applies, whence it follows that the total displacement $\{X(t) - X(s)\}$ is normally distributed. Further, suppose that the displacement $\{X(t) - X(s)\}$ depends on the length of the interval (s, t) and not on the time-point s and that $\{X(t) - X(s)\}$ has the same distribution as $\{X(t + h) - X(s + h)\}$ for all $h > 0$.

It is to be noted that here both time and space variables are continuous. The equations of the process obtained by taking limits of both time and space variables will be partial differential equations in both time and space variables. These equations, called diffusion equations, will be discussed in Sec. 5.3. In Sec. 5.2 we develop Wiener process as the continuous limit of the simple random walk.

It may be noted that there are some measure-theoretic subtleties involved in the passage from the discrete to the continuous case. Their considerations will, however, be beyond the scope of this book.

We assume that the process $\{X(t), t \geq 0\}$ is Markovian. Let the cumulative transition probability be

$$P(x_0, s; x, t) = \Pr\{X(t) \leq x \mid X(s) = x_0\}, s < t \tag{1.1}$$

and let the transition probability density p be given by

$$p(x_0, s; x, t)dx = \Pr\{x \leq X(t) < x + dx \mid X(s) = x_0\}. \tag{1.2}$$

For a homogeneous process the transition probability depends only on the length of the interval $(t - s)$ and then the transition probability may be denoted in terms of the three parameters, $x_0, x, t - s$. We denote

$$\Pr\{x \leq X(t + t_0) < x + dx \mid X(t_0) = x\} \text{ by } p(x_0, x; t)dx$$

for any t_0. The Chapman-Kolmogorov equation can be written (as the continuous state analogue of (11.1) of Ch. 3) as follows:

$$P(x_0, s; x, t) = \int d_z P(x_0, s; z, v) P(z, v; x, t).$$

In terms of transition probabilities $p(x_0, s; x, t)$, we have

$$p(x_0, s; x, t) = \int p(x_0, s; z, v) p(z, v; x, t)\, dz.$$

5.2 WIENER PROCESS

Consider that a (Brownian) particle performs a random walk such that in a small interval of time of duration Δt, the displacement of the particle to the right or to the left is also of small magnitude Δx, the total displacement $X(t)$ of the particle in time t being x. Suppose that the random variable Z_i denotes the length of the ith step taken by the particle in a small interval of time Δt and that

$$\Pr\{Z_i = \Delta x\} = p \text{ and } \Pr\{Z_i = -\Delta x\} = q, p + q = 1$$

$0 < p < 1$, where p is independent of x and t.

Suppose that the interval of length t is divided into n equal subintervals of length Δt and that the displacements $Z_i, i = 1, \ldots, n$ in the n steps are mutually independent random variables. Then $n \cdot (\Delta t) = t$ and the total displacement $X(t)$ is the sum of n i.i.d. random variables Z_i, i.e.

$$X(t) = \sum_{i=1}^{n(t)} Z_i, \quad n \equiv n(t) = t/\Delta t.$$

We have

$$E\{Z_i\} = (p - q)\Delta x \text{ and var } (Z_i) = 4pq(\Delta x)^2.$$

Hence

$$E\{X(t)\} = nE(Z_i) = t(p - q)\Delta x/\Delta t, \tag{2.1}$$

and

$$\text{var } \{X(t)\} = n \text{ var } (Z_i) = 4pqt(\Delta x)^2/\Delta t.$$

To get a meaningful result, as $\Delta x \to 0$, $\Delta t \to 0$, we must have

$$\frac{(\Delta x)^2}{\Delta t} \to \text{a limit, } (p - q) \to \text{a multiple of } (\Delta x). \tag{2.2}$$

We may suppose, in particular, that in an interval of length t, $X(t)$ has mean-value function equal to μt and variance function equal to $\sigma^2 t$. In other words, we suppose that as $\Delta x \to 0$, $\Delta t \to 0$, in such a way that (2.2) are satisfied, and per unit time

$$E\{X(t)\} \to \mu \text{ and var } \{X(t)\} \to \sigma^2 \tag{2.3}$$

From (2.1) for $t = 1$ and (2.3) we have

$$\frac{(p - q)\Delta x}{\Delta t} \to \mu; \quad \frac{4pq(\Delta x)^2}{\Delta t} \to \sigma^2. \tag{2.4}$$

The relations (2.2) and (2.4) will be satisfied when

$$\Delta x = \sigma (\Delta t)^{1/2}, \tag{2.5a}$$

$$p = \frac{1}{2}(1 + \mu(\Delta t)^{1/2}/\sigma), \quad q = \frac{1}{2}(1 - \mu(\Delta t)^{1/2}/\sigma). \tag{2.5b}$$

Now since Z_i are i.i.d. random variables, the sum $\sum_{i=1}^{n(t)} Z_i = X(t)$ for large $n(t) (= n)$, is asymptotically normal with mean μt and variance $\sigma^2 t$ (by virtue of the central limit theorem for equal components). Note that here also t represents the length of the interval of time during which the displacement, that takes place is equal to the increment $X(t) - X(0)$. We thus find that for $0 < s < t$, $\{X(t) - X(s)\}$ is normally distributed with mean $\mu(t - s)$ and variance $\sigma^2(t - s)$. Further, the increments $\{X(s) - X(0)\}$ and $\{X(t) - X(s)\}$ are mutually independent; this implies that $\{X(t)\}$ is a Markov process.

We may now define a Wiener process or a Brownian motion process as follows:

The stochastic process $\{X(t), t \geq 0\}$ is called a *Wiener process* (or a *Wiener-Einstein process* or a *Brownian motion process*) with drift μ and variance parameter σ^2, if:

(i) $X(t)$ has independent increments, i.e. for every pair of disjoint intervals of time (s, t) and (u, v), where $s \leq t \leq u \leq v$, the random variables $\{X(t) - X(s)\}$ and $\{X(v) - X(u)\}$ are independent.

(ii) Every increment $\{X(t) - X(s)\}$ is normally distributed with mean $\mu(t - s)$ and variance $\sigma^2(t - s)$.

Note that (i) implies that Wiener process is a Markov process with independent increments and (ii) implies that a Wiener process is Gaussian.

Since $\{X(t) - X(0)\}$ is normally distributed with mean μt and variance $\sigma^2 t$, the transition probability density function p of a Wiener process is given by

$$p(x_0, x; t)dx = \Pr\{x \leq X(t) < x + dx \mid X(0) = x_0\}$$

$$= \frac{1}{\sigma\sqrt{(2\pi t)}} \exp\left\{-\frac{(x - x_0 - \mu t)^2}{2\sigma^2 t}\right\} dx. \tag{2.6}$$

A Wiener process $\{X(t), t \geq 0\}$ with $X(0) = 0$, $\mu = 0$, $\sigma = 1$ is called a *standard* Wiener process.

5.3 DIFFERENTIAL EQUATIONS FOR A WIENER PROCESS

Let $\{X(t), t \geq 0\}$ be a Wiener process. We can consider the displacement in such a process as being caused by the motion of a particle undergoing displacements of small magnitude in a small interval of time. Suppose that $(t - \Delta t, t)$ is an infinitesimal interval of length Δt and that the particle makes in this interval a shift equal to Δx with probability p or a shift equal to $-\Delta x$ with probability $q = 1 - p$. Suppose that p and q are independent of x and t. Let the transition probability that the particle has a displacement from x to $x + \Delta x$ at epoch t, given that it started from x_0 at time 0, be $p(x_0, x; t) \Delta x$. Further suppose that $p(x_0, x; t)$ admits of an expansion in Taylor's series, i.e.

$$p(x_0, x \pm \Delta x; t - \Delta t) = p(x_0, x; t) - \Delta t \frac{\partial p}{\partial t} \pm \Delta x \frac{\partial p}{\partial x}$$

$$+ \frac{1}{2}(\pm \Delta x)^2 \frac{\partial^2 p}{\partial x^2} + o(\Delta t). \tag{3.1}$$

From simple probability arguments we have

$$p(x_0, x; t)\Delta x = p \cdot p(x_0, x - \Delta x; t - \Delta t)\Delta x$$

$$+ q \cdot p(x_0, x + \Delta x; t - \Delta t)\Delta x. \tag{3.2}$$

Making use of (3.1), and cancelling out the factor Δx from both sides of (3.2) we get

$$p(x_0,x; t) = p(x_0,x; t) - \Delta t \frac{\partial p}{\partial t} - \Delta x (p - q) \frac{\partial p}{\partial x}$$

$$+ \frac{1}{2} (\Delta x)^2 \frac{\partial^2 p}{\partial x^2} + o(\Delta t).$$

Divide both sides by Δt. Using (2.4) and (2.5) and taking limits as $\Delta t \to 0, \Delta x \to 0$, we get

$$\frac{\partial}{\partial t} p(x_0,x; t) = -\mu \frac{\partial}{\partial x} p(x_0,x; t) + \frac{1}{2} \sigma^2 \frac{\partial^2}{\partial x^2} p(x_0,x; t). \tag{3.3}$$

This is a partial differential equation in the variables x and t, being of first order in t and of the second order in x. The equation is known as the *forward diffusion equation* of the Wiener process. One can likewise obtain the *backward diffusion equation* of the process in the form

$$\frac{\partial}{\partial t} p(x_0,x; t) = \mu \frac{\partial}{\partial x_0} p(x_0,x; t) + \frac{1}{2} \sigma^2 \frac{\partial^2}{\partial x_0^2} p(x_0,x; t). \tag{3.4}$$

The solution of (3.3) (as well as of (3.4)) yields $p(x_0, x; t)$ as a normal density of the form given in (2.6) (see also Sec. 5.5). It may, however, be easily verified that $p(x_0, x; t)$ given by (2.6) satisfies (3.3) as well as (3.4). The equation for a Wiener process with drift $\mu = 0$ is known as the *heat equation*.

Note: The partial differential equation (3.3) [(3.4)] is known as the forward [backward] equation because it involves differentiation in x [x_0]. The reason why it is called diffusion equation is given in the next section.

It is to be noted that in Sec. 5.2 and 5.3 we have made the following assumptions:
  (i) in a small interval of time Δt, the displacement Δx is small (and that

$\Delta x = o((\Delta t)^{\frac{1}{2}})$;

  (ii) $E\{X(t)\} \to \mu t$ in the limit;
  (iii) var $\{X(t)\} \to \sigma^2 t$ in the limit.
  The quantity μ in (ii) may also be interpreted as

$$\lim_{\Delta t \to 0} \frac{E\{X(t+\Delta t) - X(t)\}}{\Delta t} = \mu. \tag{3.5}$$

This implies that the infinitesimal mean (i.e. mean over Δt) of the variance of the increment in $X(t)$ exists and is equal to a finite quantity σ^2.

For a Wiener process, μ and σ^2 are assumed to be constants, independent of t or of x (where $X(t) = x$). By considering the transition mechanism with μ and σ^2 as functions of t or of x or both t and x, we get more general processes for which the equations corresponding to (3.3) and (3.4) will also be more general. We discuss below such equations.

5.4 KOLMOGOROV EQUATIONS

Let $\{X(t), t \geq 0\}$ be a Markov process in continuous time with continuous state space. We make the following assumptions: For any $\delta > 0$,

(i) $$\Pr\{|X(t) - X(s)| > \delta \mid X(t) = x\} = o(t - s), s < t.$$

In other words, small changes occur during small intervals of time.

(ii) $$\lim_{\Delta t \to 0} \frac{E\{X(t + \Delta t) - X(t) \mid X(t) = x\}}{\Delta t}$$

$$= \lim_{\Delta t \to 0} \int_{|y - x| \leq \delta} (y - x) p(x, t; y, t + \Delta t) dy$$

$$= a(t, x).$$

In other words, the limit of the infinitesimal mean of the conditional expectation of the increment of $X(t)$ exists and is equal to $a(t, x)$, which is known as the *drift coefficient*.

(iii) $$\lim_{\Delta t \to 0} \frac{E\{[X(t + \Delta t) - X(t)]^2 \mid X(t) = x\}}{\Delta t}$$

$$= \lim_{\Delta t \to 0} \int_{|y - x| \leq \delta} (y - x)^2 p(x, t; y, t + \Delta t) dy$$

$$= b(t, x).$$

In other words, the limit of the infinitesimal mean of the variance of the increment of $X(t)$ exists and is equal to $b(t, x)$, which is known as the *diffusion coefficient*.

A Markov process $\{X(t)\}$ satisfying the above conditions is known as a *diffusion process* and the partial differential equation satisfied by its transition p.d.f. is known as *diffusion equation*. We give below the equations (for derivation see Prabhu, Blanc-Lapiérre and Fortet).

Let $\{X(t), t \geq 0\}$ be a Markov process satisfying (i), (ii) and (iii). If its transition p.d.f. $p(x_0, t_0; x, t)$ possesses continuous partial derivatives

$$\frac{\partial p}{\partial t}, \frac{\partial}{\partial x}(a(t, x)p), \frac{\partial^2}{\partial x^2}(b(t, s)p),$$

then $p(x_0, t_0; x, t)$ satisfies the forward Kolmogorov equation

$$\frac{\partial p}{\partial t} = -\frac{\partial}{\partial x}(a(t,x)p) + \frac{1}{2}\frac{\partial^2}{\partial x^2}(b(t,x)p). \tag{4.1}$$

This equation is also known as the *Fokker-Planck equation*. Suppose that p $(x_0, t_0; x, t)$ possesses continuous partial derivatives

$$\frac{\partial p}{\partial t_0}, \frac{\partial p}{\partial x_0}, \frac{\partial^2 p}{\partial x_0^2};$$

then p $(x_0, t; x, t)$ also satisfies the backward Kolmogorov equation

$$\frac{\partial p}{\partial t_0} = -a\ (t_0, x_0)\frac{\partial p}{\partial x_0} - \frac{1}{2}b\ (t_0, x_0)\frac{\partial^2 p}{\partial x_0^2}. \tag{4.2}$$

The diffusion equations for X (t) were first derived by Kolmogorov. Feller showed that under suitable restrictions the equations admit of a unique solution. Fortet established some very interesting and important properties of the solutions.

Particular case: If the process is homogeneous, then

$$p\ (t_0, x_0; t, x) = p\ (x_0, x; t - t_0),$$

and a (t, x), b (t, x) are independent of t.

If the process is additive, i.e. given that X $(t_0) = x_0$, the increment $\{X\ (t) - X\ (t_0)\}$, depends only on t_0 and t (and not on x_0), then

$$p\ (t_0, x_0; t, x) = p\ (x - x_0; t_0, t) \text{ and } a\ (t, x), b(t, x)$$

are independent of x.

The Kolmogorov equations, in these cases, can be easily deduced from the general equations.

5.5 FIRST PASSAGE TIME DISTRIBUTION FOR WIENER PROCESS

The possible realizations of a stochastic process are called *sample paths* or *trajectories*. The structure and the properties of the same paths of a Brownian motion or Wiener process are the subject matter of deep study. Without entering into the subtleties (which are beyond the scope of this work), we discuss here some results of Wiener process, using the property that the sample paths are continuous functions. We also make use of the simple but powerful '*reflection principle*' (enunciated by D. André in 1887). The principle relates to the fact that there is a one-to-one correspondence between all paths from A (a_1, a_2) to B (b_1, b_2) which touch or cross the x-axis and all paths from A' $(a_1, -a_2)$ to B (see, Feller, Vol I for details). We shall first consider the following from which the distribution of the first passage time will be derived.

5.5.1 Distribution of the Maximum of a Wiener Process

Lemma: Let $\{X(t), 0 \le t \le T\}$ be a Wiener process with $X(0) = 0$ and $\mu = 0$. Let $M(T)$ be the maximum of $X(t)$ in $0 \le t \le T$, i.e. $M(T) = \max_{0 \le t \le T} X(t)$. Then for any $a > 0$

$$\Pr\{M(T) \ge a\} = 2\Pr\{X(T) \ge a\}.$$

(This result was first obtained by Bachelier (1900).)

Proof: Consider the collection of sample path $X(t), 0 \le t \le T$ such that $X(T) \ge a$. Since $X(0) = 0$ and $X(t)$ is continuous, there exists a time τ at which $X(t)$ first attains the value a (or $X(t)$ hits first the value a). The time T_a is itself a random variable. For $t > T_a$, $X_a(t)$ given below

$$X_a(t) = \begin{cases} X(t), & t < T_a \\ 2a - X(t), & t > T_a \end{cases}$$

gives the reflection of $X(t)$ about the line $x = a$. Note that $X_a(T) \le a$, and that $M(T) \equiv \max_{0 \le t \le T} X(t) \ge a$ and $M_a(T) \equiv \max_{0 \le t \le T} X_a(t) \ge a$; further, by symmetry the sample paths $X(t)$ and $X_a(t)$ have the same probability of occurrence. From reflection principle, it follows that corresponding to every sample path $X(t)$ for which $X(T) \ge a$, there exist two sample paths such that $M(T) \ge a$. Further, its converse is also true, viz., every sample path $X(t)$ for which $M(T) \ge a$ corresponds to two sample paths $X(t)$ with equal probability, one of the paths being such that $X(T) > a$, unless $\{X(T) = a\}$, whose probability is zero. In fact, the set $\{M(T) \ge a\}$ is the union of three disjoint sets

$$\{M(T) \ge a, X(t) > a\},$$

$$\{M(T) \ge a, X(t) < a\},$$

and
$$\{M(T) \ge a, X(t) = a\}.$$

The probability of the third set is zero, while the other two are mapped onto one another by reflection about the line $x = a$ after the time T_a. Thus we have

$$\Pr\{M(T) \ge a\} = 2\Pr\{X(T) \ge a\}. \qquad \blacktriangle$$

The above gives a heuristic proof of the lemma; as already indicated, a rigorous proof involves considerations beyond the scope of this book (see Karlin and Taylor, Iosifescu and Tautu).

Let $\{X(t), t \ge 0\}$ be a Wiener process with $X(0) = 0$, $\mu = 0$ and let $M(t) = \max_{0 \le s \le t} X(s)$. Then from the lemma, we get, for $t > 0$, $\sigma = 1$,

$$\Pr \{M \, (t) \geq a\} = 2 \Pr \{X(t) \geq a\}$$

$$= \frac{2}{\sqrt{(2\pi t)}} \int_a^\infty \exp \left(-x^2/2t\right) dx \tag{5.1a}$$

$$= \frac{2}{\sqrt{(2\pi)}} \int_{a/\sqrt{t}}^\infty \exp \left(-y^2/2\right) dy \tag{5.1b}$$

(by changing the variable to $y = x/\sqrt{t}$)

$$= 2 \left\{ 1 - \frac{1}{\sqrt{(2\pi)}} \int_{-\infty}^{a/\sqrt{t}} \exp \left(-y^2/2\right) dy \right\}$$

$$= 2 \{1 - \Phi \, (a/\sqrt{t})\}, \tag{5.1c}$$

Φ being the distribution function of the standard normal variate.
By changing the variable to $s = a^2 t/x^2$, (5.1a) can be written as

$$\Pr \{M \, (t) \geq a\} = \frac{a}{\sqrt{(2\pi)}} \int_0^t s^{-3/2} \exp \left(-\frac{a^2}{2s}\right) ds, s > 0. \tag{5.1d}$$

5.5.2 Distribution of the First Passage Time to a Fixed Point

We can use the lemma to obtain the distribution of the random variable T_a, the first passage time to a fixed point a (> 0) (or the time of hitting a fixed point a first), for a Wiener process $\{X \, (t)\}$ with $X \, (0) = 0$, $\mu = 0$. The time T_a for $X \, (t)$ to hit the level a first will be less than t iff $M \, (t) = \max_{0 \leq s \leq t} X \, (s)$ in that time is at least a.

Thus for $t > 0$

$$\Pr \{M \, (t) \geq a\} = \Pr \, (T_a \leq t). \tag{5.2}$$

Hence the distribution function $F \, (t) = \Pr \{T_a \leq t\}$ is

$$F \, (t) = \frac{2}{\sqrt{(2\pi t)}} \int_a^\infty \exp \left(-x^2/2t\right) dx \tag{5.3a}$$

$$= 2 \{1 - \Phi \, (a/\sqrt{t})\} \tag{5.3b}$$

$$= \frac{a}{\sqrt{(2\pi)}} \int_0^t s^{-3/2} \exp \left(-\frac{a^2}{2s}\right) ds, s > 0. \tag{5.3c}$$

The density function of T_a is obtained by differentiating (5.3) with respect to t. Differentiation of (5.3c) (and also 5.3b) readily gives

$$f_{T_a}(t) = F'(t) = \frac{a}{\sqrt{(2\pi)}} t^{-3/2} \exp\left(-\frac{a^2}{2t}\right), \quad t > 0 \tag{5.4}$$

It may be easily verified that the Laplace transform is

$$\bar{f}(s) = \int_0^\infty e^{-st} f(t)\, dt$$

$$= \exp\{-a\sqrt{(2s)}\}.$$

It can be seen that no moment of T_a exists finitely.

Let us find the density function of $M(t)$. The distribution function is

$$G(a) = \Pr\{M(t) \le a\}$$

$$= 1 - \Pr\{M(t) > a\}$$

$$= 1 - 2\Pr\{X(t) \ge a\}$$

$$= 1 - \frac{2}{\sqrt{(2\pi t)}} \int_a^\infty \exp(-x^2/2t)\, dx \tag{5.5a}$$

$$= 1 - 2\{1 - \Phi(a/\sqrt{t})\} \tag{5.5b}$$

$$= 2\Phi(a/\sqrt{t}) - 1.$$

Differentiating (5.5a) (or 5.5b) with respect to a, we get the density function

$$g_M(a) = G'(a) = \frac{2}{\sqrt{(2\pi t)}} \exp\left(-\frac{a^2}{2t}\right), a > 0. \tag{5.6}$$

The results given above (for $\sigma = 1$) can be suitably modified for any $\sigma > 0$.

Note 1: We have obtained the distribution of T_a by using the lemma which gives a relation between the distributions of the Wiener process $X(t)$ and its maximum $M(t)$. However, the distribution of T_a can be obtained directly without bringing in the distribution of the maximum. In fact, it can be directly shown that, if $X(0) = 0$, $a > 0$, then

$$\Pr\{T_a \le t\} = 2\Pr\{X(t) \ge a\}. \tag{5.7}$$

For a proof of the above, see Prohorov and Rozanov (p. 245), who use conditional expectations to obtain the above. The distribution of the maximum $M(t)$ can then be obtained by using (5.2).

Note 2: For an alternative approach to the distribution of T_a using differential equations, see Cox and Miller (pp. 210, 220, 230), who obtain the distribution for any $\mu \geq 0$ and $\sigma > 0$. The density function $f_{T_a}(x)$ (for $X(0) = 0$, $\mu \geq 0$, $\sigma > 0$) of T_a is found to be

$$f_{T_a}(x) = \frac{a}{a\sqrt{(2\pi x^3)}} \exp\left\{-\frac{(a-\mu x)^2}{2\sigma^2 x}\right\}, x > 0; \tag{5.8}$$

and its Laplace transform is

$$\bar{f}(s) = \exp\left[(a/\sigma^2)\{+\mu - \sqrt{(\mu^2 + 2s\sigma^2)}\}\right].$$

The mean and the variance for T_a for $\mu \neq 0$ are given by

$$E\{T_a\} = a/\mu \text{ and var}\{T_a\} = a\,\sigma^2/\mu^3.$$

Note 3: The function (5.8) with $\mu > 0$ is the density function of the distribution of the first passage time of Brownian motion with a positive drift. This distribution having density function (5.8) is known as *inverse Gaussian distribution* because of the inverse relationship between the cumulant generating function of this distribution and that of normal distribution. Such a distribution was also obtained by Wald as the limiting form for the distribution of the sample size in a sequential probability ratio test. For properties of this distribution, see Johnson and Kotz (1970), and for statistical applications, see Folks and Chhikara (1978).

Example 5(a). Suppose that $\{X(t), 0 < t,\}$ is a Wiener process with $X(0) = 0$, and $\mu = 0$. Then

$$\Pr\{X(t) \leq x\} = \Pr\{X(t)/\sigma\sqrt{t} \leq x/\sigma\sqrt{t}\} = \Phi(x/\sigma\sqrt{t}).$$

Consider the process

$$Y(t) = t\,X(1/t) \text{ in } 0 < t \leq 1 \text{ with } Y(0) = 0.$$

We have

$$E\{Y(t)\} = 0 \text{ and var}\{Y(t)\} = t^2(\sigma^2/t) = \sigma^2 t.$$

Further,

$$\Pr\{Y(t) \leq y\} = \Pr\{t\,X(1/t) \leq y\} = \Pr\left\{\frac{X(1/t)}{\sigma\sqrt{(1/t)}} \leq \frac{y/t}{\sigma\sqrt{(1/t)}}\right\}$$

$$= \Phi(y/\sigma\sqrt{t}).$$

Thus $\{Y(t), 0 < t \leq 1\}$ with $Y(0) = 0$ is also a Wiener process with $\mu = 0$ and variance $\sigma^2 t$.

Example 5(b). Consider a Wiener process $\{X(t)\}$ with $X(0) = 0$. Its mean value function is μt and variance function $\sigma^2 t$. For $0 < s < t$, the covariance function is

$$C(s,t) = \text{cov } \{X(s), X(t)\} = \text{cov } \{X(s), X(s) + X(t) - X(s)\}$$

$$= \text{cov } \{X(s), X(s)\} + \text{cov } \{X(s), X(t) - X(s)\}$$

$$= \text{cov } \{X(s), X(s)\},$$

since the process has independent increments. Thus

$$C(s,t) = \text{var } \{X(s)\} = \sigma^2 s$$

(see also Exercise 2.7). The process is not covariance stationary even when $\mu = 0$.

Example 5(c). Suppose that $\{X(t), t > 0\}$ is a Wiener process with $X(0) = 0$. Its first passage time, T_a, to a has the same distribution as $1/u^2$, where u is a normal variate with mean 0 and s.d. σ/a. For, the distribution function of $1/u^2$ for $t > 0$, is

$$F(t) = \Pr \{1/u^2 \leq t\} = \Pr \{u^2/(\sigma/a)^2 \geq a^2/\sigma^2 t\}$$

$$= \Pr \{(au/\sigma) \geq (a/\sigma\sqrt{t})\} + \Pr \{(au/\sigma) \leq -(a/\sigma\sqrt{t})\}$$

$$= 1 - \Phi(a/\sigma\sqrt{t}) + \Phi(-a/\sigma\sqrt{t})$$

$$= 2(1 - \Phi(a/\sigma\sqrt{t})),$$

which is the distribution function of T_a (see equation (5.3b)). The distribution of $1/u^2$ is *inverse Gaussian*.

Example 5(d). Let $\{X(t), t \geq 0\}$ be a Wiener process with $\mu = 0$ and $X(0) = 0$. To find the distribution of T_{a+b} for $0 < a < a + b$.

Suppose that t_a is a value of T_a, i.e. $X(t)$ reaches the level a for the first time a epoch t_a. We may then consider that the process starts at (t_a, a) and reaches the level $(a + b)$, which is b units higher than a. Suppose that $T_{a+b} - T_a$ is the duration of the interval at the ends of which $X(t)$ first reaches the level a and then reaches first the level $a + b$. Then T_a and $(T_{a+b} - T_a)$ are independent random variables denoting first passage times to a and b respectively. The L.T. of the p.d.f. of T_a is

$$\overline{h}_a(s) = \exp \{-\sqrt{(2s)}/\sigma\}$$

and that of T_a is

$$\overline{h}_b(s) = \exp \{-b\sqrt{(2s)}/\sigma\}.$$

Thus the L.T. of the p.d.f. of T_{a+b} is

$$\bar{h}_a(s)\bar{h}_b(s) = \exp\{-(a+b)\sqrt{(2s)}/\sigma\} = \bar{h}_{a+b}(s).$$

Remarks: Wiener process, originally evolved to study physical phenomena, now has several interesting applications e.g. in genetics and demography. It is used for modelling storage systems of dams and inventories; for modelling financial systems (replenishment of reserves by jump process) and for modelling various economic phenomena (see, Malliaris and Brock (1982)); as well as for modelling in management (see, for example Sethi and Thompson (1981)). In queueing theory, it is applied for heavy-traffic approximation (see, for example, Kleinrock, 1976; Medhi, 1991).

5.6 ORNSTEIN-UHLENBECK PROCESS

We have seen that for a Wiener process $\{X(t)\}$, the displacement Δx, in a small interval of time Δt is also small, being of $[O(\sqrt{\Delta t})]$. The velocity which is of $O(\sqrt{\Delta t}/\Delta t) = O(1/\sqrt{\Delta t})$ tends to infinity as $\Delta t \to 0$. Thus the Wiener process does not provide a satisfactory model for Brownian motion for small values of t, although for moderate and large values of t it does so. An alternative model which holds for small t was proposed by Ornstein and Uhlenbeck in 1930. Here instead of the displacement $X(t)$, the velocity $U(t) = X'(t)$ at time t is considered.

The equation of motion of a Brownian particle can be written as

$$dU(t) = -\beta U(t)dt + dF(t), \tag{6.1}$$

where $-\beta U(t)$ represents the systematic part due to the resistance of the medium and $dF(t)$ represents the random component. It is assumed that these two parts are independent and that $F(t)$ is a Wiener process with drift $\mu = 0$ and variance parameter σ^2. The Markov process $\{U(t), t \geq 0\}$ is such that in a small interval of time the change in $U(t)$ is also small. Since $F(t)$ is a Wiener process, we have from (6.1)

$$\lim_{\Delta t \to 0} \frac{E\{U(t+\Delta t) - U(t) \mid U(t) = u\}}{\Delta t}$$

$$= -\beta u + \lim_{\Delta t \to 0} \frac{E\{\Delta F(t)\}}{\Delta t}$$

$$= -\beta u,$$

and

$$\lim_{\Delta t \to 0} \frac{\text{var}\{U(t+\Delta t) - U(t) \mid U(t) = u\}}{\Delta t}$$

$$= \lim_{\Delta t \to 0} \frac{(\Delta t)^2}{\Delta t} + \lim_{\Delta t \to 0} \frac{\text{var}\{\Delta F(t)\}}{\Delta t},$$

$$= \sigma^2.$$

In other words, the limits exist. So the process $\{U(t), t \geq 0\}$ is a diffusion process and its transition p.d.f. $p(u_0; u, t)$ satisfies the forward Kolmogorov equation (4.1) with $a(u, t) = -\beta u$ and $b(u, t) = \sigma^2$. That is, p satisfies the differential equation

$$\frac{\partial p}{\partial t} = \beta \frac{\partial}{\partial u}(up) + \frac{1}{2}\sigma^2 \frac{\partial^2 p}{\partial u^2}. \tag{6.2}$$

Let us assume that $U(0) = u$ and that as $u_0 \to \pm \infty$, $p \to 0$ and $(\partial p / \partial u) \to 0$. The solution of (6.2) gives p, the transition p.d.f. of $U(t)$. It is more convenient to consider the equation corresponding to (6.2) in terms of the characteristic function of p, i.e.

$$\phi(u_0; \theta, t) = \int_{-\infty}^{\infty} e^{i\theta u} p(u_0; u,) \, du.$$

We have

$$\int_{-\infty}^{\infty} e^{i\theta u} \frac{\partial}{\partial u}(up) \, du = e^{i\theta u} up \Big|_{-\infty}^{\infty} - \int_{-\infty}^{\infty} i\theta e^{i\theta u} up \, du$$

$$= -\theta \frac{\partial}{\partial \theta} \int_{-\infty}^{\infty} e^{i\theta u} p \, du$$

$$= -\theta \frac{\partial \phi}{\partial \theta};$$

$$\int_{-\infty}^{\infty} e^{i\theta u} \frac{\partial^2 p}{\partial u^2} = e^{i\theta u} \frac{\partial p}{\partial u} \Big|_{-\infty}^{\infty} - i\theta \int_{-\infty}^{\infty} e^{i\theta u} \frac{\partial p}{\partial u} \, du$$

$$= -i\theta \{ e^{i\theta u} p \Big|_{-\infty}^{\infty} - i\theta \int_{-\infty}^{\infty} e^{i\theta u} p \, du \}$$

$$= -\theta^2 \phi.$$

The equation (6.2) then becomes

$$\frac{\partial \phi}{\partial t} + \beta \theta \frac{\partial \phi}{\partial \theta} = -\frac{1}{2}\sigma^2 \theta^2 \phi. \tag{6.3}$$

The equation (6.3) is of Lagrange type. It can be shown that

$$\phi(u_0; \theta, t) = \exp \left\{ i\theta u_0 e^{-\beta t} - \frac{1}{4\beta}\theta^2 \sigma^2 (1 - e^{-2\beta t}) \right\}. \tag{6.4}$$

This is the characteristic function of normal distribution with $m(t) = u_0 e^{-\beta t}$ and variance function $\sigma^2(t) = \sigma^2(1 - e^{-2\beta t}/4\beta$.

In other words, the transition p.d.f. p is normal with mean value function $m(t)$ and variance function $\sigma^2(t)$ and p can be written as:

$$p(u_0; u, t) = \frac{1}{\sqrt{(2\pi\sigma^2(t))}} \exp\{-(x - m(t))^2/2\sigma^2(t)\}. \tag{6.5}$$

Thus the process $\{U(t), t \geq 0\}$ is a Gaussian process with mean value function $m(t)$ and variance function $\sigma^2(t)$. $\{U(t), t \geq 0\}$ is a Markov process but it does not possess independent increments like the Wiener process. $\{U(t), t \geq 0\}$ is known as *Ornstein-Uhlenbeck process (O-U.P.)*. For large t, $m(t) \rightarrow 0$ and $\sigma^2(t) \rightarrow \sigma^2/2\beta$, i.e. the distribution of velocity is normal with mean 0 and variance $\sigma^2/2\beta$. We thus get an equilibrium distribution and $U(t)$ is said to in statistical equilibrium. For small t, $m(t) \rightarrow u_0$ and $\sigma^2(t) \rightarrow \sigma^2 t$.

Example 6(a). Joint distribution of $U(t)$ and $U(t + \tau)$ when $U(t)$ is in equilibrium. For large t, the limiting distribution of $U(t)$ is normal with mean 0 and variance $\sigma^2/2\beta = \sigma_0^2$ (say). The conditional distribution of $U(t + \tau)$, given $U(t) = u$, is normal with mean $u e^{-\beta\tau}$ and variance $\sigma^2(\tau) = \sigma^2(1 - e^{-2\beta\tau})/2\beta$. Thus the unconditional distribution of $U(t + \tau)$ has the following density

$$h(x) = \frac{1}{\sqrt{(2\pi\sigma_0^2)}} \int_{-\infty}^{\infty} [\exp\left(-\frac{x_0^2}{2\sigma_0^2}\right) \frac{1}{\sqrt{(2\pi\sigma^2(t))}} \times$$

$$\exp\left\{-\frac{1}{2\sigma^2(t)}(x - x_0 e^{-\beta t})^2\right\}] \, dx_0$$

$$= \frac{1}{\sqrt{(2\pi\sigma_0^2)}} \exp\left(-\frac{x^2}{2\sigma_0^2}\right).$$

Thus the unconditional distribution of $U(t + \tau)$ is Gaussian (normal), has mean 0 and variance σ_0^2 and the unconditional distribution of $U(t + \tau)$ is the same as the equilibrium distribution of $U(t)$. The joint distribution of $U(t)$ and $U(t+\tau)$ has the density

$$f(x, y) = \frac{1}{\sqrt{(2\pi\sigma_0^2)}} \exp\left(-\frac{x^2}{2\sigma_0^2}\right) \cdot \frac{1}{\sqrt{(2\pi\sigma^2(t))}} \exp\left\{-\frac{1}{2\sigma^2(t)}(y - xe^{-\beta t})^2\right\}$$

$$= \frac{1}{2\pi\sigma_0^2\sqrt{(1 - c^2)}} \exp\left\{-\frac{1}{2\sigma_0^2(1 - c^2)}(y^2 - 2axy + x^2)\right\},$$

where $c = e^{-\beta t}$.

It follows that $U(t)$ and $U(t + \tau)$ have a bivariate Gaussian distribution with

$$E\{U(t)\} = E\{U(t + \tau)\} = 0,$$

$$\text{cov } \{U(t), U(t + \tau)\} = \sigma_0^2 e^{-\beta|\tau|}$$

and

$$\text{var } \{U(t)\} = \sigma_0^2.$$

The mean and the variance of $U(t)$ are finite and the covariance function cov $\{U(t), U(t + \tau)\}$ is a function of the absolute difference only. Hence $\{U(t)\}$ is covariance stationary. Again, since $\{U(t)\}$ is Gaussian, $\{U(t)\}$ is strictly stationary. Note that Wiener process is not covariance stationary.

Example 6(b). *The O-U.P. as a transformation of a Wiener process:*
Let $\{X(t), t \geq 0\}$, be a standard Wiener process. Let

$$Y(t) = \frac{1}{\sqrt{g(t)}} X(a g(t)), a > 0,$$

and let the (non-random) function $g(t)$ be positive, strictly increasing with $g(0) = 1$.

We have

$$E\{Y(t)\} = 0$$

and

$$\text{var } \{Y(t)\} = \frac{1}{g(t)} \text{ var } \{X(a g(t))\}$$

$$= \frac{1}{g(t)} \{a g(t)\} = a.$$

Since $(X(t_1), \ldots, X(t_n))$ is multivariate normal, so also is $(X(a g(t_1)), \ldots, X(a g(t_n))$, and for $\tau > 0$,

$$\text{cov } \{Y(t), Y(t + \tau)\} = \text{cov } \left\{ \frac{X(a g(t)), X(a g(t + \tau))}{(g(t) g(t + \tau))^{\frac{1}{2}}} \right\}$$

$$= \text{var } \{X(a g(t))\} / \{(g(t) g(t + \tau))\}^{\frac{1}{2}},$$

since, for the Wiener process $\{X(t)\}$, cov $\{X(t), X(t + \tau)\} = \text{var } X(t)$ (see Example 5(b)).

Now,

$$\text{var } \{X(a g(t))\} = \text{var } \{\sqrt{g}(t) Y(t)\}$$

$$= g(t) \text{ var } \{Y(t)\} = a g(t)$$

and thus

$$\text{cov } \{Y(t), Y(t + \tau)\} = a \{g(t)/g(t + \tau)\}^{\frac{1}{2}}$$

and for $t = 0$, the covariance equals $a/\{g(\tau)\}^{\frac{1}{2}}$.

The process $\{Y(t)\}$, which has finite mean and variance, will be covariance stationary provided cov $\{Y(t), Y(t + \tau)\}$ depends only on τ. Thus we must have $g(t + \tau) = g(\tau) \cdot g(t)$, and in order to satisfy this equation, $g(t)$ must be an exponential function, say, $g(t) = e^{2\beta t}$ $(\beta > 0)$. We have then

$$\text{cov }\{Y(t), Y(t+\tau)\} = a\, e^{-\beta \tau}, (\tau > 0),$$

and thus we find that for $a > 0$, $\beta > 0$

$$Y(t) = e^{-2\beta t} X(a\, e^{2\beta t})$$

is a stationary Gaussian Markov process. In other words, $Y(t)$ has the structure of an Ornstein-Uhlenbeck process.

Remarks. The concepts of O–U. process have been applied extensively in Finance, Economics and Management. O–U. process has been used as models for continuous control systems, for buffer stock control, for continuous industrial processes in chemical plants, for process control in thermal plants and so on in industrial management; for pricing in a large system of cash bonds (Abikhalil et al. (1985) and so on in financial management; as well as for short term interest rate behaviour (Vasicek, 1977) and so on in economics. For application of diffusion processes in Finance and Economics, refer to Malliaris and Brock (1982).

EXERCISES

5.1   If $X(t)$, with $X(0)$ and $\mu = 0$, is a Wiener process, show that $Y(t) = \sigma X(t/\sigma^2)$ is also a Wiener process. Find its covariance function.

5.2   If $X(t)$ with $X(0)$ and $\mu = 0$ is a Wiener process and $0 < s < t$, show that for at least one τ satisfying $s \le \tau \le t$,

$$\Pr\{X(\tau) = 0\} = (2/\pi)\cos^{-1}((s/t)^{1/2}).$$

5.3   Let $X(t)$, with $X(0) = 0$, be a standard Wiener process and let T_a be the first passage time of $X(t)$. Show that T_a and $a^2 T_1$ are identically distributed. If Z_i, $i = 1, 2, \ldots, n$ are i.i.d. as T_1 then show that $\Sigma\, Z_i/n$ and Z_1 are identically distributed.

5.4   *Additive process*: Here

$$a(t, x) = a(t) \text{ and } b(t, x) = b(t).$$

Show that the Kolmogorov forward equation can be reduced to

$$\frac{\partial p}{\partial t_1} = \frac{1}{2}\frac{\partial^2 p}{\partial x_1^2}$$

by a suitable change of variables, and that the corresponding distribution is Gaussian.

5.5   Let $X(t)$ be the displacement process corresponding to the velocity process $U(t)$ (O-U. process). Show that in equilibrium,

$$E\{X(t) - X(0)\} = 0$$

and

$$\text{var}\{X(t) - X(0)\} = \sigma^2(\beta t - 1 + e^{-\beta t})/\beta^3.$$

Deduce that,

$$\text{var}\{X(t) - X(0)\} \sim \sigma^2 t^2/2\beta \text{ for small } t$$

$$\sim \sigma^2 t/\beta^2 \quad \text{for large } t.$$

5.6   Show that if $Y(t)$ is a O-U.P. with mean value 0 and covariance function $C\{Y(s), Y(t)\} = ae^{-\beta|t-s|}$, then

$$X(t) = (a/t)^{-1/2} Y\left(\frac{1}{2}\beta \log(t/a)\right)$$

is a standard Wiener process.

5.7   Suppose that $\{U(t)\}$ is an O-U. process with $E\{U(t)\} = 0$,

$$\text{cov}\{U(s), U(t)\} = \sigma_0^2 e^{-\beta|t-s|}$$

and $X(t)$, (where $(d/dt) X(t) = U(t)$)
gives the position of the particle.

Show that the conditional distribution of $X(t)$ $(t > 0)$ given that $X(0) = x_0$ is normal with mena $E\{X(t)\} = x_0$
and var $\{X(t)\} = (2\sigma_0^2/\beta^2)(e^{-\beta t} - 1 + \beta t)$ (Prabhu (1965)).

5.8   *Inverse Gaussian Process*: Consider a Wiener process $\{X(t)\}$ with positive drift. Without loss of generality, suppose that $X(0) = 0$, $\mu = 1$, $\sigma = 1$. Then T_a, the first passage time to a given point a $(a > 0)$ has density

$$f(a, x)dx = \frac{a}{\sqrt{(2\pi x^3)}} \exp\{-(x-a)^2/2x\} \, dx, x > 0.$$

Instead of a given point a, consider a variable point t (>0). Then T_t has density

$$f(t, x) \, dx = \frac{t}{\sqrt{(2\pi x^3)}} \exp\{-(x-t)^2/2x\} \, dx.$$

This inverse Gaussian distribution has mean t and variance t. A new stochastic process having f for its distribution can now be defined.

An *inverse Gaussian process* (or *a first passage time distribution process*) is a stochastic process $\{X(t), t \geq 0\}$, with $X(0) = 0$ such that

(1) $X(t)$ has independent increments, i.e. for every pair of disjoint intervals of time (s, t) and (u, v) where $s \leq t \leq u \leq v$, the random variables $\{X(t) - X(s)\}$ and $\{X(v) - X(u)\}$ are independent.

(2) Every increment $\{X(t + h) - X(h)\}$ has an inverse Gaussian distribution with mean t and variance t.

For
$$s < t, E\{X(t) - X(s)\} = (t - s),$$

$$\operatorname{var}\{X(t) - X(s)\} = (t - s)$$

and
$$C(s, t) = \operatorname{cov}\{X(s), X(t)\} = \operatorname{var} X(s))$$

(Wasan, 1969).

5.9 *Gamma Process*: If $Y_i(t)$, $i = 1, 2, \ldots, m$ are i.i.d. Gaussian processes with zero mean value function and covariance function

$$\operatorname{cov}\{Y_i(s), Y_i(t)\} \equiv C(s, t) \text{ for all } i,$$

then $X(t) = \frac{1}{2} \sum_{i=1}^{m} Y_i^2(t)$ is a Gamma process with parameter $m/2$.

Show that

$$E\{X(t)\} = \frac{1}{2} m\, C(t, t),$$

$$\operatorname{cov}\{X(t), X(s)\} = \frac{1}{2} m (C(t, t))^2$$

and that the distributions of a Gamma process are gamma distributions.

5.10 *Negative binomial process*: In our discussion on Poisson process we consider λ as a constant or as a non-random function $\lambda(t)$ of time. It is natural to introduce some kind of random variations of λ also; we may consider λ to be a random variable or as a random function (or stochastic process) $\lambda(t)$. A Poisson process whose intensity function $\lambda(t)$ is a stochastic process is called a conditional Poisson process or a doubly stochastic Poisson process. Several studies have been made on conditional and mixed Poisson process (see also Sec. 4.3.6).

If the parameter λ of a Poisson distribution is a random variable having gamma distribution, then the mixed Poisson distribution is negative binomial. By analogy, a negative binomial process $\{N(t)\}$ is defined as a conditional Poisson process whose intensity function $\lambda(t)$ is a Gamma process. It may be remarked that the distributions of a negative binomial process are not necessarily negative binomial.

Suppose $Y_i(t)$ are i.i.d. *O-U*. processes with $E(Y_i) = 0$ and covariance function $C(s, t) = \sigma^2 e^{-\beta|t-s|}$. Then $\frac{1}{2} \sum_{i=1}^{m} Y_i^2(t)$ is a Gamma process and the conditional Poisson process, whose intensity function $\lambda(t)$ is a Gamma process, is a *negative binomial process* $\{N(t)\}$.

Examine whether the distributions of $N(t)$ are negative binomial. Show that

$$E\{N(t)\} = \frac{1}{2} m\sigma^2 t$$

and

$$\text{var}\{N(t)\} = \frac{1}{2} m\sigma^2 t + m\sigma^4 (e^{-2\beta t} - 1 + 2\beta t)/4\beta^2$$

(Barndorff-Nielsen and Yeo, 1969).

REFERENCES

F. Abikhalil, P. Dupont and J. Janseen. A mathematical model of pricing in a large system of cash bonds, *App. Stoch. Model & Data Analysis*, **1**, 55–64, (1985).

O. Barndorff-Nielson and G.F. Yeo. Negative binomial processes, *J. Appl. Prob.* **6**, 633–647 (1969).

R.N. Bhattacharya and E.C. Waymire. *Stochastic Processes With Applications*, Wiley, New York (1990).

A. Blanc-Lapièrre and R. Fortet. *Théorie des Fonctions Aléatories*, Masson et Cie., Paris (1953).

S. Chandrasekhar. Stochastic problems in physics and astronomy, *Rev. Mod. Phy.* **15**, 1–89 (1943) [Reprinted in *Selected Papers on Noise and Stochastic Processes*, (Ed. Wax, N.) 1–91, Dover, New York (1954)].

D.R. Cox and H.D. Miller. *The Theory of Stochastic Processes*, Methuen, London (1965).

J.L. Doob. *Stochastic Processes*, Wiley, New York (1953).

—— The Brownian movement and stochastic equations, *Ann. Math.* **43**, 351–369 (1942).

A. Einstein. *Investigation on the Theory of Brownian Movement*, Dover, New York (1956).

W. Feller. *An Introduction to Probability Theory and Its Applications*, Vol **I**, 3rd ed. (1968), Vol **II** (1966), Wiley, New York.

J.L. Folks and R.S. Chhikara. The inverse Gaussian distribution and its statistical application–a review, *J.R.S.S.*, **B. 40**, 263–275 (1978).

J.M. Harrison. *Brownian Motion and Stochastic Flow Systems*, Wiley, New York (1985).

R.A. Howard. *Dynamic Programming and Markov Processes*, M.I.T. Press, Cambridge, MA (1960).

M. Iosifescu and P. Tautu. *Stochastic Processes and Applications in Biology and Medicine*, **I**, Springer-Verlag (1973).

N.L. Johnson and S. Kotz. *Continuous Univariate Distributions*, Vol. I. Houghton Mifflin Co., Boston (1970).

S. Karlin and H.M. Taylor. *A First Course in Stochastic Processes*, 2nd ed. Academic Press, New York (1975).

P. Kind, R. Sh. Liptser and W.J. Runggaldier. Diffusion approximation in past dependent models and applications to option pricing. *Ann. Appl. Prob.* **1**, 379–405 (1991).

L. Kleinrock, *Queueing Systems*, Vol. **II**, Wiley, New York (1976).

P. Lévy. *Processus Stochastiques et Mouvement Brownien*, 2nd ed. Gauthier-Villars, Paris (1965).

A.G. Malliaris and W.A. Brock. *Stochastic Methods in Economics and Finance*, North-Holland (1982).

J. Medhi. *Stochastic Models in Queueing Theory*, Academic Press, Boston, & San Diego, CA (1991).

M.K. Ochi. *Applied Probability and Stochastic Processes In Engineering & Physical Sciences*, Wiley, New York (1990).

N.U. Prabhu. *Stochastic Processes*, Macmillan, New York (1965).

Yu. V. Prohorov and Yu. A. Rozanov. *Probability Theory*, Springer-Verlag (1969).

S. Sethi and G. Thompson. *Optimal Control Theory : Applications to Management Science*, Martinus Nijhoff, Boston (1981).

G.E. Uhlenbeck and L.S. Ornstein. On the theory of Brownian motion, *Physical Review*, **36**, 93–112 (1930) [Reprinted in *Selected Papers on Noise and Stochastic Processes*,(Ed. Wax, N.) Dover, New York (1954)].

D. Vasicek. An equilibrium characterization of the term structure, *J. Fin. Eco.*, **5**, 177–188 (1977).

M.T. Wasan. Sufficient conditions for a first passage time to be that of a Brownian motion, *J. Appl. Prob.* **6**, 218–223 (1969).

D. Williams. *Diffusions, Markov-process and Martingales*, Wiley, New York (1979).

E.Wong and J. Hajek. *Stochastic Processes in Engineering Systems*, Springer-Verlag, New York (1989).

CHAPTER 6

Renewal Processes and Theory

6.1 RENEWAL PROCESS

Renewal process was introduced as a generalisation of Poisson process in Section 4.3.7. There it was considered as a process in continuous time. Renewal theory has assumed great importance because of its theoretical structure as well as for its application in diverse areas. Renewal theory and renewal theoretic arguments have often been advanced in a variety of situations, such as demography, manpower studies, reliability, replacement and maintenance, inventory control, queueing, and simulation and Monte-Carlo methods and so on. For applications, reference may be made, for example, to Cox (1962), Keyfitz (1968), Bartholomew and Forbes (1979), Barlow and Proschan (1975), Crane and Lemoine (1977), Tijms (1986), Sahin (1990) and so on.

We propose to discuss the topic in greater detail, first taking renewal process in discrete time.

6.1.1 Renewal Process in Discrete Time

Consider a sequence of repeated trials with possible outcomes E_j, $j = 1, 2, \ldots$ The trials need *not* be independent; we assume that the trials can be repeated infinitely. Suppose that we are interested in a certain outcome in a trial or a pattern of outcomes in a number of trials. We denote this event by E^*. Whenever E^* occurs we say that a *renewal* has occurred; if it occurs at nth trial, we say that a renewal occurs at trial number n. Once a renewal occurs at nth trial, trials are counted thenceforth from scratch.

The interval between occurrences of two successive renewals (two successive occurrences of the pattern E^*) is called a *renewal period* of the process. Denote

$$f_n = \Pr \{E^* \text{ occurs } \textit{for the first time} \text{ at the } n\text{th trial}\} \tag{1.1}$$
$$p_n = \Pr \{E^* \text{ occurs at the } n\text{th trial (not necessarily for the first time)}\}. \tag{1.2}$$

Define

$$f_0 = 0, p_0 = 1 \tag{1.3}$$

$$F(s) = \sum_{n=0}^{\infty} f_n s^n, \quad P(s) = \sum_{n=0}^{\infty} p_n s^n \tag{1.4}$$

Now

$$f^* = \Sigma f_n \tag{1.5}$$

is the probability that the renewal $E*$ occurs at some trial in a long sequence of trials. We have $f* \leq 1$; when $f* = 1$, then $\{f_n\}$ is a proper probability distribution representing the distribution of the length of a renewal period T i.e. $P\{T = n\} = f_n$. However $\{p_n\}$ is not a probability distribution. The renewal event is termed as *persistent (recurrent)* when $f* = 1$; and *transient* when $f* < 1$.

Example 1(a). Consider a dice throwing experiment. The event $E*$ that corresponds to the pattern that a 6 occurs is a renewal event. If the die is fair, then $E*$ occurs in a trial with probability 1/6 and the probability of non-occurrence at a trial is 5/6. We have

$$f_n = \left(\frac{5}{6}\right)^{n-1} \cdot \frac{1}{6}, \quad n \geq 1, \sum_{n=1}^{\infty} f_n = 1.$$

The renewal event is persistent and the renewal period has geometric distribution.

Let T_i denote the ith renewal period, then

$$S_n = T_i + \ldots + T_n \tag{1.6}$$

If N_m denote the number of renewals in a total of m trials, then it follows that

$$N_m \geq n \Leftrightarrow S_n \leq m. \tag{1.7}$$

6.1.2 Relation Between $F(s)$ and $P(s)$

The event that $E*$ occurs at the nth trial may be a compound event such that $E*$ occurs for the first time at the rth ($r < n$) trial and again at the later trial number n (i.e. in subsequent $n - r$ trials) and thus

$$p_n = \sum_{r=1}^{n} f_r \, p_{n-r}, \quad n \geq 1. \tag{1.8}$$

The r.h.s. in a *convolution* relation $\{f_n\} * \{p_n\}$ between two sequences. Multiplying by s^n, $n = 1, 2, \ldots$

$$p_n s^n = (f_1 s)(p_{n-1} s^{n-1}) + \cdots + (f_n s^n) p_0$$

and adding

$$\sum_{n=1}^{\infty} p_n s^n = (f_1 s) \sum_{n=1}^{\infty} p_{n-1} s^{n-1} + (f_2 s^2) \sum_{n=2}^{\infty} p_{n-2} s^{n-2} + \ldots$$

or

$$P(s) - 1 = P(s)\left[\sum_{n=1}^{\infty} f_n s^n\right] = P(s) F(s)$$

Thus

$$P(s) = \frac{1}{1 - F(s)}$$

(1.9)

and

$$F(s) = \frac{P(s) - 1}{P(s)}.$$

Now from (1.9) it follows that $\sum p_n = P(1) = \frac{1}{1 - F(1)}$ is convergent if and only if $F(1) < 1$, i.e. E^* is transient. In other words for E^* is transient if and only if

$$\sum p_n = P(1) \text{ is finite.}$$

The probability that E^* ever occurs is given by

$$f^* = F(1) = \frac{\sum p_n - 1}{\sum p_n};$$

E^* is persistent if and only if $\sum p_n$ is divergent.

Definition. The renewal event E^* is said to be *periodic* if there exists an integer m (> 1) such that E^* can occur only at trials numbered $m, 2m, \ldots$ The greatest m with this property is said to be the *period* of E^*. It is said to be *aperiodic* if no such m exists.

The sequence $\{a_n\}$ is said to be periodic with period m (> 1) if $a_n = 0$, for $n \neq km$, $k = 1, 2, \ldots$ and m is the greatest integer with this property.

Definition. For a persistent and aperiodic renewal event (pattern) $F'(1) = \sum n f_n = E(T)$ is the *mean recurrence time* (i.e. mean time between two consecutive renewals or mean waiting time between two consecutive renewals). $F'(1)$ may be finite or infinite.

6.1.3 Renewal Interval

The renewal interval (period) T has the p.m.f.

$$\Pr\{T = n\} = f_n.$$

T is a proper r.v. when $\sum f_n = F(1) = 1$ with mean recurrence time $\sum n f_n = F'(1)$. T is also called the waiting time for the occurrence of the renewal E^* (or renewal period of the process). The generating function of T is $F(s) = \sum f_n s^n$.

The probability $f_n^{(2)}$ that E^* occurs for the second time at nth trial is given by

$$f_n^{(2)} = \sum_{k=1}^{n-1} f_k f_{n-k}$$

Similarly, the probability $f_n^{(r)}$ that E^* occurs for the rth time at nth trial in given by

$$f_n^{(r)} = \sum_{k=1}^{n-1} f_k f_{n-k}^{(r)} \tag{1.10}$$

Thus $\{f_n^{(r)}\}$ gives the probability distribution of

$$T^{(r)} = T_1 + \dots + T_r$$

where T_i are i.i.d. random variables distributed as T. The generating function of $\{f_n^{(r)}\}$ is given by

$$F^{(r)}(s) = \sum_n f_n^{(r)} s^n$$

$$= [F(s)]^r. \tag{1.11}$$

Putting $s = 1$, we get

$$\sum_n f_n^{(r)} = F^{(r)}(1) = [F(1)]^r = (f^*)^r.$$

$\sum_n f_n^{(r)}$ is the probability that E^* occurs *at least* r times if the process is continued indefinitely. It follows that

Pr $\{E^*$ occurs *exactly* r times if the process is continued indefinitely$\}$

$$= (f^*)^r - (f^*)^{r+1} = (f^*)^r [1 - f^*].$$

6.1.4 Generalised Form : Delayed Recurrent Event

So far we have assumed that the renewal (or recurrence interval) upto the *first* occurrence of the renewal event E^* has the same distribution as $\{f_n\}$, the recurrence interval between successive occurrences of E^*. In many situations, it is more realistic to assume that the recurrence interval $\{b_n\}$ upto the first occurrence of E^* has a distribution different from $\{f_n\}$.

The first occurrence of E^* is then called a *delayed recurrent event* while the subsequent occurrences are ordinary recurrent events. For this situation, denote

$$v_n = \Pr\{E^* \text{ occurs at } n\text{th trial}\}.$$

Suppose that the first occurrence of E^* happens at trial number k (with probability b_k) and then a renewal occurrence of E^* occurs at the subsequent $(n-k)$ trials (with probability p_{n-k}). Thus we get

$$v_n = b_n + b_{n-1}p_1 + b_{n-2}p_2 + \cdots + b_1 p_{n-1} + b_0 p_n \qquad (1.12)$$

That is,

$$v_n = \{b_n\} * \{p_n\}. \qquad (1.13)$$

Denoting $V(s) = \sum v_n s^n$, $B(s) = \sum b_n s^n$, we can write

$$V(s) = B(s) P(s) \qquad (1.14)$$

$$= \frac{B(s)}{1 - F(s)} \text{ (using (1.9))}. \qquad (1.15)$$

We now state an important relation concerning the delayed recurrent events.

Theorem 6.1. If $p_n \to \alpha$, then

$$v_n \to \alpha b \qquad (1.16)$$

where $b = B(1) = \sum b_n$
If $\sum p_n$ converges to β $(< \infty)$ then

$$\sum v_n \to b \beta \qquad (1.17)$$

Proof: Denote

$$r_k = \Pr\{\text{first renewal period} > k\}$$

$$= b_{k+1} + b_{k+2} + \cdots .$$

We can choose k sufficiently large such that $r_k < \epsilon$. Since $p_m \le 1$, we get, from (1.12),

$$b_0 p_n + b_1 p_{n-1} + \cdots + b_k p_{n-k} \le v_n$$

$$= b_0 p_n + \cdots + b_k p_{n-k} + \{b_{k+1} p_{n-(k+1)} + \cdots + b_{n-1}p_1 + b_n\}$$

$$\le b_0 p_n + \cdots + b_k p_{n-k} + \{b_{k+1} + \cdots + b_n\}$$

$$\le b_0 p_n + \cdots + b_k p_{n-k} + r_k. \qquad (1.18)$$

As $p_n \rightarrow \alpha$

$$b_0 p_n + b_1 p_{n-1} + \dots + b_k p_{n-k}$$

$$\rightarrow (b_0 + b_1 + \dots + b_k)\alpha$$

$$= (b - r_k)\alpha$$

$$> b\,\alpha - \in \alpha$$

$$> b\,\alpha - 2 \in, \text{ since } p_n < 1 \text{ implies that } \alpha \leq 1 < 2;$$

and

$$b_0 p_n + \dots + b_k p_{n-k} + r_k$$

$$\rightarrow (b - r_k)\alpha + r_k$$

$$= b\,\alpha + r_k(1 - \alpha)$$

$$< b\,\alpha + (1 - \alpha) \in$$

$$< b\,\alpha + 2 \in$$

so that from (1.18) we get

$$b\,\alpha - 2 \in \, < v_n < b\,\alpha + 2 \in .$$

Making $\in$ sufficiently small, we get

$$\lim_{n \rightarrow \infty} v_n \rightarrow b\,\alpha$$

which gives (1.16).
From (1.14) we get

$$\sum v_n = V(1) = B(1)\,P(1)$$

$$\rightarrow b\,\beta \quad (\text{as } P(1) = \sum p_n \rightarrow \beta)$$

which gives (1.17).   ▲

Corollary: If E^* is persistent, then $v_n \rightarrow \dfrac{b}{\mu}$. A rigorous proof is outside the scope of

the book. We give a rough demonstration here.

$$v_n = \lim_{s \to 1-0} (1-s) V(s)$$

$$= \lim_{s \to 1-0} B(s) \frac{1-s}{1-F(s)}$$

$$\to B(1) \frac{1}{F'(1)} = \frac{b}{\mu}. \tag{1.19}$$

Note. The above also implies that when E^* is persistent

$$p_n \to \frac{1}{\mu} \tag{1.20}$$

6.1.5 Renewal Theory in Discrete Time

We shall now consider a result of greater generality (not necessarily connected with the stochastic behaviour of recurrent events).

Suppose that $\{f_n, n = 1, 2, \ldots\}$, $\{b_n, n = 0, 1, 2, \ldots\}$ are two sequences of real numbers such that

$$f_n \geq 0, f = \Sigma f_n < \infty \tag{1.21}$$

and

$$b_n \geq 0, b = \Sigma b_n < \infty. \tag{1.22}$$

Define a new sequence $\{v_n, n = 0, 1, \ldots\}$ by the convolution relation

$$v_n = b_n + v_{n-1} f_1 + v_{n-2} f_2 + \ldots + v_0 f_n$$

$$= b_n + \sum_{r=1}^{n} f_r v_{n-r}. \tag{1.23}$$

The above defines v_n uniquely in terms of $\{b_n\}$ and $\{f_n\}$. In terms of their generating functions we get

$$V(s) = B(s) + F(s) V(s)$$

or

$$V(s) = \frac{B(s)}{1-F(s)}; \tag{1.24}$$

$F(s)$ and $B(s)$ converge at least for $0 \leq s < 1$ and if $F(s) < 1$, then $V(s)$ is a power series in s.

The sequence $\{f_n\}$ is periodic, if there exists an integer m such that $f_n = 0$ except for $n = km$.

We now state the basic theorem.

Theorem 6.2. Renewal Theorem

Suppose that the relations (1.22) hold and $\{f_n\}$ is not periodic.

(a) If $f < 1$, then $v_n \to 0$

and $\quad \Sigma v_n = \dfrac{b}{1-f}$

(b) If $f = 1$, then $v_n \to \dfrac{b}{\mu}$.

Proof: (a) From (1.24) we get, when $f < 1$

$$V(1) = \frac{B(1)}{1-f}$$

or
$$\Sigma v_n = \frac{b}{1-f}.$$

Σv_n is convergent; this implies that $v_n \to 0$.

(b) When $f = 1$, $\Sigma v_n \to \infty$.

But as in Corollary above, we can see that

$$v_n \to \frac{b}{\mu}, \quad \mu = F'(1) = \Sigma n f_n.$$

Note: (1) The above result has great importance. In a large number of applications in stochastic processes one can find that probabilities connected with a process satisfy a relation of convolution type as given in (1.23). The Renewal Theorem then enables us to get limit theorems.

(2) For rigorous proofs and other details, refer to Feller, I.

6.2 RENEWAL PROCESSES IN CONTINUOUS TIME

Renewal process as a generalisation of Poisson process was introduced in Sec. 4.3.7. Here we propose to discuss the topic in some details.

Let $\{X_n, n = 1, 2, \dots\}$ be a sequence of non-negative independent random variables. Assume that $\Pr\{X_n = 0\} < 1$, and that the random variables are identically distributed and are continuous with a distribution function $F(\cdot)$. Since X_n is non-negative, it follows that $E\{X_n\}$ exists and let us denote

$$E(X_n) = \int_0^\infty x \, dF(x) = \mu,$$

where μ may be infinite. Whenever $\mu = \infty$, $1/\mu$ shall be interpreted as 0.

Let
$$S_0 = 0, S_n = X_1 + X_2 + \cdots + X_n, \quad n \geq 1$$

and let $F_n(x) = \Pr\{S_n \leq x\}$ be the distribution function of S_n, $n \geq 1$;

$$F_0(x) = 1 \text{ if } x \geq 0 \text{ and } F_0(x) = 0 \text{ if } x < 0.$$

Definition. Define the random variable

$$N(t) = \sup\{n : S_n \leq t\}. \tag{2.1}$$

The process $\{N(t), t \geq 0\}$ is called a *renewal process* with distribution F (or generated or induced by F).

It is also customary to say that the sequence of random variables $\{S_n, n = 1, 2, \ldots\}$ (as also the sequence $\{X_n, n = 1, 2, \ldots\}$) constitutes a renewal process with distribution F.

If, for some n, $S_n = t$, then a renewal is said to occur at t; S_n gives the time (epoch) of the nth renewal, and is called nth *renewal* (or *regeneration*) epoch. The random variable $N(t)$ gives the number of renewals occurring in $[0, t]$. The random variable X_n gives the interarrival time (or waiting time) between $(n-1)$th and nth renewals. The interarrival times are independently and identically distributed. When they have common exponential distribution, we get Poisson process as a particular case.

The renewals, where X_i are i.i.d. random variables, are called *Palm flow of events*; where X_i are i.i.d. exponential the renewals are called *ordinary* or *Poisson flow of events*.

Note: Renewal processes in *more than one dimension* are also discussed in the literature. Prabhu (1980) considers a renewal process in two dimensions, associated with a random walk.

Simple Examples

One of the simplest examples of a renewal process is provided by lifetime distributions of a component such as an electric bulb which either works or fails completely. Suppose that the detection of the failure of a bulb and its replacement by a new bulb take place instantaneously and suppose that the lifetimes of bulbs are i.i.d. random variables with distribution F. We then have a renewal process with distribution F.

Consider a stage in an industrial process relating to production of a certain component in batches. Immediately on completion of production of a batch, that of another batch is undertaken. Suppose that the times taken to produce successive batches are i.i.d. random variables with distribution F. We get a renewal process with distribution F.

6.2.1 Renewal Function and Renewal Density

The function $M(t) = E\{N(t)\}$ is called the *renewal function* of the process with distribution F. It is clear that

$$\{N(t) \geq n\} \Leftrightarrow \{S_n \leq t\} \tag{2.2}$$

or $\{N(t) \geq n\}$ if and only if $\{S_n \leq t\}$.

Equivalently $\{N(t) < n\} \Leftrightarrow \{S_n > t\}$.

Theorem 6.3. The distribution of $N(t)$ is given by

$$p_n(t) = \Pr\{N(t) = n\} = F_n(t) - F_{n+1}(t) \tag{2.3}$$

and the expected number of renewals by

$$M(t) = \sum_{n=1}^{\infty} F_n(t) \tag{2.4}$$

Proof: We have

$$\Pr\{N(t) = n\} = \Pr\{N(t) \geq n\} - \Pr\{N(t) \geq n+1\}$$
$$= \Pr\{S_n \leq t\} - \Pr\{S_{n+1} \leq t\}$$
$$= F_n(t) - F_{n+1}(t).$$

Again,

$$M(t) = E\{N(t)\} = \sum_{n=0}^{\infty} n\, p_n(t)$$

$$= \sum_{n=0}^{\infty} n\, \{F_n(t) - F_{n+1}(t)\}$$

$$= \sum_{n=1}^{\infty} F_n(t)$$

$$= \sum_{n=1}^{\infty} \Pr\{S_n \leq t\}. \text{ Hence proved.} \qquad \blacktriangle$$

The relation (2.4) can be put in terms of Laplace transforms as follows:

Let $F'(x) = f(x)$ be the density function of (p.d.f.) of X_n and $g^*(s)$ denote the Laplace transform of a function $g(t)$. Then taking Laplace transform of both sides of (2.4), we get

$$M^*(s) = \sum_{n=1}^{\infty} F_n^*(s) = \frac{1}{s} \sum_{n=1}^{\infty} f_n^*(s)$$

$$= \frac{1}{s} \sum_{n=1}^{\infty} [f^*(s)]^n = \frac{f^*(s)}{s[1 - f^*(s)]}. \tag{2.5}$$

This is equivalent to

$$f^*(s) = \frac{s\,M^*(s)}{1 + s\,M^*(s)}.$$ (2.6)

These show that $M(t)$ and $F(x)$ can be determined *uniquely* one from the other.

Note that $M(t) = E\{N(t)\}$ is a *sure* function and *not* a random function or stochastic process.

Renewal Density
    The derivative $m(t)$ of $M(t)$ (i.e. $M'(t) = m(t)$) is called the *renewal density*. We have

$$m(t) = \lim_{\Delta t \to 0} \frac{\Pr\{\text{one or more renewals in } (t, t + \Delta t)\}}{\Delta t}$$

$$= \sum_{n=1}^{\infty} \lim_{\Delta t \to 0} \frac{\Pr\{n\,\text{th renewal occurs in } (t, t + \Delta t)\}}{\Delta t}$$

$$= \sum_{n=1}^{\infty} \lim_{\Delta t \to 0} \frac{f_n(t)\Delta t + o(\Delta t)}{\Delta t}$$

(assuming that $F(x)$ is absolutely continuous and $F'_n(t) = f_n(t)$)

$$= \sum_{n=1}^{\infty} f_n(t) = \sum_{n=1}^{\infty} F'_n(t)$$

$$= M'(t).$$

The function $m(t)$ specifies the mean number of renewals to be expected in a narrow interval near t.
    Note that $m(t)$ is *not* a probability density function. As $m^*(s) = \text{L.T. } \{m(t)\} = s\,M^*(s)$, it follows from (2.5) that

$$m^*(s) = \frac{f^*(s)}{1 - f^*(s)}.$$

Example 2(a). Let X_n have gamma distribution having density

$$f(x) = \frac{a^k x^{k-1} e^{-ax}}{(k-1)!}, x \geq 0.$$

$$= 0, \text{ elsewhere.}$$

Then,

$$f^*(s) = \left(\frac{a}{s + a}\right)^k$$

and the density $F'_n(x)$ of

$$S_n = X_1 + X_2 + \dots + X_n$$

has the L.T.

$$\left(\frac{a}{s+a}\right)^{nk}$$

Thus

$$F_n'(x) = \frac{a^{nk} x^{nk-1} e^{-ax}}{(nk-1)!}$$

and hence

$$F_n(x) = \int_0^x F_n'(y)\,dy$$

$$= 1 - e^{-ax} \sum_{r=0}^{nk-1} \frac{(ax)^r}{r!}, \quad n \geq 1.$$

Thus

$$p_n(t) = F_n(t) - F_{n+1}(t)$$

$$= e^{-at} \sum_{r=nk}^{(n+1)k-1} \frac{(ax)^r}{r!},$$

and using (2.5), we get

$$M*(s) = \frac{a^k}{s[(s+a)^k - a^k]}. \tag{2.7}$$

Particular Cases
(1) *Markovian case*: When $k = 1$, X_n has negative exponential distribution and the renewal process then reduces to a Poisson process. As is to be expected, we have then,

$$p_n(t) = e^{-at} \frac{(at)^n}{n!}$$

and

$$M*(s) = a/s^2 \text{ so that } M(t) = at.$$

(2) When $k = 2$, (Gamma with shape parameter 2), then from (2.7),

$$M*(s) = \frac{a^2}{s[(s+a)^2 - a^2]} = \frac{a}{2}\left[\frac{1}{s^2} - \frac{1}{s(s+2a)}\right]$$

$$= \frac{a}{2}\left[\frac{1}{s^2} - \frac{1}{2a}\left\{\frac{1}{s} - \frac{1}{s+2a}\right\}\right].$$

Inverting the L.T. we get

$$M(t) = \frac{a}{2}t - \frac{1}{4} + \frac{1}{4}e^{-2at}$$

as the expected number of renewals in an interval of time t.

For a component having for lifetime the said gamma distribution, the expected number of renewals in time t is given by the above $M(t)$.

Example 2(b). *Hyper-exponential distribution.* Let X_n have density

$$f(t) = pae^{-at} + (1-p)be^{-bt}, \quad 0 \le p \le 1,$$

$$a > b > 0. \tag{2.8}$$

Such a model may be used to describe a system which has two kinds of components—a proportion 'p' of components having, say, a high failure rate a and the remaining proportion $(1-p)$ of components having a different, say, a lower failure rate b. We have

$$f*(s) = \frac{pa}{s+a} + \frac{(1-p)b}{s+b} \tag{2.9}$$

so that

$$M*(s) = \frac{ab + s[pa + (1-p)b]}{s^2[s + (1-p)a + pb]}.$$

Writing $A = pa + (1-p)b$ and $B = (1-p)a + pb$, we get

$$M*(s) = \frac{As + ab}{s^2(s+B)}$$

$$= \frac{A}{s(s+b)} + \frac{ab}{s^2(s+B)}$$

$$= \frac{A}{B}\left\{\frac{1}{s} - \frac{1}{s+B}\right\} + \frac{ab}{B}\left[\frac{1}{s^2} - \left\{\frac{1}{s} - \frac{1}{s+B}\right\}\frac{1}{B}\right].$$

Inverting the L.T., we get

$$M(t) = \frac{A}{B}(1 - e^{-Bt}) + \frac{ab}{B}\left\{ t - \frac{1 - e^{-Bt}}{B} \right\}$$

$$= \frac{abt}{B} + C(1 - e^{-Bt}), \tag{2.10}$$

where

$$C = \left(\frac{A}{B} - \frac{ab}{B^2} \right)$$

$$= \frac{p(1-p)(a-b)^2}{B^2} \geq 0.$$

Markovian case: When $p = 1$ (or $p = 0$) the distribution of X_n reduces to negative exponential and then $C = 0$, i.e. the second term of (2.10) vanishes, so that we get $M(t) = at$ (or bt).

6.3 RENEWAL EQUATION

An integral equation can be obtained for the renewal function

$$M(t) = E\{N(t)\},$$

which gives the expected number of renewals in $[0, t]$.

Theorem 6.4. The renewal function M satisfies the equation

$$M(t) = F(t) + \int_0^t M(t-x)\, dF(x). \tag{3.1}$$

Proof: By conditioning on the duration of the first renewal X_1, we get

$$M(t) = E\{N(t)\} = \int_0^\infty E\{N(t) \mid X_1 = x\}\, dF(x).$$

Consider $x > t$; given that $X_1 = x > t$, no renewal occurs in $[0, t]$, so that

$$E\{N(t) \mid X_1 = x\} = 0.$$

Consider $0 \leq x \leq t$; given that the first renewal occurs at x ($\leq t$), then the process starts again at epoch x, and the expected number of renewals in the remaining interval of length $(t - x)$ is $E\{N(t-x)\}$, so that

$$E\{N(t)\,|\,X_1\} = 1 + E\{N(t-x)\}$$

$$= 1 + M(t-x).$$

Thus considering the above two equations, we get

$$M(t) = \int_0^t \{1 + M(t-x)\}\, dF(x)$$

$$= F(t) + \int_0^t M(t-x)\, dF(x).$$

We have thus established (3.1).   ▲

The equation (3.1) is called the *integral equation of renewal theory* (or simply *renewal equation*) and the argument used to derive it is known as '*renewal argument*'. The renewal equation is also expressed as

$$M = F + M*f.$$

The equation (3.1) can also be established as given below.

We have
$$M(t) = \sum_{n=1}^{\infty} F_n(t) = F_1(t) + \sum_{n=1}^{\infty} F_{n+1}(t)$$

$$= F(t) + \sum_{n=1}^{\infty}\left\{\int_0^t F_n(t-x)\, dF(x)\right\},$$

F_{n+1} being the convolution of F_n and $F_1 = F$. Thus, assuming the validity of the change of order of integration and summation, we get

$$M(t) = F(t) + \int_0^t \left\{\sum_{n=1}^{\infty} F_n(t-x)\right\} dF(x);$$

or,
$$M(t) = F(t) + \int_0^t M(t-x)\, dF(x).$$

It follows that $M(t) = \sum_{n=1}^{\infty} F_n(t)$ satisfies the integral equation (3.1).

The renewal equation (3.1) can be generalised as follows

$$v(t) = g(t) + \int_0^t v(t-x)\, dF(x),\, t \geq 0, \tag{3.2}$$

where g and F are known and v is unknown. The equation (3.2) is called a *renewal type equation*.

A unique solution of $v(t)$ exists in terms of g and F as can be seen from the following.

Theorem 6.5. If

$$v(t) = g(t) + \int_0^t v(t-x)\, dF(x),\ t \geq 0$$

then

$$v(t) = g(t) + \int_0^t g(t-x)\, dM(x),\tag{3.3}$$

where

$$M(t) = \sum_{n=1}^{\infty} F_n(t).$$

Proof: Taking L.T. of (3.2), we get

$$v^*(s) = g^*(s) + v^*(s)f^*(s)$$

or

$$v^*(s) = \frac{g^*(s)}{1 - f^*(s)}$$

$$= g^*(s)\left[1 + \frac{f^*(s)}{1 - f^*(s)}\right]$$

$$= g^*(s)[1 + sM^*(s)].$$

Inverting the L.T., we get

$$v(t) = g(t) + \int_0^t g(t-x)\, dM(x)$$

and the solution $v(t)$ is unique, since a function is uniquely determined by its L.T. Thus the theorem is proved. ▲

Note: (1) Renewal and renewal type equations occur in various different situations. Feller, Vol. II contains a number of examples of phenomena satisfying renewal type equations. Keyfitz (1968) considers use of renewal type equation in demography, Bartholomew (1973) in social processes, Bartholomew (1976), Bartholomew and Forbes (1979) in manpower studies, and Sahin (1990) in inventory models and so on; see also Kohlas (1982), Tijms (1986).

Note: (2) Bartholomew (*J.R.S.S.* **B 25** (1963) 432–441) has put forward a simple approximate solution of the renewal equation

$$h(t) = f(t) + \int_0^t h(t-x)f(x)\, dx$$

[where $h(t)$ is a renewal density and $f(x)$ is the p.d.f. of a r.v. X (completed length of service); this equation was encountered in manpower studies].

The simple approximate solution given by him is

$$h(t) = f(t) + F^2(t) \bigg/ \left[\int_0^t \overline{F}(x)\, dx \right]$$

where $F(\cdot)$ is the d.f. and $\overline{F}(\cdot)$ is the complementary d.f. of the r.v. X.

6.4 STOPPING TIME : WALD'S EQUATION

6.4.1 Stopping Time

Before going to renewal theorems, we discuss a special type of non-negative r.v. associated with a sequence of r.v. $\{X_i\}$ (or a stochastic process $\{X(t)\}$). Such a variable, first considered by Wald (1947) while formulating sequential analysis, is known as *r.v. independent of the future* or a *Markov time or a stopping time*. For an elaborate discussion on Markov time and its applications, see A.N. Sirjaev (1973). We consider a somewhat special case here.

Definition. An integer-valued random variable N is said to be a *stopping time* for the sequence $\{X_i\}$ if the event $\{N = n\}$ is independent of

$$X_{n+1}, X_{n+2}, X_{n+3}, \dots \text{ for all } n = 1, 2, \dots.$$

It can be shown that N is a proper random variable.

As an example of such a variable, consider a coin tossing experiment; let the outcome of the ith toss be denoted by $X_i = 1$ or 0 depending on the result being head or tail respectively; and let $\Pr(X_i = 1) = p = 1 - \Pr(X_i = 0)$ and then $E(X_i) = p$.

The sum

$$S_n = X_1 + \dots + X_n$$

denotes the cumulative number of heads in the first n tosses. Suppose that m is a given positive integer, then $N = \min\{n : S_n = m\}$ is a stopping time.

We have $E(S_N) = m$, and $E(N) = m/p$.

Consider the number $N(t)$ of renewals by time t, w.r.t. a sequence of interarrival times $\{X_i\}$. Now $N(t) = n$ whenever $S_n \leq t$ and $S_{n+1} > t$, i.e. the event $N(t) = n$ depends not only on $X_1, \dots, X_n$ but also on X_{n+1}.

Consider the variable $N(t) + 1$. Now $N(t) + 1 = n$ implies that $S_{n-1} \leq t$ and $S_n > t$, so that the event $\{N(t) + 1 = n\}$ is independent of $X_{n+1}, X_{n+2}, \dots$. Thus $N(t) + 1$ is a stopping time for the sequence $\{X_i\}$ while $N(t)$ is not.

From a corollary to Theorem 1.3 we find that: if

$$S_N = \sum_{i=1}^{N} X_i,$$

where X_i are i.i.d. random variables and N is a r.v. (independent of X_i's) having finite expectation, then

$$E\left\{\sum_{i=1}^{N} X_i\right\} = E(X_i)\,E(N).$$

The same result holds also when N is a stopping time for the sequence $\{X_i\}$. The proposition which is used in proving certain results in renewal theory is given below.

6.4.2 Wald's Equation

Let $\{X_i\}$ be a sequence of independent random variables, having the same expectation and let N be a stopping time for $\{X_i\}$ and $E(N) < \infty$, then

$$E\left\{\sum_{i=1}^{N} X_i\right\} = E(X_i)\,E\{N\}. \tag{4.1}$$

Proof: Let

$$Z_i = \begin{cases} 1 & \text{if} \quad N \geq i \\ 0 & \text{if} \quad N < i, \end{cases}$$

so that

$$\sum_{i=1}^{N} X_i = \sum_{i=1}^{\infty} X_i Z_i;$$

and thus

$$E\left\{\sum_{i=1}^{N} X_i\right\} = E\left\{\sum_{i=1}^{\infty} X_i Z_i\right\} = \sum_{i=1}^{\infty} E\{X_i Z_i\},$$

assuming the validity of the change of order of expectation and summation. Now Z_i is determined by $\{N < i\}$, i.e. by $X_1, \ldots, X_{i-1}$ and is thus independent of X_i. Thus

$$E\left\{\sum_{i=1}^{N} X_i\right\} = \sum_{i=1}^{\infty} E\{X_i\}\,E\{Z_i\}$$

$$= E\{X_i\} \sum_{i=1}^{\infty} E\{Z_i\}$$

$$= E\{X_i\} \sum_{i=1}^{\infty} \Pr\{N \geq i\}$$

$$= E\{X_i\}\,E\{N\}.$$

Hence proved. ▲

Corollary: Since $N(t) + 1$ is a stopping time for the sequence $\{X_i\}$ we have,

$$E\{S_{N(t)+1}\} = E\left\{\sum_{i=1}^{N(t)+1} X_i\right\}$$

$$= E\{X_i\}\,E\{N(t)+1\}$$

$$= E\{X_i\}.\{M(t)+1\}.$$

Remarks (1). For Wald's equation to hold, the r.v.'s X_i's need not be identically distributed but X_i's must be independent and have the same mean, i.e. $E(X_i) = E(X)$ for all i.

(2) If N is independent of $\{X_i\}$, then

$$E\left\{\sum_{i=1}^{N} X_i \,|\, N\right\} = N\, E\{X_i\}$$

and

$$E\{X_i\} = E\left[E\left\{\sum_{i=1}^{n} X_i \,|\, N\right\}\right] = E(N)E(X_i).$$

Example 4(a). Consider a sequence of independent coin tosses. The result of the nth toss is denoted by X_n: $X_n = 1$ if Head occurs and $X_n = -1$ if Tail occurs. Let $\Pr\{X_n = 1\} = p$, $\Pr\{X_n = -1\} = q$. Then

$$E\{X_n) = p - q \text{ for all } n,$$

and

$$S_n = X_1 + ... + X_n$$

gives the number of Heads minus the number of Tails in n tosses. Let

$$N = \min\{n: S_n = 1, n = 1, 2, ...\};$$

N is the first toss in which the number of Heads exceeds the number of tails by exactly 1. Here N is a stopping time and $S_N = 1$ for any value of N so that $E\{S_N\} = 1$. By Wald's theorem,

$$1 = E\{S_N\} = E\{X_i\}\, E\{N\};$$

thus

$$E\{N\} = \frac{1}{p-q} \qquad \text{when } p > q,$$

$$E\{N\} = \infty \qquad \text{when } p = q,$$

and N is defective when $p < q$.

6.5 RENEWAL THEOREMS

Poisson process (with parameter a) is a renewal process having exponential interarrival times X_n (with mean $1/a$), we have

$$M(t) = at$$

or,

$$M(t)/t = a = 1/E(X_n).$$

In the general case, the result

$$M(t)/t \to 1/\mu, \mu = E(X_n) \leq \infty$$

holds as $t \to \infty$. The result (due to Feller) is known as elementary renewal theorem.
Note: While for a Poisson process, the renewal function is *exactly linear*, for other renewal processes, it is *asymptotically linear*.
Before considering this, we consider a simple result.
Theorem 6.6. With probability 1,

$$\frac{N(t)}{t} \to \frac{1}{\mu} \text{ as } t \to \infty, \tag{5.1}$$

where
$$\mu = E(X_n) \leq \infty.$$

Proof: Consider an interval $[0, t]$; we have

$$S_{N(t)} \leq t < S_{N(t)+1} \tag{5.2}$$

Now the strong law of large numbers holds for the sequence $\{S_n\}$, so that, as $n \to \infty$,

$$\frac{S_n}{n} = \frac{X_1 + X_2 + \cdots + X_n}{n} \to E(X_n) = \mu$$

with probability 1. Again

$$\text{as } t \to \infty, N(t) \to \infty$$

with probability 1. Thus, with probability 1,

$$\frac{S_{N(t)}}{N(t)} \to \mu \text{ as } t \to \infty. \tag{5.3}$$

Similarly, with probability 1,

$$\frac{S_{N(t)+1}}{N(t)} = \frac{S_{N(t)+1}}{N(t)+1} \cdot \frac{N(t)+1}{N(t)} \to \mu \text{ as } t \to \infty. \tag{5.4}$$

Thus from the three relations, we get that, with probability 1,

$$N(t)/t \to 1/\mu \text{ as } t \to \infty. \qquad \blacktriangle$$

This theorem shows that for large t, the number of renewals per unit time converges to $1/\mu$.

We now go to the renewal theorems. The elementary renewal theorem is due to Feller (1941); here we shall follow the arguments put forward by Smith (1958).

6.5.1 Elementary Renewal Theorem

Theorem 6.7. *Elementary Renewal Theorem*. We have

$$\frac{M(t)}{t} \to \frac{1}{\mu} \quad \text{as } t \to \infty, \tag{5.5}$$

where $\mu = E(X_n) < \infty$, the limit being interpreted as 0 when $\mu = \infty$.

Proof: Let

$$N = N(t) + 1 \text{ and } S_N = t + Y(t),$$

where $Y(t)$ is the residual lifetime of the unit in use at time t. Let $\{X_i^{(j)}, j = 1, 2, \ldots\}$, be a sequence of independent realisations of the renewal process $\{X_i\}$, $\{S_n^{(j)}\}$ be the corresponding partial sums, and $N^{(j)}(t)$ the corresponding number of renewals in $[0, t]$.

Let

$$\tau = S_N, \ M_k = N^{(1)}(t) + \ldots + N^{(k)}(t),$$

$$T_k = \tau^{(1)} + \ldots + \tau^{(k)}.$$

Now T_k is the sum of $(k + M_k)$ i.i.d. random variables X_i. Thus, by the strong law of large numbers, if $\mu = E(X_i) < \infty$, then as $k \to \infty$

$$\frac{T_k}{k + M_k} = \frac{\sum_{i=1}^{k+M_k} X_i}{k + M_k} \to \mu \tag{5.6}$$

with probability 1.

By the same law, we get, as $k \to \infty$,

$$\frac{M_k}{k} = \frac{\sum_{j=1}^{k} N^{(j)}(t)}{k} \to E\{N(t)\} = M(t) \tag{5.7}$$

and

$$\frac{T_k}{k} = \sum_{j=1}^{k} \frac{\tau^{(j)}}{k} \to E(\tau) = E\{t + Y(t)\}$$

$$= t + E\{Y(t)\} \tag{5.8}$$

with probability 1. Combining (5.6), (5.7) and (5.8), we get, as $k \to \infty$

$$\frac{T_k}{k + M_k} = \frac{T_k}{k} \left[1 + \frac{M_k}{k} \right]^{-1} \to \frac{t + E(Y(t))}{1 + M(t)} \to \mu$$

or

$$t + E(Y(t)) = \mu \{1 + M(t)\}$$

i.e.
$$1 + \frac{E(Y(t))}{t} = \mu \left\{ \frac{1}{t} + \frac{M(t)}{t} \right\};$$

and now $Y(t)$ is positive, $E(Y(t))$ finite and hence as $t \to \infty$

$$\liminf_{t \to \infty} \frac{M(t)}{t} = \frac{1}{\mu}. \qquad (5.9)$$

To prove the theorem we have to show that

$$\limsup_{t \to \infty} \frac{M(t)}{t} \le \frac{1}{\mu}.$$

For doing this we define a new renewal process as follows:
Let A be a constant > 0 and for $n = 1, 2, \ldots$, and let

$$X_n^* = \begin{cases} X_n & \text{if } X_n \le A \\ A & \text{if } X_n > A \end{cases}$$

Let
$$S_n^* = \sum_{i=1}^{n} X_i^* \text{ and } N^*(t) = \sup \{ n : S_n^* \le t \},$$

$$M^*(t) = E\{N^*(t)\}.$$

Now
$$S_n^* \le S_n, \text{ and hence } N^*(t) \ge N(t)$$

and
$$M^*(t) \ge M(t).$$

Again
$$E(X_n^*) = \mu_A \le \mu \text{ and } \mu_A \to \mu, \text{ as } A \to \infty.$$

We have

$$(S_{N(t)+1}^*) \le t + A$$

$$E(S_{N(t)+1}^*) = \mu_A(M^*(t) + 1)$$

and hence

$$\limsup_{t \to \infty} \frac{M^*(t)}{t} \le \frac{1}{\mu_A}$$

and as $A \to \infty$,

$$\limsup_{t \to \infty} \frac{M(t)}{t} \le \frac{1}{\mu}. \qquad (5.10)$$

From (5.9) and (5.10) the result follows. ▲

Note: As an illustration, consider a renewal process generated by $\{X_n\}$, where X_n has gamma distribution with shape parameter $k = 2$. From Example 2(a), we find that as $t \to \infty$, $M(t)/t \to a/2$ and this is equal to $1/E\{X_n\}$.

The theorem implies that the average number of renewals per unit time, in the long run, converges to $1/\mu$; this should have intuitive appeal.

6.5.2 Applications

Example 5(a). *Age and block replacement policies*:
              (Barlow and Proschan, 1964)
The usual replacement policy implies replacement of a component, as and when it fails, by a similar new one. There are other policies besides this; the two most important replacement policies that are in use, in general, are the age and block replacement policies. Under an *age replacement policy*, a component is replaced upon failure *or* when it attains a specified age T, whichever occurs earlier. Under a *block* replacement policy, a component is replaced upon failure *and* also regularly at times $T, 2T, \ldots$. We shall call T, the *replacement interval* under these two policies. We assume, as usual, that components or units fail permanently, independently and that the detection of a failure and replacement of the failed item are instantaneous. Suppose that the successive lifetimes of the units are random variables with a common distribution function $F(x)$ having mean μ. Considering failures as renewals, the number of failures in $[0, t]$ under the three replacement policies, the usual one, the age replacement and the block replacement can be denoted by three renewal processes as indicated below:

$N(t)$ = the number of failures in $[0, t]$ for an ordinary renewal process.
$N_A(t, T)$ = the number of failures in $[0, t]$ under age replacement policy, with replacement interval T.
$N_B(t, T)$ = the number of failures in $[0, t]$ under block replacement policy, with replacement interval T.

Let the corresponding renewal functions be

$$M(t) = E\{N(t)\}$$

$$M_A(t, T) = E\{N_A(t, T)\} \tag{5.11}$$

$$M_B(t, T) = E\{N_B(t, T)\}.$$

It will be shown here (see Barlow and Proschan) that

$$\lim_{t \to \infty} \frac{N_A(t, T)}{t} = \lim_{t \to \infty} \frac{M_A(t, T)}{t} = \frac{F(T)}{\int_0^T \{1 - F(x)\}\, dx} \tag{5.12}$$

$$\lim_{t \to \infty} \frac{N_B(t,T)}{t} = \lim_{t \to \infty} \frac{M_B(t,T)}{t} = \frac{M(T)}{T}. \tag{5.13}$$

To prove (5.12) we first note that $\{N_A(t, T), t \geq 0\}$ is a renewal process with distribution

$$F_A(t) = \Pr(Y_i \leq t)$$

such that in,

$$nT \leq t < (n+1)T, n = 0, 1, 2, \dots,$$

$$1 - F_A(t) = \Pr(Y_i > t) = [1 - F(T)]^n [1 - F(t - nT)]. \tag{5.14}$$

Hence

$$\mu_A = E(Y_i) = \int_0^\infty \Pr(Y_i \geq t) \, dt$$

$$= \sum_{n=0}^\infty \int_{nT}^{(n+1)T} \{[1 - F(T)]^n [1 - F(t - nT)]\} \, dt$$

$$= \sum_{n=0}^\infty [1 - F(T)]^n \left[\int_0^T \{1 - F(x)\} \, dx \right]$$

$$= \int_0^T \{1 - F(x)\} \, dx / F(T). \tag{5.15}$$

With this value of μ_A, using Theorems 6.6 and 6.7, we at once get (5.12).

To prove (5.13) we note that after successive block replacements at $T, 2T, \dots$, the process starts anew, and if $N_{B_r}(T)$ is the number of failures in the interval $[(r-1)T, rT]$, $r = 1, 2, 3, \dots$, following block replacements, then $N_{B_r}(T) = N(T)$.

Suppose that for some specified integral value of n, $nT \leq t < (n+1)T$, i.e. for some n and $0 \leq \tau < T$,

$$t = nT + \tau,$$

then

$$N_B(t,T) = \sum_{r=1}^n N_{B_r}(T) + N(\tau)$$

$$= \sum_{r=1}^n N(T) + N(\tau).$$

We then get

$$\lim_{t \to \infty} \frac{N_B(t,T)}{t} = \lim_{n \to \infty} \frac{\sum\limits_{r=1}^{n} N(t)}{nT}$$

$$= E\{N(T)\} / T, \text{ (by the law of Large Numbers)}$$

$$= M(T)/T.$$

Thus we get (5.13).   ▲

Particular Cases

(i) *Markovian Case*: When $F(x) = 1 - e^{-x/a}$ with mean $\mu = a$, then

$$\mu_A = \int_0^T [1 - F(x)] \, dx / F(T) = a = \mu$$

and $$M(T)/T = aT/T = a = \mu,$$

as is to be expected (Explain why?).

(ii) *Limiting Case*: Suppose that μ is finite, then for large T, we have asymptotically,

$$1/\mu_A \to 1/\mu$$

and   by Theorem 6.7, $$M(T)/T \to 1/\mu.$$

It should be intuitively clear that for $T \to \infty$, the renewal processes under age and block replacement policies are both equivalent to the ordinary renewal process $N(t)$. For some interesting properties and renewal theoretical implications of these replacement policies, refer to Barlow and Proschan (1964, 1975).

Example 5(b). *Replacement on failure and block replacement*
Schwitzer (1967) considers two replacement policies: replacement on failure and block replacement at interval T (fixed) and two failure time distributions: (i) hyper-exponential, and (ii) uniform. Let m denote the number of items and F, the common lifetime d.f., with p.d.f. f and mean μ. Let r_1 be the cost of individual replacement (on failure of an item) and r_2 be the cost of block replacement per item.

The m items will be failing asymptotically at the rate of m/μ per unit time (Blackwell's theorem: Barlow and Proschan (1975), p. 51). Thus the expected cost per unit time equals $C_1 = mr_1/\mu$.

For block replacement, the expected cost per unit time equals

$$C_2 = [mr_2 + mr_1 M(t)] / t,$$

where $M(t)$, the renewal functions gives the expected number of failures by time t from the start of a new item. If F^{k*} is the k-fold convolution of F with itself, then

$$M(t) = \int\limits_0^t \sum_{k=1}^{\infty} dF^{k*}(x).$$

If $(r_2/r_1) < 1$, then one might expect the block replacement policy to be more economical than the policy of individual replacement. It turns out (as shown by Schweitzer) that block replacement policy is not desirable for exponentially distributed lifetime. For rectangular lifetime, block replacement is preferable for $(r_2/r_1) < 0.3863$.

Remarks. Various block replacement policies have been considered in the literature (see, for example, Nakagawa (1979, 1982)). Some of these are:

When a failure occurs to a component, just before the scheduled block replacement, it remains failed till the replacement time (Cox (1962), Crookes, *Opnl. Res. Qrly.* **14** (1963) 167–184).

When a failure occurs to a component just before block replacement, it is replaced by an used unit (Bhat, *J. Appl. Prob.* **6** (1969) 309–318, Tango, *I.E.E.E. Trans.* **R–28** (1979) 400–401), *J. Appl. Prob.* **15** (1978) 560–572).

Extended block replacement policy with used item constraints has been considered by Murthy and Nguyen (1983).

A policy such that an operating unit of young age is not replaced at scheduled replacement and remains in service has been considered by Berg and Epstein, *Nav. Res. Log. Qrly.* **23** (1976) 15–24. For a comparison of various policies, see, for example, Langberg, *J. Appl. Prob.* **25** (1988) 780–788.

6.5.3 Some Definitions

Definition. A non-negative random variable X is said to be a *lattice variable* (or to have a *lattice distribution*) if it takes on values nd $(n = 0, \pm 1, \pm 2, \ldots, d > 0)$,* i.e. integral multiplies of a non-negative number d. The largest d is said to be the *period* of the distribution. (Feller uses the terminology 'arithmetic' for lattice and 'span' for period but the terminology used here is more common).

Lattice variables form a special type of the class of discrete variables. When $d = 1$, a lattice variable becomes an integer-valued variable.

By saying that a distribution is *non-lattice*, we shall mean that the corresponding random variable is other than lattice.

Definition. Let $f(x)$ be a function defined on $[0, \infty)$. For fixed $h > 0, n = 1, 2, \ldots$, let

$$\overline{m}_n = \max \{f(x): \ (n-1)h \leq x \leq nh\}$$

$$\underline{m}_n = \max \{f(x): \ (n-1)h \leq x \leq nh\}$$

---

*Some authors consider X to be a lattice variable if it takes on values $c + nd$ with c arbitrary. We take $c = 0$

and let

$$\bar{\sigma} = h \sum \bar{m}_n \, , \underline{\sigma} = h \sum \underline{m}_n.$$

If both the series *converge absolutely* for every $h > 0$ and if $\bar{\sigma} - \underline{\sigma} \to 0$ as $h \to 0$ then $f(x)$ is said to be *directly Riemann integrable*.

If $f(x)$ is non-negative and non-increasing on $[0, \infty)$ ($f(\infty) = 0$) and is integrable there on in the ordinary (Riemann) sense then, $f(x)$ is *directly Riemann integrable*.

We shall now go to the other Renewal theorems.

6.5.4 Renewal Theorems (Blackwell's and Smith's)

Theorem 6.8. (Blackwell's theorem): For X_i non-lattice and for fixed $h \geq 0$

$$M(t) - M(t - h) \to h/\mu \quad \text{as} \quad t \to \infty \tag{5.16}$$

and for a lattice X_i with period d,

$$\lim_{n \to \infty} \Pr \{\text{renewal at } nd\} \to d/\mu. \tag{5.17}$$

An alternative form of the renewal theorem is as follows:

Theorem 6.9. (Smith's theorem or Key Renewal Theorem)

Let $H(t)$ be *directly Rienmann integrable* and $H(t) = 0$ for $t < 0$. If X_i is non-lattice, then

$$\int_0^t H(t - x) \, dM(x) \to \frac{1}{\mu} \int_0^\infty H(t) dt \text{ as } t \to \infty, \tag{5.18}$$

the limit being interpreted as 0 when $\mu = \infty$.

If X_i is lattice with period d, then

$$H(c + nd) \to \frac{d}{\mu} \sum_{k=0}^\infty h(c + kd). \tag{5.19}$$

We shall consider here a somewhat restricted form of the theorem as given below (Theorem 6.9A).

Theorem 6.9A. Let $H(t)$ be a non-negative, non-increasing function of $t \geq 0$ such that

$$\int_0^\infty H(t) \, dt < \infty$$

and let X_i be non-lattice. Then, as $t \to \infty$

$$\int_0^t H(t-x)\,dM(x) \to \frac{1}{\mu}\int_0^\infty H(t)\,dt, \qquad (5.20)$$

the limit being interpreted as 0 when $\mu = \infty$.

Proof: We have

$$\int_0^t H(t-x)dM(x) = \int_0^{t/2} H(t-x)dM(x) + \int_{t/2}^t H(t-x)dM(x),$$

$$= I_1 + I_2 \text{ (say)} \qquad (5.21)$$

Since $H(\cdot)$ is non-negative and nonincreasing $H(t-x) \le H(t/2)$ for $0 \le x \le t/2$ and so

$$I_1 = \int_0^{t/2} H(t-x)dM(x) \le H(t/2)\int_0^{t/2} dM(x) = H(t/2)M(t/2)$$

$$= \{(t/2)H(t/2)\}\frac{M(t/2)}{t/2}.$$

Now as $t \to \infty$,

$$\frac{M(t/2)}{t/2} \to \frac{1}{\mu} \qquad \text{(Theorem 6.7)}$$

and $(t/2)\,H(t/2) \to 0$. Thus

$$I_1 \to 0 \text{ as } t \to \infty. \qquad (5.22)$$

Let

$$J = \int_0^\infty H(t)\,dt = \sum_{n=0}^\infty \int_{nh}^{nh+h} H(t)\,dt \qquad (5.23)$$

Now

$$H(nh+h) \le H(t) \le H(nh), \text{ for } nh \le t < nh+h$$

so that

$$k\sum_{n=0}^\infty H(nh+h) \le J \le k\sum_{n=0}^\infty H(nh)$$

$$= h\,H(0) + h\sum_{n=1}^\infty H(nh).$$

Thus

$$0 \le J - k \sum_{n=1}^{\infty} H(nh) \le h\, H(0) < \epsilon$$

if

$$0 < k < \epsilon\, /H\,(0).$$

Again,

$$I_2 = \int_{t/2}^{t} H(t-x)dM(x)$$

$$= \int_{0}^{t/2} H(y)[-dM(t-y)]$$

$$= \sum_{n=0}^{N-1} \int_{nh}^{nh+h} H(y)[-dM(t-y)]$$

where N is the greatest integer contained in $t/2h$, i.e. $[t/2h] = N$.

It follows that

$$\sum_{0}^{N-1} H\,(nh+h)\,[M(t-nh)-M(t-nh-h)] \le I_2,$$

$$\le \sum_{0}^{N-1} H\,(nh)\,[M(t-nh)-M(t-nh-h)].$$

For large t, we can make

$$\left| \frac{M(t-nh)-M(t-nh-h)}{h} - \frac{1}{\mu} \right| < \epsilon$$

and

$$h \sum_{N+1}^{\infty} H(nh) < \epsilon\ .$$

Thus we find that

$$\left(\frac{1}{\mu}-\epsilon\right)(J-2\,\epsilon) < I_2 < \left(\frac{1}{\mu}+\epsilon\right)(J+\epsilon)$$

Since ϵ is arbitrary and can be made as well as we please,

$$I_2 \to \frac{J}{\mu} \text{ as } t \to \infty;$$

(5.24)

hence   as $t \to \infty$

$$I_1 + I_2 \rightarrow \frac{1}{\mu} \int\limits_0^\infty H(t)\, dt.$$

The result follows.   ▲

Remarks:
   (1). If we take

$$H(t) = 1/h, \quad 0 < t \le h$$

$$= 0, \qquad \text{otherwise;}$$

we at once get

$$M(t) - M(t-h) \rightarrow \frac{1}{\mu} \int_0^h dt = \frac{h}{\mu}$$

which is Theorem 6.8.

   (2) The two theorems 6.8 and 6.9 are in fact equivalent in the sense that one can be deduced from the other. The proof that Theorem 6.9 can be deduced from Theorem 6.8 is somewhat complicated.

6.5.5 Central Limit Theorem for Renewals

   We shall consider a CLT for the renewal process $\{N(t), t \ge 0\}$.

Theorem 6.10. Let $\{X_n, n = 1, 2, \ldots\}$ be a renewal process with distribution F, for which the mean $\mu = E(X_i)$ and variance $\sigma^2 = E\{(X_i - \mu)^2\}$ exist and are finite. Let $\{N(t), t \ge 0\}$ be the renewal process generated by F. Then

$$\lim_{t \to \infty} \left\{ \frac{N(t) - t/\mu}{\sqrt{t\sigma^2/\mu^3}} < x \right\} = \Phi(x) \tag{5.25}$$

where

$$\Phi(x) = \frac{1}{\sqrt{2\pi}} \int\limits_{-\infty}^x \exp\left(-\frac{1}{2}t^2\right) dt \ \text{ is the d.f. of the standard normal distribution.}$$

Proof: From the Central Limit Theorem applied to $S_n = X_1 + \ldots + X_n$, we find that as $n \to \infty$, S_n is asymptotically normal with mean $n\mu$ and variance $n\sigma^2$. That is

$$\lim_{n \to \infty} \Pr \left\{ \frac{S_n - n\mu}{\sigma\sqrt{n}} \le x \right\} = \Phi(x).$$

Let x be fixed and let $n \to \infty$ and $t \to \infty$ in such a way that

$$\lim_{\substack{n \to \infty \\ t \to \infty}} \frac{t - n\mu}{\sigma\sqrt{n}} \to -x.$$

that is, $t \to n\mu - x \sigma \sqrt{n}$ as $n \to \infty$, $t \to \infty$.
We have

$$\lim_{\substack{n \to \infty \\ t \to \infty}} \Pr\{S_n > t\} = \lim_{\substack{n \to \infty \\ t \to \infty}} \left\{ \frac{S_n - n\mu}{\sigma\sqrt{n}} > -x \right\}$$

$$= 1 - \Phi(-x) = \Phi(x).$$

Again since

$$\{N(t) < n\} \Leftrightarrow \{S_n > t\}$$

$$\Phi(x) = \lim_{\substack{n \to \infty \\ t \to \infty}} \Pr\{N(t) < n\}$$

$$= \lim_{\substack{n \to \infty \\ t \to \infty}} \Pr \left\{ \frac{N(t) - t/\mu}{\sqrt{(t\sigma^2/\mu^3)}} < \frac{n - t/\mu}{\sqrt{(t\sigma^2/\mu^3)}} \right\}$$

$$= \lim_{t \to \infty} \Pr \left\{ \frac{N(t) - t/\mu}{\sqrt{(t\sigma^2/\mu^3)}} < x \right\}.$$

Then $N(t)$ is asymptotically normal with mean t/μ and variance $t\,\sigma^2/\mu^3$.   ▲

6.6 DELAYED AND EQUILIBRIUM RENEWAL PROCESSES

The renewal process $\{X_i\}$ which we have considered so far is such that all the X_i's, $i = 1, 2, \ldots$, are i.i.d. r.v.'s The process is what is known as *ordinary renewal process*. We consider two generalisations of the ordinary renewal process.

6.6.1 Delayed (modified) Renewal Process

First, suppose that the first interarrival time X_1 (i.e. time from the origin upto the first renewal) has a distribution G which is different from the common distribution F of the remaining interarrival times $X_2, X_3, \ldots$, i.e. the initial distribution G is different from subsequent common distribution F. We then get what is known as a *modified* or *delayed renewal process*. Such a situation arises when the component used at $t = 0$ is not new. When $G \equiv F$, the modified process reduces to the ordinary renewal process.

Definition. Let $\{X_i, i = 1, 2, \ldots\}$ be a sequence of independent non-negative random variables such that the first interarrival time X_1 has the distribution G and the subsequent interarrivals times X_n, $n = 2, 3, \ldots$ have the identical distribution F. Let

$$S_0 = 0, S_n = \sum_{i=1}^{n} X_i \text{ and } N_D(t) = \sup \{n : S_n \leq t\}.$$

The stochastic process $\{N_D(t), t \geq 0\}$ is called a *modified*, *delayed* or a *general renewal process*.

The sequence of random variables $\{X_n, n = 1, 2, \ldots\}$, where the initial distribution G is different from subsequent common distribution F, is said to constitute a modified renewal process.

The following results can be easily obtained:
We have

$$\Pr \{N_D(t) = n\} = \Pr \{S_n \leq t\} - \Pr \{S_{n+1} \leq t\}$$

$$= G * F_{n-1}(t) - G * F_n(t)$$

so that

$$M_D(t) = E \{N_D(t)\} = \sum_{n=1}^{\infty} G * F_{n-1}(t) \qquad \qquad '(6.1)$$

and

$$M_D^*(s) = \frac{g^*(s)}{s\{1 - f^*(s)\}}, \qquad \qquad (6.2)$$

g^* and f^* being the L.T. of the p.d.f. of G and F respectively.
Again,

$$M_D(t) = \int_0^{\infty} E \{N_D(t) \mid X_1 = x\} \, dG(x)$$

$$= \int_0^{t} \{1 + M(t - x)\} dG(x)$$

$$= G(t) + \int_0^{t} M(t - x) dG(x); \qquad \qquad (6.3)$$

$M(t)$ being the renewal function of the ordinary renewal process with distribution F.

Example 6(a). Consider a modified renewal process with the first interarrival time X_1 having hyper-exponential distribution (considered in Example 2(b)) and the other interarrival times $X_2, X_3, \ldots$ having simple exponential distribution with parameter b. Such a model would be suitable when there is no defective item among replacement parts (i.e. whereas X_1 has distribution as described therein, the distribution of $X_2, X_3, \ldots$ corresponds to $p = 0$ in the distribution of X_1. Such a situation has been considered by Karmarkar, 1978; see Example 6(d)). We have

$$g(t) = pae^{-at} + (1-p)be^{-bt}, a > b > 0.$$

From (6.2) we get

$$M_D{}^*(s) = \frac{g^*(s)}{s\{1 - f^*(s)\}}$$

$$= \frac{ab + \{b + p(a - b)\}s}{s^2(s + a)}$$

$$= \frac{ab}{s^2(s + a)} + \frac{b + p(a - b)}{s(s + a)}.$$

Inverting, we get

$$M_D(t) = \frac{ab}{a}\left\{t - \frac{1 - e^{-at}}{a}\right\} + \frac{b + p(a - b)}{a}(1 - e^{-at})$$

$$= bt + \frac{p(a - b)}{a}(1 - e^{-at}).$$

$M_D(t)$ can also be obtained directly from (6.3), by noting that the renewal function $M(t)$ of the ordinary renewal process with distribution F is given by

$$M(t) = bt$$

and
$$G(t) = \int_0^t g(x)\,dx.$$

We state, without proof, the corresponding renewal theorem for modified renewal process.

Theorem 6.11. Let $\mu = E\{X_n\}$, $n = 2, 3, \ldots$. Then
(1) with probability 1,

$$M_D(t)/t \to 1/\mu \text{ as } t \to \infty. \tag{6.4}$$

(2) If F is non-lattice, then

$$\{M_D(t) - M_D(t-a)\} \to a/\mu \text{ as } t \to \infty \tag{6.5}$$

and if F is lattice with period d, then

$$\Pr\{\text{renewal at } nd\} \to d/\mu \text{ as } n \to \infty.$$

The above theorem shows that the renewal rate tends to a constant and is independent of the initial distribution G. We have, as Feller observes, an analogue to the Ergodic theorem for Markov chains.

Now $M_D(t) \to t/\mu$ raises a question: whether there exists an initial distribution G such that $M_D(t) = t/\mu$. The answer is in the affirmative. This is true in case of a kind of renewal process discussed below.

6.6.2 Equilibrium (or Stationary) Renewal Process

Now a modified renewal process with initial distribution $G \equiv F_e$, exists such that

$$M_e(t) = t/\mu,$$

or

$$M_e^*(s) = 1/\mu s^2$$

iff

$$f_e^*(s) = \frac{1 - f^*(s)}{\mu s},$$

or

$$f_e(t) = \frac{1 - F(x)}{\mu},$$

i.e.

$$F_e(t) = \int_0^t \frac{1 - F(x)}{\mu} dx. \tag{6.6}$$

The distribution having d.f. $F_e(t)$ is called equilibrium distribution of F and the corresponding renewal process is called equilibrium (or stationary) renewal process. We denote an equilibrium process by $\{N_e(t), t \geq 0\}$ and the corresponding renewal function by $M_e(t)$. Poisson process is clearly an equilibrium renewal process.

Equilibrium renewal processes, which are extremely important in applications, can arise as follows: Suppose that an ordinary renewal process, has been continuing from time $t \to -\infty$, remote from $t = 0$. If the observation of the process is started at $t = 0$, then the interarrival time X_1 to the first renewal will have distribution function $G = F_e$ as given by (6.6).

Example 6(b). We have, for all $t > 0$, and $n = 0, 1, 2, \ldots$.

$$\Pr\{N_e(t) \geq n\} \geq \Pr\{N(t) \geq n\},$$

iff

$$\frac{\int_t^\infty \{1 - F(x)\}\, dx}{1 - F(t)} \le \mu.$$

Proof: Assume that the condition holds. Since

$$\frac{1}{\mu} \int_t^\infty \{1 - F(x)\}\, dx = \frac{1}{\mu}\left[\int_0^\infty \{1 - F(x)\}\, dx - \int_0^t \{1 - F(x)\}\, dx\right]$$

$$= 1 - F_e(t),$$

we have, by hypothesis, for $t \ge 0$

$$1 - F_e(t) \le 1 - F(t)$$

or,

$$F_e(t) \ge F(t).$$

If S_n and S_n' denote the nth renewal epochs of $N(t)$ and $N_e(t)$ respectively, then, for $n \ge 1$,

$$\Pr\{N_e(t) \ge n\} = \Pr\{S_n' \le t\}$$

$$= \int_0^t F^{n-1}(t - x)\, dF_e(x)$$

$$\ge \int_0^t F_{n-1}(t - x)\, dF(x)$$

$$= \Pr\{S_n \le t\} = \Pr\{N(t) \ge n\}.$$

Further, $\Pr\{N_e(t) \ge 0\} = \Pr\{N(t) \ge 0\} = 1$, for all $t \ge 0$. Thus the result holds.

To prove the *converse*, assume that $\Pr\{N_e(t) \ge n\} \ge \Pr\{N(t) \ge n\}$ holds for $n = 0, 1, 2, \ldots$. Since quality holds for $n = 0$, we must have

$$\Pr\{N_e(t) = 0\} \le \Pr\{N(t) = 0\}$$

or,

$$1 - F_e(t) \le 1 - F(t)$$

and therefore the converse is also true.   ▲

Corollary. $M(t) \le t/\mu$ *iff* $F_e(t) \ge F(t)$.

The result given in the above example may thus be restated as:

$$\Pr\{N_e(t) \ge n\} \ge \Pr\{N(t) \ge n\}, t \ge 0, n = 0, 1, 2, \ldots.$$

as well as $M(t) \to t/\mu$, *iff* the distribution F including the renewal process is NBUE.

6.6.3 Probability Generating Function (p.g.f.) of Renewal Processes

We obtain here the p.g.f. of the renewal processes

$$\{N(t), t \geq 0\}, \{N_D(t), t \geq 0\}, \{N_e(t), t \geq 0\}.$$

We have

$$G(t, z) = \sum_{r=0}^{\infty} z^r \Pr\{N(t) = r\}$$

$$= \sum_{r=0}^{\infty} z^r \{F_r(t) - F_{r+1}(t)\}$$

$$= 1 + \sum_{r=1}^{\infty} z^{r-1}(z - 1)F_r(t). \tag{6.7}$$

The above relation holds for all the three types of renewal processes. We denote the p.g.f. of ordinary, delayed (modified) and equilibrium renewal processes by G, G_m and G_e respectively. Now the Laplace transform of the p.d.f. of F_r (for $r \geq 1$) is

$$f_r^*(s) = [f^*(s)]^r \qquad \text{(for an ordinary renewal process)} \tag{6.8a}$$

$$= g^*(s)[f^*(s)]^{r-1} \quad \text{(for a delayed renewal process)} \tag{6.8b}$$

$$= \frac{1 - f^*(s)}{\mu s}[f^*(s)]^{r-1} \quad \text{(for an equilibrium renewal process).} \tag{6.8c}$$

Denote the L.T. of the p.g.f. of the ordinary, modified and equilibrium renewal processes by

$$G^*(s, z), G_m^*(s, z), G_e^*(s, z)$$

respectively. From (6.7), we have

$$G^*(s, z) = \frac{1}{s} + \frac{1}{s}\sum_{r=1}^{\infty} z^{r-1}(z-1)[f^*(s)]^r$$

$$= \frac{1}{s} + \frac{1}{s} \cdot \frac{(z-1)f^*(s))}{1 - z f^*(s)}$$

$$= \frac{1 - f^*(s)}{s\{1 - zf^*(s)\}}; \tag{6.9}$$

$$G_m^*(s, z) = \frac{1}{s} + \frac{1}{s}\sum_{r=1}^{\infty} z^{r-1}(z-1)g^*(s)[f^*(s)]^{r-1}$$

$$= \frac{1}{s} + \frac{(z-1)g^*(s)}{s\{1 - zf^*(s)\}}$$

$$= \frac{1 - zf^*(s) + zg^*(s) - g^*(s)}{s\{1 - zf^*(s)\}}; \tag{6.10}$$

$$G_e^*(s,z) = \frac{1}{s} + \frac{1}{s}\sum_{r=1}^{\infty} z^{r-1}(z-1)\frac{1-f^*(s)}{\mu s}[f^*(s)]^{r-1}$$

$$= \frac{1}{s} + \frac{(z-1)\{1-f^*(s)\}}{\mu s^2\{1-zf^*(s)\}} \qquad\qquad (6.11a)$$

$$= \frac{1}{s} + \frac{(z-1)}{\mu s}G^*(s,z). \qquad\qquad (6.11b)$$

Inverting (6.11a), we get

$$G_e(t,z) = 1 + \frac{z-1}{\mu}\int_0^t G(u,z)\,du;$$

comparing the coefficients of z^r from both sides, we get

$$\Pr\{N_e(t) = r\} = \frac{1}{\mu}\int_0^t [\Pr\{N(u) = r-1\} - \Pr\{N(u) = r\}]\,du \qquad (6.12a)$$

$$\Pr\{N_e(t) = 0\} = 1 - \frac{1}{\mu}\int_0^t \Pr\{N(u) = 0\}\,du. \qquad (6.12b)$$

Example 6(c). *Second moment and variance of $N(t)$.* The moments of $N(t)$ can be found from the L.T. of the p.g.f. However, the following simple procedure may be employed to find the second moment $L(t)$ of $N(t)$ and thence its variance. We have

$$L(t) = E\{N(t)\}^2 = \sum_{r=0}^{\infty} r^2\{F_r(t) - F_{r+1}(t)\}.$$

Taking L.T., we get

$$L^*(s) = \sum_{r=0}^{\infty} \frac{r^2}{s}[\{f^*(s)\}^r - \{f^*(s)\}^{r+1}].$$

On simplification, we get

$$L^*(s) = \frac{f^*(s)\{1+f^*(s)\}}{s\{1-f^*(s)\}^2}$$

$$= \left[\frac{1+f^*(s)}{1-f^*(s)}\right]M^*(s)$$

which can also be written as:

$$L^*(s) = M^*(s) + \frac{(2f^*(s))}{s\{1 - f^*(s)\}} (sM^*(s)).$$

Inverting the L.T., we get

$$E\left[(N(t))^2\right] = L(t) = M(t) + 2 \int_0^t M(t - x)\, dM(x). \qquad (6.13a)$$

This gives the second moment of $N(t)$ in terms of the renewal function $M(t)$. Thus var $\{(N(t)\}$ can be obtained in terms of $M(t)$. We get

$$\text{var } \{N(t)\} = M(t) + 2 \int_0^t M(t - x)\, dM(x) - [M(t)]^2. \qquad (6.13b)$$

Example 6(d). *An application to a management problem.* For certain kinds of electronic equipments of office use (such as electronic calculators etc.) and of domestic appliances (such as T.V., refrigerators etc.) the manufacturer offers to the customer a certain kind of service contract by which the manufacturer undertakes to maintain and repair the product over a certain specified period against payment of a fee by the customer. In arriving at a decision regarding the offer of the terms of contract (including the amount of fees to be charged thereof), the manufacturer should be aware of the expected cost of future repair. A simple renewal theoretic model developed by Karmarkar (1978) is considered here. It is assumed that each failure is followed by instantaneous detection and repair.

Denote

$R(t) = $ cost associated with a repair (renewal) at time t; it is assumed that $R(t) = R$,

$c = $ discount factor for "continuous" discounting of future costs,

$f(t) = $ p.d.f. of time to failure (interarrival time X_n),

$k(t) = $ p.d.f. of $S_r (= X_1 + \ldots + X_r)$; $F_r = $ d.f. of S_r,

$V_r = $ discounted cost of rth renewal.

$g^*(\cdot) = $ L.T. $\{g(\cdot)\}$.

For a single failure followed by renewal, the expected discounted cost $E(V_1)$ is given by

$$E(V_1) = \int_0^\infty Re^{-ct} f(t)\, dt = Rf^*(c).$$

The expected discounted cost for the rth renewal is

$$E(V_r) = \int_0^\infty Re^{-ct}k_r(t)\,dt$$

$$= Rk_r*(c)$$

$$= R[f*(c)]^r.$$

For a sequence of n renewals, the total expected discounted cost is

$$E(V) \equiv E(V_1 + V_2 + \dots + V_n) = \sum_{r=1}^n R\,[f*(c)]^r$$

$$= \frac{R\,[1-\{f*(c)\}^n]f*(c)}{1-f*(c)}$$

$$\rightarrow \frac{R\,f*(c)}{1-f*(c)} \quad \text{as} \rightarrow \infty. \qquad (6.14)$$

For a sequence of infinite renewals truncated at time T the expected cost of future renewals is

$$E\{V(T)\} = \sum_{r=1}^n \int_0^T Re^{-ct}k_r(t)\,dt$$

$$= \int_0^T Re^{-ct}d\left(\sum_{r=1}^n F_r(t)\right)$$

$$= \int_0^T Re^{-ct}\,dM(t), \qquad (6.15)$$

assuming the interchangeability of summation, integration and differentiation. As $T \rightarrow \infty$,

$$E\{V(T)\} \rightarrow RcM*(c) = \frac{Rf*(c)}{1-f*(c)} \qquad (6.16)$$

which agrees (as $n \rightarrow \infty$) with (6.14).

Particular cases
   (1) Suppose that $f(t) = be^{-bt}$, then

$$E(V) = \frac{R\left[1 - \left(\frac{b}{c+b}\right)^n\right]\frac{b}{c+b}}{1 - \frac{b}{b+c}}$$

$$= \left(\frac{Rb}{c}\right)\left[1 - \left(\frac{b}{c+b}\right)^n\right]$$

$$\rightarrow \frac{Rb}{c} \quad \text{as } n \rightarrow \infty;$$

and for finite renewals truncated at T, we get

$$E\{V(T)\} = \int_0^T Re^{-ct}\, d(bt)$$

$$= \left(\frac{Rb}{c}\right)(1 - e^{-ct})$$

$$\rightarrow \frac{Rb}{c} \quad \text{as } T \rightarrow \infty.$$

(2) Suppose that $f(t) = pae^{-at} + (1-p)be^{-bt}$, $0 \le p \le 1$, $a > b > 0$, i.e. X_n has a hyper-exponential distribution. Then

$$E(V) \rightarrow R\left[\frac{pa}{c+a} + \frac{(1-p)b}{c+b}\right] \Bigg/ \left[1 - \left\{\frac{pa}{c+a} + \frac{(1-p)b}{c+b}\right\}\right]$$

$$= \frac{R\{Ac + ab\}}{c(c+B)},$$

where $\qquad A = pa + (1-p)b, B = (1-p)a + pb,$

$$E\{V(T)\} = \int_0^T Re^{-ct}\left\{\frac{ab}{B} + \left(\frac{A}{B} - \frac{ab}{B^2}\right)Be^{-Bt}\right\} dt$$

$$= \frac{Rab}{cB}(1 - e^{-cT}) + \frac{R(AB - ab)}{B(c+B)}[1 - e^{-(c+B)}].$$

(3) Suppose that

$$f(t) = \frac{\lambda^k t^{k-1} e^{-\lambda t}}{\Gamma(k)}, t \ge 0$$

so that

$$f^*(s) = \left(\frac{\lambda}{s+\lambda}\right)^k;$$

then

$$E(V) \to \frac{R\lambda^k}{(c+\lambda)^k - \lambda^k}.$$

Example 6(e): *Number of renewals in a random time.* So far we have considered the number $N(t)$ of renewals in a fixed interval $[0, t]$. Consider now that T is a random variable independent of X_i so that $[0, T]$ is a random interval; suppose that N is the number of renewals in $[0, T]$. Let $G(t, z)$ be the p.g.f. of $N(t)$ and $G(z)$ be the p.g.f. of N. If $q(t)$ is the p.d.f. of T, then it is clear that

$$G(z) = \int_0^\infty G(t,z)\, q(t)\, dt. \tag{6.17}$$

Particular cases

(1) Suppose that T has exponential distribution, with p.d.f. $q(t) = a\, e^{-at}$. Then

$$G(z) = \int_0^\infty a e^{-at} G(t,z)\, dt$$

$$= aG^*(a,z).$$

If the renewal process is ordinary, then from (6.9) we get

$$G(z) = aG^*(a,z)$$

$$= \frac{a\{1 - f^*(a)\}}{a\{1 - zf^*(a)\}}$$

$$= \frac{1 - f^*(a)}{1 - zf^*(a)},$$

which is the p.g.f. of a geometric distribution.

We have

$$\Pr\{N = r\} = [f^*(a)]^r\, [1 - f^*(a)], \quad r = 0, 1, 2, \ldots.$$

Suppose further that the renewal process is Poisson with $f(t) = be^{-bt}$; then

$$G(z) = \frac{a}{a + b - zb}$$

and
$$\Pr\{N = r\} = \left(\frac{a}{a+b}\right)\left(\frac{b}{a+b}\right)^r, \quad r = 0, 1, 2, \ldots.$$

This result was obtained in Ch. 4 (see Poisson count process).

(2) Suppose that the renewal process is Poisson but T has gamma distribution having p.d.f.

$$q(t) = a^k \, t^{k-1} \, e^{-at}/\Gamma \, (k), \, t \geq 0;$$

T can be considered as the sum of k i.i.d. exponential distributions with

p.d.f. $\qquad\qquad\qquad\qquad q(t) = a e^{-at}, \, t \geq 0.$

From the memoryless property of exponential distribution, we then get that the distribution of N is negative binomial (Pascal) with p.g.f.

$$[G \, (z)]^k = \left[\frac{a}{a+b-zb} \right]^k$$

so that

$$\Pr \{N = r\} = \binom{k+r-1}{r} \left(\frac{a}{a+b} \right)^k \left(\frac{b}{a+b} \right)^r, \, r = 0, 1, 2, \dots \; .$$

6.7 RESIDUAL AND EXCESS LIFETIMES

We discuss here two random variables of interest in renewal theory. To a given $t > 0$, there corresponds uniquely a $N \, (t)$ such that

$$S_{N(t)} \leq t < S_{N(t)+1},$$

i.e. t falls in the interval $X_{N(t)+1}$.

(i) The *residual* (or *excess*) *lifetime* of the individual alive at age t is given by the time $Y \, (t)$ from t to the next renewal epoch, i.e.

$$Y(t) = S_{N(t)+1} - t.$$

It is also called *forward recurrence time* at t.

(ii) The *spent* (or *current*) *lifetime* of the individual alive at age t is given by the time to t since the last renewal epoch, i.e.

$$Z(t) = t - S_{N(t)}.$$

It is also called *backward recurrence time* at t.

(iii) The total lifetime at t (or length of the lifetime containing t) is given by

$$Y(t) + Z(t) = S_{N(t)+1} - S_{N(t)} = X_{N(t)+1}.$$

The distributions of $Y \, (t)$ and $Z \, (t)$ are given in Section 6.7.2.

6.7.1 Poisson Process as a Renewal Process

We consider the distribution of $Y(t)$ and $Z(t)$ when $\{N(t), t \geq 0\}$ is a Poisson process with parameter μ. The residual lifetime $Y(t)$ at age t (> 0) exceeds x (≥ 0) *iff* no renewal (no occurrence of the Poisson process) takes place in the interval $(t, t + x)$. Now Poisson process has stationary increments and hence

$$\text{Pr \{no renewal in } (t, t + x]\}$$
$$= \text{Pr \{no occurrence of the Poisson process in } (0, x]\}$$
$$= \text{Pr } \{N(x) = 0\} = e^{-\mu x}.$$

Thus, the distribution of residual lifetime $Y(t)$ is given by

$$\text{Pr } \{Y(t) \geq x\} = e^{-\mu x}$$

or
$$\text{Pr } \{Y(t) \leq x\} = 1 - e^{-\mu x}; \tag{7.1}$$

These are independent of t (> 0).
This property characterizes Poisson process among all renewal processes.
The spent lifetime

$$Z(t) = t - S_{N(t)} \leq t;$$

for $x < t$, $z(t)$ exceeds x *iff* there is no renewal (no occurrence of the Poisson process) in $(t - x, t]$, the probability of this event being $e^{-\mu x}$. Thus

$$\text{Pr } \{Z(t) \leq x\} = 1 - e^{-\mu x}, 0 \leq x < t$$
$$= 1 \qquad , x \geq t;$$

That is

$$\text{Pr } \{Z(t) \leq x\} = F(x), x < t$$
$$= 1 \qquad , x \geq t \tag{7.2}$$

where $F(x) = P\{X_i \leq x\}$.

These are again independent of t. This property also characterizes Poisson process among all renewal processes.
We have

$$E\{Y(t)\} = \frac{1}{\mu}$$

$$E\{Z(t)\} = \int_0^t \text{Pr } \{Z(t) > x\} \, dx$$

$$= \int_0^t e^{-\mu x} dx$$

$$= \frac{1}{\mu}(1 - e^{-\mu t})$$

so that the expected total life equals

$$E\{Y(t) + Z(t)\} = \frac{1}{\mu}(2 - e^{-\mu t}).$$

It may be noted that the mean total lifetime is greater than $\frac{1}{\mu}$ $(= E(X_i))$, the mean length of any particular renewal interval X_i (inter-occurrence time).
As $t \to \infty$

$$E\{Y(t) + Z(t)\} \to \frac{2}{\mu};$$

that is, the mean total lifetime is *twice* as much as the mean inter renewal time. This fact appears to be paradoxical at first. The apparent paradox is known as *inspection paradox*.

6.7.2 Distribution of Y (t) and Z (t)

Theorem 6.12. We have

$$\Pr\{Y(t) \le x\} = F(t+x) - \int_0^t [1 - F(t+x-y)] \, dM(y). \tag{7.3}$$

If in addition, F is non-lattice, then

$$\lim_{t \to \infty} \Pr\{Y(t) \le x\} = \frac{1}{\mu} \int_0^x [1 - F(y)] \, dy. \tag{7.4}$$

Proof: It is clear that

$$Y(t) > x \Leftrightarrow \text{no renewals in } [t, t+x].$$

Let

$$P(t) = \Pr\{Y(t) > x\}.$$

By conditioning on X_1, we get

$$P(t) = \Pr\{Y(t) > x\}$$

$$= \int_0^\infty \Pr\{Y(t) > x \mid X_1 = y\} \, dF(y).$$

Now three situations may arise: (i) $y > t + x$, in which case it is certain that no renewal occurs in $[t, t + x]$, so that

$$\Pr \{Y(t) > x\} = \Pr \text{ (no renewal in } [t, t + x])$$

$$= 1$$

(ii) $t < y \leq t + x$, in which case one renewal (the one corresponding to $X_1 = y$) occurs in $[t, t + x]$, so that

$$\Pr \{Y(t) > x\} = \Pr \{\text{no renewal in } [t, t + x]\} = 0$$

(iii) $0 < y \leq t$, in which case the first renewal occurs in $(y, y + dy)$ and then the process restarts itself and

$$\Pr \{Y(t) > x\} = \Pr \{Y(t - y) > x\}$$

$$= P(t - y).$$

Considering the above three situations, we get

$$P(t) = \int_0^t P(t - y) dF(y) + \int_{t+x}^{\infty} dF(y)$$

or, $$P(t) = 1 - F(t + x) + \int_0^t P(t - y) dF(y),$$

which is a renewal type equation. Applying Theorem 6.5, we have

$$P(t) = 1 - F(t + x) + \int_0^t \{1 - F(t + x - y)\} dM(y), \tag{7.5}$$

which is equivalent to (7.3). Thus (7.3) is proved.   ▲

Now suppose that F is non-lattice. Then applying the Key Renewal Theorem (Theorem 6.9) with $h(t) = 1 - F(t + x)$ we get, as $t \to \infty$

$$\int_0^t \{1 - F(t + x - y)\} dM(y) \to \frac{1}{\mu} \int_0^{\infty} \{1 - F(t + x)\} dt$$

$$= \frac{1}{\mu} \int_x^{\infty} [1 - F(y)] dy.$$

Hence from (7.3), we have

$$\lim_{t \to \infty} \Pr\{Y(t) > x\} = \lim_{t \to \infty} P(t) = \frac{1}{\mu} \int_0^\infty \{1 - F(y)\}\, dy$$

or
$$\lim_{t \to \infty} \Pr\{Y(t) \le x\} = 1 - \frac{1}{\mu} \int_x^\infty \{1 - F(y)\}\, dy$$

$$= \frac{1}{\mu} \int_0^x \{1 - F(y)\}\, dy.$$

Thus (7.4) is proved. ▲

Note: Coleman (1982) gives an explicit expression for $E\left[\{Y(t)\}^r\right], t \ge 0, r = 1, 2, \ldots$ (See Exercise (6.13))

Theorem 6.13. We have

$$\Pr\{Z(t) \le x\} = \begin{cases} F(t) - \displaystyle\int_0^{t-x} \{1 - F(t-y)\}\, dM(y), & x \le t; \quad (7.6) \\[2mm] 1, & x < t \quad (7.7) \end{cases}$$

If, in addition, F is non-lattice, then

$$\lim_{t \to \infty} \Pr\{Z(t) \le x\} = \frac{1}{\mu} \int_0^x \{1 - F(y)\}\, dy. \qquad (7.8)$$

Proof: The distribution of $Z(t)$ can be obtained immediately by noting that

$$\{Z(t) > x\} \Leftrightarrow \text{no renewals in } [t - x, t]$$

$$\Leftrightarrow Y(t - x) > x,$$

so that
$$\Pr\{Z(t) > x\} = \Pr\{Y(t - x) > x\}.$$

Hence from the Theorem 6.12, we get

$$\Pr\{Z(t) \le x\} = \Pr\{Y(t - x) \le x\}$$

$$= F(t) - \int_0^{t-x} \{1 - F(t - y)\}\, dM(y), \quad x \le t$$

$$= 1, \quad x > t;$$

and if F is non-lattice, then

$$\lim_{t \to \infty} \Pr\{Z(t) \le x\} = \frac{1}{\mu} \int_0^x \{1 - F(y)\}\, dy \qquad \blacktriangle$$

Note: (1) It is to be noted that the result

$$\lim_{t \to \infty} \Pr\{Y(t) \le x\} = \frac{1}{\mu} \int_0^x \{1 - F(y)\}\, dy$$

holds also for the modified renewal process regardless of the initial distribution G.

(2) Now the R.H.S. is the distribution function $F_e \equiv G$ of the equilibrium renewal process which has a uniform renewal rate $M(t)/t = \mu$. Thus if $\mu < \infty$, then the residual lifetime $Y(t)$ has a distribution which coincides with the distribution of the process attaining uniform renewal rate.

If $\mu = \infty$, then the probability that the residual lifetime exceeds x tends to unity for all x. This should be intuitively clear.

(3) The asymptotic behaviour of the residual lifetime distribution has been investigated at length by Balkema and De Hann (1974) who obtained the possible limit distribution types.

(4) For some generalisation to multidimensional case, see Spitzer (1984).

6.7.3 Moments of the Asymptotic Distributions

Let the asymptotic distribution of $Y(t)$ be denoted F_Y and that of $Z(t)$ by F_Z. That is

$$\lim_{t \to \infty} \Pr\{Y(t) \le x\} = F_Y(x) \equiv \Pr\{Y \le x\}$$

$$\lim_{t \to \infty} \Pr\{Z(t) \le x\} = F_Z(x) \equiv \Pr\{Z \le x\}.$$

Let f_Y and f_Z denote the corresponding p.d.f.'s.

Then the L.T. $\bar{f}(\cdot)$ of $f_Y(\cdot)$ is given by

$$\bar{f}_Y(s) = \int_0^\infty e^{-st} f_Y(t)\, dt = \frac{1 - F_Y{}^*(s)}{s\mu} \qquad (7.9)$$

where

$$F_Y{}^*(s) = \int_0^\infty e^{-st}\, dF_Y(t).$$

Denote by μ_r, the rth moment of X_i (having d.f. $F = \Pr\{X_i \le x\}$). Then

$$E\{Y^r\} = \int_0^\infty x^r\, dF_Y(x) = \frac{1}{\mu} \int_0^\infty x^r [1 - F(x)]\, dx$$

$$= \frac{1}{\mu}\left[\{1-F(x)\}\right]\frac{x^{r+1}}{r+1}\Big|_0^\infty + \frac{1}{\mu}\int_0^\infty \frac{x^{r+1}}{r+1}dF(x)$$

$$= \frac{\mu_{r+1}}{(r+1)\mu}, \quad r = 1,2,3,\dots \tag{7.10}$$

Thus
$$E\{Y\} = \frac{\mu_2}{2\mu}$$

Similarly, we find

$$E\{Z^r\} = \frac{\mu_{r+1}}{(r+1)\mu}, \quad r = 1,2,3,\dots \tag{7.11}$$

For exponential X_i, $E\{Y\} = E(Z) = E\{X_i\}$.

Example 7(a). *Markovian case*: Let X_i have exponential distribution with distribution function $F(x) = 1 - e^{-ax}$
Then $M(t) = at$, and we have

$$\Pr\{Y(t) \le x\} = \{1 - e^{-a(t+x)}\} - \int_0^t \{1 - 1 + e^{-a(t+x-y)}\}\, d(ay)$$

$$= 1 - e^{-ax} \text{ for all } t.$$

Thus $Y(t)$ has the same exponential distribution. In fact this is one of the characteristics of exponential distribution.
For $x \le t$,

$$\Pr\{Z(t) \le x\} = 1 - e^{-at} - \int_0^{t-x} \{e^{-a(t-y)}\}\, d(ay)$$

$$= 1 - e^{-ax}$$

and for $x > t$,

$$\Pr\{Z(t) \le x\} = 1.$$

Thus $Z(t)$ has a truncated exponential distribution

$$\Pr\{Z(t) \le x\} = \begin{cases} F(x), & x \le t \\ 1, & x > t \end{cases}$$

For large t, $Z(t)$ will also have the same exponential distribution.
From the memoryless property of exponential distribution, it is clear that Y and Z are independent. As

$$X_{N(t)+1} = Y(t) + Z(t),$$

the renewal interval containing t is given by the convolution of the residual and spent lifetimes at t. We have

$$E\{X_{N(t)+1}\} = E\{Y_t\} + E\{Z_t\}$$

$$= \frac{1}{a} + \int_0^t \Pr\{Z(t) > x\}\, dx$$

$$= \frac{1}{a} + \frac{1}{a} \int_0^t e^{-ax}\, dx$$

$$= \frac{1}{a} + \frac{1}{a}(1 - e^{-at})$$

so that the mean total life is larger than the mean life $1/a$ (inspection paradox). For large t, the distribution of $X_{N(t)+1}$ will be gamma with mean $2/a$, i.e.

$$E\{X_{N(t)+1}\} = 2/a = 2E(X_i) \qquad \text{(see Section 7.6.1).}$$

Thus for large t, the expected duration of the renewal interval containing t will be *twice* the expected duration of an ordinary renewal interval.

6.8 RENEWAL REWARD (CUMULATIVE RENEWAL) PROCESS

Consider an ordinary renewal process generated by the sequence

$$\{X_n, n = 1, 2, ...\}$$

of interarrival times. Suppose that each time a renewal occurs, a reward is received (e.g. for the occurrence of the renewal) or a cost is incurred (e.g. for replacement of the failed component). Denote the reward or cost associated with n th renewal by

$$Y_n, n = 1, 2, ...$$

Now Y_n will usually depend on X_n, the duration of the interarrival time between $(n - 1)$ th and n th renewal; but we assume that the pairs of random variables

$$\{X_n, Y_n\}, n = 1, 2, ...$$

are independently and identically distributed. The sequence of the pairs of i.i.d. random variables

$$\{X_n, Y_n\}, n = 1, 2, \dots.$$

is said to generate a *renewal reward process* (or *compound renewal process* or *cumulative renewal process*).

Now
$$V(t) = \sum_{n=1}^{N(t)} Y_n \qquad (8.1)$$

gives the total reward earned (or cost incurred) by time t;

$$\{V(t), t \geq 0\}$$

is called the cumulative renewal process generated by $\{X_n, Y_n\}$. If X_n represents the inter demand times (interarrival times of customers in a queue) and Y_n represents the size of the demands (size of the arrival batch) then $Y(t)$ represents the total demand (total number of customers) in $(0, t]$.

Distribution of V(t)

Denote

$$F_n(x) = \Pr\{X_1 + \cdots + X_n \leq x\}$$
$$G_n(x) = \Pr\{Y_1 + \cdots + Y_n \leq x\}$$

$N(t)$ = number of occurrences (renewals) in $(0, t]$.
The d.f. of $Y(t)$ can be obtained by conditioning. We have

$$\Pr\{V(t) \leq x\} = \sum_{n=0}^{\infty} \Pr\{V(t) \leq x \mid N(t) = n\} \Pr\{N(t) = n\}$$

$$= \sum_{n=0}^{\infty} G_n(x)[F_n(t) - F_{n+1}(t)] \qquad (8.2)$$

We have from (1.19) and (1.20) of Chapter I,
$$E\{V(t)\} = M(t)E\{Y_n\} \qquad (8.3)$$

$$\text{var}\{V(t) = M(t) \text{ var}\{Y_n\} + [E\{Y_n\}]^2 \text{ var}\{N(t)\} \qquad (8.4)$$

First Passage Time

Denote the first passage time of level ω by T_ω, i.e.

$$T_\omega = \inf\{t : V(t) > \omega\}$$

Then

$$\{T_\omega \leq x\} \Leftrightarrow \{V(x) \geq \omega\}$$

so that the d.f. of T_ω equals, for $\omega \geq 0$, $x \geq 0$.

$$\Pr\{T_\omega \leq x\} = \Pr\{V(x) \geq \omega\}$$

$$= 1 - \sum_{n=0}^{\infty} G_n(\omega)[F_n(x) - F_{n+1}(x)]. \qquad (8.5)$$

The mean first passage time equals

$$E[T_\omega] = \int_0^\infty [1 - \Pr\{T_\omega \leq x\}]\, dx$$

$$= \sum_{n=0}^{\infty} G_n(\omega)\left[\int_0^\infty [\{1 - F_{n+1}(x)\} - \{1 - F_n(x)\}]\, dx\right]$$

$$= \sum_{n=0}^{\infty} G_n(\omega)[(n+1)\mu - n\mu]$$

$$= \mu \sum_{n=0}^{\infty} G_n(\omega) = \mu\left[1 + \sum_{n=1}^{\infty} G_n(\omega)\right]$$

$$= \mu[1 + R(\omega)] \qquad (8.6)$$

where $R(\omega)$ is the renewal function associated with $\{Y_n, n = 1, 2, \ldots\}$ (here the process does not refer to time).

We have the following important theorem.

Theorem 6.14. Suppose that

$$E(X) = E(X_n) \text{ and } E(Y) = E(Y_n)$$

are finite. Then

(a) with probability 1, $V(t)/t \to E(Y)/E(X)$ as $t \to \infty$

and   (b) $EV(t)/t \to E(Y)/E(X)$ as $t \to \infty$

Proof of (a): We have

$$\frac{V(t)}{t} = \frac{V(t)}{N(t)} \cdot \frac{N(t)}{t}$$

$$= \frac{\sum_{n=1}^{N(t)} Y_n}{N(t)} \cdot \frac{N(t)}{t}.$$

By the strong law of large numbers, with probability 1,

$$\frac{\sum\limits_{n=1}^{N(t)} Y_n}{N(t)} \to E\,(Y) \text{ as } t \to \infty.$$

and $\qquad N\,(t)/t \to 1/\mu = 1/E\,(X) \text{ as } t \to \infty \qquad\qquad$ (by Theorem 6.6).

Proof of (b):

We have

$$E\,\{V(t)\} = E\,\{Y\}E\,\{N(t)\}$$

$$= E\,\{Y\}M(t).$$

Thus

$$\frac{E\{V(t)\}}{t} = E(Y).\frac{M(t)}{t}$$

$$\to E(Y).\frac{1}{E(X)} \quad \text{as } t \to \infty \text{ (Theorem 6.7)}.$$

Hence the result. $\qquad\qquad\qquad\qquad\qquad\qquad\qquad\qquad\qquad$ ▲

Asymptotic expression

Denote

$$\frac{E(Y)}{E(X)} = \xi$$

and $\qquad\qquad\qquad\qquad E(V(t)) = C(t);$

$C\,(t)$ is the average expected total reward (or cost) in $[0, t]$. Then from (b) above, we get, for large t,

$$C(t) = t\,\xi.$$

A more refined asymptotic form of $C\,(t)$ has been given by Christer (1978), who obtains the asymptotic expression for $C\,(t)$ in the form

$$C(t) = t\,\xi + \zeta,$$

where ξ and ζ are functions of the parameters and decision variables of the renewal process.

Example 8(a). *Age-based renewal reward process.* Consider a non-repairable item with lifetime X_i having distribution F. Suppose that replacements take place under an age replacement policy, with replacement interval T, and that the costs of replacements upon failure and upon attainment of age T are A and B respectively, $(B < A)$. Then the associated age-based renewal reward process is given by $\{X_n, Y_n\}$, where

$$\Pr\{X_n \le x\} = F(x), x < T$$

$$= 1 \quad x \ge T$$

and
$$\Pr\{Y_n = A\} = \Pr\{X_n < T\} = F(T)$$

$$\Pr\{Y_n = B\} = \Pr\{X_n \ge T\} = 1 - F(T).$$

We have
$$E(X) = E(X_n) = \int_0^T \{1 - F(x)\}\, dx$$

and
$$E(Y) = E(Y_n) = A\, F(T) + B\{1 - F(T)\}.$$

Then applying theorem 6.14, we have, for large t,

$$\frac{C(t)}{t} \to \xi = \frac{E(Y)}{E(X)}.$$

In this case, the result

$$C(t)/t = \xi \, (t > 0),$$

can be obtained by straight forward calculation; the above provides an illustration of a reward renewal process. Christer (1978) gives an expression of the term ζ in the more refined asymptotic form

$$C(t) = t\,\xi + \zeta,$$

and this may be used to obtain better approximation in the asymptotic case.

Note: Tijms (1986, Sec. 1.5) discusses analyses of control problems in production and queueing by using renewal reward processes.

6.9 ALTERNATING (OR TWO-STAGE) RENEWAL PROCESS

The renewal processes so far considered are concerned with the occurrences of only one kind of event; the corresponding model relates to a system having only one state. For example, we considered the working of a component, the lifetime (or time to failure) being given by a sequence $\{X_n\}$ of i.i.d. random variables, on the assumption that the detection of failure and repair or replacement of the failed component take place *instantaneously*. Here the corresponding system has only one state–the working state and a renewal occurs at the termination of a working state (or failure of a component). Consider now that the detection and repair or replacement of a failed item are not instantaneous and that the time taken to do so is a random variable. The system then has two states–the working state and the repair state (during which repair of the failed component or search for a new one is under way). Here the two sequences of states–the working states and the repair (failed) state alternate. Suppose that the durations of the working states (or lifetimes or times to failure) are given by a sequence i.i.d. random variables and the durations of repair states (times taken to repair or search) are given by a sequence of i.i.d. random variables. We have then an *alternating* or *two-stage* renewal process. As can be easily imagined, a great many renewal processes occurring in applications may be described by alternating renewal processes.

Consider an alternating renewal process. Suppose that the two alternating states are denoted by 1 and 0 and that the process starts at time $t = 0$ with state 1. Suppose that the duration of the two states 1 and 0 are given by the sequence of i.i.d. random variables X_n (having a common distribution F_1) and Y_n (having a common distribution F_0) respectively. Assume that a renewal of state i occurs at each termination of state i. Suppose that X's and Y's are independent. Denote by $N_i(t)$, $i = 1, 0$, the number of renewals of state i in $[0, t]$. Then $\{N_0(t), t \geq 0\}$ is an ordinary renewal process generated by the sequence of random variables $\{X_i + Y_i\}$ having distribution $H = F_1 * F_0$ (i.e. H is given by convolution of F_1 and F_0) and $\{N_1(t), t \geq 0\}$ is a modified renewal process with initial distribution F_1 (i.e. initial interarrival time X_1) and subsequent distribution $H = F_1 * F_0$ (i.e. subsequent interarrival times $X_{i+1} + Y_i$, $i = 1, 2, \ldots$).

Denote the renewal functions by

$$M_i(t) = E\{N_i(t)\}, \quad i = 0, 1$$

Let f_0, f_1 be the p.d.f. of Y_i and X_i respectively and let $M_i^*(s), f_i^*(s)$ be the L.T.'s of the corresponding functions. We then get

$$M_1^*(s) = \frac{f_1^*(s)}{s\{1 - f_1^*(s)f_0^*(s)\}} \tag{9.1}$$

$$M_0^*(s) = \frac{f_1^*(s)f_0^*(s)}{s\{1 - f_1^*(s)f_0^*(s)\}}. \tag{9.2}$$

Theorem 6.15. For a system described by an alternating renewal process (starting with state 1 at $t = 0$) the probability that the system will be in state i at time t is given by

$$p_1(t) = M_0(t) - M_1(t) + 1 \tag{9.3}$$

and $$p_0(t) = M_1(t) - M_0(t). \tag{9.4}$$

Proof: We first note that the rth renewal epoch of the renewal process $\{N_0(t), t \geq 0\}$ is given by $S_r = \sum_{i=1}^{r} (X_i + Y_i)$. The distribution of S_r is the r-fold convolution with itself of $H = F_1 * F_0$. Thus the distribution of S_r is given by H^{r*} and has p.d.f. h_r (say).

Now, the event that the system is in state 1 at time t is equivalent to the following mutually exclusive events:

(A): the duration of the initial state 1 (in which the system started at $t = 0$) exceeds t, the probability of this event being given by

$$\Pr(A) = \Pr\{X_1 > t\} = 1 - F_1(t);$$

(B): in $(u, u + du)$, for some $u < t$, a renewal epoch S_r (for some $r = 1, 2, \ldots$) occurred and the system remained in state 1 for the remaining time $(t - u)$. The event (B) has the probability

$$\Pr(B) = \int_0^t [\sum_{r=1}^{\infty} \{\Pr(u \le S_r < u + du)\}] \Pr\{X_1 > t - u)].$$

The probability $p_1(t)$ of the event that the system is in state 1 at time t is given by

$$p_1(t) = 1 - F_1(t) + \int_0^t \sum_{r=1}^{\infty} h_r(u)\{1 - F_1(t - u)\}\, du. \tag{9.5}$$

Since L.T. of $h_r(u)$ is $h_r^*(s) = [f_1^*(s)f_0^*(s)]^r$, we have, taking Laplace transforms of (9.5),

$$p_1^*(s) = \left\{\frac{1}{s} - \frac{f_1^*(s)}{s}\right\} + \sum_{r=1}^{\infty} [f_1^*(s)f_0^*(s)]^r \left[\frac{1}{s} - \frac{f_1^*(s)}{s}\right]$$

$$= \frac{1 - f_1^*(s)}{s}\left[1 + \frac{f_1^*(s)f_0^*(s)}{1 - f_1^*(s)f_0^*(s)}\right]$$

$$= \frac{1 - f_1^*(s)}{s\{1 - f_1^*(s)f_0^*(s)\}}. \tag{9.6}$$

From (9.1) and (9.2), we find that the above can also be expressed as

$$p_1^*(s) = M_0^*(s) - M_1^*(s) + \frac{1}{s} \tag{9.7}$$

from which we get, on inversion,

$$p_1(t) = M_0(t) - M_1(t) + 1.$$

It follows that

$$p_0(t) = 1 - p_1(t)$$

$$= M_1(t) - M_0(t).$$

The theorem is thus proved.   ▲

Corollary. *Steady State Distribution: Limiting Value of $p_i(t)$*

If $E(X), E(Y)$ exist finitely, then

$$\lim_{t \to \infty} p_1(t) = \frac{E(X)}{E(X) + E(Y)}$$

$$\lim_{t \to \infty} p_0(t) = \frac{E(Y)}{E(X) + E(Y)}.$$

Proof: Let $\qquad\qquad E(X) = a, E(Y) = b.$ Then

$$f_1^*(s) = 1 - as + o(s)$$

$$f_0^*(s) = 1 - bs + o(s).$$

Substituting in (9.6), we get

$$p_1^*(s) = \frac{as + o(s)}{s\{1 - 1 + (a + b)s + o(s)\}}$$

$$= \frac{1}{s} \cdot \frac{a}{a + b} + o\left(\frac{1}{s}\right).$$

Now

$$\lim_{t \to \infty} p_1(t) = \lim_{s \to 0} s\, p_1^*(s)$$

$$= \frac{a}{a + b}$$

and thus

$$\lim_{t \to \infty} p_1(t) = \frac{E(X)}{E(X) + E(Y)}. \qquad (9.8)$$

It follows that

$$\lim_{t \to \infty} p_0(t) = \frac{E(Y)}{E(X) + E(Y)}. \qquad (9.9)$$

Note: Further generalisations of this alternating renewal process have led to interesting results. One such generalisation is that the system has N states $1, 2, \ldots i,$ $\ldots, j, \ldots, N$, such that transition from one state i to any other state j takes place with transition probability

$$p_{ij}, \quad \sum_{j=1}^{N} p_{ij} = 1,$$

and that the time spent in state i before moving to state j is a random variable whose distribution depends on both these states. We have then a semi-Markov process.

Markov renewal process is another important generalisation of the renewal process. We deal with these two topics in a separate chapter.

6.10 REGENERATIVE STOCHASTIC PROCESSES: EXISTENCE OF LIMITS

In the Corollary to Theorem 6.15, we were concerned with finding steady-state behaviour or limiting distribution of $p_i(t)$, as $t \to \infty$, $i = 0, 1$. Steady state distribution for stochastic processes in the more general case is of considerable importance. Even without finding the exact limiting distribution, it is helpful to know about the existence of the limits in stochastic processes. Renewal theorem provides a very simple and powerful tool in proving the existence of limits or of steady state for a large class of stochastic processes. The result is given below as an existence theorem.

Definition. Let $\{X(t), t \geq 0\}$ be a stochastic process having a countably infinite number of states $\{0, 1, 2, \ldots, k, \ldots\}$. Assume that there are epochs $t_1, t_2, \ldots$ at which the process probabilistically restarts from scratch, that is, from epoch t_1, the process is independent of its past and the stochastic behaviour from epoch t_1 is the same as it had from $t_0 = 0$. The same property holds for $t_2, t_3, \ldots$. Denote $T_n = t_n - t_{n-1}$, $n = 1, 2, \ldots$.

A stochastic process $\{X(t), t \geq 0\}$ is called a *regenerative process* if there exists a non-negative random variable T_1 such that

(i) the process $\{X(t + T_1), t \geq 0\}$ is independent of the process $\{X(t), t \geq s, T > s\}$ for every $s \geq 0$, and

(ii) the processes $\{X(t + T_1), t \geq 0\}$ and $\{X(t), t \geq 0\}$ have the same joint distribution.

The epochs $t_1, t_2, \ldots$ are called *regeneration epochs* (points) and the lengths $T_1, T_2, \ldots$ are called *regeneration cycles*. T_i's are i.i.d. random variables and $\{T_i, i = 1, 2, \ldots\}$ constitutes a renewal process. The renewal process is said to be embedded in $\{X(t)\}$ at the epochs $t_1, t_2, \ldots$. Every time a renewal occurs a cycle is said to be completed.

For example, a renewal process is regenerative; T_1, the time of the first renewal is the first renewal cycle. The alternating renewal process considered in Section 6.9 is regenerative, a regeneration cycle being of length Z having d.f. $H = F_1 * F_0$.

A regenerative phenomenon arising in queues in given in Example 10(a). A regenerative inventory system is discussed in section 6.11.

Theorem 6.16. Let $\{X(t)\}$ be a regenerative process with discrete state space and let

$$p_k(t) = \Pr \{X(t) = k\}.$$

Then $$\lim_{t \to \infty} p_k(t) = p_k \text{ exists such that}$$

$$p_k \geq 0 \text{ and } \sum_{k=0}^{\infty} p_k = 1.$$

Proof: Let F ($\cdot$) be the d.f. of the interarrival time

$$(t_n - t_{n-1}) = T_n \quad (t_0 = 0; t_1 \equiv T_1),$$

having mean, say, μ; and let

$$q_k(t) = \Pr\{X(t) = k, t_1 > t\}.$$

Then
$$\sum_{k=0}^{\infty} q_k(t) = \Pr\{t_1 > t\} = 1 - F(t).$$

We have, by conditioning on t_1,

$$p_k(t) = \int_0^{\infty} \Pr\{X(t) = k \mid t_1 = s\}\, dF(s)$$

$$= q_k(t) + \int_0^t p_k(t-x)\, dF(x); \tag{10.1}$$

since
$$q_k(t) = \Pr\{X(t) = k, t_1 > t\}$$

$$= \int_0^{\infty} \Pr\{X(t) = k, t_1 > t \mid t_1 = s\}\, dF(s)$$

$$= \int_t^{\infty} \Pr\{X(t) = k \mid t_1 = s\}\, dF(s).$$

From Theorem 6.5 we see that the renewal equation (10.1) has a unique solution

$$p_k(t) = q_k(t) + \int_0^t q_k(t-x)\, dM(x) \tag{10.2}$$

where $M(x)$ is the renewal function of the renewal process with distribution F.

Now $q_k(t)$ is non-negative, non-increasing and tends to 0 as $t \to \infty$, and $\int_0^{\infty} q_k(t)\, dt < \infty$.

Thus application of the Key Renewal Theorem (Theorem 6.9A) to (10.2) at once yields

$$\lim_{t \to \infty} p_k(t) \to \frac{1}{\mu} \int_0^{\infty} q_k(x)\, dx, \tag{10.3}$$

i.e. $\qquad \lim_{t \to \infty} p_k(t) = p_k$ exists. Clearly, $p_k \geq 0$,

and $\qquad \displaystyle\sum_{k=0}^{\infty} p_k = \frac{1}{\mu} \int_0^{\infty} \left\{ \sum_{k=0}^{\infty} q_k(x) \right\} dx = \frac{1}{\mu} \int_0^{\infty} \{1 - F(x)\} \, dx = 1.$ ▲

Note: Clearly, μ is the expected duration of a complete cycle, and $\int_0^{\infty} q_k(t) \, dt =$

$\int_0^{\infty} \Pr \{X(t) = k, t_1 > t\} \, dt =$ expected duration of sojourn at (or visit to) state k during

a cycle $= E(\tau_k)$, say. Thus

$$p_k = \lim_{t \to \infty} \Pr \{X(t) = k\} = \frac{E(\tau_k)}{\mu}. \tag{10.4}$$

Remark: In Theorem 6.16, it has been assumed that the state space S of $X(t)$ is discrete. This requirement is *not* necessary, and the result is valid, under certain conditions, for $X(t)$ having a continuous state space $S \equiv (-\infty, \infty)$ also.

Example 10(a). *Queueing process** with arrival and service rates λ and μ and $\lambda/\mu < 1$. Let the state of the queueing process at epoch t be given by the number in the system $X(t)$ at that epoch; then the states of the process are $i \in N = \{0, 1, 2, \ldots\}$. Suppose that time is reckoned from the epoch 0 at which an arriving customer finds the system empty with a free server and the service begins, i.e. with commencement of a busy period (with distribution, say, F_1). The busy period B is followed by an idle period I during which no customer is present in the system (I having distribution say, F_0). A busy period and an idle period alternate and together constitute a *busy cycle*; the whole process restarts itself (probabilistically) from scratch at the end of each busy cycle. The epoch of commencement of a busy period is a *regeneration point*; a busy cycle is a regeneration cycle. An idle period corresponds to the visit to state 0 while a busy period to states $i (\neq 0)$. Clearly, $q_i(t) = 1 - F_i(t)$,

$\int_0^{\infty} q_0(t) \, dt = $ Expectation of the duration of idle period $= E(I)$,

and $\qquad \int_0^{\infty} q_1(t) \, dt = $ Expectation of the duration of busy period $= E(B)$.

Thus for such a system

$$p_0 = \lim_{t \to \infty} \Pr \{X(t) = 0\} = \frac{E(I)}{E(I) + E(B)} \tag{10.5}$$

---

*See Sec. 10.1 for description.

which is a very interesting result. It is also put as

$$\frac{E(B)}{E(I)} = \frac{1-p_0}{p_0}. \tag{10.6}$$

In this example we are concerned with an alternating (two-stage) system; however the theorem is applicable to multi-stage systems also.

6.11 REGENERATIVE INVENTORY SYSTEM

6.11.1 Introduction

Consider that a store maintains a certain quantity of an item in stock. It places an order for replenishment when the stock level runs low. The inventory policy is of the (S, s) type, $S \geq 0, S > s$, where s could be negative (implying unfilled back orders). Suppose that stock at hand is continuously observed; if the stock level at a point of time falls below s, then order for replenishment is placed and stock is procured to bring the inventory upto level S (order-up-to level) and if the stock level is greater than or equal to s, then no action is taken. The inventory policy is called (S, s) policy under continuous review. We assume that the initial stock position at epoch 0 is S and that an order placed is immediately filled (that is, lead time is zero). Denote $I(t)$, the *stock (inventory) position* at time t and by $t_0 (= 0), t_1, t_2, \ldots$ the epochs at which replenishment is done and stock level is raised again to S. The epochs (or points of time) t_0, t_1, are regeneration points and $T_n = t_n - t_{n-1}, (n = 1, 2, \ldots)$ is the length of the nth regeneration cycle. The r.v.'s $T_n, n = 1, 2, \ldots$ are i.i.d. and so $\{T_n, n = 1, 2, \ldots\}$ is a renewal process embedded in the stochastic process $\{I(t), t \geq 0\}$. Denote by $C(x)$, $c(x)$ and $k(x)$, respectively, the d.f., p.d.f. and the associated renewal density of the renewal process $\{T_n, n = 1, 2, \ldots\}$. We call it *refill process*.

Denote by $X_i, i = 1, 2, \ldots$ the interdemand time, by $A(x) = \Pr\{X_i \leq x\}$ the d.f. of the i.i.d. random variables X_i, and by $\mu_a = E(X_i)$. Then $\{X_i, i = 1, 2, \ldots\}$ is a renewal process. Denote $S_n = X_1 + \ldots + X_n$ $(S_0 = 0)$ and $A_n(x) = \Pr(S_n \leq x)$. We have

$$E(S_n) = \int_0^\infty [1 - A_n(u)] \, du = n \, \mu_a \tag{11.1}$$

Denote by $N(u, u + t)$, the number of occurrences of demand in $(u, u + t]$. If u is an arrival point of a demand then $N(u, u + t) = N(t)$ and

$$\Pr\{N(t) = n\} = A_n(t) - A_{n+1}(t), \quad n = 0, 1, 2, \ldots \tag{11.2}$$

Denote by $Y_i, i = 1, 2, \ldots$ the size of the batch size of demand at the ith occurrence of demand, and by $B(x) = \Pr\{Y_i \leq x\}$, the d.f. of the i.i.d. r.v.'s Y_i. Then $\{Y_i, i = 1, 2, \ldots\}$ is a renewal process. Denote $B_n(x) = \Pr\{Y_1 + \ldots + Y_n \leq x\}, n = 1, 2, \ldots$. The renewal function of this renewal process is given by

$$R(x) = \sum_{n=1}^{\infty} B_n(x).$$ (11.3)

Assume that $A(x)$ and $B(x)$ are absolutely continuous with p.d.f. $a(x)$ and $b(x)$ respectively. The renewal density is given by $r(x) = \sum b_n(x)$. Suppose that u is an arrival point of demand; then the cumulative demand during $(u, u + t]$ is given by

$$D(t) = \sum_{i=1}^{N(t)} Y_i.$$ (11.4)

$\{D(t), t \geq 0\}$ is a cumulative renewal process having its d.f. (using (8.2))

$$\Pr\{D(t) \leq x\} = \sum_{n=0}^{\infty} B_n(x)\{A_n(t) - A_{n+1}(t)\}, t \geq 0, x \geq 0.$$ (11.5)

6.11.2 Inventory Position

$\{I(t), t \geq 0\}$, where $I(t)$ denotes the inventory position at epoch t, is a stochastic process in continuous time with continuous state space; we take $I(0) = S$. Denote the p.d.f. of $I(t)$ by $f(t, x)$, that is,

$$f(t,x)dx = \Pr\{x \leq I(t) < x + dx\}, t \geq 0, s \leq x \leq S$$ (11.6)

We have the following result.

Theorem 6.17. The distribution of $I(t)$ is given by

$$\Pr[I(t) = S] = 1 - A(t) + \int_0^t k(t-x)[1 - A(x)]\, dx$$ (11.7)

and

$$f(t,x) = \sum_{n=1}^{\infty} [A_n(t) - A_{n+1}(t)]\, b_n(S - x)$$

$$+ \sum_{n=1}^{\infty} b_n(S - x) \int_0^t k(t-x)[A_n(x) - A_{n+1}(x)]\, dx,$$ (11.8)

$$(t \geq 0, s \leq x \leq S)$$

where $k(x)$ is the renewal density associated with the renewal process $\{T_n, n = 1, 2, \ldots\}$; that is, $k(x)$ is the solution of the integral equation

$$k(u) = c(u) + \int_0^u k(u-x)c(x)\, dx$$ (11.9)

with $c(x)$ given by

$$c(x) = \sum_{k=0}^{\infty} [a_{k+1}(x) - a_k(x)] B_k (S - s). \qquad (11.10)$$

Proof: It is clear that

$$\{T_n > x\} \Leftrightarrow \{D(x) < S - s\}$$

that is,

$$\{T_n \le x\} \Leftrightarrow \{D(x) > S - s (= \Delta, \text{say})\}$$

so that (using 11.5) we get

$$C(x) = \Pr\{T_n \le x\} = \Pr\{D(x) > S - s\}$$

$$= 1 - \sum_{n=0}^{\infty} B_n(\Delta) \{A_n(x) - A_{n+1}(x)\}, x \ge 0.$$

Taking derivative, we at once get (11.10).
With the above expression of $c(x)$, we get that the renewal density $k(t)$ associated with T_n as the solution of the integral equation

$$k(u) = c(u) + \int_0^u k(u - x) c(x) dx$$

which is (11.9).
The renewal function $K(u)$ associated with $\{T_n, n = 1, 2, \ldots\}$ is given by

$$K(u) = C(u) + \int_0^u k(u - x) dC(x)$$

Now, by conditioning on the first renewal T_1 of the renewal process $\{T_n, n = 1, 2, \ldots\}$, we have

$$\Pr\{I(t) = S\} = 1 - A(t) + \int_0^t \Pr[I(t - u) = S] c(u) du$$

which is a renewal equation.
Using Theorem 6.5, we get

$$\Pr\{I(t) = S\} = 1 - A(t) + \int_0^t [1 - A(t - u)] k(u) du$$

$$= 1 - A(t) + \int_0^t k(t - x) [1 - A(x)] du$$

(changing the variable u in the integral by x, where $t - u = x$). Thus we get (11.7). Using similar renewal arguments we can get (11.8).   ▲

6.11.2.1 Asymptotic Distribution of I(t)

We shall now consider limiting distribution as $t \to \infty$.

Denote
$$\lim_{t \to \infty} I(t) = I$$

$$\lim_{t \to \infty} f(t,x) = f(x).$$

Theorem 6.18. The limiting distribution I of inventory position is given by

$$\Pr[I = S] = \frac{1}{1 + R(S - s)} \tag{11.11}$$

$$f(x) = \frac{r(S - x)}{1 + R(S - s)}, \quad x \le x \le S \tag{11.12}$$

Proof: A renewal cycle T_n is the first passage time to level $S - s$. We have, from (8.6)

$$E\{T_n\} = \mu_a [1 + R(S - s)].$$

Using the Key Renewal Theorem (Theorem 6.9), we get

$$\int_0^t k(t - x)[1 - A(x)]\, dx = \int_0^t [1 - A(t - x)]\, k(x)\, dx$$

$$\to \frac{1}{E(T_n)} \int_0^t [1 - A(u)]\, du$$

$$= \frac{\mu_a}{E(T_n)}$$

$$= \frac{1}{1 + R(S - s)}.$$

Thus, as $t \to \infty$, the second term $\to 1$, and we have (from (11.7))

$$\Pr(I = S) \to \frac{1}{1 + R(S - s)}.$$

As $t \to \infty$, we note that

$$[A_n(t) - A_{n+1}(t)] \to 0$$

and

$$\int_0^t k(t-x)[A_n(x) - A_{n+1}(x)]\, dx$$

$$= \int_0^t k(t-x)[\{1 - A_{n+1}(x)\} - \{1 - A_n(x)\}]\, dx$$

$$\rightarrow \frac{1}{E(T_n)}\left[\int_0^\infty \{1 - A_{n+1}(x)\}\, dx - \int_0^\infty \{1 - A_n(x)\}\, dx\right]$$

$$= \frac{1}{E(T_n)}[(n+1)\mu_a - n\,\mu_a]$$

$$= \frac{1}{1 + R(S - s)}.$$

Taking limit of (11.8) as $t \rightarrow \infty$, and using (11.3) we get

$$f(x) = \sum_{n=1}^\infty b_n(S-x)\left[\frac{1}{1 + R(S-s)}\right]$$

$$= \frac{r(S-x)}{1 + R(S-s)} \quad, x \geq 0, s \leq x \leq S.$$

Remarks: (1) The interdemand time has no effect on the limiting distribution of the inventory position. The latter depends entirely on the renewal function of the renewal process $\{Y_i, i = 1, 2, \ldots\}$ (where Y_i is the ith demand size).

(2) A detailed account of renewal-theoretic analysis of a single item (S, s) inventory system has been given by Sahin (1990), to which reference may be made for further results.

6.12 GENERALISATION OF THE CLASSICAL RENEWAL THEORY

The classical theory (to which we devoted our attention) deals with a sequence of *i.i.d. non-negative* r.v.'s $\{X_i\}$ and considers the asymptotic behaviour of $E\{N(t)\}$ as $t \rightarrow \infty$ (where $N(t) = \sup\{n: S_n \leq t\}$, $= \sum_{i=1}^n X_i$, $E\{N(t)\} = \sum_{n=0}^\infty \Pr\{S_n \leq t\}$). Feller (1941) obtained the elementary renewal theorem which states that as $t \rightarrow \infty$, $E\{N(t)\}/t \rightarrow 1/\mu$, where $\mu = E(X_i)$. This result of classical theory was generalised, step by step, in several directions. We refer here to some of the works of the several researchers who made very significant contribution of immense mathematical and theoretical interest. This has been done to indicate the broad areas and to draw attention to some of the works for the benefit of those who might be interested in pursuing the topic more deeply. First generalisation was concerned with sequence of r.v.'s $\{X_i\}$ which were *not* assumed to be non-negative, i.e. generalisation from distributions concentrated on the half-line to distributions on the whole line. Feller considers what he terms 'transient distribution' F on the whole line.

Important generalisation of renewal theory to various cases of transient distributions, are due, among others to Chung *et al* (*Ann. Math.* **55** (1952) 1–6, *Proc. Am. Math. Soc.* **3** (1952) 303–309), Blackwell (*Pacific J. Math.* **3** (1953) 315–320), Feller *et al.* (*J. Math. Mech.* **10** (1961) 619–624. We state below (without proof) an important *extension of Blackwell's renewal theorem.*

Let $\{X_i\}$ be a sequence of (continuous) i.i.d. random variables which assume values in the range $(-\infty < x < \infty)$ such that $0 < \mu = E(X_i) \le \infty$.

Let $U(t) = \sum_{n=0}^{\infty} \Pr\{S_n \le t\}, -\infty < t < \infty$, where $S_n = \sum_{i=1}^{n} X_i$. Then for $h > 0$,

$$U(t+h) - U(t) = \frac{h}{\mu} \quad \text{as } t \to \infty$$

$$U(t+h) - U(t) = 0 \quad \text{as } t \to -\infty.$$

It is to be noted that in this case $U(t)$ is, in general, not equal to $E\{N(t)\}$, where $N(t) = \sup\{n: S_n \le t\}$ (for details of proof reference may also be made to the books by Prabhu and Feller Vol. II).

Attention was also turned towards the nature of $\{X_i\}$. Chow and Robbins (*Ann. Math. Statis.* **34** (1963) 390–395) investigated the conditions on the joint distribution of X_i which would ensure the validity of the elementary renewal theorem,

$$\sum_{n=0}^{\infty} \Pr\{S_n \le t\}/t \to 1/\mu$$

for some positive constant μ. They relaxed the restriction of independence of the r.v.'s X_i and obtained a renewal theorem for the dependent case as well as for a sequence of non-identical r.v.'s Smith (*Pacific J. Math.* **14** (1964) 673–699) also extended the theorem to non-identically distributed r.v.'s Heyde (*Ann. Math. Statis.* **37** (1966) 699–710) considered asymptotic behaviour of more general forms like

$$\sum_{n=0}^{\infty} a_n \Pr\{S_n \le t\} \quad \text{and} \quad \sum_{n=0}^{\infty} a_n \Pr\left\{\max_{1 \le k \le n} S_k \le t\right\}$$

for a sequence of i.i.d. r.v.'s $\{X_i\}$ and for a sequence of positive constants $\{a_n\}$ such that Σa_n diverges.

Later in a series of papers Sankaranarayana *et al.* (*Pacific J. Math.* **30** (1969) 785–803, *Act. Math. Aca. Sc. Hung.* **31** (1978) 1–8, *Math. Oper. Statist. Ser. Opt.* **8** (1977) 581–590) considered suitable sequences of $\{a_n\}$ and studied the asymptotic behaviour of such functions for sequences of correlated and dependent r.v.'s, as well as for stationary discrete stochastic processes.

Renewal theory for Markov chains has been investigated by Kesten ((1974) and Athreya *et al.* (*Ann. Prob.* **6** (1978) 788–797), Keener (*Ann. Prob.*, **10** (1982) 942–955 and others). Ryan Jr. (*Ann. Prob.* **4** (1976) 656–661) has considered the multidimensional case; so also has Spitzer (1984). Chow *et al.* (*Ann. Prob.* **7** (1979) 304–318) have developed an extended renewal theory.

Neyman (1982) obtains a variant of the classical renewal theorem in which convergence is uniform on classes of r.v.'s. Renewal theorem for a urn model has been obtained by Sen (1982).

These would give an idea of the several directions in which extensions have been made. The above is an illustrative (and by no means an exhaustive) account of the many interesting and far-reaching contributions made by several workers towards generalisation and extension of the renewal theory, as well as towards applications.

EXERCISES

6.1 Show that

$$\Pr\{N_e(t) = r\} = \frac{1}{\mu} \int_0^T [\Pr\{N(u) = r - 1\} - \Pr\{N(u) = r\}]\, du$$

$$r = 1, 2, 3, \ldots$$

and
$$\Pr\{N_e(t) = 0\} = 1 - \frac{1}{\mu} \int_0^T \Pr\{N(u) = 0\}\, du.$$

6.2 Taking $f(t), g(t)$ as given in Example 6(a), obtain $M_D(t)$ by using the relation (6.3).

6.3 Show that, the renewal function of a renewal process $\{N(t), t \geq 0\}$ generated by $X_n, n = 1, 2, \ldots$ is given by

$$E\{N(t)\} = M(t) = \frac{t}{\mu} + \frac{\sigma^2 - \mu^2}{2\mu^2} + o(1),$$

where $\mu (< \infty)$ and σ^2 are the mean and the variance of X_n. Further, show that var $\{N(t)\} = (\sigma^2/\mu^3)\, t + o(t)$ (Smith, 1954).

6.4 Find the second moment $L(t) = E\{(N(t))^2\}$ for the renewal process with hyper-expotential distribution having p.d.f.

$$f(t) = pae^{-at} + (1-p)be^{-bt}, 0 \leq p \leq 1, a > b > 0.$$

Consider also 2-stage hypo-exponential distribution (Sec. 1.3.4.8).

6.5 Put forward direct probabilistic argument to show that, in the Markovian case,

$$\Pr\{N(t) \geq n\} = \Pr\{N_A(t, T) \geq n\} = \Pr\{N_B(t, T) \geq n\}$$

for all
$$t \geq 0, T \geq 0, n = 0, 1, 2, \ldots.$$

6.6   Show that, for $t \geq 0$, $T \geq 0$, $n = 0, 1, 2, \ldots$ both

$$\text{(i)} \quad \Pr\{N(t) > n\} \geq \Pr\{N_A(t,T) > n\}$$

and $$\text{(ii)} \quad \Pr\{N(t) > n\} \geq \Pr\{N_B(t,T) > n\}$$

hold *iff* F is NBU, i.e. the lifetime X of the unit satisfies

$$\Pr\{X > x + y\} \leq \Pr\{X > x\}\,\Pr\{Y > y\} \text{ for } x \geq 0,\ y \geq 0.$$

(Barlow and Proschan, 1975).

6.7   Let $R_A(t, T)$, $R_B(t, T)$ be the number of removals in $[0, t]$ under age and block replacement policies respectively. Show that

$$\Pr\{N(t) \geq n\} \leq \Pr\{R_A(t,T) \geq n\} \leq \Pr\{R_B(t,T) \geq n\}$$

for $$t \geq 0, T \geq 0, n = 0, 1, 2, \ldots.$$

(The intuitive meaning of the above should be quite clear).

(Barlow and Proschan, 1964).

6.8   Show that

$$\lim_{t \to \infty} \frac{E\{R_A(t,T)\}}{t} = \frac{1}{\displaystyle\int_0^T \{1 - F(x)\}\, dx}$$

$$\lim_{t \to \infty} \frac{E\{R_B(t,T)\}}{t} = \frac{M(t) + 1}{T}$$

and that for all $T > 0$,

$$M(t) \geq \frac{T}{\displaystyle\int_0^T \{1 - F(x)\}\, dx} - 1$$

$$\geq T/\mu - 1 \qquad\qquad \text{(Barlow and Proschan, 1964).}$$

6.9   Consider the number N of renewals in random time T having gamma distribution and p.d.f.

$$q(t) = \frac{a^k t^{k-1} e^{-at}}{\Gamma(k)},\ k = 1, 2, \ldots,\ t \geq 0$$

Show that the p.g.f. of N is given by

$$G(z) = \frac{a^k}{\Gamma(k)} \left(-\frac{\partial}{\partial s} \right)^{k-1} \left[\frac{1 - f^*(s)}{s\{1 - zf^*(s)\}} \right]_{s=a},$$

Examine the truth of the following assertion:
N has negative binomial distribution with p.g.f.

$$G(z) = \left[\frac{1 - f^*(a)}{1 - zf^*(a)} \right]^k$$

if and only if the renewal process $\{X_i\}$ is Poisson (i.e. X_i has exponential distribution with p.d.f., say, $f(t) = ce^{-ct}$).

6.10 Distribution of residual lifetime $Y(t)$: alternative form.
Show that

$$\Pr\{Y(t) \le x\} = \int_t^{t+x} \{1 - F(t + x - y)\}\, dM(y).$$

6.11 Show that the mean residual (remaining) life of an item aged x is given by

$$e(x) \equiv E[X - x \mid X > x] = \int_x^\infty \{1 - F(t)\}\, dt \, / \, [1 - F(x)]$$

where $F(x)$ is the d.f. of the lifetime X of the item. Examine the Markovian case.
Note that $x + e(x)$ is a non-decreasing function of x. [$e(x)$ corresponds to the expectation of life at age x in demography].

6.12 Show that if $F(x)$ is DFR, then $m(x)$ is monotone decreasing on $x \ge 0$ and $M(x)$ is concave. Verify the result for hyper-exponential X.
Note that $F(x)$ IFR does not imply that $m(x)$ is increasing. For monotonicity properties, refer to Brown (1980).

6.13 Show that the rth moment of the forward recurrent time $Y(t)$ of X is given (for $r = 1, 2, \ldots$) by

$$E[\{Y(t)\}^r] = E[(X - t)^r] + \int_0^t \{E[(X - u)^r - (-u)^r]\} m(t - u)\, du.$$

Coleman (1982).
Deduce the corresponding result for the asymptotic distribution of $Y(t)$.

6.14 If $Z(t)$ is the spent lifetime at time t for the process $\{X_i\}$ and $\mu = E(X_i) < \infty$, then, with probability 1

$$\frac{Z(t)}{t} \to 0 \quad \text{as} \quad t \to \infty$$

and that

$$E\left[\frac{Z(t)}{t}\right] \to 0 \quad \text{as} \quad t \to \infty.$$

6.15 Show that for the renewal reward process $\{X_n, Y_n\}$, the asymptotic form of $C(t)$ is given by

$$C(t) = t\xi + \zeta,$$

where

$$\xi = E(Y)/E(X)$$

and

$$\zeta = \frac{E(Y) - E(X^2)}{2\{E(X)\}^2} - \lim_{t \to \infty} E\{Y_{N(t)+1}\},$$

and that for the age-based renewal reward process considered in Example 8(a)

$$\zeta = -\left[\frac{2TE(X) - E(X^2)E(Y)}{2\{E(X)\}^2} - \frac{A\{T - E(X)\}}{E(X)}\right]$$

(Christer, 1978).

6.16 Find $M_i(t), i = 1, 0$ for the alternating renewal process $\{X_n, Y_n\}$ with distribution of X_n and Y_n given by

$$F_1(x) = 1 - e^{-ax}, \quad \text{and} \quad F_0(x) = 1 - e^{-bx} \text{ respectively.}$$

Show that

$$p_1(t) = \frac{b}{a+b} + \frac{a}{a+b} e^{-(a+b)t}$$

$$p_0(t) = 1 - p_1(t) = \frac{a}{a+b} - \frac{a}{a+b} e^{-(a+b)t};$$

further that

$$\lim_{t \to \infty} p_1(t) = \frac{b}{a+b} = \frac{E(X)}{E(X) + E(Y)}$$

$$\lim_{t \to \infty} p_0(t) = \frac{a}{a+b} = \frac{E(Y)}{E(X) + E(Y)}.$$

Note that $p_1(t)$ for the Markovian case was obtained earlier by using results of Markov processes (see Example 5(b), Ch. 4), and that the result giving the limiting value of $p_1(t)$ as $t \to \infty$ holds independently of the form of the distributions of X and Y (see Corollary to Theorem 6.15).

6.17 For block replacement (with replacement interval T) with costs of replacement upon failure and at intervals $T, 2T, \ldots$ being A and B respectively, find $E\{Y(t)\}/t$ as $t \to \infty$, where $Y(t)$ is the total cost incurred by time $t \geq 0$.

6.18 Prove that the (asymptotic) joint distribution of the residual lifetime $Y(t)$ and the spent lifetime $Z(t)$, is given by

$$\Pr\{Z(t) > x, X(t) > y\} \to \frac{1}{\mu} \int_{x+y}^{\infty} \{1 - F(u)\}\, du$$

$$\text{as } t \to \infty \ (x > 0, y > 0)$$

(where F denotes the distribution of the lifetime with mean μ).

6.19 Show that, for a regenerative inventory system, the mean inventory position is given by

$$E(I) = s + \frac{\overline{R}(S-s)}{1 + R(S-s)}$$

where
$$\overline{R}(t) = \int_0^t R(u)\, du.$$

Find var (I).

6.20 Show that, with the initial inventory position given by $I(0) = y$, $s < y < S$,

$$\Pr\{I(t) = y\} = 1 - A(t)$$

$$\Pr\{I(t) = S\} = \int_0^t \overline{m}(t-u)[1 - A(u)]\, du$$

where $\overline{m}(\cdot)$ is the renewal density of the delayed renewal process $\{T_n, n = 1, 2, \ldots\}$ with distribution of T_1 being different from that of T_n, $n = 2, 3, \ldots$.

Further show that the limiting distributions of $I (= \lim_{t \to \infty} I(t))$ and of $f(x)$ are the same as those given in (11.11) & (11.12) respectively (Sahin, 1990).

REFERENCES

A.A. Balkema and L.De Hann. Residual lifetime at great age, *Ann. Prob.*, **2**, 792–804 (1974).

R.E. Barlow and F. Proschan. Comparison of replacement policies and renewal theory implications, *Ann. Math. Stat.*, **35**, 577–589 (1964).

—— *Statistical Theory of Reliability and Life Testing*, Holt, Rinehart & Winston, New York (1975).

D.J. Bartholomew. Renewal theory models in manpower planning, *Symp. Proc. Series* No. 8, The Institute of Mathematics and its Applications, 57–73 (1976).

—— *Stochastic Models for Social Processes*, 2nd ed., Wiley, Chichester (1973).

—— and A.F. Forbes. *Statistical Techniques for Manpower Planning*, Wiley, Chichester (1979).

L.A. Baxter, E.M. Schauer, W.R. Blischke and D.J. McConalogue. On the tabulation of renewal function, *Technometrics* **24**, 151–152 (1982a).

—— *Renewal Tables: Tables of Functions Arising in Renewal Theory*, University of South Calif., Los Angeles (1982b).

D. Blackwell. A renewal theorem, *Duke Math. J.* **15**, 145–150 (1948).

—— Extension of a renewal theorem, *Pacific J. Math.* **3**, 315–320 (1953).

M. Brown. Bounds, inequalities and monotonicity properties for some renewal processes, *Ann. Prob.* **8**, 227–240 (1980).

A.H. Chister. Refined asymptotic costs for renewal reward processes, *J. Opns. Res. Soc.* **29**, 577–583 (1978).

R. Coleman. The moments of forward recurrence time, *Euro. J. Opnl. Res.* **9**, 181–183 (1982).

D.R. Cox. *Renewal Theory*, Methuen, London (1962).

M.A. Crane and A.J. Lemoine. *An Introduction to the Regenerative Method for Simulation Analysis*, Springer-Verlag, New York (1977).

W.Feller. *An Introduction to Probability Theory and its Applications*, Vol. II, (1966), Vol. I 3rd. ed. (1968) Wiley, New York.

A.K.S. Jardine. *Maintenance, Replacement and Reliability*, London (1973).

S. Karlin and H.M. Taylor. *A First Course in Stochastic Processes*, 2nd, ed. Academic Press, New York (1975).

U.S. Karmarkar. Future costs of service contracts for consumer durable goods, *AIIE Trans.*, **10**, 380–387 (1978).

H. Kesten. Renewal theory for Markov chains, *Ann. Prob.* **3**, 355–387 (1974).

N. Keyfitz. *An Introduction to Mathematics of Population*. Addison-Wesley, Reading, MA (1968; with revisions, 1977).

J. Kingman. *Regenerative Phenomena*, Wiley, New York (1972).

J. Kohlas. *Stochastic Methods in Operations Research*, Cambridge Univ. Press (1982).

D.N.P. Murthy and D.G. Nguyen. Replacement policies with used item constraints, *J. Opnl. Res. Soc.* **34**, 633–639 (1983).

S. Nahmias. *Production and Operations Analysis*, Irwin, Homewood, IL, USA (1989).

T. Nakagawa. A summary of block replacement policies, *R.A.I.R.O.* **13**, 351–361 (1979).

—— A modified block replacement policy with two variables. *I.E.E.E. Trans. Rel.* **R-31**, 398–400 (1982).

A. Neyman. Renewal theory for sampling without replacement. *Ann. Prob.* **10**, 464–481 (1982).

N.U. Prabhu. *Stochastic Processes*, Macmillan, New York (1965).

—— *Stochastic Storage Processes*, Springer-Verlag, New York (1980).

S.M. Ross. *Applied Probability Models with Optimization Applications*, Holden-Day, San Francisco, (1970).

—— *Stochastic Processes*, Wiley, New York (1983).

I. Sahin. *Regenerative Inventory Systems*, Springer, New York (1990).

P.J. Schweitzer. Optimal replacement policies for hyper-exponentially and uniformly distributed lifetimes, *Opns. Res.* **15**, 360–362 (1967).

P.K. Sen. A renewal theorem for a urn model, *Ann. Prob.* **10**, 835–843 (1982).

S. Sethi and G. Thompson. *Optimal Control Theory: Application to Management Science*, Martinus Nijhoff, Boston (1981).

E.A. Silver. Operations research in inventory management: A Review and Critique, *Opns. Res.* **29**, 628–645 (1981).

A.N. Sirjaev. *Statistical Sequential Analysis*, American Mathematical Society, Providence, R.I. (1973).

W.L. Smith. Renewal Theory and its Ramifications, *J.R.S.S-B.* **20**, 243–302 (1958).

F. Spitzer. A multidimensional renewal theorem in: *Probability, Statistical Mechanics and Number Theory*, Advances in Mathematics, Suppl. Studies, Vol 9 (Ed. G.C. Rota), 147–155 (1984).

H.C. Tijms. *Stochastic Modelling and Analysis: A Computational Approach*, Wiley, Chichester, U.K. (1986).

CHAPTER 7

Markov Renewal and Semi-Markov Processes

7.1 INTRODUCTION

We shall consider here some kind of generalisation of a Markov process, as well as of a renewal process. We first consider a Markov process with discrete state space, and its generalisation by way of random tranformation of the time scale of the process. Let $\{X(t)\}$ be a Markov process with discrete state space and let the transitions occur at epochs (or instants of time) t_i, $i = 0, 1, 2, \ldots$. Then the sequence $\{X_n = X(t_n +)\}$ forms a Markov chain and the time intervals $(t_{n+1} - t_n) = \tau_n$ between transitions have independent exponential distributions, the parameters of which may depend on X_n. We shall now suppose that the transitions $\{X_n\}$ of the process $X(t)$ still constitute a Markov chain, but that the transition intervals $(t_{n+1} - t_n)$ have any distribution, with parameter which may depend not only on X_n but also on X_{n+1}. The process $\{X_n, t_n\}$ (as well as $\{X_n, \tau_n\}$) is then called a *Markov renewal process* and the process $\{X(t)\}$ is known as a *semi-Markov process*. A semi-Markov process (s-M.P.) is thus a stochastic process in which changes of state occur according to a Markov chain and in which the time interval between two successive transitions is a random variable, whose distribution may depend on the state from which the transition takes place as well as on the state to which the next transition takes place. The s-M.P. provides a convenient way of describing the underlying Markov renewal process. Semi-Markov processes were first studied by Lévy and Smith, while Markov renewal processes by Pyke. Considerable interest in these topics has grown over the years, as, such generalisation of Markov process can be fruitfully applied in building models of many phenomena. Many problems in queueing, reliability and inventory theories may be approached through Markov renewal and semi-Markov processes (see Cinlar 1969, 1975 for a survey). In stochastic models for many types of social behaviour these processes have been used (see, for example, Bartholomew, 1967, Bartholomew and Forbes, 1979, Valliant and Milkovitch, 1977). For example, problems of occupational mobility, where transitions between occupations as well as intervals between transitions are taken into account, may be treated by means of semi-Markov processes. So also in some studies of problems relating to market behaviour. A semi-Markov storage problem has been studied by Senturia and Puri (1973). A semi-Markov model has been used by Kao (1974) in some context of hospital administration—in the study of the dynamics of movement of patients through various care zones (e.g. medical unit, surgical unit, intensive care unit etc.) in a hospital. Weiss *et al* (1982) present a methodology to identify and specify a semi-Markov model of hospital patient flow within a network of service facilities. The theory of birth has been developed as a Markov renewal process by Sheps (1964) (also Sheps *et al.*, 1973), and the theory has been applied to microdemography by Potter (1970) (see also Keyfitz, 1977). In fact, the importance of semi-Markov processes, in particular, and Markov renewal theory, in general, lies to a large extent in the wide domain of applications rather than in the richness of their theoretical developments.

7.2 DEFINITIONS AND PRELIMINARY RESULTS

Definitions. Let the states of a process be denoted by the set $E = \{0, 1, 2, \ldots\}$, and let the transitions of the process occur at epochs (or instants of times) $t_0 = 0, t_1, t_2, \ldots (t_n < t_{n+1})$. Let X_n denote the transition occurring at epoch t_n. If

$$\Pr\{X_{n+1} = k, t_{n+1} - t_n \leq t \mid X_0 = ., \ldots, X_n = \cdot; \ t_0, \ldots, t_n\}$$

$$= \Pr\{X_{n+1} = k, t_{n+1} - t_n \leq t \mid X_n = .\}, \tag{2.1}$$

then $\{X_n, t_n\}, n = 0, 1, 2, \ldots$ is said to constitute a *Markov renewal process* with state space E. $\{X_n, \tau_n\}$ is also used to denote the process $\{X_n, t_n\}$. Assume that the process is temporally (or time) homogeneous, that is,

$$\Pr\{X_{n+1} = k, t_{n+1} - t_n \leq t \mid X_n = j\} = Q_{jk}(t) \tag{2.2}$$

is independent of n. Define

$$p_{jk} = \lim_{t \to \infty} Q_{jk}(t) = \Pr\{X_{n+1} = k \mid X_n = j\}. \tag{2.3}$$

(Symbols $Q_{j,k}$ for Q_{jk} and $p_{j,k}$ for p_{jk} will also be used.)
The matrix $Q(t) = (Q_{jk}(t))$ is known as the *semi-Markov kernel* of the Markov renewal process $\{X_n, t_n\}$.

Proposition (1):
$\{X_n, n = 0, 1, 2, \ldots\}$ constitutes a Markov chain with state space E and t.p.m. $P = (p_{jk})$.
The continuous parameter process $\{Y(t)\}$ with state space E, defined by

$$Y(t) = X_n \quad \text{on} \quad t_n \leq t < t_{n+1} \tag{2.4}$$

is called a *semi-Markov process*. The Markov chain $\{X_n\}$ is said to be an *embedded Markov chain* of the semi-Markov process. X_n refers to the state of the process at transition occurring at epoch t_n and $Y(t)$ that of the process at its most recent transition. Consider, for example, a simple random walk performed by a particle. Suppose that $t_0, t_1, \ldots$, are the epochs when transitions take place and that X_n is the transition at the epoch t_n; also that at t_n the particle makes a move to the right (and enters the state j), at t_{n+1} to the left (enters the state $j - 1$), at t_{n+2} to the right (enters the state j), at t_{n+3} to the right (enters the state $j + 1$), at t_{n+4} moves to the right (enters the state $j + 2$), and so on.

Then

$$Y(t) = j, \qquad t_n \ \leq \ t < t_{n+1}$$
$$= j - 1, \qquad t_{n+1} \ \leq \ t < t_{n+2}$$
$$= j, \qquad t_{n+2} \ \leq \ t < t_{n+3}$$
$$= j + 1, \qquad t_{n+3} \ \leq \ t < t_{n+4}$$
$$= j + 2, \qquad t_{n+4} \ \leq \ t < t_{n+5}$$

and so on. The position of the particle is shown in Fig. 7.1.

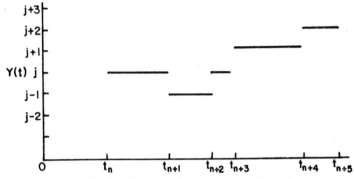

Fig. 7.1 Diagrammatic representation of $\{Y(t)\}$

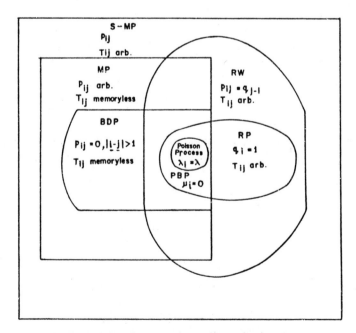

Fig. 7.2 Relationships among interesting stochastic processes
s–MP.: semi-Markov process; MP: Markov process;
BDP: birth-death process; PBP: pure birth process;
RP: renewal process; RW: random walk
[Reproduced with kind permission of John Wiley & Sons, Inc. from *Queueing Theory*, Vol 1:
Theory by Leonard Kleinrock. Copyright © 1975, by John Wiley & Sons, Inc.]

Example 2(a). A pure birth process is a Markov renewal process with

$$Q_{jk}(t) = 1 - e^{-a_j t}, \quad k = j+1$$

$$= 0, \quad \text{otherwise.}$$

Unless otherwise stated, it will be assumed in the following sections, that the state space of the process is finite having m states, $1, 2, \ldots, m$.

7.2.1 Waiting Times

Let the time spent by the system in state j before its next transition, given that the next transition is into state k, be a random variable T_{jk} having distribution function (d.f.):

$$W_{jk}(t) = \Pr\{T_{jk} \le t\} = \Pr\{t_{n+1} - t_n \le t \mid X_n = j, X_{n+1} = k\},$$

$$j, k = 1, 2, \ldots, m. \tag{2.5}$$

The random variable T_{jk} depends on the state X_n being visited and the state X_{n+1} to be entered in the very next transition. T_{jk} is the *conditional sojourn time* in state j given that the next transition is to state k. Assume that if $p_{jk} = 0$ for some j, k then $Q_{jk}(t) = 0$ for all t and $W_{jk}(t) = 1$. For $p_{jk} > 0$

$$W_{jk}(t) = Q_{jk}(t)/p_{jk}. \tag{2.6}$$

The variable T_{jk} denotes the conditional waiting time in state j given that the next state is k; and if T_j denotes the unconditional waiting time in state j, then

$$T_j = \sum_k p_{jk} T_{jk} \tag{2.7}$$

and

$$W_j(t) = \Pr\{T_j \le t\} = \sum_k p_{jk} W_{jk}(t) = \sum_k Q_{jk}(t). \tag{2.8}$$

We also have

$$M_j^{(r)} = E(T_j^r) = \sum_k p_{jk} E(T_{jk}^r) = \sum_k p_{jk} \alpha_{jk}^{(r)},$$

$$r = 1, 2, \ldots$$

whenever the moments $\alpha_{jk}^{(r)}$ of order r of T_{jk} exists. Denote $M_j^{(1)}$ by M_j and $\alpha_{jk}^{(1)}$ by α_{jk}. We have

$$\alpha_{jk} = \int_0^\infty \Pr\{T_{jk} > t\}\, dt = \int_0^\infty \{1 - W_{jk}(t)\}\, dt \qquad (2.9)$$

and $\qquad M_j = \int_0^\infty \Pr\{T_j > t\}\, dt = \int_0^\infty \{\, 1 - \sum_k Q_{jk}(t)\,\}\, dt; \qquad (2.10)$

M_j is the expected amount of time spent in state j during each visit and α_{jk} is the (conditional) expected time spent in state j during each visit given that the next state visited is k.

Note: If $Q_{jk}(t)$ is of the form

$$Q_{jk}(t) = p_{jk}\left\{1 - e^{-a_j t}\right\}$$

(equivalently if $W_{jk}(t)$ has the form of the second factor), then it can be shown that

$$\Pr\{Y(t+s) = k \mid Y(u), u \le t\} = \Pr\{Y(t+s) = k \mid Y(t)\}$$

and that

$$\Pr\{Y(t+s) = k \mid Y(t) = j\} = p_{jk}(s)$$

is independent of t. In this case the semi-Markov process becomes a time-homogeneous Markov process with t.p.m. $P = (p_{jk})$. (In fact, this is what we get in Example 2(b) given below). The Markov property which holds, in case of semi-Markov processes *only* at the epochs t_n, will hold for all t *only when* $Q_{jk}(t)$ is of the above form. Thus, in semi-Markov processes there is freedom in the distribution of the sojourn times T_{jk}, but only at the expense of the Markov property, which instead of holding for all t holds only at the jump epochs t_n.

A Markov process is a semi-Markov process but the converse is not necessarily true.

Suppose that $X(t)$ is a s-M.P. with discrete time, and that the transition intervals $(t_{n+1} - t_n)$ are equal to unity for all n. Then the sequence $\{X_n\}$, where $X_n = X(t_n +)$ reduces to a simple Markov chain.

Example 2(b). The *M/M/1* queue*. Suppose that customers arrive at a counter in accordance with a Poisson process with parameter λ and are served by a single server, the service times of customers being i.i.d. exponential random variables with mean $1/\mu$. Then $\{N(t), t \ge 0\}$, where $N(t)$ is the number of customers in the system at time t, is a s-M.P. (it is also a Markov process). The states of the s-M.P. are $S = \{0, 1, 2, \ldots\}$ Suppose that $t_0, t_1, \ldots$ are the successive instants at which changes of states (by way of arrival or departure of a unit) take place and

$$X_n = N(t_n + 0);$$

---

*For description, see Ch. 10.

$$
\begin{aligned}
Q_{0,1}(t) &= \Pr\{X_{n+1} = 1, t_{n+1} - t_n \le t \mid X_n = 0\} \\
&= \Pr\{\text{the process enters state 1 from state 0 by time } t\} \\
&= 1 - e^{-\lambda t},
\end{aligned}
$$

$$
\begin{aligned}
Q_{j,j-i}(t) &= \Pr\{X_{n+1} = j - 1, t_{n+1} - t_n \le t \mid X_n = j\}, \quad j \ge 1 \\
&= \Pr\{\text{the process enters state } j - 1 \text{ from state } j \text{ by time } t\} \\
&= \Pr\{\text{one transition (service completion or arrival) within} \\
&\qquad \text{time } t\} \times \Pr\{\text{transition is a service completion}\} \\
&= \Pr\{\min(A, B) \le t\} \times \Pr\{A > B\},
\end{aligned}
$$

where the random variables A, B denote respectively the interarrival and service times (which are independent exponential with mean $1/\lambda$ and $1/\mu$). Since $\Pr\{A > B\} = \mu/(\lambda + \mu)$ and $\min(A, B)$ is exponential with mean $1/(\lambda + \mu)$, we have

$$
Q_{j,j-1}(t) = \{1 - e^{-(\lambda+\mu)t}\} \cdot \{\mu/(\lambda+\mu)\}, j \ge 1.
$$

Similarly,

$$
Q_{j,j+1}(t) = \Pr\{\min(A, B) \le t\} \cdot \Pr\{B > A\}
$$

$$
= \{1 - e^{-(\lambda+\mu)t}\} \cdot \{\lambda/(\lambda+\mu)\}, j \ge 1
$$

Further
$$
Q_{jk}(t) = 0, \text{ for } k \ne j - 1, j + 1 \ (j, k \ge 1).
$$

Hence the transition probabilities of the embedded Markov chain $\{X_n\}$ are

$$
p_{0,1} = 1, \quad p_{jk} = 0, \quad k \ne j - 1, \ j + 1
$$

$$
p_{j,j-1} = \mu/(\lambda+\mu), \quad p_{j,j+1} = \lambda/(\lambda+\mu), \quad j \ge 1.
$$

Thus

$$
W_{0,1}(t) = \Pr\{t_{n+1} - t_n \le t \mid X_n = 0, X_{n+1} = 1\}
$$

$$
= 1 - e^{-\lambda t},
$$

and
$$
W_{j,j-1}(t) = 1 - e^{-(\lambda+\mu)t} = W_{j,j+1}(t), j \ge 1.
$$

Further, T_0, the unconditional waiting time in state 0 is the same as the interarrival time, i.e. T_0 is exponential with mean $1/\lambda$. T_j $(j \ge 1)$, the unconditional waiting time in state j, is the time required for a transition (either an arrival or service completion). Thus T_j which is equivalent to $\min(A, B)$, is exponential with mean $1/(\lambda + \mu)$. We have

$$M_0 = E(T_0) = 1/\lambda$$

$$M_j = E(T_j) = 1/(\lambda + \mu), j \geq 1.$$

Example 2(c). The *M/G/1* queue. Here we suppose that the service time distribution is general, other things being the same as in the *M/M/1* queue. We note that $\{N(t), t \geq 0\}$ is *not* a semi-Markov process, because the state of the system after a transition depends not only on the state of $N(t)$, but also on the amount of elapsed service time of the person receiving service, if any. If we consider that transitions take place only when a customer completes his service, there will be no elapsed service time. X_n will here give the number of customers in the system at the instant that nth customer departs. If $Y(t)$ denotes the number of customers left-behind in the system by the most recent departure, then $Y(t) = X_n, t_n \leq t < t_{n+1}$ and $\{Y(t)\}$ will be a semi-Markov process having $\{X_n, n = 0, 1, \ldots\}$ for its embedded Markov chain. $\{X_n, t_n\}, n = 0, 1, 2, \ldots$ is a Markov renewal process. The sequence of intervals $(t_{n+1} - t_n), n = 0, 1, 2, \ldots$ being the inter-departure times of successive units (or equivalently $\{t_n\}, n = 0, 1, 2, \ldots$) forms a renewal process.

Let $G(\cdot)$ be the d.f. of the service time γ.
We have, for $j > 0, k \geq j - 1$

$$Q_{jk}(t) = \Pr\{\gamma \leq t \text{ and } k - j + 1 \text{ units arrive during } \gamma\}$$

$$= \int_0^t \frac{(\lambda x)^{k-j+1}}{(k-j+1)!} dG(x)$$

$$\equiv q_{k-j+1}(t) \text{ (say)},$$

$$Q_{jk}(t) = 0, \quad \text{for} \quad k < j - 1,$$

and

$$Q_{0k}(t) = \Pr\{\text{service starts with the arrival of a fresh unit to an empty system at time } x \ (0 \leq x \leq t) \text{ and } k \text{ units arrive during the service time } (t - x) \text{ of that unit}\}$$

$$= \int_0^t \lambda e^{-\lambda x} q_k(t-x) dx.$$

Further,

$$p_{ij} = \lim_{t \to \infty} Q_{ij}(t)$$

correspond to the transition probabilites of the embedded Markov chain.

Note. Neuts (*Ann. Math. Stat.* **38** (1967) 759-770) considers semi-Markov formulation for an $M/G/1$ system under general bulk service rule (for a description of the rule, see Sec. 10.5.2). See Teghem *et al.* (1969) who consider such formulations for $GI/M/1$ queues as well.

Example 2(d). *Availability of a series system*

Suppose that a system consists of k components (not necessarily identical) such that the system functions if all the components function. The lifetimes (or times to failure) of the components have independent exponential distributions, the lifetime L_i of component i having parameter λ_i; the repairtime R_i of component i has d.f. $\Pr \{R_i \le t\} = F_i (t)$ $(i = 1, 2, \ldots, k)$. When one component, say component i fails, the functioning of the system fails, and the component i is immediately sent for repair; no other component fails during the repairtime R_i of i.

Let us consider that a transition (change of state) occurs when either a component fails (the system fails) or the repairtime of the failed component ends (and the system starts functioning). Let t_n, $n = 0, 1, 2, \ldots$ be the epochs (alternating epochs of failure and end of repairtime) when such transitions occur.

Let $X_n = i$ denote that the component i fails when t_n is a failure epoch and $X_n = 0$ when t_n marks the end of repairtime of the unit under repair. Thus $X_n = 0$ implies that the system is functioning; the state space of X_n is $S = \{0, 1, 2, \ldots k\}$. Then $\{X_n, t_n\}$, $n = 0, 1, 2, \ldots$ is a Markov renewal process. We have, for $i \ne 0$,

$$Q_{i0}(t) = \Pr \{X_{n+1} = 0, t_{n+1} - t_n \le t \mid X_n = i\}$$

$$= \Pr \{R_i \le t\} = F_i(t)$$

and

$$Q_{0i}(t) = \Pr \{X_{n+1} = i, t_{n+1} - t_n \le t \mid X_n = 0\}$$

$$= \Pr \{L_i \le L_j \text{ for all } j \ne i \quad \text{and} \quad L_i \le t\}$$

$$= \int_0^t \lambda_i e^{-\lambda_i x} \left\{ \prod_{j \ne i} e^{-\lambda_j x} \right\} dx$$

$$= \frac{\lambda_i}{\lambda} (1 - e^{-\lambda t}),$$

where

$$\lambda = \sum_{i=1}^k \lambda_i.$$

Futher, $Q_{ij}(t) = 0$, when i, j are both non-zero. Taking limits, we get, for $i > 0$,

$$p_{i0} = 1, \quad \text{(as is evident)}$$

and $$p_{0i} = \frac{\lambda_i}{\lambda} \quad \text{(as expected : see Section 1.3.4.3. Property (b)).}$$

Generalisation of renewal process

We now examine how Markov renewal processes can be viewed as a generalisation of renewal process. It can be shown that, for $n = 1, 2, \ldots$

$$\Pr\{t_1 - t_0 \le u_1, t_2 - t_1 \le u_2, \ldots, t_n - t_{n-1} \le u_n \mid X_0 = i_1, X_1 = i_2, \ldots, X_n = i_{n+1}\}$$

$$= W_{i_1 i_2}(u_1)\, W_{i_2 i_3}(u_2) \ldots W_{i_{n-1} i_n}(u_n). \tag{2.11}$$

In particular, if there is only one state, then the transition intervals $(t_n - t_{n-1})$ are i.i.d. non-negative random variables and $\{t_n\}$ is a renewal process.

Proposition (2):
If E consists of a single point, then the sequence $\{t_n\}$, $n = 0, 1, 2, \ldots$ constitutes a renewal process. Suppose that $j \in E$ is fixed and $S_1^j, S_2^j, \ldots$ are the successive epochs t_n for which $X_n = j$. Then for each fixed j, the sequence $\{S_n^j\}$, $n = 0, 1, 2, \ldots$ constitutes a renewal process which is ordinary if j is the initial state (with $S_0^j = t_0 = 0$) and delayed, otherwise.

Proposition (3):
To each state $j \in E$, there corresponds (a possibly delayed) renewal process $\{S_n^j\}$, $j = 1, 2, \ldots, m$ and the superposition of all these renewal processes gives the set of transition epochs $t_0, t_1, t_2, \ldots$. The fact that $\{X_n\}$ constitutes a Markov chain and the fact that $\{S_n^j\}$, $j \in E$ constitutes renewal processes provide a justification for the terminology Markov renewal Process–as a generalisation of Markov chains and renewal processes. In view of proposition (3), an alternative definition of the Markov renewal process is also given below.

Definitions. If $N_j(t)$ denotes the number of transitions into state j (renewals of state j) in $[0, t]$, then

$$N(t) = \{N_1(t), N_2(t), \ldots, N_m(t)\}, \tag{2.12}$$

is called a *Markov renewal process*.

Suppose that the process starts with state i at time 0, i.e. $X_0 = Y(0) = i$. Define

$$G_{ij}(t) = \Pr\{N_j(t) \geq 1 \mid Y(0) = i\} \tag{2.13}$$

$$= \Pr\{S_1 \leq t \mid Y(0) = i\},$$

$$f_{ij}(t) = \Pr\{Y(t) = j \mid Y(0) = i\} \tag{2.14}$$

and

$$Q_{ij}^{(n)}(t) = \Pr\{X_n = j, t_n \leq t \mid X_0 = i\}. \tag{2.15}$$

$G_{ij}(t)$ is the d.f. of the *first passage time* V_{ij} into state j given that the initial state is i. Here irrespective of the number of transitions (into other states) involved, the visit to j must be for the *first time*. G_{jj}, which gives the d.f. of the time between two occurrences of state j, may be different from G_{ij} $(i \neq j)$. Denote $E(V_{ij}) = \mu_{ij}$, when it exists. A more detailed discussion on first passage time is given in a latter section. Now $Q_{ij}^{(0)}(t) = 1$ or 0 depending on whether $j = i$ or not. Denote $Q^{(1)}$ by Q.

Proposition (4):

For $\qquad\qquad\qquad\qquad n = 0, 1, 2, \ldots$

$$Q_{ij}^{(n+1)}(t) = \sum_k \int_0^t \{dQ_{ik}(u)\} Q_{kj}^{(n)}(t-u). \tag{2.16}$$

This can be seen by considering that of the $(n+1)$ transitions, the first one takes place at some time u $(\leq t)$ and reaches some state k, and the subsequent n transitions from state k take place within time $t - u$, and end with j.

Markov renewal function. The functions

$$M(i, j, t) = E\{N_j(t) \mid X_0 = i\}, \quad j \in E \tag{2.17}$$

are known as *Markov renewal functions* corresponding to Q.

Proposition (5): We have

$$M(i, j, t) = \sum_n Q_{ij}^{(n)}(t). \tag{2.18}$$

This can be seen by noting that

$$N_j(t) = \sum_n A_n, \quad \text{where}$$

$$A_n = 1, \quad \text{if } X_n = j, \quad t_n \leq t$$

$$= 0, \quad \text{otherwise}$$

and
$$\Pr\{A_n = 1 \mid X_0 = i\} = \Pr\{X_n = j, t_n \leq t \mid X_0 = i\}$$

$$= Q_{ij}^{(n)}(t). \tag{2.19}$$

For $j = i$, the interarrival times between two occurrences of i (having d.f. $G_{ii}(t)$) induce an ordinary renewal process. Denoting the n-fold convolution of G_{ii} with itself by G_{ii}^{n*}, we can write the Markov renewal function (i.e. conditional expectation of the number of entrances to state i) as

$$M(i,i,t) = \sum_n G_{ii}^{n*}. \tag{2.20}$$

Proposition (6):
Blackwell's and Smith's renewal theorems hold for the (ordinary) Markov renewal function $M(i, i, t)$.

Classification of states, Suppose that the initial state in a Markov renewal process $\{X_n, t_n\}$ is i. The state i is said to be recurrent (transient) *iff* $G_{ii}(\infty) = 1 (< 1)$. A recurrent state is positive or null-recurrent depending on whether $\mu_{ii} <$, or $= \infty$. We have the following result.

Proposition (7):
For the Markov renewal process $\{X_n, t_n\}$, state i is recurrent (transient) *iff* state i in the embedded Markov chain $\{X_n\}$ is recurrent (transient).

7.3 MARKOV RENEWAL EQUATION

We obtain a Markov renewal equation which gives a relationship between f_{ij} and Q_{ij}.

Theorem 7.1 For all i, j and for $t \geq 0$,

$$f_{ij}(t) = \delta_{ij} h_i(t) + \sum_k \int_0^t f_{kj}(t-x) \, dQ_{ik}(x), \tag{3.1}$$

where
$$h_i(t) = 1 - \sum_k Q_{ik}(t) = 1 - W_i(t) = \Pr\{T_i > t\}.$$

Proof: We apply renewal argument; by conditioning on the time t_1 of the first transition as well as the state then entered, we get

$$f_{ij}(t) = \Pr \{Y(t) = j \mid X(0) = i\}$$

$$= \sum_k \int_0^\infty \Pr \{Y(t) = j \mid X_0 = i, X_1 = k, t_1 = x\} \, dQ_{ik}(x).$$

If $t_1 = x > t$, i.e. the first transition epoch is beyond t, then $Y(t) = Y(0) = i$. If $t_1 = x \le t$, then starting from state k (reached at first transition by time x) the process ends up with state j by the remaining time $(t - x)$. Thus

$$\Pr \{Y(t) = j \mid X_0 = i, X_1 = k, t_1 = x\} = \delta_{ij}, \quad \text{if } x > t$$

$$= f_{kj}(t - x), \quad \text{if } x \le t.$$

Hence

$$f_{ij}(t) = \sum_k \left[\int_0^t f_{kj}(t - x) \, dQ_{ij}(x) + \delta_{ij} \int_t^\infty dQ_{ik}(x) \right].$$

Since

$$\sum_k \int_0^\infty dQ_{ik}(x) = 1,$$

the second term equals

$$\delta_{ij} \left\{ 1 - \sum_k \int_0^t dQ_{ik}(x) \right\}$$

$$= \delta_{ij} \left\{ 1 - \sum_k Q_{ik}(t) \right\} = \delta_{ij} h_i(t) \qquad (3.2)$$

and thus we get (3.1).   ▲

The equation (3.1) in $f_{ij}(t)$ can be solved as a Markov renewal equation. We shall not consider this aspect here. We are interested mainly in the asymptotic behaviour of the stochastic process $\{Y(t)\}$, which will be found without explicitly finding the solution $f_{ij}(t)$ first. We shall introduce a matrix similar to the t.p.m. of a Markov chain.

7.3.1 Interval Transition Probability Matrix (i.t.p.m.)

The matrix with functions of t, $k_{ij}(t)$, $i, j = 1, 2, \ldots, m$ as elements will be denoted by $\mathbf{K}(t)$, i.e. $\mathbf{K}(t) = (k_{ij}(t))$. The matrices obtained by taking derivatives, integrals, Laplace transforms etc. (provided they exist) of the elements will be represented accordingly. For example,

$$\int_0^t \mathbf{K}(x) dx = \left(\int_0^t k_{ij}(x) \, dx \right)$$

$$\mathbf{K}*(s) = (k_{ij}^{*}(s)),$$

where $k*(.)$ denotes Laplace transform.

We shall refer to these matrices here by

$$\mathbf{F}(t) = (f_{ij}(t)),$$

$$\mathbf{C}(t) = (c_{ij}(t)), \text{ where } c_{ij}(t)\, dt = p_{ij} dW_{ij}(t) = dQ_{ij}(t)$$

$$\mathbf{A}(t) = (a_{ij}(t)), \text{ where } a_{ij}(t) = \delta_{ij}\, h_i(t). \tag{3.3}$$

The matrix $\mathbf{A}$ is a diagonal matrix. Since $f_{ij}(t) \geq 0$ and $\sum_j f_{ij}(t) = 1$ for all $i = 1, 2, \ldots, m$,

the matrix $\mathbf{F}(t)$ is a stochastic matrix; it is called *interval transition probability matrix* (i.t.p.m.).

Now (3.1) which holds for all states $i, j = 1, 2, \ldots, m$ can be put as

$$f_{ij}(t) = a_{ij}(t) + \int_0^t \sum_k c_{ik}(x) f_{kj}(t-x)\, dx, \quad i, j = 1, 2, \ldots, m \tag{3.4}$$

and in matrix form as

$$\mathbf{F}(t) = \mathbf{A}(t) + \int_0^t \mathbf{C}(x)\mathbf{F}(t-x)\, dx. \tag{3.5}$$

Denoting the convolution of $\mathbf{C}$ and $\mathbf{F}$ by $\mathbf{C}*\mathbf{F}$, we can put the same as

$$\mathbf{F}(\cdot) = \mathbf{A}(\cdot) + \mathbf{C}(\cdot)*\mathbf{F}(\cdot) \cdot \tag{3.6}$$

Taking Laplace transforms, we get

$$\mathbf{F}*(s) = \mathbf{A}*(s) + \mathbf{C}*(s)\mathbf{F}*(s). \tag{3.7}$$

Note that $\mathbf{C}$ is a somewhat special type of matrix.

7.4 LIMITING BEHAVIOUR

Asymptotic behaviour of a stochastic process is a question of great general interest. We shall now investigate such behaviour in case of a semi-Markov process. Suppose that $\{X_n, t_n\}$ is a Markov renewal process and $Y(t) = X_n$, $t_n \leq t < t_{n+1}$, is the semi-Markov process induced by $Q_{ij}(t)$. The limiting distribution of the embedded Markov chain $\{X_n\}$ is not the same as that of the semi-Markov process. If the Markov chain is irreducible and ergodic and has the t.p.m. $\mathbf{P} = (p_{ij})$, then the limiting probabilities $\lim_{n \to \infty} p_{ij}^{(n)} = v_j, j \in E$ are given as unique solutions of $v_j = \sum_k v_k p_{kj}, j \in E$, i.e.

the probability vector $\mathbf{V} = (v_1, v_2, \ldots)$ is the unique solution of $\mathbf{V} = \mathbf{VP}$.

In the ergodic theorem given below, the limiting distribution

$$\lim_{t \to \infty} f_{ij}(t) = \lim_{t \to \infty} \Pr \{Y(t) = j \mid Y(0) = i\}$$

is obtained. It also gives the relationship between the two limiting distributions.

Theorem 7.2 Suppose that $\{X_n\}$ is irreducible and ergodic and that the mean waiting time in state i, $M_i < \infty$ for any $i \in E$, then

$$\lim_{t \to \infty} f_{ij}(t) = f_j = \frac{v_j M_j}{\sum_k v_k M_k}.$$ (4.1)

Proof: Suppose that the variables T_{ij} are continuous having p.d.f. $w_{ij}(t)$. Then T_i is also continuous and has the p.d.f. $w_i(t) = \sum_j p_{ij} w_{ij}(t)$. Referring to (3.3), we get

$$c_{ij}^*(s) = p_{ij} w_{ij}^*(s), \quad a_{ij}^*(s) = \delta_{ij} h_i^*(s) = \delta_{ij} \left\{ \frac{1}{s} - \frac{w_i^*(s)}{s} \right\}.$$ (4.2)

Taking limits as $s \to 0$,

$$c_{ij}^*(s) \to p_{ij}, \quad a_{ij}^*(s) \to \delta_{ij} \frac{d}{ds} (-w_i^*(s))|_{s=0} = \delta_{ij} M_i$$ (4.3)

so that, as $s \to 0$

$$\mathbf{C}^*(s) = (c_{ij}^*(s)) \to (p_{ij}) = \mathbf{P}$$ (4.4)

and

$$\mathbf{A}^*(s) = (a_{ij}^*(s)) \to (\delta_{ij} M_i) = \mathbf{M};$$

$\mathbf{M}$ is a diagonal matrix with diagonal elements $M_1, M_2, \ldots$.

From (3.7) we get

$$[\mathbf{I} - \mathbf{C}^*(s)] \mathbf{F}^*(s) = \mathbf{A}^*(s),$$

$\mathbf{I}$ being the unit matrix.

Thus

$$s \mathbf{F}^*(s) = \mathbf{B}(s) \mathbf{A}^*(s),$$ (4.5)

where

$$\mathbf{B}(s) = s[\mathbf{I} - \mathbf{C}^*(s)]^{-1}.$$ (4.6)

Hence

$$\mathbf{B}(s)[\mathbf{I} - \mathbf{C}^*(s)] = s\mathbf{I}$$

or

$$\mathbf{B}(s) - \mathbf{B}(s)\mathbf{C}^*(s) = s\mathbf{I}.$$

Denote
$$\lim_{s \to 0} \mathbf{B}(s) = \mathbf{B} = (b_{ij});$$

then taking limits as $s \to 0$, we get

$$\mathbf{B} - \mathbf{BP} = \mathbf{0}$$

or
$$\mathbf{B} = \mathbf{BP}. \tag{4.7}$$

Thus we find that $\mathbf{B}$ satisfies the same equation as $\mathbf{V}$, i.e.

$$b_{ij} = \sum_k b_{ik} p_{kj}, \quad j = 1, 2, \dots \tag{4.8}$$

$$v_j = \sum_k v_k p_{kj}, \quad j = 1, 2, \dots. \tag{4.9}$$

The solutions of (4.9) are unique. As both the sets (4.8) and (4.9) hold, the rows of $\mathbf{B}$ and $\mathbf{V}$ must be proportional, i.e.

$$b_{ij} = d_i v_j, \text{ where } d_i \text{ are constants.}$$

Now
$$f_{ij} = \lim_{t \to \infty} f_{ij}(t) = \lim_{s \to 0} s f_{ij}^*(s);$$

writing

$$\mathbf{F} = \langle f_{ij} \rangle = \lim_{t \to \infty} (f_{ij}(t))$$

$$= \lim_{s \to 0} (s f_{ij}^*(s)) = \lim_{s \to 0} s \mathbf{F}^*(s),$$

we have from (4.5)

$$\mathbf{F} = \lim_{s \to 0} s \mathbf{F}^*(s) = \lim_{s \to 0} \mathbf{B}(s) \lim_{s \to 0} \mathbf{A}^*(s)$$

$$= \mathbf{BM}. \tag{4.10}$$

Thus
$$f_{ij} = b_{ij} M_j = d_i v_j M_j \cdot$$

Since
$$\sum_j f_{ij} = 1, d_i = 1 / \sum_j v_j M_j \text{ for all } i,$$

finally we get

$$\lim_{t \to \infty} f_{ij}(t) = f_{ij} = \frac{v_j M_j}{\sum_k v_k M_k} \cdot$$

The limits $f_{ij}(t)$ are independent of the initial state i; we may write $f_{ij} = f_j$.

To complete the proof of the theorem, we note that in case the variables are discrete, by considering generating functions instead of Laplace transforms, and proceeding on the same lines, we can obtain the result.   ▲

Particular case. When the waiting times are all exponential and $W_{ij}(t) = 1 - e^{-a_i t}$, then $\{Y(t)\}$ becomes a Markov process. We then get $M_j = 1/a_j$. Substituting this value of M_j in the expression for f_j, one can ultimately arrive at the expression [Eq. (5.19), Ch. 4]

$$\sum_j f_j a_{jk} = 0, \quad (\sum_j f_j = 1)$$

satisfied by the limiting distribution $\{f_j\}$ of an irreducible Markov process.

When the semi-Markov process reduces to an ordinary Markov chain, then $M_j = 1$ for all j and one gets $f_k = v_k$, as is to be expected. The relation $f_k = v_k$ holds also in the case when the unconditional waiting times in all the states (i.e. interoccurrence times between recurrence points) are independently distributed with the same mean $M_j = M$ for all j. It may be noted that the expression (4.1) giving f_{ij} involves only the v_j's, the limiting transition probabilities of the embedded Markov chain and the mean waiting times $M_j = \sum_k p_{jk} \alpha_{jk}$, i.e. the mean α_{jk} and no other parameter of T_{jk}.

The distinction between the quantities v_j and f_j should be carefully observed. Suppose that a process has been in operation for a sufficiently long time and that a sufficiently large number of transitions have taken place so that a steady state has been reached. Suppose that the situation at some instant is considered. The quantity f_j gives the probability that the process is in (or occupies) the state j, while v_j gives the probability that, given that the process is now making a transition, the same is to state j. Except in the cases stated above the two quantities v_j, f_j are unequal. Even in the case when the process is a Markov process in continuous time they are unequal (see Example 4(b)). We consider a numerical example below.

Example 4(a). Suppose that the embedded Markov chain $\{X_n\}$ has the t.p.m.

$$P = \begin{pmatrix} 0.5 & 0.3 & 0.2 \\ 0.2 & 0.4 & 0.4 \\ 0.1 & 0.5 & 0.4 \end{pmatrix}$$

and that the matrix of complementary distribution functions of T_{ij} is

$$W = \begin{pmatrix} e^{-2t} & e^{-5t} & e^{-t} \\ e^{-4t} & e^{-2t} & e^{-5t} \\ e^{-5t} & e^{-2t} & e^{-t} \end{pmatrix}.$$

Using $M = \sum_j p_{ij} \alpha_{ij}$, we get

$$M_1 = 0.51, M_2 = 0.33, M_3 = 0.67,$$

the limiting probabilities $v_j = \lim\limits_{n \to \infty} p_{ij}^{(n)}$ have values

$$v_1 = 0.2353, \ v_2 = 0.4118, \ v_3 = 0.3529$$

(see Example 5(b) Ch. 3).

Using (4.1), we get

$$f_1 = 0.2437, \ f_2 = 0.2760, \ f_3 = 0.4803.$$

Example 4(b). Consider the *M/M/*1 queue discussed in Example 2(b). Assume that $\lambda < \mu$, and denote

$$a = \mu/(\lambda + \mu), \quad b = \lambda/(\lambda + \mu);$$

then

$$p_{0,1} = 1, \quad p_{j,j-1} = a, \quad p_{j,j+1} = b, j \geq 1, \quad p_{jk} = 0, \quad k \neq j-1, j+1.$$

The elements of $\mathbf{V} = (v_0, v_1, v_2, \ldots)$, $\Sigma \, v_i = 1$ are the solutions of $\mathbf{V} = \mathbf{V}P$. We have

$$v_0 = a v_1$$

$$v_1 = v_0 + a v_2$$

and

$$v_k = b v_{k-1} + a v_{k+1}, \quad k = 2, 3, \ldots.$$

The solution of the last equation is given by

$$v_k = c_1 + c_2 (b/a)^k, \quad k = 1, 2, \ldots.$$

Solving in terms of v_0, we get

$$c_1 = 0 \text{ and } c_2 = v_0/b,$$

$$v_k = (1/b) \, (b/a)^k \, v_0$$

$$= \left(\frac{\lambda + \mu}{\lambda}\right)\left(\frac{\lambda}{\mu}\right)^k v_0, \quad k = 1, 2, \ldots$$

Now

$$M_0 = 1/\lambda, \ M_k = 1/(\lambda + \mu), \ k = 1, 2, \ldots$$

so that

$$v_k M_k = \frac{v_0}{\lambda(1 - \lambda/\mu)}.$$

Thus we get

$$f_{ij} = f_j = \frac{v_j M_j}{\sum_k v_k M_k}$$

$$= \left(1 - \frac{\lambda}{\mu}\right)\left(\frac{\lambda}{\mu}\right)^j, \quad j = 0, 1, 2, ...$$

This result which can be obtained, as solution of difference equations, is considered here as an illustration of application of the theory of s-M.P.

Note that $\qquad\qquad v_k > f_k, k \geq 1, v_0 < f_0.$

7.4.1 Limiting Distribution of s-M.P. and 'Recurrence Times'

There are many other distributions connected with Markov renewal processes, whose asymptotic behaviour is of considerable significance and many limit theorems have been obtained. We shall discuss here the limiting behaviour in another important case.

Suppose that $\{X_n, t_n\}$ is a Markov renewal process and that $\{Y(t)\}$ is its associated s-M.P. For each t, there exists n, such that $t_n \leq t < t_{n+1}$; then $U(t) = t - t_n$, $V(t) = t_{n+1} - t$ are respectively called the time since the last transition and the time until the next transition ('*recurrence times*'). By applying renewal type of arguments and using the renewal theorems, the limiting form of the joint distribution of $Y(t), U(t), V(t)$ can be obtained as stated below. For details of proof, refer to Pyke (1961), Cinlar (1969).

Theorem 7.3 If the process is irreducible and the state space is non-lattice, then, as $t \to \infty$

$$\Pr\{Y(t) = j, X_{n+1} = k, U(t) \geq x, V(t) > y \mid Y(0) = i\}$$

$$\to \frac{v_j}{\sum_r v_r M_r} \int_{x+y}^{\infty} p_{jk}\{1 - W_{jk}(u)\}\, du, \tag{4.11}$$

provided the integral exists.

Summing over k, we have, as $t \to \infty$

$$\Pr\{Y(t) = j, U(t) \geq x, V(t) > y \mid Y(0) = i\}$$

$$\to \frac{v_j}{\sum_r v_r M_r} \int_{x+y}^{\infty} \{1 - \sum_k Q_{jk}(u)\}\, du = \frac{f_j}{M_j} \int_{x+y}^{\infty} \Pr\{T_j > u\}\, du. \tag{4.12a}$$

Thus
$$\Pr\{U(t) \ge x, V(t) > y \mid Y(t) = j, Y(0) = i\}$$

$$\rightarrow \int_{x+y}^{\infty} \frac{\Pr\{T_j > u\}}{M_j} \, du. \tag{4.12b}$$

Putting $x = 0$ in (4.11), we have, as $t \rightarrow \infty$,

$$\Pr\{Y(t) = j, X_{n+1} = k, U(t) \ge 0, V(t) > y \mid Y(0) = i\}$$

$$\rightarrow \frac{v_j}{\sum_r v_r M_r} \int_y^{\infty} P_{jk} \{1 - W_{jk}(u)\} \, du. \tag{4.13}$$

It may be noted that when the Markov renewal process reduces to an ordinary renewal process (i.e. there is only one state), the above (4.13) yields the limiting distribution of the residual lifetime, whereas (4.12b) gives the limiting joint distribution of residual and spent lifetimes ($V(t)$ and $U(t)$ respectively).

7.5 FIRST PASSAGE TIME

We consider here some of their properties. First, we obtain a relationship between $G_{ij}(t)$ and $f_{ij}(t)$.

Theorem 7.4. For all i, j and for $t \ge 0$,

$$f_{ij}(t) = \delta_{ij} \, h_i(t) + \int_0^t f_{jj}(t - x) \, dG_{ij}(x), \tag{5.1}$$

where $h_i(t) = \Pr\{T_i > t\}$.

Proof: The event $[Y(t) = j]$, given $Y(0) = i$ can happen in the following two mutually exclusive ways (these give the first and the second term respectively on the r.h.s. of (5.1)):

(i) No transition occurs by time t; then $Y(t) = i$, and the time spent in state i exceeds t.

(ii) One or more transitions occur by time t. Suppose that the process reaches the state j for the first time by time $x < t$ and then starting from this initial state j, the process in the remaining time $t - x$, ends up again with state j. ▲

We now obtain a relation between $G_{ij}(t)$ and $Q_{ij}(t)$.

Theorem 7.5. For all states i, j and $t \geq 0$

$$G_{ij}(t) = Q_{ij}(t) + \sum_{k \neq j} \int_0^t G_{kj}(t-x)\,dQ_{ik}(x) \tag{5.2a}$$

$$= Q_{ij}(t) + \sum_k \int_0^t G_{kj}(t-x)\,dQ_{ik}(x)$$

$$- \int_0^t G_{jj}(t-x)\,dQ_{ij}(x). \tag{5.2b}$$

Proof: The first term on the r.h.s. of (5.2a) corresponds to the case of direct transition from state i to state j in the first transition within t; the second term corresponds to the case of transition from state i to k $(\neq j)$ in the first transition at time $x < t$ and then starting from the initial state k, the process, in the remaining time $(t - x)$, ending up with state j.   ▲

Laplace transform formulation. Suppose that the variables V_{ij}, T_{ij} are continuous and

$$dG_{ij}(x) = g_{ij}(x)dx \tag{5.3}$$

$$dQ_{ij}(x) = p_{ij}\,dW_{ij}(x) = p_{ij}\,w_{ij}(x)\,dx.$$

Then taking Laplace transforms of (5.1), we get

$$f_{ij}^*(s) = \delta_{ij}\,h_i^*(s) + g_{ij}^*(s)f_{jj}^*(s),$$

so that

$$f_{ij}^*(s) = \frac{h_i^*(s)}{1 - g_{ii}^*(s)}, \quad j = i \tag{5.4a}$$

$$f_{ij}^*(s) = \frac{g_{ij}^*(s)h_j^*(s)}{1 - g_{jj}^*(s)}, \quad j \neq i. \tag{5.4b}$$

Taking Laplace transform of (5.2a), we get

$$G_{ij}^*(s) = Q_{ij}^*(s) + \sum_{k \neq j} G_{kj}^*(s)\,q_{ik}^*(s). \tag{5.5}$$

Equivalently,

$$g_{ij}^*(s) = q_{ij}^*(s) + \sum_{k \neq j} g_{kj}^*(s)\,q_{ik}^*(s). \tag{5.6a}$$

Denoting $(g_{ij}^*(s)) = G^*(s)$ and the diagonal matrix $(\delta_{ij}\,g_{ii}*(s)) = G_j^*(s)$, (5.6a) can be written in matrix form as

$$G^*(s) = C^*(s) + C^*(s) G^*(s) - C^*(s) G_j^*(s). \tag{5.6b}$$

Mean first passage time. If $G_{ij}(\infty) = 1$, then

$$\mu_{ij} = E(V_{ij}) = \int_0^\infty t dG_{ij}(t) \tag{5.7}$$

is the mean first passage time and μ_{jj} is the mean recurrence time of state j.

Theorem 7.6. For all i, j

$$\mu_{ij} = M_i + \sum_{k \neq j} p_{ik} \mu_{kj}. \tag{5.8}$$

Proof: Suppose that the variables V_{ij}, T_{ij} are continuous and that the relations (5.3) hold; then $g_{ij}^*(s) = w_{ij}^*(0) = 1$

$$\mu_{ij} = -\frac{d}{ds} g_{ij}^*(s)\Big|_{s=0}$$

$$\alpha_{ij} = -\frac{d}{ds} w_{ij}^*(s)\Big|_{s=0}$$

and
$$M_i = \sum_k p_{ik} \alpha_{ik}.$$

From (5.6a) we get

$$\frac{d}{ds} g_{ij}^*(s) = p_{ij} \frac{d}{ds} w_{ij}^*(s) + \sum_{k \neq j} \left\{ p_{ik} w_{ik}^*(s) \frac{d}{ds} g_{kj}^*(s) + p_{ik} g_{kj}^*(s) \frac{d}{ds} w_{ik}^*(s) \right\}.$$

Hence
$$\mu_{ij} = p_{ij} \alpha_{ij} + \sum_{k \neq j} \{ p_{ik} \mu_{kj} + p_{ik} \alpha_{ik} \}$$

$$= \sum_k p_{ik} \alpha_{ik} + \sum_{k \neq j} p_{ik} \mu_{kj}$$

$$= M_i + \sum_{k \neq j} p_{ik} \mu_{kj}.$$

The proof, when the variables are discrete can be given in the same manner by considering the generating functions of V_{ij}, T_{ij} in place of their Laplace transforms.

▲

Note: (1) The intuitive meaning of (5.8) is clear. The mean first passage time from state i to j is the meantime required to leave the state i *plus* the sum of the product of meantimes required to go to state j from state k (for all k except $k = j$) reached at first transition and the probability that the first transition is from state i to k.

(2) Higher moments of V_{ij} can be obtained from (5.6a).

We obtain another interesting result.

Theorem 7.7. If the embedded chain $\{X_n\}$ is irreducible and ergodic, then for all i, j

$$\mu_{jj} = \sum_i v_i M_i / v_j. \qquad (5.9)$$

Proof: Since $\{X_n\}$ is irreducible ergodic, v_i are given as the unique solutions of

$$v_k = \sum_i v_i p_{ik}$$

Now multiplying both sides of (5.8) by v_i, $i = 1, 2, \ldots$ and adding, we get

$$\sum_i v_i \mu_{ij} = \sum_i v_i M_i + \sum_i \left\{ \sum_k v_i p_{ik} \mu_{kj} - v_i p_{ij} \mu_{jj} \right\}$$

$$= \sum_i v_i M_i + \sum_k \left\{ \sum_i v_i p_{ik} \right\} \mu_{kj} - \left\{ \sum_i v_i p_{ij} \right\} \mu_{jj}$$

$$= \sum_i v_i M_i + \sum_k v_k \mu_{kj} - v_j \mu_{jj} .$$

Thus
$$\mu_{jj} = \sum_i v_i M_i / v_j. \qquad \blacktriangle$$

Note: We can write (5.9) as follows:

$$\mu_{jj} = \sum_i M_i (v_i / v_j).$$

The intuitive meaning of this expression would be clear once it is noted that (v_i / v_j) is the expected number of visits to state i between two visits of state j and each visit to state i lasts a time whose mean is M_i.

Corollary 1. When the s-M.P. reduces to a simple Markov chain, then $M_j = 1$ for all j and then we get the well known result

$$\mu_{jj} = 1 / v_j.$$

Corollary 2. From (4.1) and (5.9), we get

$$f_j = M_j / \mu_{jj}, \qquad (5.10)$$

which gives f_j without directly involving v_i's.

It can be easily seen that the results derived for a finite set E will also hold good, in general, even for a set E containing a countably infinite number of points.

Example 5(a). *Alternating renewal process.* Suppose that an equipment has lifetime L with d.f. $G(t) = \Pr\{L \le t\}$. As soon it fails, it is sent for repair. The repairtime R has d.f. $F(t) = \Pr\{R \le t\}$; after repair it has again the same lifetime L. Let t_n, $n = 0, 1, 2, \ldots$ be the epochs (alternating epochs of up-state and down-state of the equipment) of transitions. $X_n = 0$ denotes that the equipment is at down-state and $X_n = 1$ that it is at up-state. Then $\{X_n, t_n\}$, $n = 0, 1, 2, \ldots$ is a Markov renewal process, having for its kernel

$$Q_{01}(t) = F(t), \quad Q_{10} = G(t), \quad Q_{00}(t) = Q_{11}(t) = 0.$$

Again $\qquad p_{01} = p_{10} = 1, \quad p_{00} = p_{11} = 0.$

It can easily be verified that

$$\mu_{00}(\mu_{11}) = \text{meantime between two consecutive down-states (up-states)}$$
$$= E(L) + E(R) \quad \text{[as is evident]}$$

Further

$$\lim_{t \to \infty} f_{i0}(t) \equiv f_0 = \frac{E(R)}{E(R) + E(L)}, \quad i = 0, 1$$

$$\lim_{t \to \infty} f_{0i}(t) \equiv f_1 = \frac{E(L)}{E(R) + E(L)}, \quad i = 0, 1.$$

[See Sec. 6.9, Ch. 4].

Example 5(b). *Availability of a series system* [see Example 2(d)].
We have

$\qquad M_i \equiv$ Meantime of the system at down-state with component i under repair

$\qquad\qquad$ = Mean repairtime of component i

$\qquad M_0 = E(R_i) = \mu_i$ (say).

$\qquad\qquad$ = Meantime that the system is at working state

$$= \frac{1}{\sum\limits_{i=1}^{k} \lambda_i} = \frac{1}{\lambda}$$

since min $(L_1, \ldots L_k)$ is exponential with parameter $\sum\limits_{i=1}^{k} \lambda_i = \lambda$.

Again
$$p_{i0} = 1, \quad p_{0i} = \frac{\lambda_i}{\lambda} \quad (i \neq 0)$$

$$p_{00} = 0, \quad p_{ij} = 0 \quad (i, j, \text{ both non-zero})$$

The chain $\{X_n\}$ is irreducible and ergodic; $V = (v_0, v_1, \ldots, v_k)$ is given as the unique solution of $VP = V$. We get

$$v_0 = \frac{1}{2}, \quad v_i = \frac{\lambda_i}{2\lambda}, \quad i = 1, 2, \ldots, k.$$

Thus, using (5.9) we get

μ_{jj} = Meantime between two consecutive failures of component j

$$= \frac{\sum\limits_{i=0}^{k} v_i M_i}{v_j} = \frac{1}{\lambda_j} + \frac{\sum\limits_{i=1}^{k} \lambda_i \mu_i}{\lambda_j}$$

and

μ_{00} = Meantime between two consecutive resumptions of operation after repair

$$= \frac{\sum\limits_{i=0}^{k} v_i M_i}{v_0} = \frac{1 + \sum\limits_{i=1}^{k} \lambda_i \mu_i}{\lambda}$$

Using (5.10) we get

$$f_0 = \left(1 + \sum\limits_{i=1}^{k} \lambda_i \mu_i\right)^{-1}$$

and

$$f_j = \lambda_j \mu_j \left(1 + \sum\limits_{i=1}^{k} \lambda_i \mu_i\right)^{-1}, \quad j \neq 0.$$

as the limiting probabilities

$$\lim \Pr\{Y(t) = j \mid Y(0) = i\}.$$

Concluding Remarks

We have covered, within the scope of this book, the most important aspects of Markov renewal and semi-Markov processes. The niceties of theoretical structures and the vast potentialities of applications have attracted a large number of researchers to these topics. The literature has grown tremendously over the last three decades, from only about 5 titles in 1960 to a fairly large number now. It is not possible to list them all here and reference may be made to Cinlar (1975) for a list of 71 titles and to the mimeographed compilation by Cheong *et al*. (1974) for an exhaustive list. See references for basic theory and for applications.

EXERCISES

7.1 The embedded Markov chain of a two-state $\{1, 2\}$ s-M.P. has t.p.m.

$$P = \begin{pmatrix} 2/3 & 1/3 \\ 1/2 & 1/2 \end{pmatrix}$$

and the matrix of the complementary distribution functions of the waiting times is

$$W(t) = \begin{pmatrix} e^{-t} & e^{-3t} \\ e^{-2t} & e^{-4t} \end{pmatrix}.$$

Find   (i)   the L.T.'s of interval transition probabilities $f_{jk}(t)$, $j, k = 1, 2$
and   (ii)   the limiting probabilities f_k, $k = 1, 2$.

7.2 *Continuation*. Find

     (i)   the first passage time densities $g_{jk}(t)$,
     (ii)   the mean recurrence times of the states 1, 2,
and   (iii) the mean of the first passage times from state j to state k, $j \neq k$.

7.3 Suppose that the employment status of certain section of the population in an area can be described by a Markov chain with two states employed (0) and unemployed (1) and that the chain has t.p.m.

$$P = \begin{pmatrix} 0.4 & 0.6 \\ 0.7 & 0.3 \end{pmatrix}.$$

Suppose that the waiting times densities are given by $w_{00}(t) = e^{-t}$, $w_{01}(t) = 2e^{-2t}$, $w_{10}(t) = te^{-t}$ and $w_{11}(t) = \frac{1}{2} t^2 e^{-t}$.

  (i)   If a person has just been employed, what is the probability that he will be unemployed after time t ?
  (ii)   If a person has just been employed, what is the p.d.f. of the first passage time to the state of unemployed? What is the mean of this first passage time?
  (iii) What is the limiting probability that he will be: (a) employed and (b) unemployed after a sufficiently long time?

7.4 *Entrance probabilities*. Let $e_{ij}(t)\, dt$ denote the probability that the process will enter state j in the interval $(t, t + dt)$ given that it entered state i at $t = 0$. Show that

$$f_{ij}(t) = \int_0^t e_{ij}(x) \, h_j (t - x) \, dx,$$

where

$$h_j(u) = \Pr \{T_j > u\}.$$

7.5   *Continuation.* If the process is irreducible and ergodic, show that $\lim_{t \to \infty} e_{ij}(t) = e_j$ exists and is independent of the initial state i; the limit e_j (called the *limiting entrance rate to state j*) is given by

$$e_j = f_j / M_j$$

and further that

$$e_j = 1/\mu_{jj}.$$

7.6   Show that the second moments

$$E(V_{ij}^2) = \mu_{ij}^{(2)} = \int_0^\infty t^2 \, dG_{ij}(t)$$

of the first passage times V_{ij} (with d.f. G_{ij}) satisfy the relation

$$\mu_{ij}^{(2)} = M_i^{(2)} + \sum_{k \neq j} p_{ik} \{2\alpha_{ij}\mu_{kj} + \mu_{kj}^{(2)}\}.$$

Use this relation to find var (V_{12}) and var (V_{21}) for Exercise 7.1.

7.7   Show that for an irreducible and ergodic process, if $_j N_i$ denotes the number of visits to state i in between two visits to state j, then

$$E \{_j N_i\} = v_i / v_j$$

also

$$E \{_j N_i\} = \mu_{jj} / \mu_{ii}.$$

7.8   Show that for an irreducible ergodic process

$$\lim_{t \to \infty} \frac{M(k,j,t)}{M(h,i,t)} \to \frac{v_j}{v_i}.$$

7.9   For a queue having exponential interarrival and service times having mean $1/a$ and $1/b$ respectively and with batch service (taking in a batch not more than c units i.e. batch of size c or the number in the queue at the time of initiation of service, whichever is less) find the expressions for $Q_{ij}(t)$ and p_{ij}.

7.10   For an $M/G/1$ queue (with exponential interarrival time having mean $1/a$ and general service time with distribution $B(x)$ and mean $1/b$) taking t_n as the epoch of the n th departure and $Y(t)$ as the number left in the system by the most recent departure, find expressions for $Q_{ij}(t)$.

7.11 Repeat the above exercise with the modification that service is in batches of size not more than c.

7.12 Examine the validity of the relation $f_j = v_j$ in case of $GI/M/1$ and $M/G/1$ queues.

REFERENCES

D.J. Bartholomew. *Stochastic Models for Social Processes*, Wiley, New York (1967) (2nd ed. (1973)).

—— and A.F. Forbes. *Statistical Techniques for Manpower Planning*, Wiley, New York (1979).

U. Narayan Bhat, *Elements of Applied Stochastic Processes*, 2nd ed. Wiley, New York (1984).

C.K. Cheong, J.H.A. de Smit and J.L. Teugels, *Bibliography on Semi-Markov Processes* (mimeographed), 4th ed., Inst. of Maths., Catholic Univ. of Louvain, Heverlee, (1974).

E. Cinlar. Markov Renewal Theory, *Adv. Appl. Prob.*, **1**, 123–187 (1969).

—— Markov Renewal Theory: A survey, *Management Science*, **21**, 727–752 (1975).

A.J. Fabens. The solution of queueing and inventory models by semi-Markov Processes, *J.R.S.S.B.* **23**, 113–127 (1961) [Correction, **24**, 455–456 (1962)].

R.A. Howard. *Dynamic Probabilistic System*, Vol. II: *Semi-Markov and Decision Processes*, Wiley, New York (1971).

E.P.C. Kao. Modeling the movement of coronary patients within a hospital by semi-Markov processes", *Opns. Res.*, **22**, 683–699 (1974).

N. Keyfitz. *Applied Mathematical Demography*, Wiley, New York (1977).

J. Kohlas. *Stochastic Methods of Operations Research* (English Edition) Cambridge University Press, Cambridge (1982).

P. Lévy. Processus semi-Markoviens, *Proc. Inter. Cong. Math.* (Amsterdam), **3**, 416–426 (1954).

M.F. Neuts. The single server queue with Poisson input and semi-Markov service times, *J. Appl. Prob.*, **3**, 202–230 (1966).

R.G. Potter. Births averted by contraception: An approach through renewal theory, *Theo. Pop. Biol.*, **1**, 251–272 (1970).

R. Pyke. Markov renewal processes: definitions and preliminary properties, *Ann. Math. Stat.*, **32**, 1231–1242 (1961a).

—— Markov renewal processes with finitely many states, *Ann. Math. Stat.*, **32**, 1243–1265 (1961b).

S.M. Ross. *Applied Probability Models with Optimization Applications*, Holden-Day, San Francisco (1970).

I. Sahin. *Regenerative Inventory Systems*, Springer-Verlag, New York (1990).

R. Schassberger. Insensivity of steady state distributions of generalized semi-Markov processes, *Ann. Prob.* Part I, **5**, 87–99 (1977); Part II: **6**, 85–93 (1978).

—— Insensivity of steady-state distributions of generalized semi-Markov processes with speeds, *Adv. Appl. Prob.* **10**, 906–912 (1979).

J. Senturia and P.S. Puri. A semi-Markov storage problem, *Adv. Appl. Prob.* **5**, 362, (1973).

J. Teghem, J. Loris-Teghem and J.P. Lambotte. *Modéles d'Attente M/G/1 et GI/M/1 á Arrivées et Services en Groupe*, Springer-Verlag, New York (1969).

M.C. Sheps. On the time required for conception, *Population Studies* **18**, 85–97 (1964).

M.C. Sheps and J.A. Menken, *Mathematical Models of Conception and Birth*, Univ. of Chicago Press, Chicago (1973).

W.L. Smith. Regenerative stochastic processes, *Proc. Roy. Soc.* A. **232**, 6–31 (1955).

S. Vajda. *Mathematics of Manpower Planning*, Wiley, Chichester (1977).

R. Valliant and G.T. Milkovitch. Comparison of semi-Markov and Markov models in a personnel forecasting application, *Decision Sciences* **8**, 465-477 (1977).

E.N. Weiss, M.A. Cohen and J.C. Hershey. An iterative estimation and validation procedure for specification of semi-Markov models with applications to hospital patient flow, *Opns. Res.* **30**, 1082–1104 (1982).

G.H. Weiss and M. Zelen. A semi-Markov model for clinical trials, *J. Appl. Prob.* **2**, 269-285 (1965).

CHAPTER 8

Stationary Processes and Time Series

8.1 INTRODUCTION

A series of observations $x(t)$, $t \in T$ made sequentially in time t constitutes a time series. Examples of data taken over a period of time are found in abundance in diverse fields such as meteorology, geophysics, biophysics, economics, commerce, communication engineering systems analysis etc. Daily records of rainfall data, prices of a commodity etc. constitute time series. The variate t denotes time, i.e., changes occur in time. But this need not always be so. For example, the records of measurements of the diameter of a nylon fibre along its length (distance) t also give a time series. Here t denotes length.

The essential fact which distinguishes time series data from other statistical data is the specific order in which observations are taken. While observations from areas other than time series are statistically independent, the successive observations from a time series may be dependent, the dependence based on the position of the observation in the series. The time t may be a continuous or a discrete variable. A general mathematical model of the time series $Y(t)$, $t \in T$ is as follows:

$$Y(t) = f(t) + X(t). \tag{1.1}$$

Here, $f(t)$ represents the systematic part and $X(t)$ represents the random part. These two components are also known as *signal* and *noise* respectively. The model is theoretical: $f(t)$ and $X(t)$ are not separately observable. While the model for $Y(t)$ gives the structure of the generating process, a set of observations (or time series data) is a realisation or a sample function of the process. The effect of time may be in both the systematic and the random parts.

A general model in which the effect of time is represented in the random part involves a stochastic process, i.e. a random function of time. The systematic part is represented by a (non-random or deterministic) function of time. The important types of evolution (in time) of the systematic part are described by the following components.

(i) *Trend*: It is a slowly varying function of time due to growth and other such permanent effects. A trend is usually represented by means of a polynomial of a low degree or a short Fourier series.

(ii) *Cyclical component*: It is a broad long-run movement, more or less regular, caused by fluctuations in activity and is quasi-periodic in appearance.

(iii) *Seasonal component*: It is composed of regular variations, caused by changing seasons, rhythmic activities etc. Cyclic and seasonal movements are usually represented by suitable periodic functions. Trigonometric functions are generally used for representation of cyclic movements. Both the systematic and random parts may involve a number of unknown parameters. The two parts are considered to be additive.

For description and analysis of components involving the systematic part, the reader may consult the standard texts on time series, such as Kendall, Kendall and Stuart, Malinvaud, Anderson, Fuller etc.

We shall discuss here only the models, where the random part involves stationary stochastic processes (real-valued). Here, we shall also confine mostly to processes in discrete time. A time series $\{x_t, t = 1, \ldots, m\}$ may be considered as a single particular realisation of the stationary stochastic process $\{X_t, t \in I\}$. We shall be concenred here with the stationary process $\{X_t, t \in I\}$ rather than with the analysis of the time series $\{x_t, t = 1, \ldots, m\}$.

As discussed in Ch. 2, a stochastic process $\{X_t, t \in I\}$ is said to be stationary in the wide-sense or covariance stationary *iff*, for all $t \in I$, if the mean value function $E\{X_t\} = m$ is independent of t and the covariance function $E\{(X_t - m)(X_{t+k} - m)\} = C_k$ is a function of the difference (in lag) $(t + k) - t = k$. The correlation function is $\rho_k = C_k/C_0$, where $C_0 = \mathrm{var}(X_t)$. The graph of the correlation function is called the *correlogram* of the stochastic process.

Some examples of stationary processes were given in Ch. 2. We shall consider below some stationary processes which are used as models for the generation of the random part of a time series.

8.2 MODELS OF TIME SERIES

8.2.1 Purely Random Process (or White Noise Process)

A completely random process $\{X_t\}$ has

$$E\{X_t\} = m, \tag{2.1}$$

which may be assumed to be 0, for simplicity, and

$$E\{X_t X_{t+k}\} = \begin{cases} \sigma^2, & k = 0 \\ 0, & k \neq 0. \end{cases} \tag{2.2}$$

The process is covariance stationary.

8.2.2 First Order Markov Process

$\{X_t\}$ has the structure

$$X_t + \alpha_1 X_{t-1} = e_t, \quad |\alpha_1| < 1, \tag{2.3}$$

where $\{e_t\}$ is purely random process, say, with $m = 0$, $\sigma = 1$. Further e_{t+1} is not involved in $X_t, X_{t-1}, \ldots$, i.e. X_t depends on $e_1, e_2, \ldots e_t$, but is independent of $e_{t+1}, e_{t+2}, \ldots$.

Multiplying both sides of (2.3) by X_{t-1} and taking expectation, we get

$$C_1 + \alpha_1 C_0 = 0,$$

or, $$\rho_1 + \alpha_1 = 0, \quad \text{i.e.} \quad \rho_1 = -\alpha_1 = a \text{ (say)}, |a| < 1.$$

Proceeding in the same way, we get

$$\rho_2 + \alpha_1 \rho_1 = 0, \quad \text{i.e.} \quad \rho_2 = (-\alpha_1)^2 = a^2$$
$$\cdots \qquad\qquad \cdots \qquad\qquad \cdots$$
$$\rho_n + \alpha_1 \rho_{n-1} = 0, \quad \text{i.e.} \quad \rho_n = (-\alpha_1)^n = a^n$$
$$\cdots \qquad\qquad \cdots \qquad\qquad \cdots$$

The process is covariance stationary. The representation (2.3) is a linear stochastic difference equation; X_t can be expressed as a formal solution of the difference equation as follows.

Denote the backward shift operator by B, i.e. $BX_t = X_{t-1}$.

Then $$B^2 X_t = X_{t-2} \quad \text{and} \quad B^k X_t = X_{t-k}, k = 1, 2, 3, \ldots.$$

Thus the difference equation can be written as

$$(1 + \alpha_1 B) X_t = e_t$$

or $$X_t = (1 + \alpha_1 B)^{-1} e_t = \sum_{k=0}^{\infty} \{(-\alpha_1)^k B^k\} e_t$$

$$= \sum_{k=0}^{\infty} \rho_1^k e_{t-k}. \tag{2.4}$$

8.2.3 Moving Average (MA) Process

Here $\{X_t\}$ is represented as

$$X_t = a_0 e_t + a_1 e_{t-1} + \cdots + a_h e_{t-h}, \tag{2.5}$$

where the a's are real constants and $\{e_t\}$ is a purely random process with mean 0 and variane σ^2. X_t can also be represented as

$$X_t = f(B) e_t, \quad \text{where} \quad f(B) = \sum_{r=0}^{h} a_r B^r;$$

when $a_h \neq 0$, $\{X_t, t \in T\}$ is called an MA process of order h.
We have $$E(X_t) = 0,$$

and $$C_k = E\{X_t X_{t+k}\}$$

$$= E\left\{a_0a_k e_t^2 + a_1a_{k+1}e_{t-1}^2 + \cdots + a_{h-k}a_h e_{t+k-h}^2 + \sum_{r \neq t} a_r a_t e_r e_t\right\}$$

$$= (a_0a_k + a_1a_{k+1} + \cdots + a_{h-k}a_h)\,\sigma^2, \quad k \leq h$$

$$= 0, \qquad\qquad\qquad\qquad k > h.$$

A necessary but *not* sufficient condition that such a moving average representation of order h is possible is $C_k = 0$ for $k > h$. We have

$$\rho_k = \frac{C_k}{C_0} = \frac{a_0a_k + a_1a_{k+1} + \cdots + a_{h-k}a_h}{a_0^2 + a_1^2 + \cdots a_h^2}, \quad \text{if } h \geq k$$

$$= 0, \text{ if } h < k. \tag{2.6}$$

The process is covariance stationary.

If, in particular, $a_k = 1/(h+1)$, $k = 0, 1, \ldots, h$, then

$$\rho_k = \frac{h-k+1}{h+1} = 1 - \frac{k}{h+1}, \quad 0 \leq k \leq h \tag{2.7}$$

$$= 0, \qquad\qquad \text{otherwise.}$$

The correlogram consists of a linearly decreasing set of points.

The representation (2.5) of X_t consists of a linear expression with a finite number of terms in $e_t, e_{t-1}, \ldots, e_{t-h}$. It is said to be of order h. The first order Markov process (2.4) can be considered as a moving average of infinite order with $\rho_k = a^{|k|}$ for all k.

Wold (1954) has a given a necessary and sufficient condition that there exists an MA model of order h corresponding to a specific set of autocorrelation coefficients. Since Wold's work, interest in this topic increased and several contributions have been made. Davies *et al.* (1974) found general inequalities involving autocorrelations of an MA process.

8.2.4 Autoregressive Process (AR Process)

The process $\{X_t\}$ given by

$$X_t + b_1X_{t-1} + b_2X_{t-2} + \cdots + b_hX_{t-h} = e_t, \quad b_h \neq 0, \tag{2.8}$$

where $\{e_t\}$ is a purely random process, with mean 0, is called an *autoregressive process of order h*.

X_t can be obtained as a solution of the linear stochastic difference equation

$$g(B)X_t = e_t, \tag{2.9}$$

where $\qquad\qquad g(B) = \sum_{r=0}^{h} b_r B^r, \quad b_0 = 1.$

Suppose that $g(B) = \Pi(1 - z_i B)$, $z_i \neq z_j$, i.e. $z_1^{-1}, \ldots, z_h^{-1}$ are the distinct roots of the equation $g(z) = 0$. Further suppose that $|z_i| < 1$ for all i, i.e. all the roots of $g(z) = 0$ lie outside the unit circle; the roots z_i of the characteristic equation $f(z) \equiv \sum_{r=0}^{h} b_r z^{h-r} = 0$ (where $f(z) = z^{-h} g(z^{-1})$) all lie within the unit circle. The complete solution of (2.9) can be written as

$$X_t = \sum_{r=1}^{h} A_r z_r^t + \frac{1}{g(B)} e_t,$$

where A_r's are constants. Now

$$\frac{1}{g(B)} e_t = \prod_{i=1}^{h} (1 - z_i B)^{-1} e_t$$

$$= \sum_{r=0}^{\infty} b_r' B^r e_t = \sum_{r=0}^{\infty} b_r' e_{t-r}, \quad b_0 = 1,$$

where b_r' are constants involving z_i's.

If the process is considered as begun long time ago, then the contribution of $\sum_{r=1}^{h} A_r z_r^t$ damps out of existence. X_t is then given by

$$X_t = \sum_{r=0}^{\infty} b_r' e_{t-r}. \tag{2.10}$$

Thus an AR Process can be represented by an MA process of infinite order.

The coefficients b_r' of e_{t-r} in the right hand side of (2.10) can also be obtained as follows. Using the expressions for X_t as given in (2.10), for $t, t-1, \ldots, t-h$ and then substituting them in (2.8), we get

$$\sum b_r' e_{t-r} + b_1 \sum b_r' e_{t-1-r} + b_2 \sum b_r' e_{t-2-r} + \cdots + b_h \sum b_r' e_{t-h-r} = e_t.$$

Equating the coefficients of $e_t, e_{t-1}, \ldots$ from both sides, we get.

$$\left.\begin{array}{r} b_0' = 1 \\ b_1' + b_1 b_0' = 0 \\ \cdots \cdots \\ b_{h-1}' + b_1 b_{h-2}' + \cdots + b_{h-1} b_0' = 0 \end{array}\right\} \tag{2.11}$$

and

$$b_r' + b_1 b_{r-1}' + \cdots + b_h b_{r-h}' = 0, \quad r = h, h+1, \ldots. \tag{2.12}$$

In other words, b'_r satisfies the difference equation (2.12) together with the initial cnditions (2.11). Thus, b'_r can be obtained by solving (2.12) with the help of (2.11).

We can put the result in the form of a theorem as follows:

Theorem 8.1. If the roots of the equation

$$f(z) \equiv z^h + b_1 z^{h-1} + \cdots + b_h = 0$$

all lie within the unit circle, then the autoregressive process

$$X_t + b_1 X_{t-1} + \cdots + b_h X_{t-h} = 0,$$

can be represented as an infinite moving average

$$X_t = \sum_{r=0}^{\infty} b'_r e_{t-r},$$

where b'_r are the roots of the difference equation

$$b'_r + b_1 b'_{r-1} + \cdots + b_h b'_{r-h} = 0, \quad r = h, h+1, \cdots.$$

subject to the initial conditions

$$b'_0 = 1, \quad b'_1 + b_1 b'_0 = 0, \quad b'_2 + b_1 b'_1 + b_2 b'_0 = 0,$$

$$\cdots \qquad \cdots$$

$$b'_{h-1} + b_1 b'_{h-2} + \cdots + b_{h-1} b'_0 = 0.$$

Let us now determine the autocorrelation function of X_t. We have $E\{X_t\} = 0$. Multiplying (2.8) by X_{t-k} and taking expectation, we find that the autocovariances satisfy the relation

$$C_k + b_1 C_{k-1} + \cdots + b_h C_{k-h} = 0, \quad k = 1, 2, \ldots.$$

Noting that
$$\rho_k = C_k / C_0,$$

we get

$$\rho_k + b_1 \rho_{k-1} + \ldots + b_h \rho_{k-h} = 0, \quad k = 1, 2, \ldots \qquad (2.13)$$

or
$$g(B)\rho_k = 0,$$

which is a difference equation in ρ_k of order h. These are known as *Yule-Walker Equations.*

Now $\rho_0 = 1$, and putting $k = 1, 2, \ldots, h - 1$ in (2.13), we get $(h - 1)$ equations in

$\rho_1, \ldots, \rho_{h-1}$ which can be solved in terms of the coefficients $b_1, \ldots, b_h$. These will constitute h initial conditions which could be used for the solution of the difference equation

$$\rho_k + b_1 \rho_{k-1} + \cdots + b_k \rho_{k-h} = 0, \quad k \geq h.$$

The difference equation in ρ_k $(k \geq h)$ can then be completely solved. Thus if the characteristic equation $f(z) = 0$, have distinct roots $z_1, \ldots, z_h$ then

$$\rho_k = \sum_{r=1}^{h} \alpha_r z_r^k, \tag{2.14}$$

where α_r's are constants which can be determined from the initial conditions. The process is evidently covariance stationary.

Note : The process $\{X_t\}$ given by (2.8) cannot be represented by (2.10) when the roots of the characteristic equation $f(z) = 0$ do not all lie within the unit circle. In this case $E\{X_t\} = \sum A_r z_r^t$ which is a function of t does not damp out even for large t and the process is non-stationary.

Let us consider the case $h = 2$.

8.2.5 Autoregressive Process of Order Two (Yule Process)

This is given by

$$X_t + b_1 X_{t-1} + b_2 X_{t-2} = e_t \tag{2.15}$$

where $\{e_t\}$ is a purely random process with mean 0.

Assume that the roots z_1, z_2 of the characteristic equation

$$z^2 + b_1 z + b_2 = 0$$

are distinct, and that both the roots lie within the unit circle, i.e. $b_2 < 1$, and $4b_2 > b_1^2$. Neglecting the contribution $\sum_{r=1}^{\infty} A_r z_r^t$ for large t, we get

$$X_t = \left[(1 - z_1 B)(1 - z_2 B) \right]^{-1} e_t$$

$$= \left[1 + (z_1 + z_2)B + (z_1^2 + z_1 z_2 + z_2^2) B^2 + \ldots \right] e_t$$

$$= \left[\frac{\sum\limits_{r=0}^{\infty} (z_1^{r+1} - z_2^{r+1})}{(z_1 - z_2)} B^r \right] e_t$$

$$= \sum_{r=0}^{\infty} b'_r e_{t-r}, \tag{2.16}$$

where
$$b'_r = (z_1^{r+1} - z_2^{r+1})/(z_1 - z_2), \text{ when } z_1 \neq z_2$$

$$= (r+1)z_1^r, \text{ when } z_1 = z_2. \tag{2.17}$$

We get $E\{X_t\} = 0$ and

$$\operatorname{var}(X_t) = \Sigma (b'_r)^2 \operatorname{var}(e_t)$$

$$= \frac{\operatorname{var}(e_t)}{(z_1 - z_2)^2} \left[\frac{z_1^2}{1 - z_1^2} - \frac{2z_1 z_2}{1 - z_1 z_2} + \frac{z_2^2}{1 - z_2^2} \right]$$

$$= \operatorname{var}(e_t) \left[\frac{1 + z_1 z_2}{(1 - z_1 z_2) \{ (1 - z_1^2)(1 - z_2^2) \}} \right] \text{ (on simplification).}$$

Finally,
$$\frac{\operatorname{var}(X_t)}{\operatorname{var}(e_t)} = \frac{1 + b_2}{(1 - b_2) \{ (1 + b_2)^2 - b_1^2 \}}.$$

The Yule-Walker equations are $\rho_k + b_1 \rho_{k-1} + b_2 \rho_{k-2} = 0, \quad k = 1, 2, \ldots,$

whence we get
$$\rho_k = \alpha_1 z_1^k + \alpha_2 z_2^k.$$

For $k = 1$,
$$\rho_1 + b_1 + b_2 \rho_1 = 0,$$

whence
$$\rho_1 = -\frac{b_1}{1 + b_2} = \frac{z_1 + z_2}{1 + z_1 z_2}; \text{ further } \rho_0 = 1.$$

Using these two values we can determine the two constants α_1 and α_2. Thus for unequal roots (real or complex) of the characteristic equation, (i.e. for $z_1 \neq z_2$)

$$\rho_k = \frac{z_1^{k+1}(1 - z_2^2) - z_2^{k+1}(1 - z_1^2)}{(z_1 - z_2)(1 + z_1 z_2)}, \quad k = 0, 1, 2, \ldots. \tag{2.18}$$

As $4b_2 > b_1^2$, the roots z_1, z_2 are complex, and we can write $z_1 = pe^{i\theta}, z_2 = pe^{-i\theta}$, $0 \leq \theta \leq 2\pi$, so that

$$p = \sqrt{(z_1 z_2)} = \sqrt{b_2}, \quad 2p \cos \theta = z_1 + z_2 = -b_1$$

or
$$\cos \theta = -b_1/2\sqrt{b_2}.$$

Putting these expressions for z_1, z_2 in (2.18), we get

$$\rho_k = \frac{p^{k+1}\{e^{i\theta(k+1)}(1 - p^2 e^{2i\theta}) - e^{-i\theta(k+1)}(1 - p^2 e^{2i\theta})\}}{p(1 + p^2)(e^{i\theta} - e^{-i\theta})}$$

$$= \frac{p^k}{1 + p^2}\left[\frac{2i\sin(k+1)\theta - 2ip^2\sin(k-1)\theta}{2i\sin\theta}\right]$$

$$= p^k\left[\sin k\theta\left(\frac{1 - p^2}{1 + p^2}\right)\cot\theta + \cos k\theta\right]$$

$$= p^k(\sin k\theta \cot\psi + \cos k\theta)$$

where
$$\tan\psi = \frac{1 + p^2}{1 - p^2}\tan\theta. \tag{2.19}$$

Thus
$$\rho_k = p^k\frac{\sin(k\theta + \psi)}{\sin\psi} \tag{2.20}$$

For equal (and real) root of the characteristic equation, (i.e. $z_1 = z_2 = z$,)

$$\rho_k = \{1 + k(1 - z^2)/(1 + z^2)\}\, z^k, \quad k = 0, 1, 2, \ldots.$$

This process is clearly covariance stationary. As $p < 1$, $\rho_r \to 0$ for large k. The correlogram is oscillatory (with period $2\pi/\theta$) and decaying in character (with damping factor p^k).

8.2.6. Autoregressive Moving Average Process (ARMA Process)

Consider a process having the representation

$$\sum_{r=0}^{p} b_r X_{t-r} = \sum_{s=0}^{q} a_s e_{t-s} \quad (b_0 = 1), \tag{2.21}$$

where $\{e_t\}$ is a purely random process with mean 0, and the roots of $f(z) = \sum_{r=0}^{p} b_r z^{p-r} = 0$ all lie within the unit circle. Such a process is called an *auto-regressive moving average process* and is denoted by ARMA (p, q). We can write

(2.21) as $g(B)X_t = h(B) e_t$, where $g(B)$ and $h(B)$ are polynomials in B of degrees p and q respectively. Then X_t can be represented as an infinite moving average $X_t = \Sigma v_r e_{t-r}$ as follows:

Put $u_t = \sum_{s=0}^{q} a_s e_{t-s}$ and let the roots z_i, $i = 1, 2, \ldots, p$ of the equation $f(z) = 0$ be distinct.

Then

$$X_t = \sum_{r=1}^{p} A_r z_r^t + \prod_{i=1}^{q} (1 - z_i B)^{-1} u_t.$$

Now as $|z_i| < 1$, we have for large t (neglecting the terms $\Sigma A_r z_r^t$)

$$X_t = \prod_{i=1}^{p} (1 - z_i B)^{-1} u_t = \sum_{r=0}^{\infty} b'_r u_{t-r} = \sum_{r=0}^{\infty} b'_r \left(\sum_{r=0}^{q} a_s e_{t-r-s} \right)$$

$$= \sum_{r=0}^{\infty} v_r e_{t-r} \text{ (say)}. \tag{2.22}$$

The v_r's, $r = 0, 1, 2, \ldots$ in (2.22) can be determined by substituting the above expression for X_t in the l.h.s. of (2.21) and then comparing the coefficients of e_{t-r} from both sides.

We have $E\{X_t\} = 0$. Multiplying (2.21) X_{t-k} ($k \geq 0$) and taking expectations, we get

$$g(B)\rho_k = \sum_{r=0}^{p} b_r \rho_{k-r} = 0, \quad k > q$$

whence

$$\rho_k = \sum_{r=1}^{p} \alpha_r z_r^k, \quad k > q, \tag{2.23}$$

α_r being constants.

8.3 TIME AND FREQUENCY DOMAIN: POWER SPECTRUM

Covariance and correlation functions of stationary processes are functions in the time domain. It is found that the study of processes in the 'time domain' is neither simple nor illuminating. An alternative approach for the study of stationary processes is through 'frequency domain'. This leads to simpler and more rewarding results. Consider a covariance stationary process in discrete time having C_k and ρ_k as covariance and correlation functions respectively.

8.3.1 Properties of Covariance and Correlation Functions

(1) The covariance (correlation) function is even, and is non-negative definite.
(2) The correlation function has the representation

$$\rho_k = \int\limits_{-\pi}^{\pi} e^{ikw}\, dF_1(w),\tag{3.1}$$

where $F_1(w)$ has the character of a distribution function and is unique.
When F_1 is absolutely continuous (and $d\,F_1(w) = f_1(w)\,dw$), one gets the inverse representation

$$f_1(w) = \frac{1}{2\pi} \sum_{k=-\infty}^{\infty} \rho_k e^{-ikw}\tag{3.2}$$

The corresponding representation of covariance function is

$$C_k = \int\limits_{-\pi}^{\pi} e^{ikw}\, dF(w)\tag{3.3}$$

where $dF(w) = C_0 dF_1(w)$; when $dF(w) = f(w)\, dw$,

$$f(w) = \frac{1}{2\pi} \sum_{k=-\infty}^{\infty} C_k e^{-ikw}.\tag{3.4}$$

The function $F_1(w)$ (so also $F(w)$) is called the *integrated spectrum*; the function $f(w)$ is known as the *spectral density function* and $f_1(w)$ as the *normalised spectral density* function of the process.

The information contained in the sequence $\{C_k\}$ is thus contained in the function $f(\cdot)$ and vice-versa. From covariances which are 'time domain' entities one gets a unique 'frequency domain' function.

The relation (3.1) is of the same form as that between characteristic function and distribution function of a r.v. As a consequence of non-negative definiteness of ρ_k, the relation (3.1) follows from a celebrated theorem due to Bochner-Khinchine.

Example 3(a). White noise process $\{X_t\}$. We have

$$E\{X_t\} = 0,$$

$$E\{X_r X_s\} = \sigma^2, \quad r = s$$

$$= 0, \quad r \neq s.$$

Thus $\qquad C_k = 0, k \neq 0, C_0 = \sigma^2.$ We have

$$f(w) = \frac{1}{2\pi} \sum_{k=-\infty}^{\infty} C_k e^{-ikw} = \frac{\sigma^2}{2\pi}, \quad -\pi \leq w \leq \pi.$$

Again, given $f(w)$, we at once get C_k from the inverse relationship (3.3).

Autocovariance and Autocorrelation Generating Functions

The autocovariance generating function (*AGF*) of a stationary process $\{X_t\}$ is the power series

$$C(z) = \sum_{k=-\infty}^{\infty} C_k z^k,$$

(considered for values of z for which $C(z)$ is convergent).
   Replacing C_k by ρ_k $(= C_k/C_0)$, one gets the autocorrelation generating function

$$S(z) = \sum_{k=-\infty}^{\infty} \rho_k z^k = C(z)/C_0 = C(z)/(\text{var } X_t).$$

The spectral density $f(w)$ can be expressed in terms of *AGF* as follows:

$$f(w) = \frac{1}{2\pi} \sum_{k=-\infty}^{\infty} C_k e^{-kw} = \frac{1}{2\pi} \{C(e^{-iw})\}$$

$$= \frac{\text{var } (X_t)}{2\pi} S(e^{-iw}).$$

Theorem 8.2. Let $Y_t = \sum_{j=0}^{\infty} a_j X_{t-j}$

where $\sum |a_j| < \infty$, be a stationary process. If $C_Y(z)$, $C_X(z)$ are the *AGF*'s and $f_Y(w)$, $f_X(w)$ are the spectral density functions of $\{Y_t\}$ and $\{X_t\}$ respectively, then

$$C_Y(z) = h(z)\,h(z^{-1})\,C_X(z)$$

and $\qquad\qquad f_Y(w) = h(e^{iw})\,h(^{-iw})\,f_X(w),$

where $\qquad\qquad h(z) = \sum_{j=0}^{\infty} a_j z^j.$

Proof : Assume that $E\{X_t\} = 0$, then $E\{Y_t\} = 0$. Denote $E\{X_t X_{t+k}\}$ by C_k and $E\{Y_t Y_{t+k}\}$ by R_k. We have

$$R_k = E\{Y_t Y_{t+k}\} = E\left\{\left(\sum_{j=0}^{\infty} a_j X_{t-j}\right)\left(\sum_{l=0}^{\infty} a_l X_{t+k-l}\right)\right\}$$

$$= \sum_{j=0}^{\infty} \sum_{l=0}^{\infty} a_j a_l C_{k+j-l}$$

and hence

$$C_Y(z) = \sum_{k=-\infty}^{\infty} R_k z^k$$

$$= \sum_{k=-\infty}^{\infty} z^k \left\{ \sum_{j=0}^{\infty} \sum_{l=0}^{\infty} a_j a_l C_{k+j-l} \right\}$$

$$= \sum_{j=0}^{\infty} \sum_{l=0}^{\infty} \sum_{k=-\infty}^{\infty} a_j a_l z^{-j+l} C_{k+j-l} z^{k+j-l}$$

$$= \left(\sum_{l=0}^{\infty} a_l z^l \right) \left(\sum_{j=0}^{\infty} a_j z^{-j} \right) \left(\sum_{s=-\infty}^{\infty} C_s z^s \right),$$

(setting $s = k+j-l$)

$$= h(z) h(z^{-1}) C_X(z).$$

Again

$$f_Y(w) = \frac{1}{2\pi} \{ C_Y(e^{-iw}) \}$$

$$= \frac{1}{2\pi} \{ h(e^{-iw}) h(e^{iw}) C_X(e^{-iw}) \}$$

$$= h(e^{-iw}) h(e^{-iw}) f_X(w). \qquad \blacktriangle$$

Note: Assume $\{X_t\}$ to be a white noise process; then the above theorem can be applied to obtain the spectral density functions of *MA*, *AR* and *ARMA* processes.

Example 3(b). *MA* process of order h: $X_t = \sum_{j=0}^{h} a_j e_{t-j}$, $\{e_t\}$ being a white noise process having variance σ^2. We get

$$f_X(w) = \frac{\sigma^2}{2\pi} \{ h(e^{iw}) h(e^{-iw}) \}$$

$$= (\sigma^2/2\pi) \{ a_0 + a_1 e^{iw} + \cdots + a_h e^{iwh} \} \{ a_0 + a_1 e^{-iw}$$

$$+ \cdots + a_h e^{-iwh} \}.$$

On simplification, we get

$$f_X(w) = (\sigma^2/2\pi) \sum_{k=-h}^{h} \left\{ \sum_{r=0}^{h-k} a_r a_{r+k} \right\} \cos kw$$

$$= (\sigma^2/2\pi) \sum_{k=-h}^{h} C_k e^{-ikw},$$

where $$C_k = E\{X_t X_{t+k}\}, \text{ as is to be expected.}$$

Example 3(c). First order autoregressive process or Markov process:
$X_t = aX_{t-1} + e_t, |a| < 1, \{e_t\}$ being a white noise process having variance σ^2.
The process can be represented as (see (2.4))

$$X_t = \sum_{j=0}^{\infty} a^j e_{t-j}.$$

We get
$$f_X(w) = \left(\frac{\sigma^2}{2\pi}\right) \{h(e^{iw}) h(e^{e-iw})\},$$

where
$$h(e^{iw}) = \sum_{j=0}^{\infty} a^j e^{iwj} = 1/(1 - ae^{iw}),$$

Thus
$$f_X(w) = \frac{1}{2\pi} \cdot \frac{\sigma^2}{(1 - ae^{iw})(1 - ae^{-iw})} = \frac{\sigma^2}{2\pi(1 - 2a\cos w + a^2)}.$$

Example 3(d). AR process of order h : $X_t + b_1 X_{t-1} + \cdots + b_h X_{t-h} = e_t$,
where $\{e_t\}$ is a white noise process having variance σ^2 and the roots of the equation
$f(z) = z^h + b_1 z^{h-1} + \cdots + b_h = 0$ all lie within the unit circle.
Applying the theorem to

$$e_t = \sum_{j=0}^{h} b_j X_{t-j},$$

we get
$$f(w) = \left(\sum_{j=0}^{h} b_j e^{ijw}\right)\left(\sum_{k=0}^{h} b_k e^{-ikw}\right) f_X(w),$$

where $f(w)$ is the spectral density of $\{e_t\}$. But $f(w) = \sigma^2/2\pi$, so that we at once get

$$f_X(w)\frac{\sigma^2}{2\pi} = \left[\left(\sum_{j=0}^{h} b_j e^{ijw}\right)\left(\sum_{k=0}^{h} b_k e^{-ikw}\right)\right]^{-1}.$$

Particular case: $h = 2$. The spectral density function $f_X(w)$ of a second order auto-regressive process

$$X_t + b_1 X_{t-1} + b_2 X_{t-2} = e_t$$

is given by

$$f_X(w) = \frac{\sigma^2}{2\pi}[(1 + b_1 e^{iw} + b_2 e^{2iw})(1 + b_1 e^{-iw} + b_2 e^{-2iw})]^{-1}$$

$$= \left(\frac{\sigma^2}{2\pi}\right)\frac{1}{1 + b_1^2 + b_2^2 + 2b_1(1 + b_2)\cos w + 2b_2 \cos 2w}.$$

Example 3(e). ARMA (p, q) process:

$$\sum_{r=0}^{p} b_r x_{t-r} = \sum_{r=0}^{q} a_r e_{t-r}, \quad b_0 = 1,$$

where $\{e_t\}$ is a white noise process with variance σ^2 and the roots of the equation $f(z) = z^h + b_1 z^{h-1} + \cdots + b_h = 0$ all lie within the unit circle. Putting

$$Z_t = \sum_{r=0}^{q} a_r e_{t-r}$$

and applying the theorem, we get

$$f_Z(w) = \frac{\sigma^2}{2\pi} \left(\sum_{r=0}^{q} a_r e^{irw} \right) \left(\sum_{s=0}^{q} a_s e^{-isw} \right),$$

where $f_Z(w)$ is the spectral density of the process $\{Z_t\}$. Again from $Z_t = \sum_{r=0}^{p} b_r X_{t-r}$, we get

$$f_Z(w) = \left(\sum_{r=0}^{p} b_r e^{irw} \right) \left(\sum_{s=0}^{q} a_s e^{-isw} \right) f_X(w).$$

Thus equating the two expressions for $f_Z(w)$, we get

$$f_X(w) = \frac{\sigma^2}{2\pi} \cdot \frac{\left(\sum\limits_{r=0}^{q} a_r e^{irw} \right) \cdot \left(\sum\limits_{s=0}^{q} a_s e^{-isw} \right)}{\left(\sum\limits_{r=0}^{p} b_r e^{irw} \right) \cdot \left(\sum\limits_{s=0}^{p} b_s e^{-isw} \right)}.$$

8.3.2 Continuous Parameter Processes

Stationarity for a process in continuous time is defined in a similar manner. A process $\{X(t), -\infty < t < \infty\}$ in continuous time is wide sense stationary *iff*, for all t, the mean value function $E\{X(t)\} = m$ is independent of t and the covariance function $E[\{(X(t) - m)\}\{(X(t+v) - m)\}] = R(v)$ is a function of the lag difference $(t+v) - t = v$. The continuous time analogue of (3.3) is given below.

$R(v)$ is positive semidefinite and has the representation

$$R(v) = \int_{-\infty}^{\infty} e^{ivw} \, dF(w), \tag{3.5}$$

where F is nondecreasing and bounded; when $dF(w) = f(w) \, dw$,

$$f(w) = \frac{1}{2\pi} \int\limits_{-\infty}^{\infty} e^{-ivw} R(v) dv. \tag{3.6}$$

[The result follows from Bochner's theorm; for its proof see Goldberg (1961)].

Example 3(f). For a process having $R(v) = ae^{-b|v|}$ the spectral density function is

$$f(w) = \frac{1}{2\pi} \int\limits_{-\infty}^{\infty} e^{-iwv} \, ae^{-b|v|} dv$$

$$= \frac{a}{2\pi} \left[\int\limits_{-\infty}^{\infty} e^{(b-iw)v} \, dv + \int\limits_{0}^{\infty} e^{-(b+iw)v} \, dv \right]$$

$$= \frac{a}{2\pi} \left(\frac{1}{b-iw} + \frac{1}{b+iw} \right)$$

$$= \frac{ab}{\pi(b^2 + w^2)}.$$

Note: We can also write

$$R(t,s) = E[\{X(t) - m\} \{X(s) - m\}]$$

as the covariance function of a process

$$\{X(t), -\infty < t < \infty\}. \quad \text{Then} \quad R(t,s) = R(s{\sim}t).$$

The covariance function satisfies the following properties
   (1) A positive constant C can be expressed as a covariance function. If we can take $X(t) = X$ to be a normal (Gaussian) r.v. with zero mean, then $E(X^2) = C$.
   (2) The sum of two covariance functions is again a covariance function.
   (3) The product of two covariance functions is again a covariance function.
For a detailed account of second-order processes and of impotant related aspects (such as linear prediction and filtering), refer to Wong and Hajek (1985).

8.4 STATISTICAL ANALYSIS OF TIME SERIES: SOME OBSERVATIONS

Time series is one of the most widely pursued areas of statistics. Problems of statistical inference in time series, though very important from the point of view of practitioners, are beyond the scope of this book. However, we propose to give here very briefly an idea of the type of problems that arise.
   The special feature is that the data are arranged with respect to time and successive observations are considered to be not independent. The problem of inference of such types of stochastic processes arising in diverse fields does not differ in principle from

that of other types of statistical data; but the existence of some dependence in successive observations often implies that the classial methods become inadequate and need adequate extension. For example, the asymptotic theory of maximum likelihood estimates does not apply in general to dependent observations. Further, independent repetitions of the process are not available; instead a single observed sequence is available and this also makes new methods necessary. Difficulty arises because of the fact that dependence has so many more possibilities *a priori* than independence, and more often than not, the distribution and the dependence of the observations representing stochastic processes may not be clearly known. The sampling theory of time series, like that of stochastic processes in general, is still in a comparatively early stage of development, and most of the results concern large samples, and small sample methods are less often available.

The inference problems centering the parametric models are concerned with the parameters of the models. For example, for the AR model

$$\sum_{r=0}^{h} b_r X_{t-r} = e_t, \ b_0 = 1,$$

the problems include estimation of the b's, the variance of e_t, the distribution of the estimates and hypotheses testing; and similarly for the MA, ARMA models. There has been considerable amount of work on estimation and the question is still engaging the attention of researchers (see Chatfield (1977), for a review of recent developments and for an exhaustive list of references).

Nonstationary models have also been extensively studied (e.g., Priestly, Rao, Granger *et al*, Koopmans *et al*; also see Box and Jenkins (1976) for building of stochastic models, specially non-stationary and their use in important areas of application).

In addition to making inferences about the statistical structure of the underlying process, the analysis of time series has two other purposes: forecasting and control. The Box-Jenkins forecasting method, based on a class of models called *integrated autoregressive moving average* (ARIMA) processes, has roused keen interest. They start with a modified type of ARMA model, such as $g\ (B)\ X_t = h\ (B)\ e_t$, where some roots (say, $d \geq 1$) of $g\ (B)$ are equal to unity and the others lie outside the unit circle. The corresponding nonstationary process (called ARIMA process) has the representation

$$f(B)(1-B)^d X_t = h(B)e_t,$$

where all the roots of $f\ (B)$ lie outside the unit circle. Introducing seasonality components, Box and Jenkins (1976, pp. 303–305) arrive at the general multiplicative seasonal model

$$f_p(B)\, F_P(B^s)(1-B)^d(1-B^s)^D X_t = h_q(B)\, H_Q(B^s)\, e_t \ ,$$

where $\{e_t\}$ is a white noise process, f_p, h_q are polynomials in degree p, q in the lag operator B and F_P, H_Q are polynomials in B^s (s being the number of seasons per year).

It is assumed that f_p, h_q have all the zeros outside the unit circle, h_q has no zeros on the unit circle and all the polynomials f_p, h_q, F_P, H_Q have unity as the constant terms. This class provides a wide range of models, both *stationary* and *non-stationary*. Box and Jenkins claim that this class is sufficiently flexible to provide a close approximation to the true model for many sets of homogeneous data. The main problem is the selection of an appropriate model. Box and Jenkins provide detailed techniques for identification of a proper model, for estimation of its parameters and for checking the adequacy of the model. Their techniques have aroused keen interest resulting in further active work and applications in several fields (including weather forecasting).

Apart from the estimation and associated problems of a parametric model there are some other statistical problems associated with a stationary time series with mean μ. Given a consecutive sample of observations $X_1, \ldots, X_T$, one is concerned with the estimation of the mean, of autocovariances or autocorrelations, of the spectral density and also with the large sample behaviour of the estimates (see Anderson (1971), Fuller (1976) for a detailed account). The problem of estimation of the spectral density function has long attracted the attention of time series researchers (Bartlett, Blackman and Tukey, Parzen, Grenander, Medhi, Priestley, Rao etc., to name a few; see Jenkins and Watts (1968), Hannan (1970) for some details). The difficulties of selecting a best estimator on theoretical grounds are still being appreciated (Bloomfield, 1976).

EXERCISES

8.1 Do the autocorrelation functions uniquely determine a moving average process? Discuss with the help of some suitable examples.

8.2 Verify that the four moving average processes

$$X_t = 1.6\,e_t - 0.8\,e_{t-1} + 0.4\,e_{t-2} - 0.2\,e_{t-3}$$
$$X_t = -0.2\,e_t + 0.4\,e_{t-1} - 0.8\,e_{t-2} + 1.6\,e_{t-3}$$
$$X_t = 0.4\,e_t - 0.2\,e_{t-1} + 1.6\,e_{t-2} - 0.8\,e_{t-3}$$
$$X_t = -0.8\,e_t + 1.6\,e_{t-1} - 0.2\,e_{t-2} + 0.4\,e_{t-3}$$

all have the first three autocorrelation coefficients equal to $-42/85, 4/17 - 8/85$ and the rest zero.

(Wold, 1954).

8.3 Find two moving average processes which have for autocorrelation $\rho_0 = 1$, $\rho_1 = 0.4$, $\rho_k = 0, k \geq 2$.

8.4 A set of autocorrelation functions $\rho_0 = 1, \rho_1, \ldots, \rho_h$ of an autoregressive process is given. Will these uniquely determine the autoregressive process?

It is given that an autoregressive process of order two has autocorrelation coefficients $\rho_0 = 0$, $\rho_1 = a$, $\rho_2 = b$. Determine the autoregressive process.

8.5 Consider the ARMA (1, 1) model

$$X_t - a_1 X_{t-1} = e_t + b_1 e_{t-1}, \quad t = 0, \pm 1, \pm 2, \ldots$$

where $|a_1| < 1$, $E(e_t) = E(e_t^2) = \sigma^2$, and $E(e_r e_s) = 0, r \neq s$.

Show that the model can be represented as an infinite moving average

$$X_t = e_t + (b_1 + a_1) \sum_{k=1}^{\infty} a_1^{k-1} e_{t-k}.$$

8.6 Let $\{X_t\}$ be an *MA* process of order h given by

$$X_t = a_0 e_t + a_1 e_{t-1} + \cdots + a_h e_{t-h}, \quad a_h \neq 0,$$

where $\{e_t\}$ is a purely random process. If the roots of the characteristic equation

$$z^h + a_1 z^{h-1} + \cdots + a_h = 0$$

all lie within the unit circle, then X_t can be expressed as an autoregressive process of infinite order

$$\sum_{r=0}^{\infty} c_r X_{t-r} = e_t$$

where the coefficients c_r satisfy the difference equation

$$c_r + a_1 c_{r-1} + \cdots + a_h c_{r-h} = 0, \quad r = h, h+1 \ldots,$$

and the initial conditions

$$c_0 = 1, c_1 + a_1 = 0, c_2 + a_1 c_1 + a_2 = 0, \ldots,$$

$$c_{h-1} + a_1 c_{h-2} + \cdots + a_{h-1} = 0. \tag{Fuller, 1976}$$

8.7 Let X_t be an ARMA (p, q) process given by

$$g(B)X_t = f(B) e_t$$

where $\{e_t\}$ is a purely random process,

$$g(B) = \sum_{r=0}^{p} b_r B^r, b_0 = 1,$$

$$f(B) = \sum_{r=0}^{q} a_r B^r, a_0 = 1$$

and the roots of $g(1/z) = 0$ all lie outside the unit circle. Show that X_t can then be represented as an *MA* process of infinite order

$$X_t = \sum_{r=0}^{\infty} c_r e_{t-r}.$$

8.8 Consider the process

$$X_t = aX_{t-1} + b + e_t,$$

where $\{e_t\}$ is a purely random process with variance unity. Further assume that the process was started at time 0 and that $X_0 = 0$.
Show that

$$X_t = \left(\frac{1-a^t}{1-a}\right) b + \sum_{s=1}^{t} a^{t-s} e_s.$$

Also show that

$$E\{X_t\} = \left(\frac{1-a^t}{1-a}\right) b$$

and

$$C_k = a^k \left(\frac{1-a^{2t}}{1-a^2}\right)$$

Deduce that, if the series was started long ago, and if $|a| < 1$

$$C_k = a^k / (1 - a^2).$$

8.9 Let

$$X_t + a_1 X_{t-1} + a_2 X_{t-2} = Y_t,$$

where

$$Y_t + b_1 Y_{t-1} + b_2 Y_{t-2} = e_t,$$

and $\{e_t\}$ is a purely random process and let the roots of $m^2 + a_1 m + a_2 = 0$ and $n^2 + b_1 n + b_2 = 0$ be less than unity in absolute value. Obtain the spectral density function of $\{X_t\}$.

8.10 Let $Z_t = X_t + Y_t$, where $\{X_t\}$ and $\{Y_t\}$ are two stationary processes with spectral density $f_X(w)$ and $f_Y(w)$ respectively. Then show that $\{Z_t\}$ is stationary and has spectral density $f_Z(w) = f_X(w) + f_Y(w)$. Hence find the spectral density of

$$Z_t = X_t + e_t$$

where $X_t + a X_{t-1} = v_t$, and $\{e_t\}$ and $\{v_t\}$ are two independent white noise processes.

8.11 Show that for a process having $R(v) = \frac{1}{8} e^{-|v|} (3 \cos v + \sin |v|)$, the spectral density is given by

$$f(w) = \frac{1}{4\pi} \cdot \frac{4+v^2}{4+v^4}.$$

8.12 *Narrow-band process*: A continuous state-space and continuous parameter process $\{X(t), -\infty < t < \infty\}$ is called a narrow-band process if

$$X(t) = A(t) \cos\{wt + e(t)\}$$

where w is a constant and $A(t)$, $e(t)$ are continuous state-space and continuous parameter processes such that $0 \le A(t) < \infty$ and $0 \le e(t) \le 2\pi$.

Show that the spectral density function of the process $\{X(t), -\infty < t < \infty\}$ is concentrated sharply in the neighbourhood of the frequency w.

REFERENCES

O.D. Anderson. *Time Series Analysis and Forecasting: The Box-Jenkins Approach*, Butterworths, London (1976).

T.W. Anderson. *The Statistical Analysis of Time Series*, Wiley, New York, (1971).

M.S. Bartlett. *An Introduction to Stochastic Processes*, Camb. Univ. Press London, (1954) [3rd ed., 1978].

M.S. Bartlett and J. Medhi. On the efficiency of procedures for smoothing periodograms with continuous spectra, *Biometrika*, **42**, 143–150 (1955).

R.B. Blackman and J.W. Tukey. *The Measurement of Power Spectra*, Dover, New York, (1959).

P. Bloomfield. *Fourier Analysis of Time Series :An Introduction*, Wiley, New York, (1976).

G.E.P. Box and G.M. Jenkins. *Time Series Analysis: Forecasting and Control*, Holden-Day, San Francisco (1970) [Revised ed. (1976)].

C. Chatfield. Some recent developments in time series analysis, *J.R.S.S.A.*, **140**, 492–510 (1977).

———, *The Analysis of Time series: An Introduction*, 2nd ed., Chapman & Hal, New York (1980).

N. Davies, M.B. Pata and M.G. Frost. Maximum autocorrelations for moving average processes, *Biometrika*, **66**, 295–336 (1974).

W.A. Fuller. *Introduction to Statistical Time Series*, Wiley, New York, (1976).

R.R. Goldberg. *Fourier Transform*, Camb. Univ. Press, London (1961).

C.W.J. Granger, M. Hatanka and M.J. Morris. Time Series modelling and interpretation, *J.R.S.S.A.* **139**, 246–257 (1976).

C.W.J. Granger and P. Newbold. *Forecasting Economic Time Series*, Academic Press, New York (1977).

U. Grenander and M. Rosenblatt. *Statistical Analysis of Stationary Time Series*, Wiley, New York, (1957)

E.J. Hannan. *Time Series Analysis*, Methuen, London, (1960).

———, *Multiple Time Series*, Wiley, New York, (1970).

M.G. Kendall. *Time Series*, C.Griffin, London, (1974).

M.G. Kendall and A. Stuart. *The Advanced Theory of Statistics*, Vol. 3, C.Griffin, London, (1966).

G.M. Jenkins and D.G.Watts. *Spectral Analysis and its Application*, Holden-Day, San Francisco (1968).

L.H. Koopmans. *The Spectral Analysis of Time Series*, Academic Press, New York, (1974).

B. Malinvaud. *Statistical Methods in Econometrics*, North Holland, Amsterdam, (1970).

J. Medhi. A note on the properties of a test procedure for discrimination between two types of spectra of a stationary process, *Skand. Akt.* **20**, 6–13 (1959).

J. Medhi and T.Subba Rao. Sequential and non-sequential test procedures for discrimination between discrete, continuous and mixed spectra of a stationary time series, *J. Ind. Stat. Ass.*, **5**, 1–8, (1967).

D.C. Montogomery and L.A. Johnson. *Forecasting and Time Series Analysis*, McGraw Hill, New York (1977).

M.K. Ochi. *Applied Probability and Stochastic Processes in Engineering & Physical Sciences*, Wiley, New York, 1990.

S.M. Pandit and S.M. Wu. *Time Series and Systems Analysis with Applications*, Wiley, New York (1983).

E. Parzen. An approach to time series Analysis, *Ann. Math. Stat.*, **32**, 951–989 (1961).

M.B. Priestley. The analysis of stationary processes with mixed spectra, **I & II** *J.R.S.S.B.*, **24**, **I**: 215–233, **II**: 511–523 (1962).

————, Evolutionary spectra and nonstationary process, *J.R.S.S.B.*, **27**, 204–237 (1965).

————, *Spectral Analysis and Time Series, Vol I. Univariate Series*, Academic Press, New York (1981).

H. Tong and M.B. Priestley. On the analysis of bivariate nonstationary processes, *J.R.S.S.* Series **B**. 153–166 (1973).

T. Subba Rao. *Problems of statistical analysis in Time Series*, Doctoral Dissertation, Gauhati University, India, (1966).

————, On the cross periodogram of a stationary Gaussian vector process, *Ann Math. Stat.* **38**, 593–597 (1967).

H. Wold. *A Study in the Analysis of Stationary Time Series*, 2nd ed. Almqvist & Wiksells, Stockholm, (1954).

————, *Bibliography on Time Series and Stochastic Process*, Oliver and Boyd, Edinburgh, (1965).

E. Wong and B. Hajek. *Stochastic Processes in Engineering Systems*, Springer-Verlag, New York (1985).

CHAPTER 9

Branching Processes

9.1 INTRODUCTION

The history of the study of branching processes dates back to 1874, when a mathematical model was formulated by Galton and Watson for the problem of 'extinction of families'. The model did not attract much attention for a long time; the situation gradually changed and during the last 50 years much attention has been devoted to it. This is because of the development of interest in the applications of probability theory, in general, and also because of the possibility of using the models in a variety of biological, physical, and other problems where one is concerned with objects (or individuals) that can generate objects of similar kind; such objects may be biological entities, such as human beings, animals, genes, bacteria and so on, which reproduce similar objects by biological methods or may be physical particles such as neutrons which yield new neutrons under a nuclear chain reaction or in the process of nuclear fission.

We consider first the discrete time case. Suppose that we start with an initial set of objects (or individuals) which form the 0-th generation-these objects are called *ancestors*. The offsprings reproduced or the objects generated by the objects of the 0-th generation are the 'direct descendants' of the ancestors, and are said to form the 1st generation; the objects generated by these of the 1st generation (or the direct descendants of the 1st generation) form the 2nd generation, and so on, the direct descendants of the rth generation form the $(r + 1)$st generation. The number of objects of the rth generation ($r = 0, 1, 2, ...$) is a random variable. We assume that the objects reproduce independently of other objects, i.e. there is no interference.

Definition. Let the random variables X_0, X_1, X_2, ... denote the sizes of (or the numbers of objects in) the 0th, 1st, 2nd, ..., generations respectively. Let the probability that an object (irrespective of the generation to which it belongs) generates k similar objects be denoted by p_k, where $p_k \geq 0$, $k = 0, 1, 2, ..., \sum_k p_k = 1$.

The sequence $\{X_n, n = 0, 1, 2, ...,\}$ constitutes a *Galton-Watson branching process* (or simply a *G.W. branching process*) with *offspring distribution* $\{p_k\}$.

The process is also called *Bienayame-Galton-Watson process,* in recognisation of an even earlier work by Bienayame.

Our interest lies mainly in the probability distribution of X_n and the probability that $X_n \to 0$ for some n, i.e. the probability of ultimate extinction of the family.

Note 1: Unless otherwise stated, we shall assume that $X_0 = 1$, i.e. the process starts with a single ancestor.

Note 2: The sequence $\{X_n\}$ forms a Markov chain with transition probabilities

$$p_{ij} = \Pr\{X_{n+1} = j \mid X_n = i\}, i, j = 0, 1, 2, ...$$

It is however not always easy to specify p_{ij}.

Note 3: The generating functions are very useful in the study of branching processes.

9.2 PROPERTIES OF GENERATING FUNCTIONS OF BRANCHING PROCESSES

Another definition given is as follows:

A Galton-Watson process is a Markov chain $\{X_n, n = 0, 1, 2, \ldots\}$ having state space N (the set of non-negative integers), such that

$$X_{n+1} = \sum_{r=1}^{X_n} \zeta_r, \tag{2.1}$$

where ζ_r are i.i.d. random variables with distribution $\{p_k\}$.

Let

$$P(s) = \sum_k \text{Pr}\ \{\zeta_r = k\} s^k = \sum_k p_k s^k \tag{2.2}$$

be the p.g.f. of $\{\zeta_r\}$ and let

$$P_n(s) = \sum_k \text{Pr}\ \{X_n = k\} s^k, \quad n = 0, 1, 2, \ldots \tag{2.3}$$

be the p.g.f. of $\{X_n\}$.

We assume that $X_0 = 1$; clearly $P_0(s) = s$ and $P_1(s) = P(s)$. The r.v.'s X_1 and ζ_r both give (the same) offspring distribution.

Theorem 9.1. We have

$$P_n(s) = P_{n-1}(P(s)) \tag{2.4}$$

and $$P_n(s) = P(P_{n-1}(s)). \tag{2.5}$$

Proof: We have, for $n = 1, 2, \ldots$

$$\text{Pr}\{X_n = k\} = \sum_{j=0}^{\infty} \text{Pr}\ \{X_n = k \mid X_{n-1} = j\} \cdot \text{Pr}\{X_{n-1} = j\}$$

$$= \sum_{j=0}^{\infty} \text{Pr}\ \left\{\sum_{r=1}^{j} \zeta_r = k\right\} \cdot \text{Pr}\ \{X_{n-1} = j\}$$

so that,

$$P_n(s) = \sum_{k=0}^{\infty} \text{Pr}\ \{X_n = k\} s^k$$

$$= \sum_{k=0}^{\infty} s^k \left[\sum_{j=0}^{\infty} \Pr \left\{ \sum_{r=1}^{j} \zeta_r = k \right\} \Pr \left\{ X_{n-1} = j \right\} \right]$$

$$= \sum_{j=0}^{\infty} \Pr \left\{ X_{n-1} = j \right\} \left[\sum_{k=0}^{\infty} \Pr \left\{ \zeta_1 + \zeta_2 + \ldots + \zeta_j = k \right\} s^k \right].$$

The expression within the square brackets, being the p.g.f. of the sum $\zeta_1 + \ldots + \zeta_j$ of j i.i.d. random variables each with p.g.f $P(s)$, equals $[P(s)]^j$. Thus

$$P_n(s) = \sum_{j=0}^{\infty} \Pr \left\{ X_{n-1} = j \right\} [P(s)]^j$$

$$= P_{n-1}(P(s)).$$

Thus we get (2.4). Putting $n = 2, 3, 4, \ldots$, we get, when $X_0 = 1$,

$$P_2(s) = P_1(P(s)) = P(P(s)), \quad P_3(s) = P_2(P(s)), \quad P_4(s) = P_3(P(s))$$

and so on. Iterating (2.4) we get

$$P_n(s) = P_{n-1}(P(s)) = P_{n-2}(P(P(s)))$$

$$= P_{n-2}(P_2(s)). \tag{2.6}$$

For $n = 3$, $P_3(s) = P_1(P_2(s)) = P(P_2(s)).$

Again iterating (2.6), we get

$$P_n(s) = P_{n-3}(P(P_2(s)) = P_{n-3}(P_3(s)),$$

and for $n = 4$, $P_4(s) = P_1(P_3(s)) = P(P_3(s)).$

Thus $P_n(s) = P_{n-k}(P_k(s)), \quad k = 0, 1, 2, \ldots, n,$

and for $k = n - 1$,

$$P_n(s) = P_1(P_{n-1}(s)) = P(P_{n-1}(s)).$$

Thus we get (2.5).   ▲

Note that even when $X_0 = i \neq 1$, the relation (2.5) holds but (2.4) does not hold.

9.2.1 Moments of X_n

Theorem 9.1 could be used to find the moments of X_n. We have

$$P'(1) = E(\zeta_r) = E(X_1) = m \text{ (say).}$$

Theorem 9.2. If $m = E(X_1) = \sum\limits_{k=0}^{\infty} k p_k$, and $\sigma^2 = \text{var}(X_1)$ then

$$E\{X_n\} = m^n \tag{2.7}$$

and

$$\text{var}(X_n) = \frac{m^{n-1}(m^n - 1)}{m - 1}\sigma^2, \quad \text{if } m \neq 1 \tag{2.8}$$

$$= n\sigma^2, \qquad\qquad \text{if } m = 1.$$

Proof: Differentiating (2.4), we get

$$P'_n(s) = P'_{n-1}(P(s))P'(s)$$

whence

$$P'_n(1) = P'_{n-1}(1)P'(1) = mP'_{n-1}(1)$$

and on iterating

$$P'_n(1) = m^2 P'_{n-2}(1)$$

$$\cdot \quad \cdot \quad \cdot$$

$$\cdot \quad \cdot \quad \cdot$$

$$= m^{n-1}P'(1) = m^n.$$

Thus

$$E(X_n) = P'_n(1) = m^n.$$

Differentiating (2.5) twice and proceeding in a similar fashion, one can find the second moment $P_n''(1)$, and thus the variance of X_n in the form (2.8). ▲

One can likewise proceed to get higher moments of X_n.

Alternatively, the mean and the variance of X_n can be obtained by noting that X_{n+1} is the sum of a random number of i.i.d. random variables, and using standard formulas.

Let $m \neq 1$. We can use the Corollary to Theorem 1.3 to find $E(X_{n+1})$.

Since $X_{n+1} = \sum\limits_{r=1}^{X_n} \zeta_r$, we have

$$E(X_{n+1}) = E(\zeta_r) E(X_n) = mE(X_n).$$

The solution of the difference equation is given by

$$E(X_n) = Cm^n, \quad n = 1, 2, 3, \ldots.$$

Since $\qquad$ $E(X_1) = E(\zeta_r) = m, \quad C = 1. \quad$ Thus $E(X_n) = m^n$.

Using the result given in relation (1.20), Ch. 1, we get

$$\text{var}\,(X_{n+1}) = E(X_n)\,\text{var}\,(\zeta_r) + [E(\zeta_r)]^2\,\text{var}\,(X_n) \qquad (2.9a)$$

$$= m^n \sigma^2 + m^2 \text{var}\,(X_n) \qquad (2.9b)$$

We can find var (X_n) from (2.9b) either by induction or by solving the non-homogeneous difference equation. We employ the latter method.

A particular solution of the difference equation (2.9b) is given by

$$\text{var}\,(X_n) = \frac{\sigma^2 m^n}{m - m^2}$$

and a general solution of the homogeneous equation corresponding to (2.9b) is given by var $(X_n) = A\,(m^2)^n$, where A is a constant. Thus the complete solution of (2.9b) is given by

$$\text{var}\,(X_n) = A\,(m^2)^n + \frac{\sigma^2 m^n}{m - m^2} \qquad (2.10)$$

Noting that var $(X_1) = \sigma^2$, we get $A = \sigma^2/[m\,(m-1)]$ so that (2.10) yields

$$\text{var}\,(X_n) = \frac{m^{n-1}(m^n - 1)}{m - 1}\,\sigma^2, \quad n = 1, 2, \dots.$$

The result holds for all n and $m \neq 1$. By taking limit as $m \to 1$, one gets the corresponding result for $m = 1$.

When $m = 1$, var $(X_n) = n\,\sigma^2$.

When $m = 1$, the variance of X_n increases linearly and when $m > 1$ ($m < 1$) it increases (decreases) geometrically with n.

According as $m < 1, = 1,$ or > 1, the Galton-Watson process is referred to as *subcritical, critical* or *supercritical* respectively.

We now come to the problem originally posed by Galton.

9.3 PROBABILITY OF EXTINCTION

Definition. By extinction of the process it is meant that the random sequence $\{X_n\}$ consists of zeros for all except a finite number of values of n. In order words, extinction occurs when Pr $\{X_n = 0\}$, for some value of n. Clearly, if $X_n = 0$ for $n = m$, then $X_n = 0$ for $n > m$; also Pr $\{X_{n+1} = 0 \mid X_n = 0\} = 1$.

Theorem 9.3. If $m \le 1$, the probability of ultimate extinction is 1. If $m > 1$, the probability of ultimate extinction is the positive root less than unity of the equation

$$s = P(s). \tag{3.1}$$

Proof: Let $q_n = \Pr\{X_n = 0\}$, i.e. q_n is the probability that extinction occurs at or before the nth generation. Clearly $q_n = P_n(0)$, $q_1 = P_1(0) = P(0) = p_0$ and from (2.5)

$$q_n = P(q_{n-1}). \tag{3.2}$$

If $p_0 = 0$, then $q_1 = 0$, $q_2 = 0, ...$, i.e. if the probability of no offspring is zero, extinction can never occur. If $p_0 = 1$, then $q_1 = 1$, $q_2 = 1, ...$, i.e. if the probability of no offspring is one then the extinction is certain to occur right after the 0th generation. So we confine ourselves to the case $0 < p_0 < 1$.

As $P(s)$ is a strictly increasing function of s, $q_2 = P(q_1) = P(p_0) > P(0) = q_1$. Assuming that $q_n > q_{n-1}$ we get $q_{n+1} = P(q_n) > P(q_{n-1}) = q_n$ and by induction $q_1 < q_2 < q_3 ....$ The monotone increasing sequence $\{q_n\}$ is bounded above by 1. Hence q_n must have a limit $\lim_{n \to \infty} q_n = q$ (say), $0 \le q \le 1$; q is the probability of ultimate extinction. From (3.2) it follows that q satisfies $q = P(q)$, i.e. q is a root of the equation (3.1),

$$s = P(s).$$

We now investigate further about the root. First, we show that q is the smallest positive root of (3.1). Let s_0 be an arbitrary positive root of (3.1). Then $q_1 = P(0) < P(s_0) = s_0$ and assuming that $q_m < s_0$, we get $q_{m+1} = P(q_m) < P(s_0) = s_0$ and by induction $q_n < s_0$ for all n. Thus $q = \lim_{n \to \infty} \le s_0$, which implies that q is the smallest positive root of (3.2).

For this, we consider the graph of $y = P(s)$ in $0 \le s \le 1$; it starts with the point $(0, p_0)$ and ends with the point $(1, 1)$; the curve lying entirely in the first quadrant, is convex as $P(s)$ is an increasing function. So the curve $y = P(s)$ can intersect the line $y = s$ in at most two points, one of which is the end point $(1, 1)$, i.e. the equation (3.1) has at most two roots, one of which is unity. Two cases are now to be considered (see Figs. 9.1 and 9.2).

Case I. The curve $y = P(s)$ lies entirely above the line $y = s$; in this case $(1, 1)$ is the only point of intersection, i.e. unity is the unique root of $s = P(s)$ so that $q = \lim_{n \to \infty} q_n = 1$. Then

$$P(1) - P(s) = 1 - P(s) \le 1 - s,$$

so that

$$\lim_{s \to 0} \frac{P(1) - P(s)}{1 - s} \le 1, \quad \text{i.e.} \quad P'(1) \le 1.$$

Thus

$$\lim_{n \to \infty} q_n = 1 \quad \text{when } P'(1) = m \le 1.$$

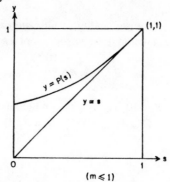

Fig. 9.1 Graphical determination of the roots of $s = P(s)$ [$m \leq 1$]

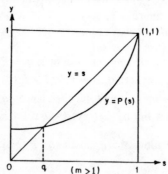

Fig.9.2 Graphical determination of the roots of $s = P(s)$ [$m > 1$]

Case II. The curve $y = P(s)$ intersects $y = s$ at another point $(\delta, P(\delta))$ such that $\delta = P(\delta)$, $\delta < 1$, i.e. there is another root of (3.1) namely $\delta < 1$; the curve $y = P(s)$, being convex, lies below the line $y = s$ in $(\delta, 1)$, and above $y = s$ in $(0, \delta)$, i.e. $P(s) < s$ in $\delta < s < 1$ and $P(s) > s$ in $0 < s < \delta$. Then $q_1 = P(0) < P(\delta) = \delta$ and assuming that $q_m < \delta$, we get $q_{m+1} = P(q_m) < P(\delta) = \delta$ and by induction $q_n < \delta$ for all n. Hence $\lim\limits_{n \to \infty} q_n = \delta$, so that $q = \delta < 1$.

Now by the mean value theorem considered in the interval $[\delta, 1]$, there is a value ξ in $\delta < \xi < 1$ such that $P'(\xi) = \frac{P(1) - P(\delta)}{1 - \delta} = 1$ and as the derivative is monotone $P'(1) > 1$.

Thus we find that q is the root less than unity of (3.1) when $m = P'(1) > 1$. We have thus proved the theorem completely.   ▲

Note: That q is a root of $s = P(s)$ can also be seen by conditioning on the number of offsprings of the first generation. We have

$$q = \Pr\{\text{ultimate extinction}\}$$

$$= \sum_{k=0}^{\infty} \Pr\{\text{ultimate extinction} \mid X_1 = k\} \cdot \Pr\{X_1 = k\}.$$

Given that $X_1 = k$, the population will be extinct *iff* each of the k families started by members of the 1st generation becomes extinct. It is assumed that families behave independently; hence

$$\text{Pr } \{\text{ultimate extinction} \mid X_1 = k\} = q^k.$$

Thus

$$q = \sum_{k=0}^{\infty} q^k \, p_k = P(q).$$

Theorem 9.4. Whatever be the value of $E(X_1) = m$, we have, as $n \to \infty$, $\lim \text{Pr } \{X_n = 0\} = q$ and $\lim \text{Pr } \{X_n = k\} = 0$, for any finite positive integral k.

Proof: We first show that $\lim_{n \to \infty} P_n(s) = q$, from which the above result will follow.

Consider the case $m \leq 1$, when $P(s) = s$ has the unique root $q = 1$. In $0 \leq s \leq q$, $P(s) \leq P(q) = q$, and $P_2(s) \leq P_2(q) = P(P_1(P)) = P(q) = q$. Assuming that $P_m(s) \leq q$, we get $P_{m+1}(s) \leq q$ and by induction $P_n(s) \leq q$ for all n. Again $P_n(s) \geq P_n(0) = q_n$; thus $q_n \leq P_n(s) \leq q$.

Hence

$$\lim_{n \to \infty} P_n(s) = q, \quad 0 \leq s \leq q.$$

Consider the case $m > 1$, when q is the root less than 1 of $P(x) = x$. In $q < s < 1$, the curve $y = P(x)$ lies below the line $y = x$, and $q < P(s) < s < 1$. Again $P_2(s) = P(P_1(s)) = P(P(s)) > P(q) = q$. Assuming that $P_m(s) > q$, we get $P_{m+1}(s) > q$, so that, by induction $P_n(s) > q$ for all n. Again $P_2(s) = P_1(P(s)) < P_1(s)$ and assuming that $P_m(s) < P_{m-1}(s)$, we get $P_n(s) < P_{n-1}(s)$ for all n.

Thus in

$$q < s < 1,$$

$$q < P_n(s) < P_{n-1}(s) < \ldots$$

so that

$$\lim_{n \to \infty} P_n(s) \geq q.$$

Suppose, if possible, that $\lim_{n \to \infty} P_n(s) = \alpha > q$, then $P(\alpha) < \alpha$, and $\lim_{n \to \infty} P_{n+1}(s) = \lim_{n \to \infty} P(P_n(s))$, and we get a contradiction which is due to our supposition that $\alpha > q$. Thus

$$\lim_{n \to \infty} P_n(s) = q.$$

So, whatever be the value of $E(X_1) = m$, $\lim_{n \to \infty} P_n(s) = q$ is independent of s (for all $s < 1$). In other words, for all $s < 1$,

$$\lim_{n \to \infty} \sum_{k=0}^{\infty} \text{Pr } \{X_n = k\} s^k = q.$$

This implies that the coefficients of s^k for $k \geq 1$ (except, possibly, for infinitely large k) all tend to 0, while the constant term tends to q.

Thus, as $n \to \infty$,

$$\text{Pr } \{X_n = k\} \to 0, \text{ for finite positive integral } k,$$

and

$$\text{Pr } \{X_n = 0\} \to q.$$

Since $P_n(1) = 1$, we have , as $n \to \infty$,

$$\text{Pr } \{X_n \to \infty\} \to 1 - q. \tag{3.3}$$

Note: The above result also follows from the general theory of Markov chains applied to the chain $\{X_n\}$, for which each of the states $k = 1, 2, 3, \ldots$ is transient while the state 0 is absorbing. We shall now consider another interesting result.

Theorem 9.5. We have, for $r, n = 0, 1, 2, \ldots$

$$E\{X_{n+r} \mid X_n\} = X_n m^r. \tag{3.4}$$

Proof : For $r = 1$ and $n = 0, 1, 2, \ldots,$ we have

$$E\{X_{n+1} \mid X_n\} = E\left\{\sum_{k=1}^{X_n} \zeta_k \mid X_n\right\} = \sum_{k=1}^{X_n} E\{\zeta_k\}$$

$$= mX_n.$$

Assume that (3.4) holds for $r = k$; then $E\{X_{n+k} \mid X_n\} = X_n m^k$. Noting the Markov nature of $\{X_n\}$, we get

$$E\{X_{n+k+1} \mid X_n\} = E[E\{X_{n+k+1} \mid X_{n+k}, X_{n+k-1}, \ldots, X_n\} \mid X_n]$$

$$= E[E\{X_{n+k+1} \mid X_{n+k}\} \mid X_n]$$

$$= E\{mX_{n+k} \mid X_n\} = m(X_n m^k\}$$

$$= X_n m^{k+1},$$

so that the result holds for $r = k + 1$.

Thus by induction, we have the result. ▲

9.3.1 Asymptotic Distribution of X_n

Another variable of interest is $W_n = X_n / m^n$, $n = 0, 1, 2, \ldots$; $\{W_n\}$ forms a Markov chain. We have $E\{W_n\} = 1$ and for $m > 1$,

$$E\{W_n^2\} = \frac{1}{m^{2n}} E(X_n^2) = \frac{1}{m^{2n}} \left\{ m^{2n} + \frac{m^{n-1}(m^n-1)\sigma^2}{(m-1)} \right\}$$

$$= 1 + \frac{\sigma^2}{m^2-m}(1 - m^{-n}).$$

Dividing both sides of (3.4) by m^{n+r}, we get

$$E\{W_{n+r} \mid W_n\} = W_n \tag{3.5}$$

and since $\{W_n\}$ is also a Markov chain,

$$E\{W_{n+r} \mid W_n, W_{n-1}, \ldots, W_0\} = E\{W_{n+r} \mid W_n\} = W_n. \tag{3.6}$$

It follows that $\{W_n, n \geq 0\}$ is also a martingale; further W_n, being non-negative, is a non-negative martingale.

Limiting distribution of X_n

One can now apply the martingale convergence theorem for the convergence of W_n. Thus we get that, with probability one,

$$\lim_{n \to \infty} W_n \quad \text{exists and is finite.}$$

Two cases arise:
 (i) $m > 1$:
then in order that W_n converges, X_n must go to ∞ at an exponentially fast rate of m (so that

$$\frac{X_n}{m^n} \to \text{a finite limit}).$$

 (ii) $m \leq 1$:
then as $m_n \to 0$, and W_n converges; we have, with probability one,

$$X_n \to 0 \quad \text{as} \quad n \to \infty.$$

This implies ultimate extinction in the subcritical and critical cases.

9.3.2 Examples

We now consider some simple examples; to fix the ideas numerical values have been taken.

Example 3(a). Let p_k, $k = 0, 1, 2$ be the probability that an individual in a generation generates k offsprings. Then $P(s) = p_0 + p_1 s + p_2 s^2$, and $P_2(s), P_3(s)$ can be calculated by simple algebra. The probability of extinction is one if $m \leq 1$; if $m > 1$, it is given by the root less than 1 of $s = P(s)$. Suppose that $p_0 = 2/3$, $p_1 = 1/6$, and $p_2 = 1/6$; then $m = 1/2 < 1$. The equation $s = P(s)$ becomes $s^2 - 5s + 4 = 0$ with roots 1 and 4; the probability of extinction is 1. Suppose that $p_0 = 1/4$, $p_1 = 1/4$, $p_2 = 1/2$; then $m = 5/4 > 1$; the equation $s = P(s)$ has the roots $1/2$ and 1. The root $1/2$ (smaller than unity) gives the probability of extinction. Note that the probability of extinction is p_0/p_2 or 1 according as $p_0 < p_2$ or $p_0 \geq p_2$ and also that $p_0 < $ (or $\geq$) p_2 *iff* $m > $ (or $\leq$) 1.

Example 3(b). Let the probability distribution of the number of offsprings generated by an individual in a generation be Poisson with mena λ, i.e. $P(s) = e^{\lambda(s-1)}$. It can be easily seen that the graph of $P(s)$ in $0 \leq s \leq 1$ (i.e. between the points $(0, e^{-\lambda})$ and $(1, 1)$) is convex, and that the curve of $y = P(s)$ always lies above $y = s$ when $\lambda \leq 1$, there being no other root of $s = P(s)$ except unity in $(0, 1)$; the probability of extinction is then 1. When $\lambda > 1$, the curve $y = P(s)$ intersects $y = s$ in another point whose s-coordinate has a value < 1 and the probability of extinction will be this value of s. For example, if $\lambda = 2$, it can be seen that $s = e^{2(s-1)}$ has a root approximately equal to 0.2 which is smaller than 1, and the probability of extinction is $q \approx 0.2$.

Example 3(c). Let the distribution of the number of offsprings be geometric with $p_k = b(1-b)^k$, $k = 0, 1, 2, \ldots$ ($0 < b < 1$). Then $m = (1-b)/b$ and $P(s) = b/(1-s(1-b))$. The equation $s = P(s)$ has the roots 1 and $b/(1-b)$. If $m \leq 1$, then the probability of extinction is 1; if $m > 1$, the root $b/(1-b) < 1$, and the probability of extinction is equal to the root $b/(1-b)$.

Example 3(d). Let $p_k = b\,c^{k-1}$, $k = 1, 2, \ldots$, $0 < b, c, b + c < 1$ and $p_0 = 1 - \sum_{k=1}^{\infty} p_k$.

Then $m = b/(1-c)^2$. We have

$$P(s) = 1 - \frac{b}{1-c} + \frac{bs}{1-cs}. \tag{3.7}$$

The quadratic equation $s = P(s)$ has the roots

$$1 \quad \text{and} \quad \frac{1-(b+c)}{c(1-c)} = s_0 \text{ (say)}.$$

If $m = 1$, then $s_0 = 1$ and the probability of extinction is 1; if $m > 1$, $s_0 < 1$, and the probability of extinction is $q = s_0$ (< 1).

This model was applied in a series of interesting papers by Lotka to find out the probability of extinction for American male lines of descent. The values estimated by him (in 1939) from census figures of 1920 give $b = 0.2126$, $c = 0.5893$ ($m \approx 1.25 > 1$) and the probability of extinction $q = s_0 = 0.819$.

Note: It is not always possible to put the generating functions $P_n(s)$ in closed form. The generating functions $P(s)$ obtained in Examples 3(c) and 3(d) are of interesting forms: they may be considered as particular cases of the more general fractional linear form (or general bilinear form)

$$P(s) = \frac{\alpha + \beta s}{\gamma + \delta s}, \quad \alpha\delta - \beta\gamma \neq 0.$$

When $P(s)$ is the above form, $P_n(s)$ is also of the same form

$$P_n(s) = \frac{\alpha_n + \beta_n s}{\gamma_n + \delta_n s}$$

where $\alpha_n, \beta_n, \gamma_n, \delta_n$ are functions of $\alpha, \beta, \gamma, \delta$.

Further, it may be noted that the equation $s = P(s)$ (where $P(s)$ is of fractional linear form) has two finite solutions 1 and s_0 and that $s_0 <, =, >$ or 1 according as $m = P'(1) >, =,$ or < 1.

Example 3(e). $P_n(s)$ for Lotka's model considered in Example 3(d) above. For any two points u, v, we get

$$\frac{P(s) - P(u)}{P(s) - P(v)} = \frac{s - u}{s - v} \frac{1 - cv}{1 - cu}.$$

Put $u = s_0$, $v = 1$, then $P(s_0) = s_0$, $P(v) = 1$, so that

$$\frac{P(s) - s_0}{P(s) - 1} = \frac{s - s_0}{s - 1} \frac{1 - c}{1 - cs_0}$$

whence

$$\frac{1 - c}{1 - cs_0} = \left\{\frac{P(s) - s_0}{s - s_0}\right\} \Big/ \left\{\frac{P(s) - 1}{(s - 1)}\right\}.$$

Let $m \neq 1$, then taking limits of the right hand side as $s \to 1$, we get

$$\frac{1 - c}{1 - cs_0} = \frac{1}{m}.$$

Hence

$$\frac{P(s) - s_0}{P(s) - 1} = \left(\frac{1}{m}\right)\frac{s - s_0}{s - 1}.$$

Thus

$$\frac{P_2(s) - s_0}{P_2(s) - 1} = \frac{P(P(s)) - s_0}{P(P(s)) - 1} = \left(\frac{1}{m}\right)\frac{P(s) - s_0}{P(s) - 1}$$

$$= \left(\frac{1}{m^2}\right)\frac{s - s_0}{s - 1}$$

and on iteration

$$\frac{P_n(s) - s_0}{P_n(s) - 1} = \left(\frac{1}{m^n}\right)\frac{s - s_0}{s - 1}, \quad n = 1, 2, \ldots.$$

Solving for $P_n(s)$, we get

$$P_n(s) = 1 - m^n\left(\frac{1 - s_0}{m^n - s_0}\right) + \frac{m^n\left(\frac{1 - s_0}{m^n - s_0}\right)^2 s}{1 - \left(\frac{m^n - 1}{m^n - s_0}\right)s}, \quad m \neq 1. \tag{3.8}$$

If $m = 1$, then $s_0 = 1$, and $P(s) = \dfrac{c + (1 - 2c)s}{1 - cs}$

and

$$P_n(s) = \frac{nc + (1 - c - nc)s}{(1 - c + nc) - ncs}. \tag{3.9}$$

Limiting Results

Suppose that $m = 1$, then $\sigma^2 = 2c/(1 - c)$,

$$n \Pr\{X_n > 0\} = n\{1 - P_n(0)\} = \frac{n(1 - c)}{1 - c + nc}$$

and

$$\lim_{n \to \infty} n \Pr\{X_n > 0\} = \frac{1 - c}{c} = \frac{2}{\sigma^2} \quad \text{(see Theorem 9.8(a)).}$$

Suppose that $m < 1$, then $s_0 > 1$ and

$$\lim_{n \to \infty} m^{-n}\Pr\{X_n > 0\} = \lim_{n \to \infty} m^{-n}\{1 - P_n(0)\}$$

$$= \lim_{n \to \infty} \frac{1 - s_0}{m^n - s_0} = \frac{s_0 - 1}{s_0}.$$

Again

$$\sum_k \Pr\{X_n = k \mid X_n > 0\}s^k = \frac{P_n(s) - P_n(0)}{1 - P_n(0)}$$

$$= 1 - \frac{1 - P_n(s)}{1 - P_n(0)}$$

$$= \frac{\left(\frac{1-s_0}{m^n - s_0}\right)s}{1 - \left(\frac{m^n - 1}{m^n - s_0}\right)s}.$$

Thus $\qquad \lim_{n \to \infty} \sum_k \Pr\{X_n = k \mid X_n > 0\}s^k = s\left(1 - \frac{1}{s_0}\right)\left(1 - \frac{s}{s_0}\right)^{-1}$

and $\qquad \lim_{n \to \infty} \Pr\{X_n = k \mid X_n > 0\} = \left(1 - \frac{1}{s_0}\right)\left(\frac{1}{s_0}\right)^{k-1}, \quad k = 1, 2, \dots.$

In other words, for large n, the distribution of $\{X_n\}$, given $X_n > 0$, is geometric with mean $s_0/(s_0 - 1)$ and p.g.f.

$$b(s) = \frac{(1 - 1/s_0)s}{1 - s/s_0}.$$

It can be easily verified that $b(s)$ satisfies the equation

$$b(P(s)) = mb(s) + 1 - m \quad \text{(see Theorem 9.9).} \qquad \blacktriangle$$

9.4 DISTRIBUTION OF THE TOTAL NUMBER OF PROGENY

Let X_n denote the size of nth generation, $n = 0, 1, 2, \dots$, and $X_0 = 1$. Then the random variable

$$Y_n = \sum_{k=0}^{n} X_k = 1 + X_1 + \dots + X_n \qquad (4.1)$$

denotes the total number of progeny, i.e. the number of descendants upto and including the nth generation and also including the ancestor.

Theorem 9.6. The p.g.f. $R_n(s)$ of Y_n satisfies the recurrence relation

$$R_n(s) = sP(R_{n-1}(s)), \qquad (4.2)$$

$P(s)$ being the p.g.f. of the offspring distribution.

Proof: Let $Z_n = X_1 + \dots + X_n$ and $G_n(s)$ be its p.g.f. Then $R_n(s) = s\, G_n(s)$. We have

$$G_n(s) = \sum_{k=0}^{\infty} \Pr\{Z_n = k\}s^k.$$

Now by conditioning on the size X_1 of the 1st generation, we get

$$\Pr\{Z_n = k\} = \sum_{i=0}^{\infty} \Pr\ \{\text{total number of descendants in the succeeding } (n-1)$$

generations following the 1st is $k - i / X_1 = i\}$.$\Pr\{X_1 = i\}$.

If the process starts with one ancestor then the probability of having r descendants in succeeding m generations is the coefficient of s^r in $G_m(s)$; and if it starts with i ancestors then the probability of having r descendants in the succeeding m generations will be the coefficient of s^r in $[G_m(s)]^i$. Thus

$$\Pr\{Z_n = k\} = \sum_{i=0}^{\infty} [\text{coefficient of } s^{k-i} \text{in } \{G_{n-1}(s)\}^i] p_i$$

$$= \sum_{i=0}^{\infty} p_i [\text{coefficient of } s^k \text{in } \{sG_{n-1}(s)\}^i]$$

$$= \text{coefficient of } s^k \text{ in } \sum_{i=0}^{\infty} p_i \{sG_{n-1}(s)\}^i$$

$$= \text{coefficient of } s^k \text{ in } P(sG_{n-1}(s)).$$

Thus
$$G_n(s) = \sum_{k=0}^{\infty} \Pr\{Z_n = k\} s^k = P(sG_{n-1}(s)),$$

whence
$$R_n(s) = sG_n(s) = sP(R_{n-1}(s)).$$

Hence the theorem.   ▲

From the recurrence relation (4.2) it is theoretically possible to calculate $R_1(s)$, $R_2(s)$, ... We are however interested in the asymptotic behaviour of $R_n(s)$ for large n.

Theorem 9.7. We have

$$\lim_{n \to \infty} R_n(s) = H(s)$$

where $H(s) = \sum_{k=0}^{\infty} \rho_k s^k$ is the generating function of the sequence of non-negative

numbers ρ_k.

The function $H(s)$ satisfies the relation

$$H(s) = sP(H(s)), \quad 0 < s < 1; \tag{4.3}$$

further $H(s)$ is the unique root of the equation

$$t = sP(t) \tag{4.4}$$

such that $H(s) \le \xi$, where ξ is the smallest positive root of $x = P(x)$ and that

$$H(1) = \Sigma \, \rho_k = \xi.$$

Proof: We have, for $0 < s < 1$,

$$R_2(s) = sP(R_1(s)) < sP(s) = R_1(s)$$

and assuming that $R_m < R_{m-1}(s)$, we get

$$R_{m+1}(s) = sP(R_m(s)) < sP(R_{m-1}(s)) = R_m(s)$$

and hence by induction $R_n(s) < R_{n-1}(s)$ for all $n > 0$. Thus for $s < 1$, $\{R_n(s)\}$ is a monotone decreasing sequence bounded below. Hence $\lim_{n \to \infty} R_n(s) = H(s)$ exists. From the continuity theorem of p.g.f.'s, it follows that $H(s)$, being the limit of a sequence of p.g.f.'s, is the generating function of a sequence of non-negative numbers ρ_k such that $H(1) = \Sigma \, \rho_k \le 1$.

Taking limit of (4.2), we get

$$H(s) = sP(H(s)), \qquad 0 < s < 1,$$

i.e. for some fixed s (in $0 < s < 1$), $H(s)$ is a root of the equation

$$t = sP(t).$$

For fixed $s < 1$, $y = sP(t)$ is a convex function of t and the graph of $y = sP(t)$ intersects the line $y = t$ in at most two points. Let ξ be the smallest positive root of $x = P(x)$; clearly $\xi \le 1$. The function $t - sP(t)$ is negative for $t = 0$ and positive for $t = \xi$, and remains positive for values of t between ξ and 1. Thus $t = sP(t)$ has exactly one root between 0 and ξ and no root between ξ and 1. The unique root of $t = sP(t)$ equals $H(s)$ and thus $H(s) < \xi$. Clearly $H(1)$ is a root of $x = P(x)$ and since ξ is the smallest root of this equation, $H(1) = \xi$. Thus the theorem is completely established.   ▲

Note: $H(1) = 1$ or < 1 depending on whether $1 + X_1 + X_2 + ..$ is finite or not (with probability one); and $H(1) = 1$ whenever $m \le 1$.

9.5 CONDITIONAL LIMIT LAWS

9.5.1 Critical Processes

In Example 3(e) we obtained some limiting results. We shall obtain here some general results. Consider a critical (i.e. with $m = 1$) G.W. process. The probability of extinction is 1. Thus from Theorem 9.4, we get $\Pr\{X_n \to 0\} = 1$. We also have $\text{var}(X_n) = n\,\sigma^2 \to \infty$. The distribution of X_n, given that $X_n > 0$, is of considerable interest.

Lemma. For a G.W. process with $m = 1$ and $\sigma^2 < \infty$, we have

$$\lim_{n \to \infty} \frac{1}{n} \left\{ \frac{1}{1 - P_n(s)} - \frac{1}{1-s} \right\} \to \frac{\sigma^2}{2} \tag{5.1}$$

uniformly in $0 \le s < 1$.

Proof: Let $0 \le s < 1$ and $P'''(1) < \infty$. Using Taylor's expansion of $P(s)$ in the neighbourhood of 1, we get

$$P(s) = s + \frac{\sigma^2}{2}(1-s)^2 + r(s)(1-s)^2, \tag{5.2}$$

where $r(s) \to 0$ as $s \to 1$.

Thus

$$\frac{1}{1 - P(s)} - \frac{1}{1-s} = \frac{P(s) - s}{(1-s)(P(s) - s)}$$

$$= \frac{1-s}{1 - P(s)} \left\{ \frac{\sigma^2}{2} + r(s) \right\}$$

$$= \left\{ \frac{\sigma^2}{2} + r(s) \right\} \left[1 - (1-s) \left\{ \frac{\sigma^2}{2} + r(s) \right\} \right]^{-1}$$

$$= \frac{\sigma^2}{2} + R(s), \tag{5.3}$$

where $R(s) \to 0$ as $s \to 1$, and R is bounded. Again using (4.6), we get

$$P_2(s) = P(P(s)) = P(s) + \frac{\sigma^2}{2}(1 - P(s))^2 + r(P(s))(1 - P(s))^2$$

so that

$$\frac{1}{1 - P_2(s)} - \frac{1}{1 - P(s)} = \frac{\sigma^2}{2} + R(P(s)) \tag{5.4}$$

and

$$\frac{1}{2} \left\{ \frac{1}{1 - P_2(s)} - \frac{1}{1-s} \right\} = \frac{\sigma^2}{2} + \frac{1}{2} \{ R(s) + R(P(s)) \}.$$

Iterating one gets

$$\frac{1}{n} \left\{ \frac{1}{1 - P_n(s)} - \frac{1}{1-s} \right\} = \frac{\sigma^2}{2} + \frac{1}{n} \sum_{k=0}^{n-1} R(P_k(s)).$$

Since $P_n(0) \le P_n(s) \le 1$ and $P_n(0) \to 1$ from the left, the convergence of $P_n(s) \to 1$ is uniform. Hence the lemma.   ▲

We shall now use the lemma to establish the following interesting limit laws (of which (a) is due to Kolmogorov and (b), (c) are due to Yaglom).

Theorem 9.8. If $m = 1$, $\sigma^2 < \infty$,
then

(a) $\lim\limits_{n \to \infty} n \Pr \{X_n > 0\} = \dfrac{2}{\sigma^2}$

(b) $\lim\limits_{n \to \infty} E\left\{\dfrac{X_n}{n} \Big| X_n > 0\right\} = \dfrac{\sigma^2}{2}$

(c) $\lim\limits_{n \to \infty} \Pr\left\{\dfrac{X_n}{n} > u \mid X_n > 0\right\} = \exp\left(-\dfrac{2u}{\sigma^2}\right)$, $u \geq 0$.

Proof: (a) $n \Pr \{X_n > 0\} = n \{1 - P_n(0)\}$

$$= \left[\frac{1}{n}\left\{\frac{1}{1 - P_n(0)} - 1\right\} + \frac{1}{n}\right]^{-1}.$$

Thus from the lemma (taking $s = 0$), we get

$$\lim_{n \to \infty} n \Pr \{X_n > 0\} = \lim_{n \to \infty}\left[\frac{\sigma^2}{2} + \frac{1}{n}\right]^{-1} = \frac{2}{\sigma^2}.$$

(b) We have

$$1 = E\{X_n\} = E\{X_n \mid X_n > 0\} \cdot \Pr \{X_n > 0\} + 0 \cdot \Pr(X_n = 0)$$

so that

$$E\{X_n \mid X_n > 0\} = \frac{1}{\Pr\{X_n > 0\}} = \frac{n\sigma^2}{2}, \text{ (from (a))}.$$

Thus $\lim\limits_{n \to \infty} E\left\{\dfrac{X_n}{n} \Big| X_n > 0\right\} = \dfrac{\sigma^2}{2}.$

(c) Let $u > 0$, and

$$dF(u) = \Pr\left\{u \leq \frac{X_n}{n} < u + du \mid X_n > 0\right\};$$ (5.5)

then taking L.T. (see equation (3.1a) Ch. 1), we get

$$\int_0^\infty \exp(-\alpha u)\, dF(u) = E\left\{\exp\left(-\frac{X_n}{n}\right) \Big| X_n > 0\right\}.$$ (5.6)

Now $\quad E\left\{\exp\left(-\dfrac{X_n}{n}\right)\right\} = E\left\{\exp\left(-\dfrac{X_n}{n}\right)\Big| X_n > 0\right\}.\Pr\{X_n > 0\} + 1.\Pr\{X_n = 0\}$

and since $\quad P_n(s) = E\{s^{X_n}\}$ is the p.g.f. of X_n, we get

$$P_n\left(\exp\left(-\frac{\alpha}{n}\right)\right) = E\left\{\exp\left(-\frac{\alpha X_n}{n}\right)\Big| X_n > 0\right\}\{1 - P_n(0)\} + P_n(0).$$

Thus

$$E\left\{\exp\left(-\frac{\alpha X_n}{n}\right)\Big| X_n > 0\right\} = \frac{P_n(\exp(-\alpha/n)) - P_n(0)}{1 - P_n(0)}$$

$$= 1 - \frac{1 - P_n(\exp(-\alpha/n))}{1 - P_n(0)}. \tag{5.7}$$

Now as $n \to \infty$

$$n\left\{1 - P_n\left(\exp\left(-\frac{\alpha}{n}\right)\right)\right\} \to \frac{2}{\sigma^2} \text{ (from (a))}$$

and from the basic lemma (because of uniform convergence), we get

$$\frac{1}{n\{1 - P_n(\exp(-\alpha/n))\}} = \frac{1}{n}\left\{\frac{1}{1 - P_n(\exp(-\alpha/n))} - \frac{1}{1 - \exp(-\alpha/n)}\right\}$$

$$+ \frac{1/n}{1 - \exp(-\alpha/n)} \to \frac{\sigma^2}{2} + \frac{1}{\alpha}.$$

Thus from (5.6) and (5.7), we get, as $n \to \infty$

$$\int_0^\infty \exp(-\alpha u)\,dF(u) \to 1 - \frac{\sigma^2/2}{\sigma^2/2 + 1/\alpha}.$$

$$= \frac{1}{1 + \alpha\sigma^2/2}.$$

Since L.T. of $\dfrac{2}{\sigma^2}\exp\left(-\dfrac{2u}{\sigma^2}\right)$ is $\dfrac{1}{1 + \alpha\sigma^2/2}$,

we have

$$\lim_{n \to \infty} \Pr\left\{ u \le \frac{X_n}{n} < u + du \mid X_n > 0 \right\} = \frac{2}{\sigma^2} \exp\left(-\frac{2u}{\sigma^2}\right),$$

which establishes the exponential limit law.   ▲

9.5.2 Subcritical Processes

Theorem 9.9. (*Yaglom's Theorem*). For a Galton-Watson process with $m < 1$,

$$\lim_{n \to \infty} \Pr\{X_n = j \mid X_n > 0\} = b_j, \quad j = 1, 2, \ldots \tag{5.8}$$

exists, and $\{b_j\}$ gives a probability distribution whose p.g.f.

$$B(s) = \sum_{j=1}^{\infty} b_j s^j \text{ satisfies the equation}$$

$$B(P(s)) = m B(s) + 1 - m \tag{5.9}$$

i.e, $$1 - B(P(s)) = m(1 - B(s)). \tag{5.9a}$$

Further, $\sum_{j=1}^{\infty} j b_j = 1/\phi(0)$, where $\phi(0) = \lim_{n \to \infty} \Pr\{X_n > 0\}/m^n$.

Proof: Using Taylor's expansion around $s = 1$, we get

$$P(s) = 1 - m(1-s) + (1-s) r(s), \ 0 \le s \le 1 \tag{5.10}$$

or, $$\frac{1 - P(s)}{1 - s} = m - r(s). \tag{5.10a}$$

Consider the function $r(s)$ in $0 \le s \le 1$; we have $r(0) = m - (1 - p_0) \ge 0$ and $\lim_{s \to 1-0} r(s) = 0$. Further as $P(s)$ is a convex function

$$P'(s) \le \frac{1 - P(s)}{1 - s},$$

so that $$r'(s) = (1-s)^{-1}\left\{ \frac{1 - P(s)}{1 - s} - P'(s) \right\} \le 0.$$

Thus $r(s)$ is monotone decreasing, is bounded above by m and $r(s) \to 0$ as $s \to 1$. Replacing s by $P_{k-1}(s)$ in (5.10a), we get

$$\frac{1-P_k(s)}{1-P_{k-1}(s)} = m\{1 - r(P_{k-1}(s))/m\}. \tag{5.11}$$

Putting $k = 1, 2, 3, \ldots, n$ and taking products of both sides, we get

$$\frac{1-P_n(s)}{1-s} = m^n \prod_{k=0}^{n-1} \{1 - r(P_k(s))/m\}. \tag{5.12}$$

Since $0 \le r/m \le 1$, the sequence $\left\{ \frac{1-P_n(s)}{m^n(1-s)} \right\}$ is monotone decreasing in n, and we have

$$\lim_{n \to \infty} \frac{1-P_n(s)}{m^n(1-s)} = \phi(s) \ge 0.$$

Putting $s = 0$, we get

$$\phi(0) = \lim_{n \to \infty} \frac{1-P_n(0)}{m^n} = \lim_{n \to \infty} \frac{\Pr\{X_n > 0\}}{m^n}. \tag{5.13}$$

Let $b_{jn} = \Pr\{X_n = j \mid X_n > 0\}$ and $B_n(s) = \sum_{j=1}^{\infty} b_{jn} s^j$ be the p.g.f. of $\{b_{jn}\}$. Then

$$B_n(s) = \frac{P_n(s) - P_n(0)}{1 - P_n(0)} = 1 - \frac{1 - P_n(s)}{1 - P_n(0)} \tag{5.14}$$

$$= 1 - (1-s) \prod_{k=0}^{n-1} \frac{1 - r(P_k(s))/m}{1 - r(P_k(0))/m} \quad \text{(from (5.12))}.$$

Since $P_k(s) \ge P_k(0)$, and $r(s)$ is monotone decreasing, $r((P)_k(s)) \le r(P_k(0))$, and so each factor of the product on the r.h.s. expression is larger than 1. Thus $B_n(s)$ is monotone decreasing and tends to a limit $B(s)$ as $n \to \infty$, i.e.

$$B(s) = \sum_{j=1}^{\infty} b_j s^j, \text{ where } b_j = \lim_{n \to \infty} b_{jn} = \lim_{n \to \infty} \Pr\{X_n = j \mid X_n > 0\}.$$

Clearly $B(0) = 0$.

Now,

$$B(P_k(0)) = \lim_{n \to \infty} B_n(P_k(0))$$

$$= \lim_{n \to \infty} \left\{ 1 - \frac{1 - P_n(P_k(0))}{1 - P_n(0)} \right\}$$

$$= 1 - \lim_{n \to \infty} \left\{ \frac{1 - P_k(P_n(0))}{1 - P_n(0)} \right\}$$

$$= 1 - \lim_{s \to 1} \frac{1 - P_k(s)}{1 - s}$$

(since $m < 1, P_n(0) \to 1$ as $n \to \infty$)

$$= 1 - m^k$$

(taking limit of (5.12) as $s \to 1$).

It follows that $\lim B\,(P_k(0)) \to 1$. As $m < 1, P_k\,(0) \to 1$ as $k \to \infty$
Hence $B\,(s) \to 1$ as $s \to 1$. Thus $B\,(s)$ is the p.g.f. of $\{b_j\}$.

Further

$$\sum_{j=1}^{\infty} jb_j = B'(1) = \lim_{k \to \infty} \frac{1 - B\,(P_k(0))}{1 - P_k(0)}$$

$$= \lim_{k \to \infty} \frac{m^k}{1 - P_k(0)}$$

$$= \lim_{k \to \infty} \frac{m^k}{\Pr\{X_k > 0\}} \to \frac{1}{\phi(0)}.$$

Again, from (5.14)

$$B_n(P(s)) = 1 - \frac{1 - P_n(P(s))}{1 - P_n(0)}$$

$$= 1 - \frac{1 - P_{n+1}(s)}{1 - P_{n+1}(0)} \cdot \frac{1 - P_{n+1}(0)}{1 - P_n(0)} \qquad (5.15)$$

$$\lim_{n \to \infty} \frac{1 - P_{n+1}(s)}{1 - P_{n+1}(0)} = \lim_{n \to \infty} (1 - B_{n+1}(s)) = 1 - B(s),$$

$$\lim_{n \to \infty} \frac{1 - P_{n+1}(0)}{1 - P_n(0)} = \lim_{s \to 1} \frac{1 - P(s)}{1 - s} = m.$$

Hence taking limits of both sides of (5.15), we get

$$B(P(s)) = 1 - (1 - B(s))m$$

$$= mB(s) + 1 - m.$$

Thus the theorem is proved.   ▲

Remarks: **1.** The above simplified proof which does not involve moment restriction is due to Joffe (1967).

The equation (5.9a) is known as a modified *Schröder functional equation*. It is not always easy to obtain $B(s)$ from it for given $P(s)$. In example 3(e) we obtained $B(s)$ directly.

The mean $\sum\limits_{k=1}^{\infty} k b_k = 1/\phi(0)$ of the limiting distribution is finite *iff*

$$E\{X_1 \log X_1\} = \sum_{k=1}^{\infty} P_k\{k \log k\} < \infty \quad \text{or} \quad p_0 = 1.$$

2. The limiting behaviour which has customarily been studied through probability generating functions and their functional iterates, has now been studied also through the martingale convergence theorem; the latter is more revealing on the nature of the process. See, for example, Heyde (1970) and Grey (1980).

Before closing the discussion on G.W. processes we mention a few interesting innovations introduced in the basic process.

9.6 GENERALISATIONS OF THE CLASSICAL GALTON-WATSON PROCESS

9.6.1 Branching Processes with Immigration

Theorem 9.4 states that for a G.W. process, $\Pr\{X_n \to 0\} = q$ and $\Pr\{X_n \to k\} = 0$, for finite k (and so $\Pr\{X_n \to \infty\} = 1 - q$), q being the probability of extinction. Further, $q = 1$ for critical and subcritical processes. Thus left to themselves G.W. populations either die out or grow without limits. Immigration from outside into a critical or subcritical process could have stabilising effect on the population size. Apart from this aspect, immigration by itself is interesting from the point of view of theory and applications. Galton-Watson processes with immigration (briefly called *G.W.I. processes*) often arise in applications in such areas as traffic theory, statistical machanics, genetics, neurophysiology etc.

Consider a G.W. process with offspring distribution $\{p_k\}$ (having p.g.f. $P(s)$ and mean $P'(1) = m$). (The process will be called underlying G.W. process.) Suppose that at time n, i.e. at the time of birth of nth generation there is an immigration of Y_n objects into the population, and that Y_n, $n = 0, 1, 2, \ldots$ are i.i.d random variables with p.g.f.

$$h(s) = \sum_{j=0}^{\infty} \Pr\{Y_n = j\} s^j = \sum h_j s^j,$$

i.e. with probability h_j, j immigrants enter the nth generation and contribute to the next generation in the same way as others already present do. The numbers of immigrants into successive generations are independent and all objects reproduce independently of each other and of the immigration process. The distribution $\{h_j\}$ will be called immigrant distribution. Let $a = h'(1)$ be the mean of this distribution. Let $X_{(n)}$ be the number of objects at the n th generation and let

$$P_{(n)}(s) = \sum_{j=0}^{\infty} \Pr\{X_{(n)} = j\}\, s^j$$

be its p.g.f. The sequence $\{X_{(n)}, n = 0, 1, 2, \ldots\}$ defines a G.W.I process. The sequence is a Markov chain whose one-step transition probabilities are given by

$$p_{ij} = \text{coeff. of } s^j \text{ in } h(s)\, [P(s)]^i, \quad i, j \in N.$$

Clearly,
$$P_{(n)}(s) = h(s)\, P_{(n-1)}\, (P(s)). \tag{5.16}$$

If
$$\lim_{n \to \infty} P_{(n)}(s) = F(s) \text{ exists, then one gets}$$

$$F(s) = h(s)\, F\, (P(s)) \tag{5.17}$$

i.e. the limit, when it exists, satisfies the above functional equation. Now the question arises: when does the limit exist, and when does it define the p.g.f. of a proper probability distribution or when does $\{X_{(n)}\}$ have a proper limit distribution?

9.6.1.1 Subcritical G.W.I. process

Corresponding to Yaglom's result for subcritical G.W. processes, Heathcote (1965) obtained the following analogue for G.W.I. processes.

Consider a Galton-Watson process with immigration, having offspring distribution $\{p_j\}$ (with p.g.f. $P(s)$) and immigrant distribution $\{h_j\}$ (with p.g.f. $h(s)$). If $m < 1$, $a = h'(1) < \infty$, then, $\lim_{n \to \infty} \Pr\{X_{(n)} = j\} = d_j$, exists for $j \in N$, and the p.g.f. $F(s) = \sum_j d_j\, s^j$ satisfies the functional equation (5.17).

Further,

$$F(1) = 1 \text{ iff } \sum_{j=1}^{\infty} h_j \log j < \infty.$$

It may be noted that, when

$$F(s) = \frac{1 - B(s)}{1 - s}$$

and
$$h(s) = \frac{1 - P(s)}{m(1 - s)}$$

then the functional equation (eqn. (5.9)) which appears in Yaglom's theorem

$$B(P(s)) = mB(s) + 1 - m,$$

reduces to the functional equation (5.17). Thus, ordinary subcritical G.W. processes may always be regarded as forming a subclass of G.W.I. processes (at least as regards Yaglom's conditional limit theorem).

The result for the critical process, due to Seneta (1970), may be stated as follows.

9.6.1.2 *Limiting Distribution of* $X_{(n)} / n$ *(Critical G.W.I. Process)*

For a Galton-Watson process with immigration having $m = 1$, $\sigma^2 < \infty$, $a < \infty$, the random variable $X_{(n)} / n$ converges in distribution, as $n \to \infty$, to a random variable having gamma density,

$$\frac{u^{\alpha-1} e^{-u/\beta}}{\Gamma(\alpha) \, \beta^{\alpha}}, \quad u \ge 0,$$

where $\qquad \alpha = 2a/\sigma^2, \ \beta = \sigma^2/2.$

If $a = h'(1) = \sigma^2/2$, i.e. $\alpha = 1$, then the gamma density reduces to an exponential density, such that

$$\lim_{n \to \infty} \ \Pr\left\{\frac{X_{(n)}}{n} > u\right\} \to \exp\left\{-\frac{2u}{\sigma^2}\right\};$$

again from Theorem 9.8(c) we get

$$\lim_{n \to \infty} \ \left\{\frac{X_{(n)}}{n} > u \mid X_n > 0\right\} \to \exp\left\{-\frac{2u}{\sigma^2}\right\}, \quad u \ge 0,$$

for the G.W. process $\{X_{(n)}\}$ without immigration.

Thus, in the critical case the effect of the conditioning of non-extinction $\{X_n > 0\}$ is the same as immigrating into the process at an average rate of $a = \sigma^2 / 2$.

Detailed studies on processes with immigration have been made by several authors (e.g. Heathcote, Seneta, Foster, Yang etc.). In Jagers' book (1975) as well as in Sankaranarayanan's (1989, Ch. 6), one can find treatment of some of them besides proof etc. of the above results.

9.6.2 Processes in Varying and Random Environments

It has been assumed that the offspring distribution $\{p_k\}$ remains the same in all the generations and that objects reproduce independently of others. It was Good who first pointed out the need to consider G.W. processes under varying environments. The assumption that the distribution $\{p_k\}$ remains the same in all the generations, was then replaced by the assumption that offspring distribution for the n th generation is of the form $\{p_{nk}\}$; several interesting results on processes in *varying environments* were obtained, among others, by Fearn (1972) and Jagers (1974) etc.

Smith and Wilkinson (1969) introduced yet another aspect. They postulated that the offspring distribution for each generation is *randomly chosen* from a class of all reproduction laws and the resulting processes are said to be in *random environments*. The topic has been pursued in great details in a number of papers by Wilkinson (1969), Athreya and Karlin (1970, 1971), Kaplan (1972) and others. See also Sankaranarayanan (1989, Ch. 5).

The models under these assumptions have wider scope for applications.

9.6.3 Multitype Galton-Watson Process

A natural extension of Galton-Watson process is concerned with the case where the population consists of a finite number of types of objects. Such processes are known as *multi-type Galton-Watson processes*. Mode (1971) deals exclusively and extensively with such processes, and discusses several models for applications in various areas.

Suppose that a population of individuals (or objects) originates with a single ancestor and that there are k (finite) types of individuals. Let $p^{(i)}(r_1, r_2, \ldots, r_k)$ be the probability that an individual of type i produces r_j offsprings of type j, $j = 1, 2, \ldots, k$. We introduce the vector notation:

$$\mathbf{r} = (r_1, r_2, \ldots, r_k),$$

$$\mathbf{s} = (s_1, s_2, \ldots, s_k),$$

$$\mathbf{e}_i = (\delta_{1i}, \delta_{2i}, \ldots, \delta_{ki}).$$

Let $f^{(i)}(\mathbf{s}) = \sum p^{(i)}(\mathbf{r}) s^{r_1} \ldots s^{r_k}$ be the p.g.f. of $\{p^{(i)}(\mathbf{r})\}$. Let

$$\mathbf{X}_n = (X_n^{(1)}, X_n^{(2)}, \ldots, X_n^{(k)})$$

represent the population size of k types in the nth generation.

Let m_{ij} be the expected number of offspring of type j produced by an object of type i. Then

$$m_{ij} = E\{X_1^{(j)} \mid \mathbf{X}_0 = \mathbf{e}_i\}$$

$$= \frac{\partial f^{(i)}}{\partial s_j}(1, 1, \ldots, 1), \quad i, j = 1, 2, \ldots, k.$$

The matrix $M = (m_{ij})$ is the matrix of moments. Let $f_n^{(i)}(\mathbf{s})$ denote the p.g.f. of the distribution of the number of objects in n th generation starting from one object of type i. A result analogous to (2.5) is as follows:

$$f_n^{(i)}(\mathbf{s}) = f^{(i)}(f_{n-1}^{(1)}(\mathbf{s}), f_{n-1}^{(2)}(\mathbf{s}) \ldots, f_{n-1}^{(k)}(\mathbf{s})),$$

$$f_0^{(i)}(\mathbf{s}) = s_i, \quad i = 1, 2, \ldots, k; \quad n = 0, 1, 2, \ldots.$$

The population becomes extinct when $\mathbf{X}_n = 0$ for some $n \geq 1$. Let

$$q_n^{(i)} = \Pr\{\mathbf{X}_n = 0 \mid \mathbf{X}_0 = \mathbf{e}_i\}$$

and
$$q^{(i)} = \lim_{n \to \infty} q_n^{(i)};$$

i.e., $q^{(i)}$ denotes the probability of extinction given that the process started with one ancestor of type i. Let $q = (q^{(i)}, q^{(2)}, \ldots, q^{(k)})$.

We confine ourselves to the case where the matrix of moments is of a certain type.

Let M be a $k \times k$ matrix with non-negative elements such that for some positive integer n, all the elements of M^n are strictly positive. Then there exists a positive eigenvalue ρ of M, which is greater than the absolute value of any other eigenvalue of M; ρ is also called the spectral radius of M.

When M is the matrix (m_{ij}), this eigenvalue ρ in the multitype G.W. case plays the role of the mean m of offspring distribution in the simple G.W. case. A multitype G.W. process is said to be subcritical, critical or supercritical depending on $\rho < 1$, $= 1$, or > 1. We now state, without proof, the *multitype analogue of* Theorem 9.3:

Let the matrix of moments M be positively regular and let ρ be its spectral radius.
   (a) If $\rho \leq 1$, then $q_i = 1$, $i = 1, 2, \ldots, k$.
   (b) If $\rho > 1$, then $q_i < 1$, $i = 1, 2, \ldots, k$.
   (c) In either case, $q = (q^{(i)}, q^{(2)}, \ldots, q^{(k)})$,
is the smallest positive solution of the vector equation

$$q = f(q).$$

i.e. $q^{(i)}$ is the smallest positive root of $q^{(i)} = f^{(i)}(q)$ for $i = 1, 2, \ldots, k$.

For proof and other results, the reader is referred to Athreya and Ney (1972) and Mode (1971). Multitype branching processes in a random environment have been considered, for example, by Tanny (1978, 1981) and Cohn (1989).

9.6.4 Controlled Galton-Watson Branching Process

In order to obtain more realistic models of population growth, two kinds of *controlled branching processes* have been considered in the literature. Sevast'yanov and Zubkov (1974), Yanev (*Theory Prob. Appl.* **20**, 421-428, 1975) and others consider a control on the number of reproducing objects; Levy (1979), Klebaner (1983, 1984), Küster (1983, 1985) and others consider the situation where the offspring distribution depends on the number of objects in the generation considered: population size-dependent (state-dependent) offspring distributions, instead of a single offspring distribution for all the generations, are introduced. Küster (1985) examines asymptotic growth of such processes and gives some criteria under which a state-dependent Galton-Watson process behaves like the classical supercritical Galton-Watson process. For details, see the above papers and the references cited therein.

Note. Many other generalisations and variations of the classical Galton-Watson process are studied. For example, *branching processes with disasters* (see Kaplan *et al.* 1973, Sankaranarayanan 1989, Ch. 7).

To study interacting particle systems, *annihilating branching process* is considered: it is a kind of random walk such that each particle places offspring on neighbouring sites but when two new particles occupy the same site, they annihilate each other. See, for example, Sudbury (1990).

Studies on generalisations are continually going on giving rise to many interesting models to suit realistic situations in diverse areas.

9.7 CONTINUOUS-TIME MARKOV BRANCHING PROCESS

So far we have been considering branching processes in discrete time; an object after one unit of time produces similar objects according to offspring distribution $\{p_k\}$. The lifetime of an object is one unit of time. Now we proceed to consider a generalisation such that the life-time of objects are i.i.d. random variables. Instead of the process $\{X_n, n = 0, 1, 2 \ldots\}$ we shall consider the process $\{X(t), t \geq 0\}$, where $X(t)$ equals the number of objects at time t. The process $\{X(t), t \geq 0\}$ may or may not be Markovian. In this section we shall confine to the Markovian case such that $\{X(t), t \geq 0\}$ will be a Markov process in continuous time with discrete state space. Such processes were considered in Sec. 4.5. We shall recall some of the basic results concerning such processes. We first note that the waiting times between changes of states (here lifetimes of objects) have independent exponential distributions. Let the process be time homogeneous.

Denote the transition probability by

$$p_{ij}(t) = \Pr\{t + t_1 = j \mid X(t_1) = i\},$$

and, in particular

$$p_{ij}(t) = \Pr\{X(t) = j \mid X(0) = i\}, j = 0, 1, \ldots, t \geq 0.$$

The transition probabilities $p_{ij}(t)$ satisfy the Chapman-Kolmogorov equation

$$p_{ij}(t + T) = \sum_{k=0}^{\infty} p_{ik}(t)\, p_{kj}(T).$$

We shall assume here that $X(0) = 1$, i.e. the process starts with a single ancestor at time 0.

Let the probability that an object will split producing j objects during a small interval $(t, t+h)$ be denoted by $a_j h + o(h)$. The constants a_j (assumed to be independent of time) are the transition densities

$$a_j = \lim_{h \to \infty} \frac{p_{1j}(h)}{h}, \quad j \neq 1$$

$$a_1 = \lim_{h \to \infty} \frac{p_{11}(h) - 1}{h}$$

so that

$$p_{1j}(h) = a_j h + o(h), \quad j = 0, 2, 3, \ldots$$

$$p_{11}(h) = 1 + a_1 h + o(h). \tag{7.1}$$

We have $\qquad a_j \geq 0, \quad j = 0, 2, 3, \ldots$ and $\sum_{j=0}^{\infty} a_j = 0.$

Now objects act independently, and in one small interval, only one particle, on the average, will split, so that

$$\Pr\{X(t+h) = n + k - 1 \mid X(t) = n\} = na_k h + o(h), \quad k = 0, 2, 3, \ldots$$

and $\qquad \Pr\{X(t+h) = n \mid X(t) = n\} = 1 + na_1 h + o(h).$

Let $F(t, s)$ be the p.g.f. of $\{p_{1j}(t)\}$. We get

$$F(t,s) = \sum_{j=0}^{\infty} p_{1j}(t)s^j = \sum_{j=0}^{\infty} \Pr\{X(t) = j \mid X(0) = 1\} s^j \tag{7.2}$$

$$= E\{s^{X(t)}\},$$

and $\qquad\qquad F(0, s) = s. \tag{7.2a}$

Definitions. *A Markov branching process* $\{X(t), t \geq 0\}$ is a continuous-time Markov chain having for state space the set of non-negative integers $N = \{0, 1, 2, \ldots\}$ (and) whose transition probabilities $p_{ij}(t)$ satisfy the relation

$$\sum_{j=0}^{\infty} p_{ij}(t)s^j = [\sum_{j=0}^{\infty} p_{1j}(t)s^j]^i = [F(t,s)]^i, \quad i \geq 0, |s| \leq 1. \tag{7.3}$$

It is the property (7.3) that characterises and distinguishes a Markov branching process from other continuous time Markov chains with state space N (discussed in Sec. 4.5). The equation (7.3) is implied in the postulate that individuals act independently.

Further generalisation has been obtained by extending the concept of Markov renewal process, defined in Sec. 7.2.

If the Markov chain $\{X_n\}$ is also a branching process, then the (bivariate) process $\{X_n, t_n\}$, $n = 0, 1, 2, \ldots$ is known as a *Markov renewal branching process*. The size X_{n+1} of the $(n+1)$st generation depends not only on the size X_n of the nth generation, but also on the lifetime $t_{n+1} - t_n$ of the nth generation.

Neuts (*Duke Math. J.* **36** (1969) 215–231; *Adv. Appl. Prob.* **2** (1970) 110–149; see also Neuts, (1974)) has studied such processes and considered applications to a number of queueing models.

We shall now obtain a continuous-time analogue of the relation (2.2) satisfied by a Galton-Watson process.

Theorem 9.10. We have

$$[F(t+T,s)]^i = [F(t,F(T,s))]^i, \, i \geq 0 \tag{7.4}$$

and in particular, $\quad F(t+T,s) = F(t,F(T,s)). \tag{7.4a}$

Proof: Using the relation (7.2) and the Chapman-Kolmogorov equation, we get

$$[F(t+T,s)]^i = \sum_{j=0}^{\infty} p_{ij}(t+T) \, s^j, \, i \geq 0$$

$$= \sum_{j=0}^{\infty} s^j \{ \sum_{k=0}^{\infty} p_{ik}(t) \, p_{kj}(T) \}$$

$$= \sum_{k=0}^{\infty} p_{ik}(t) \{ \sum_{j=0}^{\infty} p_{kj}(T) \, s^j \}$$

$$= \sum_{k=0}^{\infty} p_{ik}(t) \{ F(T,s) \}^k$$

$$= [F(t,F(T,s))]^i,$$

and, in particular,

$$F(t+T,s) = F(t,F(T,s)).$$

We shall now proceed to obtain two differential equations (forward (7.6) and backward (7.7)), satisfied by $F(t,s)$; these can be obtained from the Chapman-Kolmogorov forward and backward differential equations respectively. We shall, however, use a slightly different method.

With the given initial conditions, the differential equations can be solved in terms of the transition densities a_j. Let

$$u(s) = \sum_{j=0}^{\infty} a_j \, s^j \tag{7.5}$$

be the generating function of the sequence $\{a_j\}$.

Theorem 9.11. We have

$$\frac{\partial}{\partial t} F(t,s) = u(s) \frac{\partial}{\partial s} F(t,s) \tag{7.6}$$

and $\quad \dfrac{\partial}{\partial t} F(t,s) = u(F(t,s)). \tag{7.7}$

Proof: From (7.1) and (7.2) we have

$$F(h,s) = \sum_{j=0}^{\infty} p_{1j}(h) s^j$$

$$= \{1 + a_1 h + o(h)\} s + \sum_{j \neq 1} \{a_j h + o(h)\} s^j$$

$$= s + h \sum_{j=0}^{\infty} a_j s^j + o(h) = s + hu(s) + o(h). \qquad (7.8)$$

Thus, using (7.4a), (7.8) and Taylor's expansion for a function of two variables, we get

$$F(t+h,s) = F(t,F(h,s))$$

$$= F(t, s + hu(s) + o(h))$$

$$= F(t,s) + hu(s)\frac{\partial}{\partial s} F(t,s) + o(h)$$

so that

$$\lim_{h \to 0} \frac{F(t+h,s) - F(t,s)}{h} = u(s)\frac{\partial}{\partial s} F(t,s) + \lim_{h \to 0} \frac{o(h)}{h},$$

whence we get (7.6). Again, using (7.4a) and (7.8), we get

$$F(h+t,s) = F(h,F(t,s))$$

$$= F(t,s) + hu(F(t,s)) + o(h).$$

Thus

$$\lim_{h \to 0} \frac{F(h+t,s) - F(t,s)}{h} = u(F(t,s)) + \lim_{h \to 0} \frac{o(h)}{h}$$

whence we get (7.7).   ▲

Each of the equations (7.6) and (7.7) is subject to the initial condition $F(0, s) = s$ (eqn. (7.2a)).

Solution of the equation. $F(t, s)$ can be obtained by solving either the forward equation (7.6) or the backward equation (7.7) using of the initial condition $F(0, s) = s$.

Noting that $\dfrac{d}{dz} \displaystyle\int_0^z f(x)dx = f(z)$, and using $F(0,s) = s$, it can be seen that the equation (7.7) is equivalent to

$$\int_s^{F(t,s)} \frac{dx}{u(x)} = t, \qquad (7.9)$$

provided that $u(x) \neq 0$ in $F(0, s) = s \leq x \leq F(t, s)$. Thus $F(t, s)$ can be obtained theoretically from (7.9).

The forward equation (7.6) together with $F(0, s) = s$ leads to an identical solution.

Example 7 (a). *Simple linear birth and death process*

Here

$$\text{Pr } \{X(t+h) = n+k-1 \,|\, X(t) = n\} = n \,\lambda h + o(h), \qquad k = 2$$

$$= n\mu h + o(h), \qquad k = 0$$

$$= 1 - n(\lambda+\mu)h + o(h), \quad k = 1$$

$$= o(h), \qquad k \geq 3,$$

i.e. $a_0 = \mu, a_1 = -(\lambda+\mu), a_2 = \lambda, a_k = 0 \qquad k \geq 3.$

Thus we have

$$u(s) = \lambda s^2 - (\lambda+\mu)s + \mu.$$

From (7.9), we have

$$t = \int_s^{F(t,s)} \frac{dx}{u(x)} = \int_s^{F(t,s)} \frac{dx}{\lambda(x-1)(x-\mu/\lambda)} = \frac{1}{(\mu-\lambda)} \log \frac{(x-\mu/\lambda)}{(x-1)} \Bigg|_s^{x=F(t,s)}$$

and on simplification, we get

$$F(t,s) = \frac{\mu(s-1) - (\lambda s - \mu) \, e^{(\mu-\lambda)t}}{\lambda(s-1) - (\lambda s - \mu)e^{(\mu-\lambda)t}}. \tag{7.10}$$

Expanding $F(t, s)$ as a power series in s, we can get $p_{ik}(t)$ as coefficient of s^k, $k = 0$, $1, 2, \ldots$ [see, Sec 4.4 under 'Linear growth process'].

Probability of extinction: It may be noted that

$$p_{10}(t) = \text{Pr } \{X(t) = 0 \,|\, X(0) = 1\} \equiv F(t,0)$$

is a non-decreasing function of t. The probability of extinction q is defined by

$$q = \lim_{t \to \infty} p_{10}(t). \tag{7.11}$$

Theorem 9.12. The probability of extinction q is the smallest root in $[0, 1]$ of the equation $u(s) = 0$; further, $q = 1$ $(q < 1)$ *iff* $u'(1) \leq (>) 0$.

Proof: From (7.7) we find that $p_{10}(t)$ satisfies the equation

$$\frac{d}{dt} p_{10}(t) = u(p_{10}(t)). \tag{7.12}$$

Intergrating, we have

$$p_{10}(t) = \int_0^t u(p_{10}(y))\, dy, \tag{7.13}$$

since

$$p_{10}(0) = 0.$$

We have $u(1) = 0$ and $u(0) \geq 0$. If $u(0) = a_0 = 0$, then $p_{10}(0) = 0$ and $q = 0$; then 0 is the smallest non-negative root of $u(s) = 0$, and the theorem holds trivially. Let $u(0) = a_0 > 0$ and q^* be the smallest root of the equation $u(s) = 0$ in $0 \leq s \leq 1$. In other words, $u(s) = 0$ for $s = q^*$, while $u(s) > 0$ in $0 \leq s < q^*$. Now $p_{10}(t)$ is a non-decreasing function of t and so from (7.12) we find $u(p_{10}(t)) \geq 0$, for $t > 0$ i.e. $u(p_{10}(t))$ is non-negative for all $t > 0$. Hence $p_{10}(t) \leq q^*$ for all $t > 0$. Thus

$$q = \lim_{t \to \infty} p_{10}(t) \leq q^*;$$

further $u(q) \geq 0$. Suppose, if possible, that $u(q) > 0$, then while $p_{10}(t)$ lies in $[0, 1]$, by the continuity of $u(x)$,

$$\int_0^t u(p_{10}(y))\, dy \to \infty \quad \text{as} \quad t \to \infty;$$

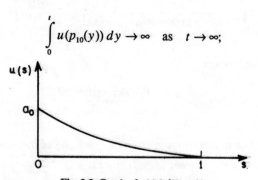

Fig. 9.3 Graph of $u(s)$ $[u'(1) \leq 0]$

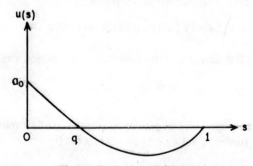

Fig 9.4 Graph of $u(s)$ $[u'(1) > 0]$

as $\lim_{t \to \infty} p_{10}(t)$ is finite, the above leads to a contradiction of (7.13), which is due to the

supposition that $u(q) > 0$. Hence we must have $u(q) = 0$. As $q \le q^*$ and as q^* is the smallest root of $u(s) = 0$, we must have $q = q^*$. We have $0 \le q = q^* \le 1$. Since $u(0) \ge 0$, $q = 1$ *iff* the function $u(s)$ is a decreasing function at $s = 1$, so that $u'(1) \le 0$, if it exists; on the other hand $q < 1$ *iff* $u'(1) > 0$. Hence the theorem.

▲

Example 7(a) (contd.). In case of linear birth and death process
$$u(s) = \lambda (s - 1)(s - \mu/\lambda).$$
Thus $q = 1$ *iff* $\lambda \le \mu$, i.e. *iff* $u'(1) \le 0$.

9.8 AGE DEPENDENT BRANCHING PROCESS: BELLMAN-HARRIS PROCESS

In the preceding section we assumed that the lifetimes of objects (particles, individuals, organisms) are exponential random variables. Here we shall generalise this further; we shall consider that the lifetimes have general, and not necessarily exponential, distributions.

Suppose that an object (ancestor) at time $t = 0$ initiates the process. At the end of its lifetime it produces a random number of direct descendants having offspring distribution $\{p_k\}$ (with p.g.f. $P(s)$). We assume, as before, that these descendants act independently of each other and that *at the end of its lifetime*, each descendant produces its own descendants with the same offspring distribution $\{p_k\}$, and that the process continues as long as objects are present. Suppose that the lifetimes of objects are i.i.d. random variables with d.f. G (which is also independent of the offspring distribution). Let $\{X(t), t \ge 0\}$ be the number of objects alive at time t. The stochastic process $\{X(t), t \ge 0\}$ is known as an *age-dependent* (or *general time*) branching process. Such a process is also known as a *Bellman-Harris process*, after Bellman and Harris who first considered such a process in 1948. We shall consider here Bellman-Harris type age-dependent process. An age-dependent process is, in general, not Markovian. For a detailed account, refer to Sankaranarayanan (1989, Ch. 4).

9.8.1 Generating Function

Theorem 9.13. The generating function

$$F(t,s) = \sum_{k=0}^{\infty} \Pr\{X(t) = k\} s^k \qquad (8.1)$$

of an age-dependent branching process $\{X(t), t \ge 0\}$; $X_0 = 1$ satisfies the integral equation

$$F(t,s) = [1 - G(t)] s + \int_0^t P(F(t-u,s)) \, dG(u). \qquad (8.2)$$

Proof: To find $\Pr\{X(t) = k\}$, we shall condition on the lifetime T at which the ancestor dies bearing i offsprings. We have

$$\Pr\{X(t) = k\} = \int_0^\infty \Pr\{X(t) = k \mid T = u\}\, dG(u)$$

$$= \int_0^t \Pr\{X(t) = k \mid T = u\}\, dG(u)$$

$$+ \int_0^\infty \Pr\{X(t) = k \mid T = u\}\, dG(u).$$

In case of the second term, $u > t$. Given that $T = u$, the number of objects at time t is then still 1 (the ancestor) and the expression under the square brackets equals δ_{1k}. Thus the second term yields $\delta_{1k}\{1 - G(t)\}$.

In case of the first term $u \leq t$, the ancestor dies at time $u \leq t$, leaving $i\ (\in N)$ direct descendants: the probability of this is $p_i\, dG(u)$; and further these i descendants (who independently initiate processes at time u) leave k objects in the remaining time $t - u$: the probability of this event is equal to the coefficient of s^k in the expansion of $[F(t - u, s)]^i$ as a power series in s. Thus we have,

$$\Pr\{X(t) = k\} = \{1 - G(t)\}\,\delta_{1k}$$

$$+ \int_0^t \sum_{i=0}^\infty p_i\, dG(u)\, \{\text{coeff. of } s^k \text{ in the expansion of } [F(t - u, s)]^i\}.$$

Multiplying both sides by s^k, $k = 0, 1, 2 \ldots$ and summing over k, we find that the l.h.s. equals $F(t, s)$; and that the first term on the r.h.s. equals $\{1 - G(t)\}s$ and the second term equals

$$\int_0^t \left[\sum_{i=0}^\infty p_i\left\{\sum_{k=0}^\infty [\text{coeff. of } s^k \text{ in } (F(t - u, s))^i]\, s^k\right\}\right] dG(u)$$

$$= \int_0^t \left[\sum_{i=0}^\infty p_i(F(t - u, s))^i\right] dG(u) = \int_0^t P(F(t - u, s))\, dG(u).$$

Hence the theorem.   ▲

Note: The integral equation (8.2) cannot be easily solved in the general case. For such finer questions such as existence and uniqueness of solutions of (8.2) and concerning properties of F, see Harris (1963) or Athreya and Ney (1972).

Particular case: Exponential lifetime. Let $G(t) = 1 - e^{-bt}$, then (8.2) yields

$$F(t,s) = se^{-bt} + b^{-bt} \int_0^t P(F(v,s)) e^{bv} dv$$

whence

$$\frac{\partial F}{\partial t} = b[P(F(t,s)) - F(t,s)], \tag{8.3}$$

which is equivalent to the backward differential equation (7.7). If $F(t, 1) = 1$, then $\{X(t), t \geq 0\}$ becomes a Markov branching process. Thus, a Bellman-Harris branching process with exponentially distributed lifetime is a Markov branching process.

It can also be seen that when G is a unit step function, (8.2) yields the functional iteration formula for $P(s)$ in the discrete case.

We now turn to the expectation $M(t) = E\{X(t)\}$ and its asymptotic behaviour. $M(t)$ satisfies an integral equation (8.4) which can be obtained rigorously from (8.2). We shall obtain (8.4) directly by using arguments similar to those used to obtain (8.2). Further, we shall show that under certain conditions, $M(t)$ is asymptotically exponential.

Theorem 9.14. The expectation $M(t) = E\{X(t)\}$ of an age-dependent process $\{X(t), t \geq 0,\}, X_0 = 1$ satisfies the integral equation

$$M(t) = [1 - G(t)] + m \int_0^t M(t - u) dG(u), \tag{8.4}$$

where $m = P'(1)$ is the mean of the offspring distribution. Further, if $m = 1$, then $M(t) = 1$ is a solution of (8.4). Suppose that α is the unique positive root of

$$\int_0^\infty e^{-\alpha u} dG(u) = \frac{1}{m}. \tag{8.5}$$

If $m > 1$ and G is non-lattice, then, as $t \to \infty$,

$$e^{-\alpha t} M(t) \to \frac{m-1}{\alpha m^2 \int_0^\infty u e^{-\alpha u} dG(u)}. \tag{8.6}$$

Proof: By conditioning on the lifetime T of the ancestor, we get

$$M(t) = \int_0^\infty E\{X(t) | T = u\} dG(u) \tag{8.7}$$

If $u > t$ then $E\{X(t) \mid T = u\} = 1$.

It $u \leq t$, then the ancestor lives for time u at the end of which, it leaves i direct descendants with probability p_i; each of these direct descendants initiates a process, the number of objects of such a process having the same distribution as $X(t - u)$. Thus for $u \leq t$,

$$E\{X(t) \mid T = u\} = \sum_{i=0}^{\infty} i p_i E\{X(t - u)\}$$

$$= \{M(t - u)\}\, m.$$

Using the above argument we see that (8.7) at once yields (8.4).

Obviously, when $m = 1$, $M(t) = 1$ is a solution of (8.4). Suppose that $m > 1$, and that α exists. Denote

$$G_\alpha(s) = m \int_0^s e^{-\alpha y}\, dG(y);$$

this implies that

$$dG_\alpha(s) = m\, e^{-\alpha s}\, dG(s). \tag{8.8}$$

Multiplying (8.4) by $e^{-\alpha t}$ and writing $v(t) = e^{-\alpha t} M(t)$, we get

$$v(t) = e^{-\alpha t}[1 - G(t)] + \int_0^t v(t - u)\, dG_\alpha(u), \tag{8.9}$$

which is a renewal type equation.

Using Theorem 6.3 (Ch. 6) we find that (8.9) has the unique solution given by

$$v(t) = e^{-\alpha t}\{1 - G(t)\} + \int_0^t e^{-\alpha(t-u)}\{1 - G(t - u)\}\, dm_\alpha(u),$$

where $m_\alpha(u)$ is of the form indicated in the theorem.

By using the Key Renewal Theorem (Theorem 6.9), we get, as $t \to \infty$

$$e^{-\alpha t} M(t) = v(t) \to \frac{\displaystyle\int_0^\infty e^{-\alpha t}\{1 - G(t)\}\, dt}{\mu},$$

where

$$\mu = \int_0^\infty u\, dG_\alpha(u) = m \int_0^\infty u e^{-\alpha u}\, dG(u).$$

Integrating by parts the integral in the numerator, we have

$$\left| \int_0^\infty e^{-\alpha t}\{1 - G(t)\}\, dt = -\{1 - G(t)\}\frac{e^{-\alpha t}}{\alpha}\right|_{t=0}^{t=\infty} - \left(\frac{1}{\alpha}\right)\int_0^\infty e^{-\alpha t}\, dG(t)$$

$$= \frac{1}{\alpha} - \frac{1}{\alpha m} \quad \text{(using (8.5))}.$$

Thus, as $t \to \infty$

$$e^{-\alpha t}M(t) \to \frac{m - 1}{\alpha m^2 \displaystyle\int_0^\infty u e^{-\alpha u}\, dG(u)}.$$

Hence the theorem. ▲

Note : A root α of the equation (8.5)

$$\int_0^\infty e^{-\alpha u}\, dG(u) = \frac{1}{m},$$

when it exists, is known as a *Malthusian parameter* ; the constant α represents the geometric rate of increase postulated by Malthus (1766-1834). It is to be noted that when $m \geq 1$, there exists a unique positive root α; if $m = 1$, $\alpha = 0$. If $m < 1$, there may or may not be a real root α and if it exists $\alpha < 0$.

Extinction probability is defined by

$$q = \Pr\{X(t) = 0, \text{ for all sufficiently large } t\}.$$

We state, without proof, the following theorem; for proof and for further results see Athreya and Ney (1972).

Theorem 9.15. If the extinction probability q is 0, then $F(t, 0) = 0$. If $q > 0$, then for each $t \geq 0$, $F(t, 0) < q$, and $F(t, 0)$ is a non-decreasing function of t ; further $\displaystyle\lim_{t \to \infty} F(t, 0) = q$.

9.9 GENERAL BRANCHING PROCESSES

In the Bellman-Harris model it is assumed that an object produces offsprings *at the end of its lifetime*. The offspring distribution is independent of the mother's lifetime.

A model allowing *real* age-dependence was introduced in 1964 by Sevast'yanov, who introduced the possibility of dependence between mother's lifetime and offspring distribution. He considered the generalisation that an object produces offsprings not necessarily at the end of its liftetime but *at randomly chosen instants during its lifetime*. For results concerning Sevast'yanov type of branching processes, the reader is referred to the books by Jagers (1975) and Sevast'yanov (1971, in Russian; also to his 1964, 1968 papers).

9.9.1 Crump-Mode-Jagers Process

A generalisation of Bellman-Harris process type as well as of Sevast'yanov type process has been obtained independently by Crump and Mode (1968, 1969) and by Jagers (1969), by way of introduction of a random process $\{N(x)\}$ representing the number of offspring born to an individual in the age-interval $(0, x)$. This is known as *Crump-Mode-Jagers (C-M-J) branching process*. (For details, see Jagers, 1975, Ch. 6 and Sankaranarayanan, 1989, Sec. 4.8). C-M-J model has attracted attention of researchers. Cohn (1985) considers a martingale approach to supercritical C-M-J process.

9.9.2 Some Extensions and Generalisations of Continuous Time Processes

In Sec. 9.5, we considered conditioned limit laws of the G.W. processes. Towards the end of the section we indicated some of the extensions of the classical G.W. processes, such as G.W. processes with immigration, G.W. Processes in varying and random environments, multitype G.W. processes etc. Such extensions have also been done in the case of continuous time processes. Continuous time multitype processes such as multitype Bellman-Harris process and multitype age-dependent processes have been studied in great details. Continuous time branching processes in varying and random environments, processes with immigration ((Pakes, 1971, 1974, 1979), and processes with disasters (Kaplan, 1975) and others) have received considerable attention. The study of the conditioned limit laws in various cases has been a topic of great interest. Functional equation technique has led to new results in the theory of *Jirina processes*, viz. branching processes in discrete time on a continuous state space (Seneta).

9.10. SOME OBSERVATIONS

9.10.1 Inference Problems

Of late, statistical inference (estimation and tests) for branching processes has attracted attention. There has been considerable interest on the estimation of the (mean and variance of the) offspring distribution and of the Malthusian growth parameter based on a complete record of the process during a time period; so also on the estimation of means of offspring and immigrant distributions for G.W.I. processes.

Mention may be made of the works of Heyde and Seneta (*J. Appl. Prob.* 9(1972) 235–256), Dion (1974, 1975), Heyde (*Adv. Appl. Prob.* 3 (1974) 421–433), Heyde (*Biometrika* 62 (1975) 49–55), Quine (*Ann. Prob.* 4 (1976) 319–325), Basawa and Scott (*Biometrika* 63 (1976) 531–536), Athreya and Keiding (*Sankhya* A 39 (1977)

101–123) Jagers, Pakes, Venkataraman and Nathi, and others. For a survey of theoretical literature, refer to Dion and Keiding (1978); see also Sankaranarayanan (1989, Ch. 9) for an account on the estimation problems.

9.10.2 Concluding Remarks

Applications of branching processes have been made in several fields like biology, medicine, epidemiology, demography and also in nuclear physics and chemistry. The books by Mode, Jagers, Srinivasan, Keyfitz, Sankaranarayanan, and Iosifescu and Tautu discuss many interesting models of branching processes in such areas, besides covering theoretical developments.

Some interesting suggestions for further research on application to demography are indicated in Mode (1978).

It is not possible to give here references of even the most important works; a limited number of references given here may be of use to the readers.

We have come a long way from the day when Francis Galton (1822-1911) first formulated the basic problem of branching processes in *Educational Times*, London, 1st April 1873, as under:

Problem 4001; 'A large nation, of whom we will concern ourselves with adult males, N in number, and who each bear separate surnames, colonise a district. Their law of population is such that, in each generation, a_0 per cent of the adult males have no male children who reach adult life; a_1 have one such child, a_2 have two, and so on up to a_5, who have five.

Find (1) what proportion of surnames will have become extinct after r generations; and (2) how many instances there will be of the same surname being held by m persons'.

The theory of branching process was formally initiated in the following year by Galton and Watson (1874). As interest in the topic grew, the names of Galton and Watson got firmly entrenched in the literature. In 1972, Heyde and Seneta (*Biometrika* **59** (1972) 680-683) drew attention to the formulation of and somewhat elaborate discussion of the topic, earlier in 1845, by I.J. Bienayame and the Galton-Watson process came be known also as Bienayame-Galton-Watson process.

Some of the luminaries whose wide interests at one time or another touched on problems of branching processes are Euler in mathematics, Lotka in demography, Fisher in statistics and mathematical genetics and Kolmogorov in probability. For a historical account, see Jagers (1975).

Branching processes have today emerged as a very rich field of research both from the point of view of theoretical structures and of applications that extend to several directions.

EXERCISES

9.1 If $\{X_n, n = 0, 1, 2, \ldots\}$, $X_0 = 1$ is a branching process with offspring distribution p.g.f. $P(s)$, show that $\{X_{rk}, r = 0, 1, \ldots\}$, where k is a fixed positive integer, is also a branching process.

9.2   Suppose that each individual at the end of one unit of time produces either k (where k is a fixed positive integer ≥ 2) or 0 direct descendants with probability p or q respectively. Considering the root of the equation $P(s) = s$, verify that the probability of ultimate extinction is less than one. Find this probability for

$k = 3$, when $p = q = \dfrac{1}{2}$.

9.3   If the offspring distribution in a G.W. process is geometric $\{q^k p\}$, $k = 0, 1, 2, \ldots$, show that

$$P_n(s) = \frac{p(q^n - p^n) - (q^{n-1} - p^{n-1})pqs}{(q^{n+1} - p^{n+1}) - (q^n - p^n)qs}, \quad p \neq q$$

$$= \frac{n - (n-1)s}{(n+1) - ns}, \quad p = q.$$

9.4   Find the variance of X_n (i.e. the number of objects in nth generation) in a G.W. process from the relation $P_n = P(P_{n-1}(s))$.

9.5   Show that the m.g.f. $\varphi_n(s) = E\{e^{-sW_n}\}$, $n = 1, 2, \ldots$, satisfies the relation

$$\varphi_{n+1}(ms) = P[\varphi_n(s)].$$

9.6   Show that the p.g.f. of the conditional distribution of X_n, given $X_n > 0$, is

$$\frac{P_n(s) - P_n(0)}{1 - P_n(0)}.$$

Find $\Pr\{X_n = r \mid X_n > 0\}$ when $\Pr\{\text{number of offspring} = k\} = \left(\dfrac{1}{2}\right)^{k+1}$, $k = 0$,

$1, 2, \ldots$

9.7   For a Galton-Watson process, show that the expected number of total progeny is given by

$$E(Y_n) = (1 - m^{n+1}) / (1 - m), \quad \text{if } m \neq 1$$

$$= n + 1, \quad \text{if } m = 1$$

and that, if $m < 1$, $E(Y_n) \to \dfrac{1}{1-m}$ as $n \to \infty$.

Show that

$$\text{var}(Y_n) = (2n+1)(n+1)\, n\sigma^2 / 6, \quad \text{if } m = 1.$$

9.8 Show that for a binary splitting process (i.e. with offspring distribution $p_0 = q, p_2 = p = 1 - q$), if $p < \dfrac{1}{2}$

$$H(s) = \frac{1 - \sqrt{4s^2 pq}}{2sp}$$

(where $H(s) = \lim\limits_{n \to \infty} R_n(s)$, $R_n(s) = E\{s^{Y_n}\}$, and Y_n = total progeny up to nth generation). Deduce that

$$\Pr\{Y_\infty = 2k\} = 0$$

$$\Pr\{Y_\infty = 2k - 1\} = \frac{(-1)^{k-1}}{2p} \binom{\frac{1}{2}}{k} (4pq)^k.$$

9.9 Find the limiting distribution of the total progeny in a G.W. process with geometric offspring distribution $\{q^k p\}$, $k = 0, 1, 2, \ldots$.

9.10 Consider a G.W. process with geometric offspring distribution $\{q^k p\}$, $k = 0, 1, 2, \ldots$. Verify that when $m = 1$

$$\lim_{n \to \infty} n \Pr\{X_n > 0\} = \frac{2}{\sigma^2}.$$

When $m < 1$ $(q < p)$, verify that Yaglom's theorem holds. Find

$$b_k = \lim_{n \to \infty} \Pr\{X_n = k \mid X_n > 0\}, \quad k = 1, 2, \ldots$$

and verify that $B(s) = \sum\limits_{k=1}^{\infty} b_k s^k$ satisfies the equation

$$B(P(s)) = mB(s) + 1 - m.$$

9.11 For a G.W. process with $m < 1$, $\sigma^2 < \infty$, if $B(s)$ is the the p.g.f. of the distribution $\{b_k\}$ defined by (5.8), show that $B''(1) < \infty$. Further, show that the variance σ_b^2 of $\{b_k\}$ is given by

$$\sigma_b^2 = \frac{\zeta \sigma^2}{m - m^2} - \zeta^2$$

where

$$\zeta = \sum_{k=1}^{\infty} k b_k = \frac{1}{\phi(0)}.$$

9.12   *Time to extinction in a G.W. process*

Show that the distribution of time to extinction T (which is the first passage time to state 0) is given by

$$r_n = \Pr\{T = n\} = P_n(0) - P_{n-1}(0), \quad n = 1, 2, \ldots$$

Find r_n when

(i) $p_k = q^k p, k = 0, 1, 2, \ldots$

(ii) $p_k = (1-a)(1-c)\,c^{k-1}, k = 1, 2, \ldots$

$$= a, \qquad\qquad\quad k = 0.$$

9.13   Suppose that at the end of exponential lifetime (having parameter a) each particle splits into $k \,(\ge 2)$ particles. Show that the generating function $F\,(t, s)$ of $\{X\,(t) \,|\, X_0 = 1\}$ is given by

$$F(t,s) = se^{-at}\,[1 - \{1 - e^{-a(k-1)t}\}\,s^{k-1}]^{-1/(k-1)}$$

9.14   Find var $(X\,(t))$, where $\{X\,(t)\}$, is a Markov branching process with $X\,(0) = 1$ and having $u\,(s)$ as the generating function of the transition densities.

9.15   Consider an age-dependent process with exponential lifetime given by $G\,(t) = 1 - e^{-bt}$. Suppose that at the time of its death an object splits into two similar objects. The process then reduces to the pure birth process considered by Yule. Obtain the differential equation for the generating function of the process. Solving the equation, show that

$$\Pr\{X(t) = k\} = e^{-bt}\,(1 - e^{-bt})^{k-1}, k = 1, 2, \ldots$$

9.16   Starting with an age-dependent process show that the differential equation satisfied by a simple linear birth and death process reduces to

$$\frac{\partial F}{\partial t} = \lambda F^2 - (\lambda + \mu)\,F + \mu,$$

where λ, μ are the birth and death rates respectively.

REFERENCES

S.Asmussen and S.Hering. *Branching Processes*, Birkhausser, Boston (1983)

K.B.Athreya. Some results on multitype continuous time Markov branching processes, *Ann Math. Stat.*, **39**, 347–357 (1968).

—— and S.Karlin. Branching processes with random environments, *Proc. Ann Math. Soc.*, **76**. 865–870 (1970).

—— On branching processes in random environments; I: Extinction probabilities, and II: Limit theorems, *Ann Math. Stat.*, **42**, 1499–1520 and 1843–1858 (1971).

K.B.Athreya and P.E.Ney. *Branching Processes*, Springer-Verlag, New York (1972).

R.E.Bellman and T.E.Harris. On the theory of age dependent stochastic branching processes, *Proc. Nat. Ac. Sc.*, **34**, 601–604 (1948).

—— On age-dependent binary branching processes, *Ann. Math.*, **55** 280–295 (1952).

H. Cohn. A martingale approach to supercritical (CMJ) branching processes, *Ann. Prob.* **13**, 1179–1191 (1985).

—— On the growth of the multitype supercritical branching process in a random environment, *Ann. Prob.* **17**, 1118–1123 (1989).

K.S. Crump and C.J. Mode. A general age-dependent branching process, I & II, *Math. Anal. & Appl.* **24**, 494–508 (1968) & **25**, 8–17 (1969).

J.P.Dion. Estimation of the mean and the initial probabilities of a branching process, *J. Appl. Prob.* **11**, 687–694 (1974).

—— Estimation of the variance of a branching process, *Ann. Statist.* **3**, 1183–1187 (1975).

—— and Keiding. Statistical estimation in : *Branching Processes* (Eds. A.Joffe and P.Ney) 105–140, Marcel Dekker, New York (1978). [It contains a list of 72 References].

D.H. Fearn. Galton-Watson processes with generation size dependence, *Proc. Sixth Berkeley Symp. Math. Stat. and Prob.*, **IV**, 159–172 (1972).

F.Galton and H.W. Watson. On the probability of extinction of families *J. of Anthr. Inst.*, **6**, 138–144 (1874).

D.R.Grey. A new look at convergence of branching processes, *Ann. Prob.* **8**, 377–380 (1980).

T.E.Harris. *The Theory of Branching Processes*, Springer-Verlag, Berlin (1963).

C.R.Heathcote. A branching process allowing immigration, *J. Roy. Stat. Soc.* **B.**, **27**, 138–143 (1965); Corrections and comments, **B 28**, 213–217 (1966).

C.C.Heyde. Extension of a result of Seneta for the supercritical branching processes, *Ann. Math. Statist.* **41**, 739–742 (1970).

J.M.Holte. Extinction probability for a critical general branching process, *Stoch. Proc. & Appl.* **2**, 303–309 (1974).

—— Critical multitype branching processes, *Ann. Prob.* **10**, 482–495 (1982).

M. Iosifescu and P. Tautu. *Stochastic Processes and Applications in Biology and Medicine*, Vols. **1** & **2**, Springer-Verlag, Berlin (1972).

P. Jagers. A general stochastic model for population development, *Scand. Actua. J.* **52**, 84–103 (1969).

—— Galton-Watson processes in varying environments, *J. Appl. Prob.*, **11**, 174–178 (1974).

—— *Branching Processes with Biological Applications*, Wiley, New York, (1975).

A.Joffe. On the Galton-Watson processes with mean less than one, *Ann. Math. Stat.*, **38**, 264–266 (1967).

—— and P.Ney (Eds.). *Branching Processes*, Marcel Dekker, New York (1978)

S.Karlin and H.M. Taylor. *A First Course in Stochastic Processes*, 2nd ed., Academic Press, New York, (1975).

N. Kaplan. A theorem on composition of random probability generating functions and applications to branching processes with random environments, *J. Appl. Prob.* **9**, 1–12 (1972).

—— A.Sudbury and T.Nilsen. A branching process with disasters, *J. Appl. Prob.*, **12**, 47–59 (1975).

N.Keyfitz. *Introduction to the Mathematics of Population*, Addison-Wesley, Reading, Mass, (1968; with revisions 1977).

F.C. Klebaner. Population size dependent branching process with linear rate of growth, *J. Appl. Prob.* **20** 242–250 (1983).

—— On population size-dependent branching processes *Adv. Appl. Prob.* **16**, 30–55 (1984).

P. Küstar. A symptotic growth of controlled Galton-Watson processes, *Ann. Prob.* **13**, 1157–1178 (1985).

J.B. Levy. Transience and recurrence of branching processes with state-dependent branching processes with an immigration component, *Adv. Appl. Prob.* **11**, 73–92 (1979).

C.J.Mode. *Multitype Branching Processes*, Elsevier, New York (1971).

—— Review of P.Jager's book *Branching Processes with Biological Applications*, Wiley (1975) in : *Ann. Prob.* **6**, 701–704 (1978).

M.F.Neuts. The Markov renewal branching processes in : *Mathematical Methods in Queueing Theory* (Ed. A.B.Clarke), Lecture Notes in Ec. & Math. Stat. 98, Springer-Verlag, New York (1974).

P.Ney. Critical branching processes, *Adv. Appl. prob.* **6**, 434–445 (1974).

A.G.Pakes. Branching processes with immigration, *J. Appl. Prob.*, **8**, 32–34 (1971).

—— On the critical G.W. processes with immigration, *J. Austr. Math. Soc.*, **13**, 277–290 (1972).

—— On Markov branching processes with immigration, *Sankhya*, A **37**, 129–138 (1974).

—— Limit theorems for the simple branching processes allowing immigration: I, the case of finite offspring mean, *Adv. Appl. Prob.* **11**, 31–62 (1979).

—— and N.L.Kaplan On the subcritical Bellman-Harris process with immigration. *J. Appl. Prob.*, **11**, 652–668 (1974).

G.Sankaranarayanan. *Branching Process and Its Estimation Theory*, Wiley Eastern, New Delhi (1989).

E.Seneta. Functional equations and the Galton-Watson process, *Adv. Appl. Prob.*, **1**, 1–42 (1969).

—— An explicit limit theorem for the critical Galton-Watson process with immigration, *J. Roy. Stat. Soc.*, **B**, **32**, 149–152 (1970).

—— On recent theorems concerning the super-critical Galton-Watson process, *Ann. Math. Statist* **39**, 2098–2102 (1968).

B.A.Sevast'yanov. Age-dependent branching processes, *Theo. Prob. Appl.*, **9**, 521–537 (1964).

—— Limit theorems for age-dependent branching processes, *Theo. Prob. Appl.*, **13**, 237–259 (1968).

—— *Branching Processes* (in Russian), Nauka, Moscow (1971).

—— and A.M. Zubkov. Controlled branching processes, *Theo. Prob. Appl.* **19**, 14–24 (1974).

W.L.Smith. Necessary conditions for almost sure extinction of a branching process with random environment, *Ann. Math. Stat.*, **39**, 2136–2140 (1968).

—— and W.E.Wilkinson. On Branching processes in random environments, *Ann Math. Stat.*, **40**, 814–827 (1969).

S.K.Srinivasan. *Stochastic Theory and Cascade Processes*, Elsevier, New York, (1969).

—— and K.M.Mehata. *Stochastic Processes*, Tata McGraw-Hill, New Delhi, (1976).

A. Sudbury. The branching annihilating process: An interacting particle system, *Ann. Prob.* **18**, 581–601 (1990)

D. Tanny. Normalizing constants for branching processes in random environments, *Stoch. Proc. Appl.* **6**, 201–211 (1978).

—— On multitype branching processes in random environments, *Adv. Appl. Prob.* **13**, 464–497 (1981).

W.C.Torrez. The birth and death chain in a random environment: instability and extinction theorems, *Ann. Prob.*, **6**, 1026–1043 (1978).

K.N.Venkataraman and K.Nathi. A limit theorem on sub-critical Galton-Watson processes with immigration, *Ann. Prob.* **10**, 1069–1074 (1982a).

—— Some limit theorems on a supercritical simple Galton-Watson process, *Ann. Prob.* **10**, 1975–1978 (1982b).

W.E.Wilkinson. On calculating extinction probabilities for barnching processes in random environments *J. Appl. Prob.* **6**, 478–492 (1969).

N.M.Yanev. On the statistics of branching processes, *Theo. Prob. & Appl.* **20**, 612–622 (1975).

Y.S.Yang. On branching processes allowing immigration, *J. Appl. Prob.*, **9**, 24–31 (1972).

CHAPTER 10

Stochastic Processes in Queueing and Reliability

10.1 QUEUEING SYSTEMS: GENERAL CONCEPTS

The queueing theory had its origin in 1909, when A.K. Erlang* (1878-1929) published his fundamental paper relating to the study of congestion in telephone traffic. The literature on the theory of queues and on the diverse areas of its applications have grown tremendously over the years. The standard works and some of the other publications specifically referred to in our discussions are listed in the references. For complete references, the standard works may be consulted. See Prabhu (1987) for a list of books and survey papers on queues and Prabhu (1986) for analytical innovations in the theory of queues. We consider here only the fundamental aspects of the theory.

A queue or waiting line ('*file d'attente*' in French and '*Warteschlange*' in German) is formed when units (or customers, clients) needing some kind of service arrive at a service channel (or counter) that offers such facility. A queueing system can be described by the flow of units for service, forming or joining the queue, if service is not immediately available, and leaving the system after being served (or sometimes without being served). The basic features which characterise a system are; (i) the input, (ii) the service mechanism, (iii) the queue discipline and (iv) the number of service channels.

By units, we mean those demanding service, e.g. customers at a bank counter or at a reservation counter, calls arriving at a telephone exchange, vehicular traffic at a traffic intersection, machines for repair before a repairman, airplanes waiting for take-off (or landing) at a busy airport, merchandise waiting for shipment at a yard, computer programmes waiting to be run on a time-sharing basis etc.

The *input* describes the manner in which units arrive and join the system. The system is called a *delay* or *loss system* depending on whether a unit who, on arrival, finds the service facility occupied, joins or leaves the system. The system may have either a limited or an unlimited capacity for holding units. The source from which the units come may be finite or infinite. A unit may arrive either singly or in a group. The interval between two consecutive arrivals (or between two consecutive arrival groups, in case of bulk arrivals) is called the *interarrival time* or *interval*.

---

*A.K. Erlang, a Danish engineer, is appropriately called the father of queueing theory. Erlang was also one of the early workers in operations research. Like O.R. consultants of present day, he assisted the engineers in the solution of problems of physical and mathematical character, his method being Socratic in nature.
For a short account on Erlang, see Saaty (1957) and for a detailed account of his life and work, refer to Brockmeyer *et al.* (1948).

The service mechanism describes the manner in which service is rendered. A unit may be served either singly or in a batch. The time required for servicing a unit (or a group, in case of batch service) is called the *service time*.

The queue discipline indicates the way in which the units form a queue and are served. The usual discipline is *first come first served* (FCFS) or *first in first out* (FIFO), though sometimes, other rules, such as, last come first served or random ordering before service are also adopted. A more realistic service discipline, called *processor-sharing*, is considered in computer science literature; this envisages that if there are m jobs, each receives service at the rate of $1/m$.

The system may have a *single channel* or *a number of parallel channels* for service.

The interarrival and service times may be deterministic or chance-dependent. The case when both the interarrival and service times are deterministic is trivial. We shall be generally concerned with chance-dependent interarrival and service times, and the theory will be essentially stochastic. When chance-dependent, the interarrival times between two consecutive arrivals are assumed to be i.i.d. random variables; the service times of units are also assumed to be i.i.d. random variables. Further the two sets of r.v.'s are also taken to be independent.

The mean arrival rate, usually denoted by λ, is the mean number of arrivals per unit time. Its reciprocal is the mean of the interarrival time distribution. The mean service rate, usually denoted by μ, is the mean number of units served per unit time, its reciprocal being the mean service time. In a single channel system, the ratio

$$\rho = \frac{\lambda}{\mu}$$

is called the *traffic intensity*. Though dimensionless, it is expressed in *erlangs* (in honour of A.K. Erlang). The definition of traffic intensity is suitably modified in case of loss systems and in case of bulk or batch queueing systems.

10.1.1 Queueing Processes

The following random variables or families of random variables that arise in the study provide important measures of performance and effectiveness of a stochastic queueing system.

(1) The number $N(t)$ in the system at time t, i.e. the number at time t waiting in the queue including those being served, if any.

(2) The busy period which means the duration of the interval from the moment the service commences with arrival of an unit at an empty counter to the moment the server becomes free for the first time.

(3) The waiting time in the queue (system), i.e. the duration of time a unit has to spend in the queue (system); also the waiting time W_n of the nth arrival.

(4) The virtual waiting time $W(t)$, i.e. the interval of time a unit would have to wait in the queue, were it to arrive at the instant t.

One needs to have their complete probabilistic description. It is clear that $\{N(t), t \geq 0\}$, $\{W(t), t \geq 0\}$, $\{W_n, n \geq 0\}$ are stochastic processes, the first two being in continuous time and the third one in discrete time. It will be seen that some of the queueing processes that we would come across are Markovian and some are semi-Markovian. From some of the non-Markovian processes that arise, Markov chains can be extracted at suitable regeneration points and semi-Markov processes can be constructed there from. The theory of Markov chains and semi-Markov processes thus plays an important role in the study of queueing processes.

10.1.2 Notation

A very convenient notation designed by Kendall (1951) to denote queueing systems has been universally accepted and used. It consists of a three-part descriptor $A/B/C$, where the first and the second symbols denote the interarrival and service time distributions respectively, and the third denotes the number of channels or servers. A and B usually take one of the following symbols:

M : for exponential (Markovian) distribution
E_k : for Erlang-k distribution
G : for arbitrary (General) distribution
D : for fixed (Deterministic) interval

Thus, by an $M/G/1$ system is meant a single channel queueing system having exponential interarrival time distribution and arbitrary (general) service time distribution. By $M/G/1/K$ is meant the same system with the fourth descriptor R denoting that the system has a limited holding capacity K.

10.1.3 Steady State Distribution

$N(t)$, the number in the system at time t and its probability distribution, denoted by

$$p_n(t) = \Pr\{N(t) = n \mid N(0) = \cdot\}$$

are both time dependent. For a complete description of the queueing process we need consider transient or time-dependent solutions. It is often difficult to obtain such solutions. Further, in many practical situations, one needs to know the behaviour in steady state, i.e. when the system reaches an equilibrium state, after being in operation for a pretty long time.

It is easier and convenient to determine

$$p_n = \lim p_n(t) \quad \text{as } t \to \infty$$

provided the limit exists. It is necessary to know the condition for the existence of the limit in the first place. This will be discussed in due course. When the limit exists, it is said that the system has reached *equilibrium* or *steady state* and the problem then boils down to finding the steady state solutions.

It is to be noted that, there can be several stochastic processes having the same equilibrium distribution; it is necessary therefore to consider transient behaviour of the system as well. See Whitt (1983), who points to the limitations of steady state distribution. The transient behaviour is also important in connection with cost-benefit analysis of an operating system.

10.1.4 Some General Relationships in Queueing Theory

There are certain useful statements and relationships in queueing theory which hold under fairly general conditions. Though rigorous mathematical proofs of such relations are somewhat complicated, intuitive and heuristic proofs are simple enough and have been known for long (Morse (1947)). It has been argued (Krakowski, (1973) also Stidham (1974)) that conservation methods could very well be applied to supply proofs of some of these relations. Conservation principles have played a fundamental role in physical and engineering sciences as well as in economics etc. Similar principles may perhaps be applied in obtaining relations for queueing systems in steady state. Some such relations are given below. The most important one is

$$L = \lambda W \tag{1.1}$$

where λ is the arrival rate, L is the expected number of units in the system and W is the expected waiting time in the system in steady state. A rigorous proof of the relation has been given by Little (1961) and so the relation is known as *Little's* formula. This result, of great generality, is independent of the form of interarrival and service time distributions, and holds under some very general conditions.

Denote the expected number in the queue and the expected waiting time in the queue in steady state by L_Q and W_Q respectively. These are related by a similar formula:

$$L_Q = \lambda W_Q. \tag{1.2}$$

(see also Brumelle (1972) for a generalisation of (1.1)).

A relation which holds for $G/G/1$ queue in steady state is

$$\lambda = (1 - p_0)\mu \quad \text{or} \quad p_0 = 1 - \rho \cdot \tag{1.3}$$

This follows from the *principle of coustomer conservation*, which states that, in equilibrium, the rate of arrival equals the rate of departure [for a $GI/G/1$ queue, the rate of departure is $\mu(1-p_0)$].

In a $GI/G/k$ queue, the expected number of idle servers is $k(1 - \rho)$ (Harris, 1974).

There are other relations, which hold under somewhat restrictive conditions. For example, for a system with Poisson input, the Pollaczek-Khinchine formula holds. We shall prove this formula later.

10.1.5 Little's Formula : $L = \lambda W$

The result has been known to hold good much earlier; however, since Little (1961) gave, for the first time, a rigorous proof, it is known as *Little's formula* or *result*. The publication of Little's paper (in which certain conceptually difficult mathematical properties were used and under which the result holds), evoked wide interest and the topic has been taken up later by several researchers (Eilon, Jewell, Stidham: see References) with a view to simplify the hypothesis and to obtain broad conditions under which the result holds. Eilon gave a simple demonstration; Jewell's proof is based on renewal theory. Stidham (1974) gave a simple rigorous proof for the validity of the result on the assumption that the average values L and W exist and are finite. Romalhoto *et al.* (1983) give an extensive lists of references of work on this topic; they also indicate applications to various situations.

One essential condition is that the system is in steady-state (or in equilibrium).

We give below some simple intuitive justification for the validity of the result, as discussed by Foster.

I. Since W is the average waiting time (say, in the system) of a unit, the average rate of departure per unit is $1/W$. The average number of units in the system being L, the average rate of departure (for the aggregate of all units) is L/W. Now, since the system is in equilibrium, this average departure rate must be equal to the average arrival rate, which is λ. Thus $\lambda = L/W$.

II. Suppose that, for every unit of time that a unit spends in the system, the system is entitled to a reward (or is required to pay a penalty) of 1 unit (of money). Then the total expected reward (or penalty) equals the average number of units in the system, that is, L. Again, the average number of arrivals is λ and the average time that a unit spends in the system is W; hence the total expected reward (or penalty) equals λW. Thus $L = \lambda W$.

Remarks: (1) The result is of great universality and is very general. It holds for all queueing systems, subject to some very broad conditions.

(2) The result connects L (customer average) with W (time average). More intresting results connecting such averages have recently been obtained. Glynn and Whitt (1989) consider an extension and new such relations of the form $H = \lambda G$. See also Brumelle (1971).

(3) The first-order moment formula has been extended to cover the case of higher-order moments for some queueing systems (Brumelle, 1972).

10.2 THE QUEUEING MODEL *M/M/1*: STEADY STATE BEHAVIOUR

This single server model envisages Poisson input and exponential service time with FCFS queue discipline. Such a queue is also known as the *simple queue* (or *Poisson queue* or *simple Markovian queue*).

The arrivals occur in accordance with a Poisson process with intensity λ (say), i.e. the probability that an arrival occurs in an infinitesimal interval of length h is $\lambda h + o(h)$, while that of more than one arrival is $o(h)$. This is equivalent to the statement that the distribution of the interarrival times is exponential having p.d.f. $a(t) = \lambda e^{-\lambda t}$. The distribution of the service times is exponential, with parameter μ (say) having p.d.f. $b(t) = \mu e^{-\mu t}$. In other words, the probability that one service is completed in an interval of infinitesimal length h is $\mu h + o(h)$, while the probability of more than one completion of service is $o(h)$. It is assumed that interarrival times as well as service times are stochastically independent. We shall call such a queue an *M/M/1* queue with parameters λ, μ (or a simple queue with parameters λ, μ). The mean interarrival time is $1/\lambda$ and the mean service time is $1/\mu$. The ratio $\rho = \lambda/\mu$ is the *traffic intensity* (also *equal to load factor* or *utilization factor*).

Let $N(t)$ be the number in the system (those in the queue including the one being served, if any) at instant t (≥ 0); then $\{N(t), t \geq 0\}$ is a Markov process in continuous time with denumerable number of states $\{0, 1, 2, \ldots\}$. Here transitions take place only to two neighbouring states. This is a type of birth and death process.

The arrivals here can be thought of as "births" and the service completions as "deaths". The birth and death process with $\lambda_n = \lambda$, $n \geq 0$ and $\mu_n = \mu$, $n \geq 1$, $\mu_0 = 0$, is also known as *immigration-emigration* process (see Sec. 4.4).

Let $\Pr\{N(t) = n \mid N(0) = \bullet\} = p_n(t), n \geq 0$. Using the same probabilistic arguments (or directly from equations (4.6) and (4.4) of Sec. 4.4) we at once get

$$p_o'(t) = -\lambda p_o(t) + \mu p_1(t) \tag{2.1}$$

$$p_n'(t) = -(\lambda + \mu)p_n(t) + \lambda p_{n-1}(t) + \mu p_{n+1}(t), n \geq 1. \tag{2.2}$$

First, let us assume the existence of a "steady-state". Then as t tends to infinity, $p_n(t)$ tends to a limit p_n, independent of t. The equations of steady-state probabilities $\Pr(N = n) = p_n$ can be obtained by putting $p_n'(t) = 0$ and replacing $p_n(t)$ by p_n in (2.1) and (2.2). We get

$$0 = -\lambda p_0 + \mu p_1 \tag{2.3}$$

$$0 = -(\lambda + \mu)p_n + \lambda p_{n-1} + \mu p_{n+1}, n = 1, 2, \ldots \tag{2.4}$$

The above equations are called *balance* (or *equilibrium* or *conservation flow*) *equations*. The same equations for an Markovian queueing system in steady state can also be obtained from the following principle:

Rate Equality Principle. In the steady state no build up occurs at any state. As such the two rates, i.e. the *rate of flow out of* an arbitrary state k and the *rate of flow into* the same state k ($k = 0, 1, 2, \ldots$) are equal.

In equilibrium (i.e. in steady state) the rate of flow out of state 0 equals λp_0 and the rate of flow into state 0 equals μp_1, thus

$$\lambda p_0 = \mu p_1, \tag{2.3a}$$

which is equation (2.3).
Again, for $k \neq 0$,

      rate of flow out of state k equals $(\lambda + \mu)p_k$

      rate of flow into state k equals $\lambda p_{k-1} + \mu p_{k+1}$

so that

$$(\lambda + \mu)p_k = \lambda p_{k-1} + \mu p_{k+1}, \tag{2.4a}$$

which is equation (2.4) (with k in place of n).

The *state-transition-rate* diagram is given in Fig. 10.1 from where the truth of (2.3a) and (2.4a) can be easily verified.

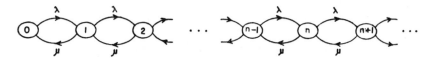

Fig 10.1 State-transition-rate diagram for *M/M/*1 queue

Note: (1) In the state-transition-rate diagram of continuous parameter Markov chain, the arcs are labelled with state transition rates, whereas in case of the graph of discrete parameter Markov chain, the arcs are labelled with conditional probabilities.

(2) State transition rate diagrams are helpful in case of Markovian queues.

10.2.1 Steady-State Solution

Dividing by μ, and writing ρ for λ/μ, the above equations can be written as

$$p_1 - \rho p_0 = 0 \qquad (2.3b)$$

$$p_{n+1} - (1+\rho)p_n + \rho p_{n-1} = 0, n = 1, 2.... \qquad (2.4b)$$

The methods of solution of these difference equations are considered in Example 2(c), Sec. A.2.4 (Appendix). The solution is given by

$$p_n = p_0 \rho^n, n \geq 0. \qquad (2.5)$$

Now in order that $\{p_n\}$ is a probability distribution, we must have

$$\sum_{n=0}^{\infty} p_n = 1 \quad \text{or,} \quad p_0 \sum_{n=0}^{\infty} \rho^n = 1.$$

The infinite geometric series converges to the sum $1/(1-\rho)$ *iff* $\rho < 1$.
For existence of steady state solution, ρ *must be less than* 1. We have then

$$p_n = (1-\rho)\rho^n, n \geq 0. \qquad (2.6)$$

The distribution $\{p_n\}$ of the random variable N, the number in the system in steady state, is geometric (see also Example 4(b) Ch. 7, where the distribution was obtained by using the theory of semi-Markov processes).

The probability that the system is empty is $p_0 = 1 - \rho$ and that the system is not empty is given by $\Pr\{N \geq 1\} = 1 - p_0 = 1 - (1-\rho) = \rho$.
We have

$$E\{N\} = \frac{\rho}{1-\rho} = \frac{\lambda}{\mu - \lambda}$$

and
$$\operatorname{var}(N) = \frac{\rho}{(1-\rho)^2} = \frac{\lambda\mu}{(\mu-\lambda)^2}.$$

The following table gives values of $E\{N\}$, var (N) for some values of ρ:

ρ	0.1	0.3	0.5	0.7	0.8	0.9	0.95	0.99
$E\{N\}$	0.11	0.43	1.00	2.33	4.00	9.00	19.00	99.00
var (N)	0.12	0.61	2.00	7.78	20.00	90.00	380.00	9900.00

Note: $\rho < 1$ (or equivalently $\lambda < \mu$) is a necessary condition for the existence of steady state, and that the queue discipline has no effect on the distribution of N. The queue discipline will have to be taken account of in obtaining the waiting time distributions.

10.2.2 Waiting Time Distributions

10.2.2.1 Waiting Time in the Queue

We assume that the queue discipline is FCFS. Let the random variable W_Q denote the time spent in waiting in the queue (or '*queueing time*') of a test unit. Now $W_Q = 0$, if at the instant of arrival of the test unit there is no unit in the system, the probability for which is $p_0 = 1 - \rho$. Given that there are n (≥ 1) units in the system, $W_Q = S_n$, where

$$S_n = v'_1 + v_2 + \cdots + v_n,$$

v'_1 being the residual service time of the customer being served and $v_2, v_3, \ldots v_n$ being the service times of the units waiting in the queue at the instant of arrival of the test unit. Now v'_1, the residual of an exponential distribution with mean $1/\mu$ is again an exponential distribution with the same mean. Thus S_n is the sum of n i.i.d. exponential distributions and is a variable with gamma distribution having p.d.f.

$$\frac{\mu^n x^{n-1} e^{-\mu x}}{\Gamma(n)}, x \geq 0$$

(see Theorem 1.10).

Hence we have, for $x > 0$,

$$w_q(x)\, dx = \Pr\{x \leq W_Q < x + dx\}$$

$$= \sum_{n=1}^{\infty} \Pr\{x \leq W_Q < x + dx \mid \text{the test unit finds } n \text{ in the system}\}$$

$$\times \Pr\{\text{the test unit finds } n \text{ in the system}\}$$

$$= \sum_{n=1}^{\infty} \left\{ \frac{\mu^n x^{n-1} e^{-\mu x}}{\Gamma(n)} dx \right\} p_n$$

$$= \mu e^{-\mu x}(1-\rho)\rho \left\{ \sum_{n=1}^{\infty} \frac{(\mu\rho x)^{n-1}}{(n-1)!} \right\} dx$$

$$= \mu\rho(1-\rho)e^{-\mu(1-\rho)x} dx, x > 0.$$

Thus the p.d.f. of the distribution of W_Q is given by

$$w_q(x) = p_0 = 1 - \rho, \quad x = 0$$

$$= \mu\rho(1 - \rho)e^{-\mu(1-\rho)x}, \quad x > 0. \tag{2.7}$$

The L.T. $\quad w_q^*(\alpha) = \int\limits_0^\infty e^{-\alpha x} w_q(x)dx = \dfrac{(1-\rho)(\alpha+\mu)}{\alpha - \lambda + \mu}.$

The distribution function $F(x)$ of W_Q is given by

$$F(x) = 1 - \rho, \quad x = 0$$

and

$$F(x) = \int\limits_0^x w_q(t)dt$$

$$= 1 - \rho e^{-\mu(1-\rho)x}, \quad x \geq 0. \tag{2.8}$$

The above gives the (unconditional) distribution of the waiting time of a test unit. The probability that the unit has not to wait is $1 - \rho$ and that it has to wait is ρ. We have

$$E(W_Q) = -\frac{d}{d\alpha} w_q^*(\alpha)\, |_{\alpha=0}$$

$$= \frac{\rho}{\mu(1-\rho)} = \frac{\lambda}{\mu(\mu-\lambda)}, \tag{2.9a}$$

$$E(W_Q^2) = \frac{d^2}{d\alpha^2} w_q^*(\alpha)\, |_{\alpha=0},$$

and

$$\mathrm{var}(W_Q) = \frac{\rho(2-\rho)}{\mu^2(1-\rho)^2}. \tag{2.9b}$$

Now the p.d.f. of the (conditional) distribution of the waiting time in the queue, denoted by $_cW_Q$, given that the test unit has to wait, is given by

$$f_c(x)dx = \Pr\{x \leq {}_cW_Q < x + dx\}, x > 0$$

$$= \Pr\{x \leq W_Q < x + dx \mid \text{the test unit has to wait}\}$$

$$= \frac{\mu\rho(1-\rho)e^{-\mu(1-\rho)x}}{\rho} dx = \mu(1-\rho)e^{-\mu(1-\rho)x}dx, \quad x > 0.$$

The distribution of $_cW_Q$ is exponential (with mean $1/\mu\,(1-\rho)$) while that of W_Q is modified exponential.

10.2.2.2 Waiting Time in the System (or *Response Time* or *Sojourn Time*)

The random variable 'time spent in the system' (or *sojourn time*) by a unit includes the service time of the unit besides its queueing time. This variable is given by $W_s = W_Q + S$, where W_Q is the queueing time and S is the service time, and its distribution can be obtained by convolution. We give below a derivation similar to the one given for W_Q.

Now the test unit has to wait in the system even if the system is empty ($n = 0$), the time spent being equal to his service time. Given that there are n (≥ 0) units in the system, its waiting time is $W_s = S_{n+1}$, where $S_{n+1} = v'_1 + v_2 + \cdots + v_{n+1}$, v_{n+1} being its own service time.

Thus the p.d.f. $w_s(x)$ of W_s (for $x > 0$) is given by

$$w_s dx = \Pr\{x \leq W_s < x + dx\}$$

$$= \sum_{n=o}^{\infty} \Pr\{x \leq S_{n+1} < x + dx \mid \text{given that}$$

there are n units in the system$\} \times \Pr\{N = n\}$.

$$= \left\{ \sum_{n=o}^{\infty} \mu \frac{(\mu x)^n}{\Gamma(n+1)} e^{-\mu x} dx \right\} \{(1-\rho)\rho^n\}$$

$$= \mu(1-\rho)e^{-\mu(1-\rho)x} dx. \tag{2.10}$$

Thus the W_s has an exponential distribution with mean $1/\{\mu(1-\rho)\} = 1/(\mu - \lambda)$ and variance $1/\{\mu(1-\rho)\}^2$.

The distribution function is

$$F(x) = \Pr\{W_s \leq x\} = \int_o^x w_S(x) dx = 1 - e^{-\mu(1-\rho)x}, \quad x > 0. \tag{2.10b}$$

Note: (1) S_{n+1} is the random (geometric) sum of $(n+1)$ i.i.d. exponential variables. That its distribution will be again exponential was also shown in Example 3(h), Ch. 1.
(2) It can be easily verified that Little's formula holds.
(3) The average response time $E\{W_s\}$ is often taken as a measure of system performance. It depends both on λ and μ. Even though ρ may be large (near 1), $E\{W_s\}$ can be small if $1/\mu$ is small.

Example 2(a). The arrivals at a counter in a bank occur in accordance with a Poisson process at an average rate of 8 per hour. The duration of service of a customer has exponential distribution with a mean of 6 minutes. Find the probability that an arriving customer: (i) has to wait on arrival, (ii) finds 4 customers in the system and (iii) has to spend less than 15 minutes in the bank. Estimate also the fraction of the total time that the counter is busy.

Here $\lambda = 8/hr., \quad \mu = 10/hr., \quad \rho = 0.8.$

The probability that the customer (i) has to wait is given by $\Pr\{N \geq 1\} = \rho = 0.8$, and (ii) finds 4 customers is given by $p_4 = (1 - \rho)\,\rho^4 = (0.2)(0.8)^4 = 0.08192$.

The probability that he has to spend less than 15 minutes ($= 0.25$ hr.) is given by

$$F(0.25) = 1 - \exp\{-\mu(1-\rho)(0.25)\} \qquad \text{(from 2.10b)}$$

$$= 1 - \exp\{-10(0.2)(0.25)\} = 1 - e^{-0.5} = 0.3935.$$

The utilization factor $\rho = 0.8$ gives an estimate of the fraction of the total time that the counter is busy.

10.2.3 Queue with Limited Waiting Space : The Queueing Model M/M/1/K

In the $M/M/1$ model, it is assumed that the system can accommodate any number of units. In practice this may seldom be the case. We have thus to consider the situation such that the system has limited waiting space and can hold a maximum number of K units (including the one being served). The analysis is similar. Equation (2.1) will hold but (2.2) will hold only for $1 \leq n \leq (K - 1)$. Further, there will be another relation. Suppose that the number of units in the system in an interval $(0, t)$ is K, then the probability that there will be K units in the interval $(0, t + h)$ is given by

$$p_K(t+h) = p_K(t)(1 - \mu h) + p_{K-1}(\lambda h)(1 - \mu h) + o(h)$$

which gives, as $h \to 0$,

$$p_K{}'(t) = -\mu p_K(t) + \lambda p_{K-1}(t)$$

The balance equations can be easily written down from the state-transition-rate diagram (Fig 10.2) for this model.

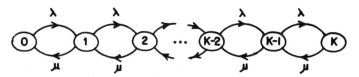

Fig. 10.2 State-transition-rate diagram of $M/M/1/K$ model

The steady state balance equations are

$$\lambda p_o = \mu p_1$$

$$(\lambda + \mu)p_n = \lambda p_{n-1} + \mu p_{n+1}, \ 1 \leq n \leq K - 1 \qquad (2.11)$$

$$\mu p_K = \lambda p_{K-1}.$$

From the first two equations we get, as in (2.5) ($\rho = \lambda/\mu$)

$$p_n = p_0 \rho^n, 0 \leq n \leq K - 1.$$

From the last one, $p_K = \rho p_{K-1} = p_0 \rho^K$, so that

$$p_n = p_0 \rho^n \quad \text{for} \quad n = 0, 1, ..., K.$$

Using $\quad\quad \sum_{n=0}^{K} p_n = 1$, we get

$$p_0 = \frac{1-\rho}{1-\rho^{K+1}}, \quad \rho \neq 1$$

$$= \frac{1}{K+1}, \quad \rho = 1.$$

Finally, we have, for $n = 0, 1, \ldots K$,

$$p_n = \frac{(1-\rho)\rho^n}{1-\rho^{K+1}}, \quad \rho \neq 1$$

$$= \frac{1}{K+1}, \quad \rho = 1. \tag{2.12}$$

The distribution is truncated geometric for $\rho \neq 1$ and uniform for $\rho = 1$.
The mean number of units in the system L_K is given by

$$L_K = \sum_{n=0}^{K} n p_n.$$

For $\rho = 1$,
$$L_K = \frac{K(K+1)/2}{K+1} = \frac{K}{2}.$$

For $\rho \neq 1$,
$$L_K = \frac{(1-\rho)\rho}{1-\rho^{K+1}} \sum_{n=0}^{K} n\rho^{n-1}$$

$$= \frac{(1-\rho)\rho}{1-\rho^{K+1}} \sum_{n=0}^{K} \frac{d}{d\rho}(\rho^n)$$

$$= \frac{(1-\rho)\rho}{1-\rho^{K+1}} \cdot \frac{d}{d\rho}\left\{\sum_{n=0}^{K}\rho^n\right\}$$

$$= \frac{(1-\rho)\rho}{1-\rho^{K+1}} \cdot \frac{1-(K+1)\rho^K + K\rho^{K+1}}{(1-\rho)^2} \tag{2.13}$$

It is to be noted that since the finite series $\sum_{n=0}^{K} \rho^n$ has a sum for all values of ρ ($\rho < 1$

and $\rho \geq 1$), the steady state solution exists even for $\rho \geq 1$. Some of the results for $M/M/1$ model (or $M/M/1/\infty$ model) can be obtained by taking limits of the corresponding expressions as $K \to \infty$. But then this could be done only when $\rho < 1$.

Note (1) The effective arrival rate (i.e. number of customers joining the system per unit time) is $\lambda(1 - p_K)$ and the effective rejection rate (i.e. number of customers diverted from the system per unit time) is $\lambda - \lambda (1 - p_K) = \lambda p_K$.

(2) The effective rate of departure of customers (after receiving service) is $\mu(1 - p_0)$.

(3) The server utilisation factor is $1 - p_0 = \rho(1 - p_K)$ ($\neq \rho$).

(4) The model has been applied in several interesting situations (see, for example, De Vany, *J. Pol. Ec.* **84** (1976) 523–541, Naor, *Econometrica* **37** (1969) 15–23, Rue & Rosenshine *Nav. Res. Log. Qrly.* **28** (1981) 489–495.

10.2.3.1 Two-stage Cyclic Queue

Consider a cyclic queueing model with K circulating jobs and with exponential servers, the server at station A having rate λ, and the server at station B having rate μ. Having been served at station B, the job goes to station A with probability q_1, or goes directly again to station B with probability q_0 (= $1 - q_1$). Let the system be in steady state.

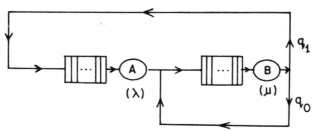

Fig. 10.3 A two-stage cyclic queue

Let p_n be the probability that number of jobs at station B (in queue or in service) is n. The state transition rate diagram is given in Fig. 10.4

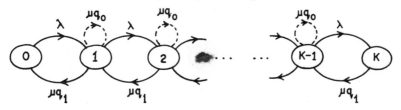

Fig. 10.4 State transition rate diagram of two-stage cyclic queue

Now, for $n = 0$,

Rate of flow out of state 0 equals λp_0

Rate of flow into state 0 equals $\mu q_1 p_1$;

thus
$$\lambda p_0 = \mu q_1 p_1. \tag{2.14a}$$

For n ($1 \le n \le K - 1$),

Rate of flow out of state n equals $\lambda p_n + \mu q_1 p_n$

Rate of flow into state n equals $\lambda p_{n-1} + \mu q_1 p_{n+1}$;

thus

$$(\lambda + \mu q_1)p_n = \lambda p_{n-1} + \mu q_1 p_{n+1} \tag{2.14b}$$

For $n = K$,

Rate of flow out of state K equals $\mu q_1 p_K$

Rate of flow into state K equals λp_{K-1};

thus

$$\mu q_1 p_K = \lambda p_{K-1}. \tag{2.15}$$

Note that the flow from a state into itself does not enter into the above relations (of flow into or flow out of that state).

We thus have the same equations (2.11) as in an $M/M/1/K$ queue with μq_i in place of μ.

Writing $\quad \rho = \lambda / \mu q_1$, we get

$p_n = \Pr \{n$ jobs in station $B, K-n$ in station $A\}$

$$= \frac{(1-\rho)\rho^n}{1-\rho^{K+1}}, \qquad \rho \neq 1$$

$$= \frac{1}{K+1}, \qquad \rho = 1$$

with

$$p_0 = \frac{1-\rho}{1-\rho^{K+1}}, \qquad \rho \neq 1$$

$$= \frac{1}{K+1}, \qquad \rho = 1$$

The utilization factor of server B equals

$$U_0 = 1 - p_0.$$

The average throughout equals $\mu U_0 p_0$.

Particular case: When $q_1 = 1$ $(q_0 = 0)$, the job from server B goes to server A and then $\rho_1 = \rho$.

The $M/M/1/K$ queue with limited waiting room can then be used as a model of two-stage cyclic queue.

Note: The above model of cyclic queue serves as a model of a multiprogramming system with server A as the I/O unit and server B as the CPU with a fixed number of jobs (or terminals). The above model is simple in the sense that the servers are assumed to be independent exponential.

Taking into consideration other types of servers (say Coxian) other models are obtained.

10.3 TRANSIENT BEHAVIOUR OF *M/M/1* MODEL

We have now to solve the equations (2.1) and (2.2) in transient state. A number of methods have been put forward for their solutions: the spectral method of Lederman and Reuter (1956); the method of generating function of Bailey (1956); the combinatorial method of Champernowne (1956). The methods are given in all advanced level books on Queueing theory (e.g. Prabhu, Saaty, Gross and Harris, Cohen, Syski, etc.). We consider Conolly's (1958) difference equation technique and also outline Bailey's (1956) generating function method.

After about three decades, there has been a resurgence of interest in the transient analysis of *M/M/1* queueing system. Fresh light has been thrown on the transient behaviour through such analyses. Reference may be made, for example, to Pegden and Rosenshine (1982), Hubbard *et al.* (1986), Parthasarathy (1987), Parthasarathy and Sharafali (1989), Abate and Whitt (1987,1988), Syski (1988), Sharma (1990); see also Medhi (1991).

While most of the authors consider continuous time framework, transient analysis in discrete time has also been receiving attention; see, for example, Mohanty and Panny (1990), Kanwar Sen and Jain (1990), Kanwar Sen *et al.* (1990), Chaudhry and Agarwal (1992).

10.3.1 Difference Equation Technique

For obtaining the solution in steady state the number in the state at the initial stage is of no consequence. Here the knowledge of the initial state is needed.

We assume, for simplicity, that the number of units at the instant $t = 0$ is 0. Let

$$p_n(t) = \Pr\{N(t) = n \mid N(0) = 0\}, \quad p_0(0) = 1, \quad p_n(0) = 0, \quad n \neq 0.$$

We refer to the approach to solution of the equations (3.13) and (3.14) as considered in Example 3(c), Appendix A.3.

Let $f_n(s)$ be the L.T. of $p_n(t)$. Then from (2.1) and (2.2), We get

$$(s + \lambda) f_0(s) = 1 + \mu f_1(s) \tag{3.1}$$

and

$$(s + \lambda + \mu) f_n(s) = \lambda f_{n-1}(s) + \mu f_{n+1}(s), n \geq 1. \tag{3.2}$$

The solution of the equation (3.2) is

$$f_n(s) = A \alpha^n + B \beta^n, n \geq 1, \tag{3.3}$$

where A, B are functions of s (and not of n) and α, β are the roots of the characteristic equation

$$\mu x^2 - (s + \lambda + \mu) x + \lambda = 0. \tag{3.4}$$

We are now to find the values of A and B.

Writing $K(s) \equiv K = \{(s+\lambda+\mu)^2 - 4\lambda\mu\}^{1/2}$, we get

$$\frac{\alpha}{\beta} \equiv \frac{(s+\lambda+\mu) \pm K}{2\mu} \quad (\alpha \equiv \alpha(s), \beta \equiv \beta(s)). \tag{3.4a}$$

It can be seen that, for all Re $(s) > 0$ and real and positive λ, μ,

$$|\alpha| > 1 \quad \text{and} \quad |\beta| < 1 \text{(see note below)}.$$

Now $\qquad \sum_{n=0}^{\infty} p_n(t) = 1$ and hence $\sum_{n=0}^{\infty} f_n(s) = \frac{1}{s}$.

Since $|\alpha| > 1$, the convergence of $\sum_{n=0}^{\infty} f_n(s) = \sum_{n=0}^{\infty} (A\alpha^n + A\beta^n)$

implies that $A = 0$. Thus $f_n(s) = B\beta^n$, $n \geq 1$. Choose B such that this holds for $n = 0$ also.

Suppose that $f_0(s) = B$, so that (3.3) is satisfied for $n = 0$ also; then

$$f_n(s) = f_0(s)\beta^n, n \geq 0. \tag{3.5}$$

Using $\sum_{n=0}^{\infty} f_n(s) = 1/s$, we get

$$f_0(s) = \frac{1}{s+\lambda-\mu\beta} = \frac{1-\beta}{s}.$$

Thus $\qquad f_n(s) = f_0(s)\beta^n = \frac{(1-\beta)\beta^n}{s}; \tag{3.5a}$

$p_n(t)$ can be obtained by taking the inverse Laplace transform of $f_n(s)$. This is done by using some properties of Laplace transforms and by taking recourse to ingenious and somewhat complicated algebraic manipulations (see Gross and Harris, for details). We proceed to find the probability Pr $\{N(t) \geq n\}$. From (3.5a) we get

$$\sum_{r=n}^{\infty} f_r(s) = \frac{1-\beta}{s} \sum_{r=n}^{\infty} \beta^r = \frac{1-\beta}{s} \cdot \frac{\beta^n}{1-\beta} = \frac{\beta^n}{s}$$

$$= (\lambda/\mu)^n \alpha^{-n}/s, \tag{3.6}$$

$$\equiv L\{A_n(t)\}, n \geq 0,$$

where $\qquad A_n(t) \equiv \text{Pr } \{N(t) \geq n\} = p_n(t) + p_{n+1}(t) + \cdots.$

The function $A_n(t)$ can be obtained by finding the inverse L.T. of (3.6).

From the tables of L.T., we find that the L.T. of $\left\{\dfrac{n}{t}I_n(at)\right\}$ is

$$\left\{\frac{a}{s+\sqrt{(s^2-a^2)}}\right\}^n .$$

Using the translation property, the L.T. of

$$a_n(t) \equiv \left(\frac{\lambda}{\mu}\right)^{n/2} e^{-(\lambda+\mu)t} \left\{\frac{n}{t}I_n\left(2t\sqrt{\lambda\mu}\right)\right\} \tag{3.7}$$

is found to be

$$\left(\frac{\lambda}{\mu}\right)^{n/2}\left\{\frac{2\sqrt{(\lambda\mu)}}{(s+\lambda+\mu)+\sqrt{\{(s+\lambda+\mu)^2-4\lambda\mu\}}}\right\}^n = \left(\frac{\lambda}{\mu}\right)^n \alpha^{-n}$$

and hence the L.T. of $\displaystyle\int_0^t a_n(u)\,du$

is $\hspace{6cm} (\lambda/\mu)^n \alpha^{-n}/s .$

Thus the derivative of $A_n(t)$ is $a_n(t)$, given by (3.7).

The steady state probability $p_n = \lim_{t\to\infty} p_n(t)$ can be obtained by taking limit of $p_n(t)$ obtained after inverting $f_n(s)$. It can also be obtained easily by using the initial value theorem of Laplace transform.

Now $\displaystyle\lim_{s\to 0}\beta(s) \equiv \lim_{s\to 0}\frac{(s+\lambda+\mu)-\sqrt{\{(s+\lambda+\mu)^2-4\mu\lambda\}}}{2\mu}$

$$= \begin{cases} \dfrac{(\lambda+\mu)-(\mu-\lambda)}{2\mu} = \rho, & \text{when } \lambda < \mu \\[2mm] \dfrac{(\lambda+\mu)-(\lambda-\mu)}{2\mu} = 1, & \text{when } \lambda \geq \mu \end{cases} \tag{3.8}$$

Thus $\displaystyle p_n = \lim_{t\to\infty}p_n(t) = \lim_{s\to 0}sf_n(s)$

$$= \lim_{s\to 0}(1-\beta)\beta^n$$

$$= \begin{cases} (1-\rho)\rho^n, & \rho < 1 \\ 0, & \rho \geq 1. \end{cases}$$

The result $p_n = 0, \rho \geq 1$ may be interpreted by saying that when the traffic intensity is greater than or equal to 1, there is ultimately zero probability that the system contains a finite number of units. One would intuitively expect that the queue size would grow larger and larger in such cases.

Another fact to be noted is that, when $\rho = 1$, the states are persistent and null having infinite recurrence time.

Note: That $|\alpha| > 1$ and $|\beta| < 1$ can be best demonstrated by applying:

Rouche's Theorem. If $f(z)$ and $g(z)$ are functions analytic inside and on a closed contour C, and if $|g(z)| < |f(z)|$ on C, then $f(z)$ and $f(z) + g(z)$ have the same number of zero (s) inside C.

(For a proof of this theorem, reference may be made to any standard book on complex-variable theory.)

Here we take C to be $|z| = 1, f(z) = (\lambda + \mu + s)z$, $g(z) = \lambda + \mu z^2$, and Re $(s) > 0$. We have, on $|z| = 1$,

$$|g(z)| = |\lambda + \mu z^2| \leq \lambda + \mu < |s + \lambda + \mu| = |(s + \lambda + \mu)z| = |f(z)|.$$

Now $f(z) = (s + \lambda + \mu)z$ has only one zero inside $|z| = 1$ and so $\mu z^2 + \lambda - (s + \lambda + \mu)z = 0$ also has only one root inside $|z| = 1$. This equation has two roots α, β of which β has the smaller modulus. Thus $|\beta| < 1$ and $|\alpha| > 1$.

A more sophisticated approach not involving generating function and Rouche's theorem has been given by Neuts (1979).

10.3.2 Method of Generating Function

To cover the more general case, let us suppose that $N(0) = i$, i.e. the process starts with i units at time $t = 0$; and we denote

$$\Pr\{N(t) = n \mid N(0) = i\} \text{ by } p_{in}(t) \text{ so that } p_{ij}(0) = \delta_{ij}$$

(so far we assumed that $N(0) = 0$; putting $i = 0$ and writing $p_n(t)$ for $p_{in}(t)$ when $i = 0$ we can get back to the case last studied). In place of the difference equations (2.1) and (2.2) we shall now get

$$p'_{i0}(t) = -\lambda p_{i0}(t) + \mu p_{i1}(t) \tag{3.9}$$

$$p'_{in}(t) = -(\lambda + \mu)p_{in}(t) + \lambda p_{in-1}(t) + \mu p_{in+1}(t), n = 1, 2, \ldots. \tag{3.10}$$

Let
$$P(z,t) = \sum_{n=0}^{\infty} p_{in}(t)z^n$$

be the generating function of $\{p_{in}(t)\}$. Then $P(z, 0) = z^i$. Let

$$f_{in}(s) \equiv L\{p_{in}(t)\}$$

and

$$F(z,s) \equiv L\{P(z,t)\}.$$

Multiplying (3.10) by z^n and adding for $n = 1, 2, \ldots$, the resulting equation with (3.9) we get

$$\sum_{n=0}^{\infty} p'_{in}(t)z^n = -(\lambda+\mu) \sum_{n=0}^{\infty} p_{in}(t)z^n + \mu p_{i0}(t)$$

$$+ \lambda \sum_{n=1}^{\infty} p_{i\,n-1}(t)z^n + \mu \sum_{n=0}^{\infty} p_{i\,n+1}(t)z^n$$

or $\quad \dfrac{\partial}{\partial t} P(z,t) = -(\lambda+\mu)P(z,t) + \mu p_{i0}(t) + \lambda z P(z,t)$

$$+ (\mu/z)\{P(z,t) - p_{i0}(t)\}$$

or $\quad z\dfrac{\partial}{\partial t}P(z,t) = \{\lambda z^2 - (\lambda+\mu)z + \mu\}P(z,t) - \mu(1-z)p_{i0}(t). \qquad (3.11)$

Taking L.T. and noting that

$$L\left\{\frac{\partial}{\partial t}P(z,t)\right\} = sF(z,s) - P(z,0)$$

$$= sF(z,s) - z^i,$$

we get

$$szF(z,s) - z^{i+1} = \{\lambda z^2 - (\lambda+\mu)z + \mu\}F(z,s) - \mu(1-z)f_{i0}(s)$$

or $\qquad F(z,s) = \dfrac{\mu(1-z)f_{i0}(s) - z^{i+1}}{\lambda z^2 - (\lambda+\mu+s)z + \mu}. \qquad (3.12)$

This involves $f_{i0}(s)$ which can be evaluated as follows. The denominator of the r.h.s. has two zeros, say, ξ and η,

where $\qquad \dfrac{\xi}{\eta} \equiv \dfrac{(s+\lambda+\mu) \pm K}{2\lambda}; \quad K \equiv K(s)$

further, $|\xi| < 1$, $|\eta| > 1$. Note that the roots of the equation $\lambda z^2 - (\lambda+\mu+s)z + \mu = 0$ are the reciprocals of the roots of the equation (3.4). Thus ξ, η are the reciprocals of α, β, i.e. $\xi = 1/\alpha$, $\eta = 1/\beta$, and thus $|\xi| < 1$ and $|\eta| > 1$. It can be shown that $\lambda z^2 - (\lambda+\mu+s)z + \mu = 0$ has only one zero $z = \xi$ in the unit circle. Since $F(z, s)$ converges in the region $|z| = 1$, the zeros in the unit circle of the numerator and the denominator of the r.h.s. of (3.12) must coincide, so that $z = \xi$ must be a zero of the numerator also. Hence

$$f_{i0}(s) = \xi^{i+1} / \mu(1-\xi)$$

and finally

$$F(z,s) = \frac{(1-z)\xi^{i+1} - z^{i+1}(1-\xi)}{\lambda(1-\xi)(z-\xi)(z-\eta)}. \tag{3.13}$$

Expanding the r.h.s. of (3.13) in power of z, $f_{in}(s)$ can be obtained as the coefficient of z^n.   ▲

Particular case: When $i = 0$, i.e., $N(0) = 0$, we get

$$F(z,s) = \frac{1}{\lambda(1-\xi)(\eta-z)} = \frac{1}{\lambda(1-1/\alpha)(1/\beta-z)} \tag{3.14}$$

$$= \frac{\alpha\beta}{\lambda(\alpha-1)} \sum_{n=0}^{\infty} \beta^n z^n = \frac{(1-\beta)}{s} \sum_{n=0}^{\infty} \beta^n z^n$$

$$\text{(since } \mu(\alpha-1)(\beta-1) = -s).$$

Hence, $$f_n(s) = f_{0n}(s)$$

$$\equiv \text{coeff. of } z^n \text{ in } F(z,s)$$

$$= \frac{(1-\beta)\beta^n}{s}, \quad \text{as in (3.5a)}.$$

Note: Pegden and Rosenshine (*Mgmt. Sci.* **28** (1982) 821–828) consider a two-dimensional state model; the state of the system is denoted by (i, j), where i is the number of arrivals and j is the number of departures until time t (with corresponding probability $p_{ij}(t)$). Expressions for $p_{ij}(t)$ have been obtained.

10.3.3. Busy Period : Zero-avoiding State Probabilities

A busy period is defined as the interval of time commencing at the instant 0 when a unit arrives at an empty counter and terminating at the instant when the server becomes free for the first time. Let the length of the interval which is a random variable be T, and $\{N^*(t)\}$ be the stochastic process denoting the number of units present at the instant t during the busy period. We have Pr $\{N^*(0) = 1\} = 1$ and the duration of the busy period is the first instant t for which $N^*(t) = 0$. Let

$$q_n(t) = \Pr\{N^*(t) = n \mid N^*(0) = 1\}, \tag{3.15}$$

be the *zero-avoiding state probabilities*.

Then $q_1(0) = 1$, $q_n(0) = 0$, $n \geq 2$. For $n \geq 2$, $q_n(t)$ will satisfy the same differential equations as $p_n(t)$, i.e. the equations (2.2) will hold also for $q_n(t)$ for $n = 2, 3, \ldots$, i.e.

$$q'_n(t) = -(\lambda + \mu)q_n(t) + \lambda q_{n-1}(t) + \mu q_{n+1}(t). \tag{3.16}$$

As the term $q_0(t)$ will not occur the equations corresponding to $n = 1$ and $n = 0$ will be as follows:

$$q_1'(t) = -(\lambda + \mu)q_1(t) + \mu q_2(t) \tag{3.17}$$

$$q_0'(t) = \mu q_1(t). \tag{3.18}$$

Let $g_n(s)$ be the L.T. of $q_n(t)$. Then taking L.T. of (3.16) we get, for $n \geq 2$,

$$s g_n(s) - q_n(0) = -(\lambda + \mu)g_n(s) + \lambda g_{n-1}(s) + \mu g_{n+1}(s)$$

or $\qquad \mu g_{n+1}(s) - (s + \lambda + \mu)g_n(s) + \lambda g_{n-1}(s) = 0.$

The solution of this equation is (as in (3.3))

$$g_n(s) = A\alpha^n + B\beta^n, \quad n \geq 2, |\alpha| > 1, |\beta| < 1.$$

We can choose the constants A and B so that they yield correct values of $g_1(s)$ and $g_0(s)$. Again, $\sum_n g_n(s)$ must converge so that A is equal to zero. Thus putting $n = 1$, we get $g_1(s) = B\beta$ so that

$$g_n(s) = g_1(s)\beta^{n-1}, \quad n \geq 1. \tag{3.19}$$

Taking L.T. of (3.17) we find

$$s g_1(s) - q_1(0) = -(\lambda + \mu)g_1(s) + \mu\beta g_1(s),$$

whence $\qquad g_1(s) = \dfrac{1}{s + \lambda + \mu - \mu\beta} = \dfrac{\beta}{\lambda}$ (as β satisfies (3.4)).

Thus $\qquad g_n(s) = g_1(s)\beta^{n-1} = \beta^n / \lambda = \rho^n / \lambda\alpha^n, \, n = 1, 2, \ldots \tag{3.20}$

Inverting the L.T., we get, for $n = 1, 2, \ldots$

$$q_n(t) = \frac{n\rho^{(n/2)}}{\lambda t} e^{-(\lambda+\mu)t} I_n(2t\sqrt{\lambda\mu}). \tag{3.21}$$

Let $b(t)$ be the p.d.f. of the busy period T. then, for small δt,

$$b(t)\delta(t) = \Pr\{t \leq T < t + \delta t\}$$

equals the probability that the busy period terminates in $(t, t + \delta t)$. It is clear that this probability equals the probability that there are $j \ (\geq 1)$ units in the system at the instant t and that the services of these j units are completed in $(t, t + \delta t)$. Thus

$$b(t)\delta t = \sum_{j=1}^{\infty} [\Pr\{N^*(t) = j \mid N^*(0) = 1\}$$

$$\Pr \times \{j \text{ units complete service in} (t, t + \delta t)\}]$$

$$= q_1(t)\mu\delta t + \sum_{j=2}^{\infty} q_j(t) (o(\delta t))$$

so that

$$b(t) = \mu \, q_1(t)$$

$$= \frac{1}{t}\rho^{-\frac{1}{2}} e^{-(\lambda + \mu)t} I_1(2t\sqrt{\lambda\mu}). \tag{3.22}$$

(using(3.21)).

The L.T. of the p.d.f. of T is

$$b^*(s) = L\{b(t)\} = L\{\mu \, q_1(t)\} = \mu \, g_1(s) = \beta/\rho. \tag{3.23}$$

Using (3.8) we get,

$$\Pr\{T < \infty\} = \lim_{s \to 0+} b^*(s) = 1, \qquad \text{if} \qquad \rho < 1$$

$$= 1/\rho, \qquad \text{if} \qquad \rho \geq 1,$$

indicating that there is non-zero probability that the busy period lasts for ever when $\rho \geq 1$.

We now find the moments of T. We have for $\rho < 1$,

$$\frac{d}{ds}\beta(s) = \frac{1}{2\mu}\{1 - 2(s + \lambda + \mu) \, / \, 2K\} = -\frac{\beta}{K}, (K \equiv K(s))$$

$$\frac{d^2}{ds^2}\beta(s) = \frac{\beta}{K^2} + \frac{\beta}{K^3}(s + \lambda + \mu) = \frac{\lambda}{K^3}$$

Thus

$$E(T) = -\frac{d}{ds}b^*(s)\bigg|_{s=0} = \frac{\beta}{\rho K}\bigg|_{s=0} = \frac{1}{\mu - \lambda} = \frac{1}{\mu(1 - \rho)}, \tag{3.24}$$

$$E(T^2) = \frac{d^2}{ds^2} b*(s)\Big|_{s=0} = \frac{2\lambda}{\rho} \cdot \frac{1}{(\mu-\lambda)^3} = \frac{2}{\mu^2(1-\rho)^3}, \qquad (3.25)$$

and $$\text{var}\,(T) = \frac{1+\rho}{\mu^2(1-\rho)^3}.$$

The coefficient of variation $C_V = \dfrac{(\text{var}\,T)^{\frac{1}{2}}}{E(T)} \times 100$

$$= \frac{(1+\rho)^{\frac{1}{2}}}{(1-\rho)^{\frac{1}{2}}} \times 100$$

increases as ρ increases, while it is constant for W_s. The following table gives $E\,(T)$ and var (T) for some values of λ, μ ($\mu = 1$, $\lambda = \rho$).

$\lambda\,(=\rho)$	0.2	0.4	0.5	0.8	0.9	0.95	0.995
$E(T)$	1.25	1.67	2	5	10	20	200
var (T)	2.34	6.48	12	225	1900	15600	1596×10^4

It is clear that C_V is large for large ρ, i.e. var (T) is large compared to $E\,(T)$; as such the expected value does not seem to have much significance for large ρ.

Note: (1) *Idle* (or *free*) *period I* is the duration of the interval from the instant the server becomes free for the first time to the instant of the next arrival when he becomes engaged. This duration being the residual portion of the interarrival time is again exponential with the same mean $1/\lambda$, since the interarrival distribution is exponential with mean $1/\lambda$. This holds for all queueing systems with Poisson arrivals.
From Example 10(a), Ch. 6, we have

$$\frac{E(T)}{E(I)} = \frac{1-p_0}{p_0} = \frac{\rho}{1-\rho}$$

whence

$$E(T) = \frac{1}{\mu(1-\rho)}.$$

(2) We have considered the initial busy period, i.e. the period initiated by one unit. The distribution of the busy period γ_j initiated by j units is given by the convolution $\tau_j = \sum_{i=1}^{j} T_i$, where T_i's are i.i.d. random variables distributed as T considered above.

(3) The concept of busy period is due to the French mathematician Emile Borel (1871-1956), who obtained the joint bivariate distribution of T and N, the number served during the busy period T for $M/D/1$ model. Various approaches are available for obtaining the distribution of T as well as the joint distribution of T and N in more general cases. Distributions of T and N can be obtained as marginal distributions of the joint distribution. The trivariate distribution of T, N and M, where M denotes the maximum queue length attained during the busy period T, has also been engaging attention of researchers.

(4) A busy period and the idle period following the former together constitute a *busy cycle* for any *queueing system*. Let time be reckoned from the epoch at which a unit arrives to an empty system and service commences. Let Z_i, $i = 1, 2, \ldots$ be the lengths of successive busy cycles, and N_i be the number served during the ith cycle. The random variables Z_i and N_i are mutually independent and are indentically distributed (N_i as N); but Z_i and N_i are not independent.

$\{Z_i\}$ and $\{N_i\}$, $i = 1, 2, \ldots$ are renewal processes.

Example 3(a). *First passage time distribution.* For an $M/M/1$ queue in transient state, let τ be the random variable denoting the time taken for the system to *fall* from the initial state a to another state $b(< a)$ for the first time. Let $\{X(t)\}$ be the process denoting such states of the system, given that $X(0) = a$, and let

$$q_n(t) = \Pr\{X(t) = n \mid X(0) = a\}. \tag{3.26}$$

Using similar arguments it can be easily shown that the state probabilities satisfy the following equations:

$$q_n'(t) = -(\lambda + \mu)q_n(t) + \lambda\, q_{n-1}(t) + \mu q_{n+1}(t), \ n \geq (b+2)$$

$$q'_{b+1}(t) = -(\lambda + \mu)q_{b+1}(t) + \mu q_{b+2}(t)$$

$$q'_b(t) = \mu\, q_{b+1}(t). \tag{3.27}$$

The p.d.f. $f(t)$ of the first passage time distribution from a to b is given by $f(t) = \mu q_{b+1}(t) = q'_b(t)$.

Putting $a = 1$, $b = 0$, we can get back to initial busy period distribution.

Waiting time process. Let $W(t)$ be the time required to serve all the units present in the system at the instant t, given that $W(0) = 0$. Then, if $N(t)$ is the number present at instant t, $(N(0) = 0,)$ and

$$W(t) = \begin{cases} 0, & \text{if } N(t) = 0 \\ v_1' + v_2 + \ldots + v_{N(t)}, & \text{if } N(t) > 0 \end{cases}$$

where v_1' is the residual service time of the unit being served at the instant t, and v_2, $\ldots, v_{N(t)}$ those of the waiting units at the instant t. $\{W(t), t \geq 0\}$, which is known as *virtual waiting time*, is a Markov process (with continuous state space). Given $W(0) = 0$, then proceeding as in the case of waiting time in the system in steady state it can be seen that its probability element

$$f(x,t)dx = \Pr\{x \leq W(t) < x + dx\}, 0 \leq x < \infty, 0 \leq t < \infty \text{ is given by}$$

$$f(x,t)dx = \sum_{n=0}^{\infty} \left\{ \mu \frac{(\mu x)^n}{\Gamma(n+1)} e^{-\mu x} dx \right\} p_{n+1}(t).$$

Its L.T. can be put in elegant closed form (see Prabhu (1965)).

10.4 BIRTH AND DEATH PROCESSES IN QUEUEING THEORY : MULTICHANNEL MODELS

10.4.1 Birth and Death Processes

Let us first consider a birth and death process with state dependent birth and death rates λ_n and μ_n respectively. Let $N(t)$ be the number present at the instant t, and

$$p_n(t) = \Pr\{N(t) = n \mid N(0) = \cdot\}$$

It was shown in Ch. 4 that $\{N(t), t \geq 0\}$ is a Markov process with denumerable state space $\{0, 1, 2, \ldots\}$ and that the forward Kolmogorov equations of the process are :

$$p_0'(t) = -\lambda_0 p_0(t) + \mu_1 p_1(t)$$

$$p'_n(t) = -(\lambda_n + \mu_n)p_n(t) + \lambda_{n-1}p_{n-1}(t) + \mu_{n+1}p_{n+1}(t), \quad n = 1, 2, \ldots$$

(see equations (4.4) and (4.6), Ch. 4).

We proceed to investigate the steady state solutions. Assume that such solutions exist, then $\lim_{t \to \infty} p_n(t) = p_n = \Pr\{N = n\}$, N being the random variable giving the number of units. Putting $p_n(t) = p_n$ and $p'_n(t) = 0$, we get from the above the following difference equations in steady state:

$$0 = -\lambda_0 p_0 + \mu_1 p_1 \tag{4.1}$$

$$0 = -(\lambda_n + \mu_n)p_n + \lambda_{n-1}p_{n-1} + \mu_{n+1}p_{n+1}, \; n = 1, 2, \ldots \tag{4.2}$$

Alternatively, we can obtain the steady state balance equations by using the rate-equality principle.

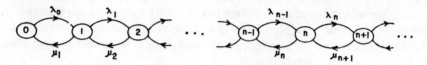

Fig. 10.5 State-transition-rate diagram of birth-death process.

From the state-transition-rate diagram (Fig. 10.5) it is clear that,

$$\text{for} \quad n = 0, \qquad \lambda_0 p_0 = \mu_1 p_1$$

and $\qquad$ for $\quad n > 0, \quad (\lambda_n + \mu_n)p_n = \lambda_{n-1}p_{n-1} + \mu_{n+1}p_{n+1},$

(which are the equations (4.1) and (4.2)).
From (4.2), we have for $n = 1, 2, \ldots,$

$$\mu_{n+1}p_{n+1} - \lambda_n p_n = \mu_n p_n - \lambda_{n-1}p_{n-1}$$

$$= \mu_{n-1}p_{n-1} - \lambda_{n-2}p_{n-2} \ (\text{putting } n-1 \text{ for } n)$$

$$\cdots \qquad \cdots$$

$$= \mu_1 p_1 - \lambda_0 p_0$$

$$= 0 \qquad (\text{from (4.1)}),$$

so that

$$p_{n+1} = \frac{\lambda_n}{\mu_{n+1}} p_n$$

$$= \left(\frac{\lambda_n}{\mu_{n+1}}\right)\left(\frac{\lambda_{n-1}}{\mu_n}\right) p_{n-1}$$

$$\cdots \qquad \cdots$$

$$= \frac{\lambda_n \lambda_{n-1} \ldots \lambda_0}{\mu_{n+1}\mu_n \ldots \mu_1} p_0$$

or

$$p_n = \prod_{k=1}^{n} \frac{\lambda_{k-1}}{\mu_k} p_0, \quad n = 1, 2, \ldots. \qquad (4.3)$$

Since $\sum_{n=1}^{\infty} p_n$ must be unity, i.e.

$$\left\{1 + \sum_{n=1}^{\infty} \prod_{k=1}^{n} \frac{\lambda_{k-1}}{\mu_k}\right\} p_0 = 1. \qquad (4.4)$$

Hence a necessary and sufficient condition for the existence of a steady state is the convergence of the infinite series $\sum_{n=1}^{\infty} \prod_{k=1}^{n} \frac{\lambda_{k-1}}{\mu_k}$ (Karlin and McGregor (1957)). When it converges, p_0 can be found from (4.4).

The queueing processes arising in some of the standard models can be considered as birth and death processes and the steady state solutions can be obtained easily by using the above. In the *M/M/*1 model considered in Sec. 10.2, $\lambda_n = \lambda$, $n = 0, 1, 2, \ldots$, and $\mu_n = \mu$, $n = 1, 2, \ldots$ Putting these values in (4.3) and (4.4) we at once get the steady state solutions.

10.4.2 The Model *M/M/s*

Here we consider a queueing model with s ($1 \leq s \leq \infty$) servers or channels *in parallel* and having identical input and service time distributions (as in the model *M/M/*1). In other words, the present model (*M/M/s*) considers a Poisson process with parameter λ as its input process and has, for each of the s channels, i.i.d. exponential service time distribution with mean rate μ. If n ($\leq s$) channels are busy, the number of services completed in the whole system is given by a Poisson process with mean $n\mu$ and the time between two successive service completions is exponential with mean $1/n\mu$; whereas if n ($> s$) channels are busy, the time between two successive service completions is exponential with mean $1/s\mu$. If $N(t)$ is the number present in the system at the instant t, the transition densities are as follows:

$$a_{n,n+1}dt \equiv \Pr\{N(t+dt) = n+1 \mid N(t) = n\} = \lambda dt + o(dt)$$

$$a_{n,n-1}dt \equiv \Pr\{N(t+dt) = n-1 \mid N(t) = n\} = \mu_n dt + o(dt)$$

$$a_{n,m}dt \equiv \Pr\{N(t+dt) = m \mid N(t) = n\} = +o(dt), m \neq n-1, n+1$$

where
$$\mu_n = n\mu, \quad \text{if} \quad 0 \leq n \leq s$$
$$= s\mu, \quad \text{if} \quad n \geq s. \tag{4.5}$$

Thus $N(t)$ is a birth and death process with constant arrival (birth) rate $\lambda_n = \lambda$ and state-dependent service (death) rates as given in (4.5).

Let $p_n(t) = \Pr\{N(t) = n \mid N(0) = .\}$, and let the steady state solutions exist. The state-transition-rate diagram is given in Fig. 10.6.

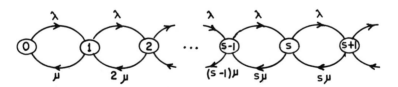

Fig. 10.6 State transition-rate diagram for an *M/M/s* model

Putting the values of λ_n and μ_n in (4.3) and (4.4) we get p_0 and p_n, $n = 1, 2, \ldots$, as follows. Denote $\lambda/s\mu$ by ρ.

For $n \leq s$,

$$p_n = \frac{\lambda \cdot \lambda \ldots \lambda}{(\mu)(2\mu)\ldots(n\mu)} p_0 = \frac{(\lambda/\mu)^n}{n!} p_0 = \frac{\lambda}{n\mu} p_{n-1}, \tag{4.6a}$$

and, for $n \geq s$

$$p_n = \frac{\lambda \cdot \lambda \ldots n \text{ factors}}{\{(\mu)(2\mu)\ldots(s\mu)\} \{(s\mu) \cdot (s\mu)\ldots(n-s)\text{factors}\}} p_0$$

$$= \frac{\lambda^n}{s!\mu^s s^{n-s}\mu^{n-s}} p_0 = \frac{(\lambda/\mu)^n}{s!s^{n-s}} p_0 = \rho^{n-s} \frac{(\lambda/\mu)^s}{s!} p_0$$

$$= \rho^{n-s} p_s. \tag{4.6b}$$

The condition $\sum_{n=0}^{\infty} p_n = 1$ gives

$$p_0^{-1} = 1 + \sum_{n=1}^{s-1} \frac{(\lambda/\mu)^n}{n!} + \sum_{n=s}^{\infty} \frac{(\lambda/\mu)^n}{s!s^{n-s}} = \sum_{n=0}^{s-1} \frac{(\lambda/\mu)^n}{n!} + \frac{s^s}{s!} \sum_{n=s}^{\infty} \left(\frac{\lambda}{s\mu}\right)^n.$$

For existence of steady state solutions, the series $\sum_{n=s}^{\infty} \left(\frac{\lambda}{s\mu}\right)^n$ must converge, and for this to happen $\lambda/s\mu$ must be less than 1. Then

$$p_0^{-1} = \sum_{n=0}^{s-1} \frac{(\lambda/\mu)^n}{n!} + \frac{(\lambda/\mu)^s}{s!(1 - \lambda/s\mu)}. \tag{4.7}$$

Thus the distribution of N, when $\rho = \lambda/s\mu < 1$ is given by (4.6), where p_0 is given by (4.7).

Note: (1) In this s-channel model, $\rho = \lambda/s\mu$ is the *utilization factor*.
(2) When $s = 1$, we get the corresponding result for $M/M/1$.
(3) The solutions p_n satisfy the following recurrence relations

$$p_n = \frac{1}{n}\left(\frac{\lambda}{\mu}\right)^n p_{n-1} = \rho\left(\frac{s}{n}\right) p_{n-1}, \quad n = 1, 2, \ldots, s-1$$

$$= \frac{1}{s}\left(\frac{\lambda}{\mu}\right)^n p_{n-1} = \rho p_{n-1}, \quad n = s, s+1, \ldots$$

For $n \leq s$, $\{p_n\}$ behaves as a Poisson distribution and for $n > s$ as a geometric distribution [when n *is finite*].

(4) The probability that an arriving unit has to wait is given by

$$C(s, \lambda/\mu) = \Pr\{N \geq s\} = \sum_{n=s}^{\infty} p_n$$

$$= \frac{(\lambda/\mu)^s}{s!(1-\rho)} p_0 = \frac{p_s}{1-\rho}. \tag{4.8}$$

This is known as *Erlang's second (or C) formula*. Extensive tables are available.

(5) The expected number in the system $E\{N\} = \sum_{n=s}^{\infty} n p_n$ can be directly evaluated by finding the sum of the series. Likewise $E\{Q\}$, the expected number in the queue can be found. Here we shall obtain them by making use of Little's formula.

(6) The results of $M/M/\infty$, for $\rho < 1$, are useful for approximating $M/M/s$ results. For approximations, see Hokstad [*Opns. Res.* **26** (1978) 510–523].

10.4.2.1 Waiting Time Distributions

We assume that the queue discipline is FCFS. The distribution of W_Q can be obtained in the same way as in $M/M/1$. We use the same notations as in Sec. 10.2.2.1.

Consider a test unit. It has not to wait, if it finds, on its arrival, less than or equal to $(s-1)$ units in the system. Thus

$$\Pr\{W_Q = 0\} = \sum_{n=0}^{s-1} p_n$$

$$= 1 - \sum_{n=s}^{\infty} p_n$$

$$= 1 - \frac{(\lambda/\mu)^s}{s!(1-\rho)} p_0 = 1 - \frac{p_s}{1-\rho} \quad \text{(using (4.8))}.$$

The test unit has to wait, i.e. $W_Q > 0$ when it finds, on arrival, that the number in the system is $n \geq s$, of which s are in the service channels and $(n-s)$ in the queue. Thus, it will have to wait for the completion of $(n-s+1)$ services. As $n \geq s$, s-channels are always in operation before its service commences, and as such the time between two successive completions is exponential with mean $1/s\mu$. Thus S_{n-s+1}, being the sum of $(n-s+1)$ i.i.d. exponential variates with mean $1/s\mu$, is a gamma variate with p.d.f.

$$\frac{(s\mu)(s\mu x)^{n-s} e^{-s\mu x}}{(n-s)!}.$$

Hence $w_q(x)dx = \Pr\{x \le W_Q < x + dx\}, x > 0$

$$= \sum_{n=s}^{\infty} \Pr\{x \le S_{n-s+1} < x + dx \mid N = n\}.\Pr\{N = n\}$$

$$= \sum_{n=s}^{\infty} \frac{(s\mu)(s\mu x)^{n-s}}{(n-s)!} e^{-s\mu x} p_n dx$$

or $w_q(x) = \dfrac{(s\mu)e^{-s\mu x}}{s!} p_0 \sum_{n=s}^{\infty} \dfrac{(s\mu x)^{n-s}}{n(n-s)!} \dfrac{r^n}{s^{n-s}}$, $(r = \lambda/\mu)$

$$= \frac{(s\mu)e^{-s\mu x}}{s!} p_0 r^s \sum_{m=0}^{\infty} \frac{(r\mu x)^m}{m!}, \quad \text{(putting } n-s=m)$$

$$= \frac{s\mu(\lambda/\mu)^s}{s!} e^{-\mu x\{s - (\lambda/\mu)\}} p_0$$

$$= s\mu\, e^{-x(s\mu - \lambda)} \frac{(\lambda/\mu)^s}{s!} p_0$$

$$= s\mu\, e^{-x(s\mu - \lambda)} p_s .$$

Thus $w_q(x) = 1 - \dfrac{(\lambda/\mu)^s}{(s-1)!(s-\lambda/\mu)} p_0 = 1 - \dfrac{(\lambda/\mu)^s}{s!(1-\rho)} p_0,\quad x = 0$

$$= \frac{(\lambda/\mu)^s}{(s-1)!} \mu e^{-(s\mu-\lambda)x} p_0, \quad x > 0 \tag{4.9}$$

and $E\{W_Q\} = (W_Q = 0)\Pr\{W_Q = 0\} + \displaystyle\int_{\varepsilon}^{\infty} x w_q(x)dx \quad \text{as } \varepsilon \to 0.$

Now $\displaystyle\int_{\varepsilon}^{\infty} x w_q(x)dx = \int_{\varepsilon}^{\infty} \frac{(\lambda/\mu)^s}{(s-1)!} p_0 \mu\{x\, e^{-(s\mu-\lambda)x}\} dx$

$$\to \left\{ \frac{(\lambda/\mu)^s}{(s-1)!} \mu p_0 \right\} \frac{1}{(s\mu-\lambda)^2} = \frac{(\lambda/\mu)^s}{s\mu(s!)(1-\rho)^2} p_0 .$$

Thus $E\{W_Q\} = \dfrac{(\lambda/\mu)^s}{s\mu(s!)(1-\rho)^2} p_0 = \dfrac{p_s}{s\mu(1-\rho)^2} = \dfrac{\Pr\{N \ge s\}}{s\mu(1-\rho)} . \tag{4.10}$

Corollary. The expected waiting time in the system is given by

$$E\{W_s\} = E\{W_Q\} + 1/\mu.$$

Using Little's formula, we get

$$E\{L_Q\} = \lambda E\{W_Q\} = \frac{\rho(\lambda/\mu)^s}{s!(1-\rho)^2} p_0 \tag{4.11}$$

$$= \frac{\rho}{(1-\rho)^2} p_s .$$

and $\qquad E\{N\} = \lambda E\{W_s\} = \frac{\lambda}{\mu} + \frac{\rho(\lambda/\mu)^s}{s!(1-\rho)^2} p_0 . \qquad (4.12)$

If the random variables, number of busy and idle channels, are denoted by B and I respectively, and the number in the queue by Q, then $N = Q + B$ and $I + B = s$,

whence we get $\qquad E\{B\} = E\{N\} - E\{Q\} = \lambda/\mu = s\rho$

and $\qquad E\{I\} = s - \lambda/\mu = s(1-\rho).$

Note: 1. $E\{I\} = s(1-\rho)$ is a fairly general result and holds for any *GI/G/s* system (see Harris, 1974).

Note 2. It can be easily shown that $\Pr\{W_Q > t\} = C(s, \lambda/\mu) \exp\{-(1-\rho) s\mu t\}$; further that the conditional distribution of W_Q given that $W_Q > 0$ is exponential with parameter $(1-\rho)s\mu$.

Note 3. Suppose that instead of having *homogeneous servers* (viz. all with i.i.d. exponential distribution with rate μ) *heterogeneous servers* having independent exponential service times with rates μ_i, $i = 1, 2, \ldots, s$, are considered. It can be seen that the waiting time density is, for $x > 0$, given by

$$\omega_q(x) = [\{\sum_{i=1}^{s} \mu_i\} \exp\{-x(\sum_{i=1}^{s} \mu_i - \lambda)\}] p_s;$$

and $\qquad E\{W_Q\} = \dfrac{\Pr\{N \geq s\}}{\sum \mu_i (1 - \lambda/\sum \mu_i)} .$ $\qquad$ (Gumbel).

Example 4(a). Patients arrive at the outpatient department of a hospital in accordance with a Poisson process at the mean rate of 12 per hour, and the distribution of time for examination by an attending physician is exponential with a mean of 10 minutes. What is the minimum number of physicians to be posted for ensuring a steady state distribution? For this number, find (i) the expected waiting time of a patient prior to being examined and (ii) the expected number of patients in the out-patient department. How many physicians, on an average, will remain idle?

Here $\lambda = 12$/hr., $\mu = 6$/hr., so that $r = \lambda/\mu = 2$. For steady state, $\rho = r/s < 1$, i.e. $2 < s$, the minimum number of physicians must be 3. We take $s = 3$, then $\rho = 2/3$. From exercise 10.7, we find that

$$E\{W_Q\} = \left[s\mu(1-\rho)\left\{1 + (1-\rho)\frac{E(s,r)}{p(s,r)}\right\} \right]^{-1}$$

Where
$$p(s,r) = \frac{e^{-r}r^s}{s!}, \quad E(s,r) = \sum_{k=0}^{s-1}\frac{e^{-r}r^k}{k!}.$$

From Poisson probability tables $p\,(3,2) = 0.180$, $E\,(3,2) = 0.677$. Extensive tables of Poisson probability function and distribution function are available (Molina (1942)).

We get (for $s = 3$)
$$E\{W_Q\} = \frac{2}{27}\,\text{hr} = 4.4\ \text{min.}$$

$$E\{W_S\} = E\{W_Q\} + 1/\mu = 13/54\ \text{hr} = 14.4\ \text{min.}$$

$$E\{N\} = \lambda E\{W_S\} = 26/9$$

and
$$E\{I\} = s(1-\rho) = 3(1-2/3) = 1.$$

Remarks

Transient-state distribution of M/M/s queue

This was first studied by Saaty (1961) and later on by Jackson and Henderson (1966) and recently by Parthasarathy and Sharafali (1989).

Busy-period distribution was first considered by Chaudhry and Templeton (1973) and later by Sharafali and Parthasarathy (1989).

For details, refer to the above and also to Medhi (1991).

10.4.3 The Model $M/M/\infty$

Here the number of channels or servers are infinite so that, on arrival, all the units are taken into service and there is no queue. The situation is more or less similar to that obtaining in a *self-service store*.

For the birth and death process associated with this, $\lambda_n = \lambda$, $\mu_n = n\mu$, for $n = 0, 1, 2,....$ The state transition-rate diagram is given in Fig. 10.7.

Fig. 10.7 Stae transition-rate diagram for an $M/M/\infty$ model

The steady state solution can be obtained from those of $M/M/s$ by letting s tend to infinity. From (4.6) and (4.7) we at once get the steady state probabilities,

$$p_n = \frac{(\lambda/\mu)^n}{n!}p_0, \quad n = 1, 2, ...,$$

where p_0 is obtained by using $\sum\limits_{n=0}^{\infty} p_n = 1$; we get

$$p_0^{-1} = \sum_{n=0}^{\infty} \frac{(\lambda/\mu)^n}{n!} = e^{\lambda/\mu}.$$

Thus
$$p_n = \frac{(\lambda/\mu)^n}{n!} e^{-\lambda/\mu}, n = 0, 1, 2, \dots. \tag{4.13}$$

The distribution of the number N in the system is Poisson with mean λ/μ. The distribution of the number of busy channels is also the same. The average number of busy channels is λ/μ.

As the series $\sum\limits_{n=0}^{\infty} (\lambda/\mu)^n/n!$ converges for all λ/μ, the steady state solutions exist

whether or not $\lambda < \mu$.

Note. The average number of busy channels in a *GI/G/∞* system is also λ/μ, as in the case of *M/M/s* queue.

10.4.4 The Model *M/M/s/s*: Loss System

This model envisages that a unit, who finds, on arrival, that all the s-channels are busy, leaves the system without waiting for service. This is called a (s-channel) *loss* system and was first investigated by Erlang. This model was also examined earlier (see Examples 5(d) and 5(g), Ch. 4).

For this birth and death process, we have

$$\lambda_n = \lambda, \quad \mu_n = n\mu, n = 0, 1, 2, \dots, s-1$$

$$\lambda_n = 0, \quad \mu_n = s\mu, \quad n \geq s. \tag{4.14}$$

The state transition-rate diagram is given in Fig 10.8

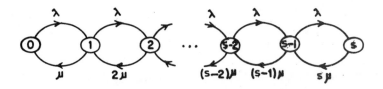

Fig. 10.8 State transition-rate diagram for *M/M/s/s* model

From (4.3) and (4.4) we get the steady state probabilities

$$p_n = \frac{(\lambda/\mu)^n}{n!} p_0 , \quad n = 0, 1, \ldots, s$$

and

$$\left\{ 1 + \sum_{k=1}^{s} \frac{(\lambda/\mu)^k}{k!} \right\} p_0 = 1 \text{ or } p_0 = \left[\sum_{k=0}^{s} \frac{(\lambda/\mu)^k}{k!} \right]^{-1}.$$

Thus

$$p_n = \frac{(\lambda/\mu)^n/n!}{\sum_{k=0}^{s} \frac{(\lambda/\mu)^k}{k!}} , \quad n = 0, 1, 2, \ldots, s. \tag{4.15}$$

The probability that an arriving unit is lost to the system (which is the same as that all the the channels are busy) is given by

$$p_s = \frac{(\lambda/\mu)^s/s!}{\sum_{k=0}^{s} (\lambda/\mu)^k/k!}. \tag{4.16}$$

The formula (4.16) is known as *Erlang's loss formula* or *blocking formula* and is denoted by $B(s, \lambda/\mu)$, while (4.15) is known as *Erlang's first formula* (or simply *Erlang's formula*, the corresponding distribution being truncated Poisson).

Note: (1) Attempts have been made since Erlang's time to generalise Erlang's results. Mention may be made of the works of Pollaczek, Palm, Kosten, Fortet, Sevast'yanov and Takacs (cf. his (1969) paper). It has been shown that Erlang's formula (4.15) holds for *any* distribution of service time (having mean $1/\mu$) provided the input is Poisson (with parameter λ), i.e. it holds for the model $M/G/s/s$ (loss system).

Brumelle (1978) considers a generalisation to state dependent arrival and service rates. See also Wolff and Wrightson (*J. Appl. Prob.* **13** (1976) 628-632) for an extension of the loss formula.

Fakinos (1980) shows that the blocking formula holds also in case of heterogeneous servers provided each server has the mean service time $1/\mu$. It is also shown that in equilibrium the epochs at which customers (who are served, who balked or who are rejected) depart from the system form a Poisson process.

(2) If $N(t)$ is the number of busy channels at instant t, then $\{N(t), t \geq 0\}$ is a Markov process, provided the service time is exponential. But the process is not Markov when the service time is not exponential; nevertheless, the loss formula holds.

(3) The steady state solutions p_n exist whether or not $\lambda < \mu$.

(4) The results of the model $M/M/\infty$ can be obtained by taking limits, as $s \to \infty$, of the corresponding results for $M/M/s/s$.

(5) For further properties of Erlang's loss formula, see Jagerman (1974).

(6) Harel (1991) obtains sharp bounds and approximations of the Erlang delay and loss formulas.

(7) Dietrich *et al.* (*Teletraffic Engineering Manual*, Standard Electrik, 1966) give tables of values of p_s for range of values s and λ/μ. See also Descloux (1962) for tables.

Example 4(b). *System with limited waiting time.* Apart from limitation of space, a second kind limitation which arises in problems dealing with practical situations need consideration. For example, a long distance telephone call may be booked to be put through within a limited time (say, during the office hours). If the call does not mature by that time, the call is cancelled and is lost. Similarly, perishable goods can often be stored for a limited time. In these cases, if a unit does not receive service, immediately on arrival, the unit waits but waits only for a limited time, on the expiry of which it is lost.

We assume that the system has Poisson input with rate λ, has s-channels with i.i.d. exponential service times each with mean $1/\mu$, that there is no limitation as regards waiting space. Further, we assume that the waiting time of a unit is a random variable having exponential distribution with mean $1/v$.

The resulting process is Markov with state space $\{0, 1, 2, \dots\}$, and can be described by a birth and death process. Here $\lambda_n = \lambda$ for $n = 0, 1, \dots, \mu_n = n\mu$ for $n = 0, 1, \dots, s$. For $n > s$, the transition from state n to state $(n - 1)$ in an infinitesimal interval h may occur in two mutually exclusive ways: either one of the s busy servers terminates his service (with probability $s\mu h + o(h)$) or the waiting time of one of the $(n - s)$ units waiting in the queue expires (with probability $(n - s) vh + o(h)$). Thus

$$\mu_n = s\mu + (n - s)v, \quad n = s + 1, \quad s + 2, \dots.$$

Putting the values of λ_n, μ_n in (4.3), we get

$$p_n = \frac{(\lambda/\mu)^n}{n!} p_0, \quad n \leq s$$

$$= \frac{\lambda^n p_0}{s! \mu^s \displaystyle\prod_{r=s+1}^{n} \{s\mu + (r - s)v\}}, \quad n > s$$

and using $\displaystyle\sum_{n=0}^{\infty} p_n = 1$, we get

$$p_0 = \left[\left\{ \sum_{k=1}^{s} \frac{1}{k!} \left(\frac{\lambda}{\mu} \right)^k \frac{\lambda^n}{\mu^s(s!)} \right\} \sum_{n=s+1}^{\infty} \frac{\lambda^n}{\displaystyle\prod_{r=s+1}^{n} \{s\mu + (r - s)v\}} \right]^{-1}.$$

10.4.5 The Model with Finite Input Source : [*M/M/s/ /m*]

We have assumed so far that the source or the population from which the input comes is infinite. We now consider the case where the source of input is not infinite but is finite and consists of *m* units. For example, such a situation occurs in practice in servicing of automatic machines. Suppose that there are m such machines, which normally do not need human attention except when there is a breakdown. In case of failure or breakdown of a machine, it is serviced by one of the s ($\leq m$) repairmen available. If all the s repairmen are engaged in repairing machines already down, the machine that is down at that time waits for repair till a repairman becomes free. We have a finite input source consisting of m machines and s ($\leq m$) repairmen.

Following Kleinrock, we shall denote the model by $M/M/s/\ /m$.

Suppose that, if a machine is in working state at the instant t, the probability that it will breakdown in the infinitesimal interval $(t, t + h)$ is $\lambda h + o(h)$ and that, when a machine is being serviced, the probability that the servicing is completed in an infinitesimal interval of length h is $\mu h + o(h)$. Let $N(t)$ be the number of machines not working (or down for repair) at the instant t. We have a birth and death process with parameters

$$\lambda_n = (m - n)\lambda, \qquad n = 0, 1, 2, ..., m - 1$$
$$= 0, \qquad n \geq m$$

and
$$\mu_n = n\mu, \qquad n = 0, 1, 2, ..., s - 1 \qquad (4.17)$$
$$= s\mu, \qquad n \geq s.$$

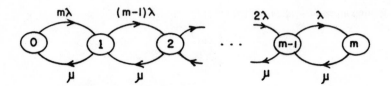

Fig. 10.9 State transition-rate diagram for $M/M/1/\ /m$ model

Using (4.3), we get in steady state,

$$p_n = \frac{(m)(m - 1)(m - 2)...(m - n + 1)\lambda^n}{n!\mu^n} p_0$$

$$= \frac{m!}{(m - n)!n!}\left(\frac{\lambda}{\mu}\right)^n p_0$$

$$= \binom{m}{n}\left(\frac{\lambda}{\mu}\right)^n p_0, \quad \text{for } n = 0, 1, 2, ..., s - 1, s \qquad (4.18)$$

and
$$p_n = \frac{m!}{(m - n)!s!s^{n-s}}\left(\frac{\lambda}{\mu}\right)^n p_0, \quad \text{for } n = s + 1, ..., m \qquad (4.19)$$

where p_0 is obtained by using $\sum_{n=0}^{m} p_n = 1$; we get

$$p_0 = \left[\sum_{n=0}^{s-1} \binom{m}{n} \left(\frac{\lambda}{\mu} \right)^n + \sum_{n=s}^{m} \frac{m!}{(m-n)! s! s^{n-s}} \left(\frac{\lambda}{\mu} \right)^n \right]^{-1}. \tag{4.20}$$

The probability that the number of machines in working order is r ($\leq m$) is given by p_{m-r}.

The model has been used as an answer to various important questions of practical interest arising out of *machine interference problems* (or vehicle fleet operation problems). Some such questions relate to determination of the optimal number of repairmen for a given set of machines (number of crew for a given fleet of vehicles), the maximum length of time during which all the m machines are working (m vehicles are plying) and so on. These have been the subject matter of investigation of several researchers (e.g. Palm, Naor, Benson and Cox; see also Feller (1968) p. 465).

In a paper, Levhari and Shesinski (1970) used such a model in a somewhat broader perspective: in the study of microeconomic production function. On the assumption that the machines produce a constant flow of output when operating, and that the breakdown and servicing of machines conform to the model of machine interference studied, they obtain the steady state probabilities of the system as well as the relation between inputs and output and fit the same to a Cobb-Douglas production function.

An application to a management problem has been considered by Montgomery (*Mgmt. Sci.* 1969, 323) and to electronics by Koenigsberg (*Int. J. Elec.* **48** (1980) 83; see also Cooper (1981) for application to time-sharing computer systems.

Wong (1979) considers waiting time distribution for this model.

Remarks: In Sec. 10.4.1 we have considered generalised birth and death processes, with state dependent mean arrival and service rates λ_n and μ_n. On the basis of the results indicated there, queueing models with *special choices* of λ_n and μ_n have been analysed in subsequent subsections. Conolly (1975), Conolly and Chan (1977) consider generalised birth and death queueing process, with state dependent rates λ_n and μ_n, as they, are. They assume the condition for the existence of steady state, and define a measure of traffic intensity in terms of 'effective' service time and interarrival interval. The results obtained show surprising similarity (with some of those of *M/M/1* and with Little's formula). The results also indicate potentiality of application to a wider range of systems.

Example 4(c). *Transient solution of M/M/1/1 model.* Here $\lambda_0 = \lambda$, $\mu_0 = \mu$ and $\lambda_1 = 0$, $\mu_n = \mu$, $n \geq 1$ (see eqn. (4.14)): if $N(t)$ denotes the number in the system at time t, then $\Pr\{N(t) = n\} = p_n(t) = 0$ for all $n > 1$, i.e. we are concerned with only $p_0(t)$ and $p_1(t)$ such that $p_0(t) + p_1(t) = 1$. The forward Kolmogorov equations of the model then become

$$p_0'(t) = -\lambda p_0(t) + \mu p_1(t)$$

$$p_1'(t) = -\mu p_1(t) + \lambda p_0(t).$$

Writing $p_1(t) = 1 - p_0(t)$ in the first equation we get $p_0'(t) + (\lambda + \mu)p_0(t) = \mu$. The solution of this first order linear differential equation with constant coefficients is given by

$$p_0(t) = Ce^{-(\lambda+\mu)t} + \frac{\mu}{\lambda+\mu}, \quad \text{where} \quad C \quad \text{is a constant.}$$

Given the initial distribution $p_i(0) = \Pr\{N(0) = i\}$, we get

$$p_0(t) = p_0(0)e^{-(\lambda+\mu)t} + \frac{\mu}{\lambda+\mu}\{1 - e^{-(\lambda+\mu)t}\}.$$

Similarly,

$$p_1(t) = p_1(0)e^{-(\lambda+\mu)t} + \frac{\lambda}{\lambda+\mu}\{1 - e^{-(\lambda+\mu)t}\}.$$

The steady state solutions are

$$p_0 = \lim_{t \to \infty} p_0(t) = \mu/(\lambda+\mu)$$

and

$$p_1 = \lim_{t \to \infty} p_1(t) = \lambda/(\lambda+\mu),$$

irrespective of whether the value of $\rho = \lambda/\mu < 1$ or not.
(These agree with p_n ($n = 0, 1$) given by (4.15).)

Assume that the initial distribution is identical with the steady-state distribution so that $p_0(0) = p_0 = \mu/(\lambda+\mu)$, $p_1(0) = p_1 = \lambda/(\lambda+\mu)$. Then we find that *for all $t \geq 0$,*

$$p_0(t) = \frac{\mu}{\lambda+\mu} = p_0, \quad p_1(t) = \frac{\lambda}{\lambda+\mu} = p_1.$$

That is, if the process is in equilibrium (steady state) initially, then it will be always (for all $t \geq 0$) in steady state. This is true for any other '*ergodic*' system.

In case of $M/M/s/s$ model, if the system is in equilibrium initially, i.e. $\Pr\{N(0) = n\} = p_n(0) = p_n, 0 \leq n \leq s$, where p_n's are steady-state probabilities (given by relation (4.15)), then $\{N(t), t \geq 0\}$ becomes a stationary process for which $N(t)$ has the same distribution *for all $t \geq 0$,* i.e. $p_n(t) = \Pr\{N(t) = n\} = p_n$ (given by (4.15)) *for all $t \geq 0$* (see Takacs (1969)).

10.5   NON-BIRTH AND DEATH QUEUEING PROCESSES: BULK QUEUES

We have seen that the queueing processes in the standard models considered in the earlier sections are birth and death processes and are, of course, Markovian. We examine here queueing processes arising out of some non-birth and death processes but which are nevertheless Markovian. The Chapman-Kolmogorov equations can be written and the analysis can be done in a similar manner.

The basic model considered is of the type of an $M/M/1$ queue with parameters λ, μ, but with group arrivals and/or batch service.

10.5.1 Group Arrivals System: $M^{(X)}/M/1$

Assume that the arrival streams form a Poisson process with parameter λ and that the number of units arriving at an arrival instant is a random variable X with p.f. $c_k = \text{Pr}\,\{X = k\}$, $k = 1, 2, \ldots$. This is equivalent to saying that, in an infinitesimal interval $(t, t + h)$, the probability that a group of k arrives is $\lambda c_k h + o\,(h)$. Assume here that service takes place singly and that the distributions of the service times are independent exponential with mean $1/\mu$. It is clear that the process is still Markovian but is a non-birth-death process, as transitions may occur to states, not necessarily neighbouring. We denote the model by $M^{(X)}/M/1$. Assume that $N\,(0) = 0$; we get

$$p_0'(t) = -\lambda p_0(t) + \mu p_1(t) \tag{5.1}$$

$$p_n'(t) = -(\lambda + \mu)p_n(t) + \mu p_{n+1}(t)$$

$$+ \sum_{k=1}^{n} \lambda c_k p_{n-k}(t), n \geq 1. \tag{5.2}$$

Let $C\,(s) = \sum_{k=1}^{\infty} c_k s^k$ be the p.g.f. of X. Since the mean arrival rate is $\lambda\,E\,(X) = \lambda C\,(1)$,

the utilization factor can be defined as

$$\rho = \lambda E(X)/\mu = \lambda C'(1)/\mu. \tag{5.3}$$

The state transition rate diagram is given in Fig 10.10
Assuming that the steady state solutions exist, we get

$$\lambda p_0 = \mu p_1 \tag{5.4}$$

$$(\lambda + \mu)p_n = \mu p_{n+1} + \lambda \sum c_k p_{n-k}, \quad n \geq 1. \tag{5.5}$$

Let $\qquad P(s) = \sum_{n=0}^{\infty} p_n s^n$ be the p.g.f. of $\{p_n\}$.

Multiplying (5.5) by s^n, for successive values of $n = 1, 2, 3, \ldots$ and adding them to (5.4), we get

$$0 = -\lambda P(s) - \mu(P(s) - p_0) + (\mu/s)(P(s) - p_0)$$

$$+ \lambda \sum_{n=1}^{\infty} \sum_{k=1}^{n} c_k p_{n-k} s^n. \tag{5.6}$$

Noting that the last term equals $\lambda \sum_{k=1}^{\infty} c_k s^k \{ \sum_{n=k}^{\infty} p_{n-k} s^{n-k} \} = \lambda C(s)P(s)$,

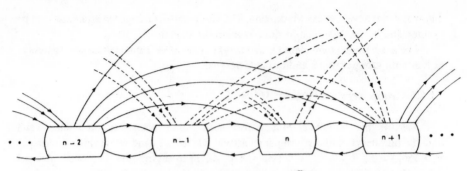

Fig. 10.10 State transition-rate diagram for $M^{(X)}/M/1$ model

we get
$$P(s) = \frac{\mu(1-s)p_0}{\mu(1-s) - \lambda s(1 - C(s))}.$$
(5.7)

To find p_0, we use $P(1) = 1$. Applying L'Hôpital's, rule, we get

$$1 = \lim_{s \to 1} P(s) = \frac{-\mu p_0}{-\mu + \lambda C'(1)}$$

or
$$p_0 = 1 - \rho.$$
(5.8)

Thus
$$P(s) = \frac{\mu(1-s)(1-\rho)}{\mu(1-s) - \lambda s(1 - C(s))}.$$
(5.7a)

Particular case: M/M/1. This model corresponds to single arrivals, i.e. in groups of size one only. Then $c_1 = 1, c_k = 0, k > 1$ and

$$P(s) = (1 - \rho)/(1 - \rho s).$$

10.5.2 Bulk Service Systems

10.5.2.1 Bulk Service Rules

There could be a number of a policies or rules according to which batches for bulk service may be formed. The following are the types of bulk service policies or rules frequently discussed in the literature.

(1) Bailey (1954) (also Downton, 1955) considers that units are served in batches of not more than b (say). If, immediately after the completion of a service, the server finds more than b units waiting, he takes a batch of b units for service while others wait; if he finds r units ($0 \le r \le b$), he takes all the r units in a batch for service. Bloemena (1960), Jaiswal (1960) Neuts (1967) consider the same rule with the restriction that $r \ne 0$ ($1 \le r \le b$), i.e. the service facility stops until a unit arrives. This rule will be called the *usual bulk service rule*, while Bailey's rule will be its *modified* type (it will be also called *bulk service rule with intermittently available server*) Jaiswal points out that the distribution of the queue length for the modified rule can

be obtained from that of the usual rule.

(2) The rule with a fixed batch size k has been considered by Fabens (1961), Takacs (1962) and others. In this case the server waits until there are k units, and then serves all the k units in a batch. If there are more than k waiting when the server becomes free, he takes a batch of k for service, while others wait.

(3) Neuts (1967) considers this rule: If, immediately after the completion of a service, the server finds less than a units present, he waits until there are a units, whereupon he takes the batch of a units for service; if he finds a or more but at most b, he takes them all in the batch and if he finds more than b, he takes in the batch for service b units, while others wait. The batch takes a minimum of a units and a maximum of b units. This rule will be called *general bulk service rule* as rules under (1) and (2) can be taken as particular cases.

(4) Bhat (1963) considers the rule that the number taken in a batch is a random variable Y. The corresponding Markov model is denoted by $M/M^{(Y)}/1$.

There are actual situations where one rule seems to be better suited or to be more appropriate than the others. Examples are given in standard textbooks and relevant research publications.

10.5.3 Poisson Queue with General Bulk Service Rule: [$M/M(a,b)/1$ Model]

Assume that the arrivals occur single in accordance with a Poisson process with parameter λ. The units are served in batches according to a general bulk service rule such that the service starts only when a minimum number of units say, a is present in the queue, the maximum service capacity being, say b. The service times of batches of size s ($a \le s \le b$) are assumed to have independent identical exponential distribution with mean $1/\mu$. For simplicity, the service time distribution is assumed to be independent of the batch size s. The queueing system is denoted by $M/M(a,b)/1$.

Denote by state $(0, q)$, the state that the service channel is idle and there are q ($0 \le q \le a - 1$) units waiting in the queue and by the state $(1, n)$, the state that the service channel is busy and there are n ($n \ge 0$) units waiting in the queue. Let

$$p_{0,q}(t) = \Pr \{\text{that at time } t, \text{ the system is in state}(0, q)\},$$

$$q = 0, 1, 2, \ldots, a - 1,$$

$$p_{1,n}(t) = \Pr \{\text{that at time } t, \text{ the system is in state}(1, n)\},$$

$$n = 0, 1, 2, \ldots$$

Assume that the steady-state distribution exists. Denote

$$\lim_{t \to \infty} p_{0,q}(t) = p_{0,q}$$

$$\lim_{t \to \infty} p_{1,n}(t) = p_{1,n}.$$

The state transition-rate diagram is given in Fig. 10.11

The steady-state probability equations are :

$$(\lambda + \mu)p_{1,n} = \lambda p_{1,n-1} + \mu p_{1,n+b}, \quad n = 1, 2, \ldots, \tag{5.9}$$

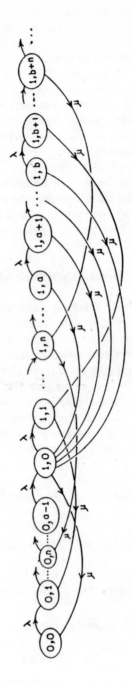

Fig. 10.11 State transition-rate diagram for *M/M(a. b)/*1 model

$$(\lambda + \mu)p_{1,0} = \lambda p_{0,a-1} + \mu \sum_{s=a}^{b} p_{1,s} \tag{5.10}$$

$$\lambda p_{0,0} = \mu p_{1,0} \tag{5.11}$$

$$\lambda p_{0,n} = \lambda p_{0,n-1} + \mu p_{1,n}, \quad n = 1, 2, \ldots, a-1. \tag{5.12}$$

Note that (5.12) will not occur when $a = 1$. First we solve (5.9).

The difference equation (5.9) can be written as

$$h(E)\{p_{1,n}\} = 0, n = 0, 1, 2\ldots \tag{5.13}$$

such that the characteristic equation is

$$h(z) \equiv \mu z^{b+1} - (\lambda + \mu)z + \lambda = 0. \tag{5.14}$$

Suppose that $\qquad f(z) \equiv -(\lambda + \mu)z \quad$ and that $\quad g(z) \equiv \mu z^{b+1} + \lambda,$

then for $|z| = 1$, $|g(z)| < |f(z)|$ and hence by Rouche's theorem, $f(z)$ and $f(z) + g(z)$ will have the same number of zeros inside $|z| = 1$. Since $f(z)$ has only one zero inside $|z| = 1$, there will be only one zero of $f(z) + g(z) = h(z)$ inside $|z| = 1$, and this zero is real.

Denote this root of $h(z) = 0$ by r, $0 < r < 1$, and the other b roots by $r_1, \ldots, r_b$, $(|r_i| \geq 1)$. Note that $h(z)$ does not depend on a.

Denote the traffic intensity by $\rho = \lambda/b\mu$. From (5.14), we get

$$b\rho = \frac{\lambda}{\mu} = \frac{r(1 - r^b)}{1 - r} = r + r^2 + \ldots + r^b. \tag{5.15}$$

Now we can write the solution of (5.9) as:

$$p_{1,n} = A r^n + \sum_{i=1}^{b} A_i r_i^n, \quad n = 0, 1, 2, \ldots$$

Since $\sum_{n=0}^{\infty} p_{1,n} < 1$, we must have $A_i = 0$ for $i = 1, 2, \ldots, b$ so that,

$$p_{1,n} = A r^n$$

$$= (p_{1,0})r^n$$

$$= (\lambda/\mu)r^n p_{0,0} \quad \text{(from (5.11))}$$

$$= \left(\frac{1 - r^b}{1 - r}\right) r^{n+1} p_{0,0}. \tag{5.16}$$

Using (5.16), we get from (5.10)

$$p_{0,a-1} = \frac{1 - r^a}{1 - r} p_{0,0} \tag{5.17}$$

and finally putting $n = 1, 2, \ldots, a - 1$ in (5.12), and using (5.16) and (5.17) we get, recursively for $q = a - 2, a - 3, \ldots, 2, 1$,

$$P_{0,q} = \frac{1 - r^{q+1}}{1 - r} P_{0,0} \tag{5.18}$$

and because of (5.17), (5.18) holds for $q = 1, 2, \ldots, a - 1$.

Since $\sum_{q=0}^{a-1} P_{0,q} + \sum_{n=0}^{\infty} P_{1,n} = 1$, we have

$$P_{0,0} = \left[\frac{a}{1-r} + \frac{r^{a+1} - r^{b+1}}{(1-r)^2} \right]^{-1}. \tag{5.19}$$

▲

Thus (5.16), (5.18), (5.19) give the steady state probabilities. One can obtain the expected number in the queue.

Particular case: Fixed size bulk service; $M/M(k, k)/1$ $[M/M^k/1]$.
Here $a = b = k$. Let P_n denote the steady state probability that the number in the *system* is n. Then

$$P_o = P_{0,0} = \frac{1-r}{k},$$

$$P_q = P_{0,q} = \frac{1 - r^{q+1}}{1 - r} P_0 = (1 - r^{q+1})/k, \quad q = 1, 2, \ldots, k - 1$$

$$P_{m+k} = P_{1,m} = (\lambda/\mu) P_0 r^m = \rho(1 - r) r^m, \qquad \rho = \lambda/(k\mu),$$
$$m = 0, 1, 2, \ldots,$$

i.e. $$P_n = \begin{cases} (1 - r^{n+1})/k, & n = 0, 1, 2, \ldots, k - 1 \\ \rho(1 - r) r^{n-k} \lambda = P_k r^{n-k} & n = k, k + 1, \ldots \end{cases} \tag{5.20}$$

10.5.3.1 Distribution of the waiting time for the system M/M $(a, b)/1$ (Medhi, 1975)

Assume that the system is in steady state. Let the random variable T denote the waiting time in the *queue* (queueing time) for an arriving (test) unit.
Let

$v(t)$	be the p.d.f. of T,
$f(\alpha, k; t)$	be the p.d.f. of gamma distribution with parameters α, k, i.e.

$$f(\alpha, k; t) = \alpha^k t^{k-1} e^{(-\alpha t)} / \Gamma(k), \quad t > 0, \quad k = 1, 2, \ldots,$$

$\Gamma_x(\alpha, k)$ be the incomplete gamma function, i.e.

$$\Gamma_x(\alpha, k) = \int_0^x f(\alpha, k; t) dt = 1 - \sum_{s=0}^{k-1} e^{-\alpha x} (\alpha x)^s / s!.$$

An arriving (test) unit may find the system in one of the following states:

(i) $(0, a - 1)$

(ii) $(0, q)$, $\quad 0 \leq q \leq a - 2$

$$\left.\begin{array}{l} \text{(iii) } (1, n), \quad a - 1 \leq m \leq b - 1 \\ \text{(iv) } (1, n), \quad 0 \leq m \leq a - 2 \end{array}\right\} \quad \begin{array}{l} n = kb + m \\ k = 0, 1, 2, \ldots \end{array} \tag{5.21}$$

In case of (i), the arriving unit does not wait; the probability of zero delay is thus $p_{0,a-1}$. In all other cases the unit has to wait, the probability of blocking (or delay) is thus $1 - p_{0,a-1}$.

In case of (ii), the arriving unit has to wait the arrival of $(a - 1 - q)$ units and the time needed for $(a - 1 - q)$ arrivals has a gamma distribution with parameters $\lambda, a - 1 - q$.

In cases of (iii), the arriving unit has to wait for the completion of services of $(k + 1)$ batches; the time required for this has a gamma distribution with parameters $\mu, (k + 1)$.

In cases of (iv), the arriving unit has to wait till either the services of $(k + 1)$ groups are completed or $(a - 1 - m)$ units arrive, whichever occurs later. This duration which is given by the maximum of two gamma variates may be denoted by the random variable Z, i.e.

$$Z = \max \{\text{gamma variate with parameters } \lambda, a - 1 - m;$$
$$\text{gamma variate with parameters } \mu, k + 1\}. \tag{5.22}$$

The distribution function $Fz(t)$ and the p.d.f. $h_z(t)$ of Z are given by

$$F_z(t) = \Pr \{Z \leq t\} = \Gamma_t(\lambda, a - 1 - m)\Gamma_t(\mu, k + 1),$$

and
$$h_z(t) = F'_z(t)$$

$$= f(\lambda, a - 1 - m; t)\Gamma_t(\mu, k + 1)$$

$$+ \Gamma_t(\lambda, a - 1 - m)f(\mu, k + 1; t). \tag{5.23}$$

It follows that the p.d.f. $v(t)$ of T is given by

$$v(t) = \sum_{q=0}^{a-2} p_{0,q} f(\lambda, a - 1 - q; t)$$

$$+ \sum_{k=0}^{\infty} \sum_{m=a-1}^{b-1} p_{1,kb+m} f(\mu, k + 1; t). \tag{5.24}$$

$$+ \sum_{k=0}^{\infty} \sum_{m=0}^{a-2} p_{1,kb+m} h_z(t), \quad 0 < t < \infty.$$

(This is obtained by conditioning on the number of units found by the test unit on arrival).

It may be verified that

$$\int_0^\infty v(t)dt = 1 - p_{0,a-1},$$

as is to be expected.

Substituting the values of $p_{0,q}p_{1,n}$ and the expressions for f and h in (5.24), we get an exact expression for the p.d.f. v (t) as follows:
Denote

$$\sum_{k=0}^{m-1} z^k/k! \equiv e(m, z); \quad e^{-z} \cdot e(m, z) \equiv E(m, z), m \geq 1. \tag{5.25}$$

Then

$$\sum_{q=0}^{a-2} r^q f(\lambda, a - q - 1; t) = \lambda r^{a-2} \exp(-\lambda t) e(a - 1, \lambda t/r).$$

Thus the first number on the right hand side (r.h.s) of (5.24) becomes

$$\lambda p_{0,0}[E(a - 1, \lambda t) - r^{a-1} \exp(-\lambda t) e(a - 1, \lambda t/r)]/(1 - r). \tag{5.26}$$

The second member on the r.h.s. of (5.24) is equal to

$$(\lambda/\mu)p_{0,0}\left[\left\{\sum_{m=a-1}^{b-1} r^m\right\}\left\{\mu \exp(-\mu t) \sum_{k=0}^\infty (\mu t r^b)^k/k!\right\}\right]$$

$$= [\lambda p_{0,0}(r^{a-1} - r^b) \exp\{-\mu(1 - r^b)t\}]/(1 - r). \tag{5.27}$$

The third member on the r.h.s. of (5.24) consists of two terms corresponding to those of (5.23); the first of these becomes

$$(\lambda/\mu)p_{0,0}\left[\left\{\sum_{m=0}^{a-2} r^m f(\lambda, a - 1 - m; t)\right\}\left\{\int_0^t \sum_{k=0}^\infty (r^b)^k f(\mu, k + 1; x) dx\right\}\right]$$

$$= \lambda p_{0,0}[r^{a-1} \exp(-\lambda t) e(a - 1, \lambda t/r)]$$

$$\times [1 - \exp\{-\mu(1 - r^b)t\}]/(1 - r), \tag{5.28}$$

and the second becomes

$$(\lambda/\mu)p_{0,0}\left[\sum_{k=0}^\infty (r^b)^k f(\mu, k + 1; t)\right]\left[\sum_{m=0}^{a-2} r^m \Gamma_t (\lambda, a - 1 - m)\right]$$

$$= \lambda p_{0,0} \left[\exp \left\{ -\mu(1 - r^b)t \right\} \right] \left[\sum_{m=0}^{a-2} r^m \left\{ 1 - \sum_{q=0}^{a-2-m} \frac{e^{-\lambda t}(\lambda t)^q}{q!} \right\} \right]$$

$$= \lambda p_{0,0} \left[\exp \left\{ -\mu(1 - r^b)t \right\} \right] \left[\frac{1 - r^{a-1}}{1 - r} - \frac{E(a-1, \lambda t)}{1 - r} \right.$$

$$\left. \left[+ \frac{r^{a-1} \exp(-\lambda t) e(a-1, \lambda t/r)}{1 - r} \right] . \right. \tag{5.29}$$

Hence adding the expressions of the r.h.s of (5.26) to (5.29), we get

$$v(t) = [\lambda p_{0,0}/(1 - r)] \left[(1 - r^b) \exp \left\{ -\mu(1 - r^b)t \right\} \right.$$

$$\left. + E(a-1, \lambda t) \{ 1 - \exp(-\mu(1 - r^b)t) \} \right], \quad 0 < t < \infty. \tag{5.30}$$

The p.d.f. of the conditional waiting time distribution, given that a customer has to wait (i.e. an arriving unit who finds the service channel busy or idle with less than $a - 1$ waiting) is given by

$$w(t) = v(t) / \{ 1 - p_{0,a-1} \}. \tag{5.31}$$

▲

10.5.3.2 Moments of the distribution of the waiting time T

To find the moments of T, we note that, for positive integral values of k and for $a \geq 2$,

$$J_1(k) \equiv \int_0^\infty t^k \exp \left\{ -\mu(1 - r^b)t \right\} dt = \frac{\Gamma(k+1)}{\{\mu(1 - r^b)\}^{k+1}}, \tag{5.32}$$

$$J_2(k) \equiv \int_0^\infty t^k E(a-1, \lambda t) dt$$

$$= \sum_{s=0}^{a-2} \int_0^\infty t^k (\lambda t)^s e^{-\lambda t} dt / s!$$

$$= \sum_{s=0}^{a-2} (s+1)(s+2)...(s+k) / \lambda^{k+1}, \tag{5.33}$$

$$J_3(k) \equiv \int_0^\infty t^k E(a-1, \lambda t) \exp \left\{ -\mu(1 - r^b)t \right\} dt$$

$$= \sum_{s=0}^{a-2} (s+1)(s+2)...(s+k) r^{s+k+1} / \lambda^{k+1} \tag{5.34}$$

The expected waiting time in the queue $E(T)$ is given by

$$E(T) = \int\limits_0^\infty tv(t)\,dt = \frac{\lambda p_{0,0}}{1-r}[(1-r^b)J_1(1)+J_2(1)-J_3(1)]$$

$$= \frac{\lambda p_{0,0}}{1-r}\left[\frac{1}{\mu^2(1-r^b)}+\frac{a(a-1)}{2\lambda^2}+\frac{ar^{a+1}(1-r)-r^2(1-r^a)}{\lambda^2(1-r)^2}\right].$$

$$(5.35)$$

It can be verified that Little's formula holds.
Higher moments can be obtained in the same manner.

10.5.4 Particular case: [M/M(1,b)/1]

For the usual bulk service rule,

$$p_{0,0} = \frac{(1-r)}{1-r+(\lambda/\mu)} = \frac{(1-r)^2}{(1-r)^2+r(1-r^b)},$$

$$p_{1,n} = \frac{(1-r)(1-r^b)}{(1-r)^2+r(1-r^b)}r^{n+1}, \quad n = 0,1,2,....$$

The waiting time density is given by

$$v(t) = [\lambda p_{0,0}/(1-r)]\,[(1-r^b)\exp\{-\mu(1-r^b)\,t\}],\qquad (5.36)$$

and the expected waiting time by

$$E(T) = \frac{r}{\mu[(1-r)^2+r(1-r^b)]}.\qquad (5.37)$$

The conditional waiting time distribution of only those who wait is exponential with mean $1/\mu\{1-r^b\}$.

10.5.4.1 Expected Busy Period

Let B and I denote a busy period and idle period respectively. The busy and idle periods alternate and constitute a busy cycle. Using the result given in Example 10 (a) Ch. 6, we get

$$\frac{E(B)}{E(I)} = \frac{1-p_{0,0}}{p_{0,0}}.$$

Here $E(I) = 1/\lambda$, and thus

$$\lambda E(B) = \frac{r(1 - r^b)}{(1 - r)^2} = \frac{\lambda/\mu}{1 - r} \text{ (from(5.15))},$$

so that $E(B) = \frac{1}{\mu(1 - r)}$.

For busy period analysis of $M/M(a, b)/1$ system refer to Medhi (1991).

Putting $b = 1$ (when $r = \rho$), we get the corresponding results for $M/M/1$.

10.5.5 Particular case: $M/M\,(a, \infty)/1$.

For the infinite server capacity case,

$$r = \lambda/(\lambda + \mu)$$

and the steady-state probabilities are obtained by putting $r^b = 0$ in (5.16) to (5.18). One can obtain expressions for $v\,(t)$ and $E\,(T)$.

10.5.6 Remarks

Subsequent to Neuts (1967), who introduced the general bulk service rule for single channel systems, such models have been studied by Teghem *et al.* (1969), Medhi *et al.* (1972), Medhi (1975, 1979), Borthakur (1975, 1976), Neuts (1979), Sim, Templeton *et al.* and Bertsimas and Papaconstantinou (1988). Arora (1964), Ghare (1968), Cromie *et al.* (1976), Chaudhry *et al.*, Weiss (1979), Powell (1985) Powell and Humblet (1986), and others examine some similar bulk service models. These serve as models for mass transportation.

10.6 NETWORK OF MARKOVIAN QUEUEING SYSTEM

10.6.1 Introduction

It often happens that a unit requiring service has to go to a number of service stations (or areas), instead of just one. There is a tendency of queues not to occur in isolation. For example, a bank customer may have to go (and, if necessary , to queue up) before a ledger clerk with his check and then to go (and, if necessary, queue up) before a cashier to receive payment. A customer in a supermarket goes to a shopping area (where merchandise is kept) and then goes to the check-out area; here the shopping area may be likened to an infinite server queue and the check-out area may be likened to a queue with a finite number of servers. These are simple examples of a unit going from one service area to another. More general types of such systems are encountered in practice; there may be a number of service areas and a unit may be not required to go for service to each area and may have to go from one area to some other area (not necessarily the next one in order). For example, in a hospital, a patient may have to go for service to a number of service areas (hospital departments) and not necessarily to all the areas. These examples are given to show how a network of queues arises. Since the publication of Jakson's paper in 1957, this topic has been receiving a good deal of attention. Networks of queues are often used to model complex computer systems as well as telecommunication systems. Here we introduce the topic briefly. Readers interested in the topic are referred to the works of Lemoine (1977, 1978), Disney and König (1985), Gelenbe and Pujolle (1987), Schwartz (1987), Walrand (1988), also to Medhi (1991) for recent references.

10.6.2 Queues in Series (or Tandem)

Before considering a general network, we discuss queues in series (or tandem), a simple one with 2 service stations such that a unit arrives from outside to the first station, receives service there and proceeds to the second one and after receiving service there departs from the system. It is assumed that there are infinite queueing space in each station.

We state (without proof) below an important result on the output of a Markovian queueing system.

Burke's Theorem. For an *M/M/c* system in steady state, the output process is also Poisson with the same rate as the input process. In other words, in an M/M/c queue in steady state, if the interarrival times are i.i.d. exponential with parameter λ then the interdeparture times are also i.i.d. exponential with the same parameter λ.

Consider now, a system in series with 2 service stations such that the arrival process to the first station is Poisson with parameter λ, the service times in the two stations are i.i.d. exponential with rates μ_1 and μ_2 respectively and with ample space before each station. The arrival process to the second station (being the output process from the first) will be again a Poisson with parameter λ. The system can be denoted by

$$M/M/1 \longrightarrow \cdot/M/1$$

In steady state, if (n_1, n_2) denote the number in the system in the two stations, and if $P(n_1, n_2)$ denote its probability, then it can be seen that

$$P(n_1, n_2) = \left[\rho_1(1 - \rho_1)^{n_1}\right]\left[\rho_2(1 - \rho_2)^{n_2}\right] \tag{6.1}$$

$$n_1, n_2 = 0, 1, 2, ..$$

where $\qquad \rho_i = \lambda/\mu_i$, $i = 1, 2.$

The result can easily be extended to cover a series system with $k(\geq 2)$ stations and with $c_i(\geq 1)$ exponential servers in the ith station.

Consider a series system with Poisson input (with rate λ) to the first station and with c_i exponential servers in the ith station. The input to the 2nd (as well as to subsequent stations), in steady state, will be Poisson with the same rate λ (Burke's theorem) so that the ith station will behave as an $M/M/c_i$ queue. Denote by $P(n_1, n_2, ..., n_k)$ the steady-state probability that there are n_i in the system at the ith station, $i = 1, 2, ..., k$. Then it can be easily seen that

$$P(n_1, n_2, ..., n_k) = \prod_{i=1}^{k} p_i(n_i) \qquad (6.2)$$

where $\quad p_i(n_i) = \quad$ Pr {there are n_i in the system in an $M/M/c_i$ queue in equilibrium, with arrival rate λ }.

We note that the joint probability of the queue lengths (numbers in the systems) is obtained in a simple *product form*: it can be put as the product of marginal distributions of each queue.

Each service station or area, which may consists of a number of servers, is called a *node*.

10.6.3 Jackson Network

We shall now consider a more general network with k nodes. Consider a Markovian system (network) such that arrivals from outside to jth node occur according to a Poisson process with parameter λ_j, and that there are c_j exponential servers at node j. Suppose that a unit completing service at node j goes to the next node i with probability p_{ji} or eventually leaves the system from node j with probability $q_j = 1 - \sum_{i \neq j} p_{ji}$, $j = 1, 2, ..., k$.

We may include two additional nodes, node O (*source*) representing the external source from which arrivals to the nodes occur (external input to a node in the network) and node $k + 1$ (*sink*) representing eventual departure from a node in the network to exterior (exterior output) after receiving service at the node. Thus, we have $k + 2$ nodes in all, with k service nodes. Arrivals from external source to node j occur in accordance with an independent Poisson process with parameter λ_j and all arrivals from external source to the nodes in the network constitute a Poisson process with parameter $\lambda = \sum_{j=1}^{k} \lambda_j$. We may write

$$\lambda_j = \lambda p_{0j} \left(\sum_{j=1}^{k} p_{0j} = 1 \right);$$

p_{0j} is the proportion of the total average external input going to node j. Again, $p_{j0} = 0$, $j = 1, 2, ..., k + 1$, $p_{k+1\,j} = 0, j = 0, 1, ..., k$, $p_{k+1\,k+1} = 1$, and $p_{jj} = 0$. The network described above is known as a *Jackson network*; it is called an *open network* as external input and external outputs are included (p_{0j}'s and p_{j0}'s are not all zeros for all j's, $j = 1, 2, ..., k$).

The matrix $P = (p_{ji})$, $i, j = 0, 1, ..., k + 1$ is a stochastic matrix; its submatrix $P_1 = (p_{ji})$, $i, j = 1, 2, ..., k$ is not stochastic.

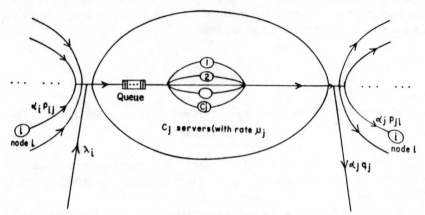

Fig 10.12 Diagrammatic representation of node j in an open Jackson network

Denote the node j visited (by a unit) by Y_j; if we denote

$$\Pr\{Y_i \mid Y_j\} = \Pr\{Y_i \mid Y_j, Y_l, \cdots\} = p_{ji}$$

then $\{Y_i\}$ is a Markov chain with t.p.m. $P = (p_{ji})$ and state space $S = \{0, 1, \ldots, k+1\}$, $k+1$ being an absorbing state. The matrix P is called a *routing matrix*.

A Jakson network without external input and external output is known as a *closed Jackson network*. For such a network, a fixed number of units circulate within the network system with nodes $1, 2, \ldots, k$; such a network has no source (node 0) nor sink (node $k+1$). We have $p_{0j} = 0$ and $q_j = 0_1, j = 1, 2, \ldots, k$. The matrix $P = (p_{ji})$, $j, i = 1, 2, \ldots, k$ is the same as its submatrix P_1 and is stochastic .

As an example of closed Jackson network, one may consider a computer system in which a fixed number of jobs circulate within the network.

A closed network is called *cyclic* if, further

$$
\begin{aligned}
p_{ji} &= 1, & i = j+1, & \quad 1 \le j \le k-1 \\
p_{k1} &= 1, & p_{ji} = 0, & \quad \text{other wise}
\end{aligned}
\tag{6.3}
$$

In a cyclic network, a unit goes from one node to the next node in order (from node j to node $j+1$) and from the last node k goes to the first node 1 and it circulates within the network.

10.6.4 Open Jackson Network

Consider an open network. Denote by α_i, the total average rate of arrival to node i; α_i is equal to average arrival rate from outside λ_i (external input to node i) *plus* the average arrival rate from other nodes $j(\ne i)$ from inside the network. Thus

$$\alpha_i = \lambda_i + \sum_{j=1}^{k} \alpha_j \, p_{ji}, \quad i, j = 1, 2, \ldots, k \tag{6.4}$$

or in matrix notation

$$\alpha = \lambda + \alpha P_1 \tag{6.5}$$

where $\alpha = (\alpha_1, \cdots, \alpha_k), \lambda = (\lambda_1, \cdots, \lambda_k), P_1 = (p_{ij})$ is a $(k \times k)$ matrix (a sub matrix of P where $i, j = 1, 2, ..., k$).

The above equations can also be written as

$$e_i = p_{0i} + \sum_{j=1}^{k} e_j \, p_{ji} \quad \text{(with } e_i = \alpha_i / \lambda) \tag{6.6}$$

The equations given above are called *traffic equations (or flow balance equations)*. The existence of their solution is a necessary condition for existence of steady state distribution for the system.

We state below Jackson's theorem.

10.6.4.1 Jackson's Theorem for Open Network

Denote by $(n_1, \ldots, n_k)$ the state of the network system (in equilibrium) that there are n_i units at the ith node and by $P(n_1, \ldots, n_k)$ the corresponding steady state probability. Assume that solutions of the traffic equations exist and that $\rho_i = \alpha_i / \mu_i < 1, i = 1, 2, ..., k$. Then the steady state distribution $P(n_1, ..., n_k)$, when it exists, is given by

$$P(n_1, ..., n_k) = A \prod_{i=1}^{k} \frac{(\rho_i)^{n_i}}{f_i(n_i)} \tag{6.7}$$

where

$$f_i(n_i) = \begin{cases} n_i!, & n_i \leq c_i \\ (c_i!)(c_i)^{n_i - c_i}, & n_i > c_i \end{cases} \tag{6.8}$$

and

$$A = \left[\sum_{n_i \geq 0} \prod_{i=1}^{k} \frac{(\rho_i)^{n_i}}{f_i(n_i)} \right]^{-1} \tag{6.9}$$

is the normalising constant. We have $P(n_1, \cdots n_k) = \prod_{i=1}^{k} p_i(n_i)$, where $p_i(n_i)$ is the

probability that there are n_i units in the system in an $M/M/c_i$ queue with arrival rate α_i. We omit the proof.

Jackson's proof uses usual probability arguments to obtain balance equations. An alternative proof has been given by Lemoine (1977).

Note: (1) The result has a surprisingly simple *product form*. Each node behaves *as if it were* an independent $M/M/c$ queue. The network can be decomposed into k number of such queues.

(2) A necessary and sufficient condition for existence of the solution is that $A > 0$.

(3) The result can be extended to cover more general cases where the external arrival rate to a node and the service rate at a node are state dependent: the service rate at a node depends on the number of units at that particular node and the external arrival rate on the total number present in the network.

10.6.5 Closed Jackson Network

Gordon and Newell (1967) consider a closed network such that there is neither external input nor external output. Then $\lambda_i = 0$, i.e. $p_{0i} = 0$, $q_i = 0$, $i = 1,2,...,k$. The matrix $P = (p_{ij})$ is a $(k \times k)$ stochastic matrix; and $P \equiv P_i$. Assume that P is an irreducible matrix so that the states of the corresponding Markov chain are all persistent.

Since, for a closed network $\lambda = 0$, the traffic equation in matrix form reduces to

$$\alpha = \alpha P \text{ [since } P_1 \equiv P] \tag{6.10}$$

Taking the routing matrix P to be irreducible it can be seen that α is the unique stationary distribution of a Markov chain having t.p.m. P.

We state (without proof) below a theorem for closed network

Gordon and Newell's Theorem for Closed Network

Consider a closed network with k nodes and a fixed number N units that circulate through the network according to a routing matrix P. The node i has c_i exponential servers each with rates μ_i.

Let $\alpha = (\alpha_1, ..., \alpha_k)$ be any non-zero solution of the equation $\alpha = \alpha P$ and let $\rho_i = \alpha_i/\mu_i < 1$.

Then the steady state probability of the system state $(n_1, \cdots, n_k)$, $\sum n_i = N, 0 \leq n_i \leq N$ is given by

$$P(n_1, \cdots, n_k) = A_c \prod_{i=1}^{k} \frac{(\rho_i)^{n_i}}{f_i(n_i)} \tag{6.11}$$

where $f_i(n_i)$ are as given by (6.8), and the normalising constant A_c is given by

$$A_c = \left[\sum_{\substack{0 \leq n_i \leq N \\ \Sigma n_i = N}} \prod_{i=1}^{k} \frac{(\rho_i)^{n_i}}{f_i(n_i)} \right]^{-1} \tag{6.12}$$

We have $P(n_1, \cdots, n_k) = \prod_{i=1}^{k} p_i(n_i)$, where $p_i(n_i)$ is the probability that there are n_i units in the system in an $M/M/c_i$ queue. But the distributions in different nodes are not independent since $\sum n_i = N$.   ▲

Note: 1. The result is of a product form; the similarity and difference between the results may be noted

2. If each node has only one server, i.e. $c_i = 1$ for all i, then $f_i(n_i) = 1$.

10.6.5.1 Cyclic Network

Using (6.3) we get as solution of $\alpha = \alpha P$

$$\rho_i = \frac{\mu_{i-1}}{\mu_i} \rho_{i-1}$$

$$\cdots$$

$$= \frac{\mu_1}{\mu_i} \rho_1, \; i = 1, 2, \cdots, k.$$

and thus, for $\sum n_i = N$,

$$P(n_1, \cdots, n_k) = A_1 \prod_{i=1}^{k} \left(\frac{\mu_1}{\mu_i} \right)^{n_i} \tag{6.13}$$

where

$$A_1 = \left[\sum_{\substack{0 \le n_i \le N \\ \sum n_i = N}} \prod_{i=1}^{k} \left(\frac{\mu_1}{\mu_i} \right)^{n_i} \right]^{-1} \tag{6.14}$$

10.6.6 Concluding Remarks

A more general network model has been considered by Baskett, Chandy, Muntz and Palacios (1975), where new service disciplines as well as a number of classes of jobs are considered. Readers may refer to the references given in Sec. 10.6.1.

10.7 NON-MARKOVIAN QUEUEING MODELS

So long we have been discussing queueing processes which are either birth and death or non-birth and death processes. They are in either case Markovian and the theory of Markov chains and processes could be applied in their studies. We shall now consider models where the distributions of the interarrival time or the service time do not possess the memoryless property, i.e. are not exponential. The process $\{N(t)\}$ giving the state of the system or system size at time t will then be no longer Markovian; however, the analysis of the process can be based on an associated process which is Markovian. Two techniques are generally being used for this purpose. Kendall (1951) uses the concept of regeneration point (due to Palm) by suitable choice of regeneration points and extracts, from the process $\{N(t)\}$, Markov chains in discrete time at those points. This is known as the technique of imbedded Markov chains. The second important technique due to Cox (1953) (see also Keilson and Kooharian (1960)) and known as *supplementary variable technique*, involves inclusion of such variable(s).

We discuss below Kendall's method.

10.7.1. Queues with Poisson Input : Model *M/G/1*

Assume that the input process is Poisson with intensity λ and that the service times are i.i.d. random variables having an arbitrary (general) distribution with mean $1/\mu$. Denote the service time by v, its d.f., by $B\,(t)$, its p.d.f., when it exists, by $b\,(t)(= B'(t))$, and its L.T. by

$$B^*(s) = \int_0^\infty e^{-st}\, d\dot{B}\,(t).$$

Let t_n, $n = 1, 2, \ldots$, $(t_0 = 0)$ be the nth departure epoch, i.e. the instant at which the nth unit completes his service and leaves the system. These points t_n are the regeneration points of the process $\{N\,(t)\}$. The sequence of points $\{t_n\}$ forms a renewal process. $N\,(t_n + 0)$, the number in the system immediately after the nth departure has a denumerable state space $\{0, 1, 2, \ldots\}$. Write $N\,(t_n + 0) \equiv X_n$, $n = 0, 1, \ldots$ and denote by A_n, the random variable giving the number of units that arrive during the service time of the nth unit. Then

$$
\begin{aligned}
X_{n+1} &= X_n - 1 + A_{n+1}, & \text{if } X_n \geq 1 \\
&= A_{n+1}, & \text{if } X_n = 0
\end{aligned}
\tag{7.1}
$$

Now the service times of all the units have the same distribution so that $A_n \equiv A$ for $n = 1, 2, \ldots$ We have

$$\Pr\{A = r \mid \text{service time of a unit is } t\} = \frac{e^{-\lambda t}(\lambda t)^r}{r!}$$

and so

$$k_r \equiv \Pr\{A = r\} = \int_0^\infty \frac{e^{-\lambda t}(\lambda t)^r}{r!}\, dB(t), \quad r = 0, 1, 2, \ldots,
\tag{7.2}$$

gives the distribution of A, the number of arrivals during the service time of a unit. The probabilities

$$p_{ij} = \Pr\{X_{n+1} = j \mid X_n = i\}$$

are given by

$$
\begin{aligned}
p_{ij} &\equiv k_{j-i+1}, & i \geq 1, j \geq i - 1 \\
&= 0 &, \quad i \geq 1, j < i - 1
\end{aligned}
\tag{7.3}
$$

$$p_{0j} \equiv p_{ij} = k_j, \quad j \geq 0.$$

The relations (7.3) clearly indicate that $\{X_n, n \geq 0\}$ is a Markov chain having t.p.m.

$$P = (p_{ij}) = \begin{pmatrix} k_0 & k_1 & k_2 & \cdots \\ k_0 & k_1 & k_2 & \cdots \\ 0 & k_0 & k_1 & \cdots \\ \cdots & \cdots & \cdots \\ \cdots & \cdots & \cdots \end{pmatrix}. \tag{7.4}$$

As every state can be reached from every other state, the Markov chain $\{X_n\}$ is irreducible. Again as $p_{ii} \geq 0$, the chain is aperiodic. It can also be shown that, when the traffic intensity $\rho = \lambda/\mu < 1$, the chain is persistent, non-null and hence ergodic. We can then apply the ergodic theorem of Markov chain (Theorem 3.11).

10.7.2. Pollaczek-Khinchine Formula

The limiting probabilities

$$v_j = \lim_{n \to \infty} p_{ij}^{(n)}, \quad j = 0, 1, 2, \ldots$$

exist and are independent of the initial state i. The probabilities $V = (v_0, v_1, \ldots)$, $\sum v_j = 1$, are given as the unique solutions of

$$V = VP.$$

Let $K(s) = \sum k_j s^j$ and $V(s) = \sum v_j s^j$ denote the p.g.f. of the distributions of $\{k_j\}$ and $\{v_j\}$ respectively.

We have
$$K(s) = \sum_{j=0}^{\infty} k_j s^j = \sum_{j=0}^{\infty} s^j \left\{ \int_0^{\infty} \frac{e^{-\lambda t}(\lambda t)^j}{j!} \, dB(t) \right\}$$

$$= \int_0^{\infty} e^{-(\lambda - \lambda s)t} \, dB(t)$$

$$= B^*(\lambda - \lambda s). \tag{7.5}$$

Hence
$$E(A) = K'(1) = -\lambda B^{*(1)}(0) = \lambda/\mu = \rho. \tag{7.6}$$

Now $V = VP$ gives an infinite system of equations. Multiplying the $(k + 1)$st equation by s^k, $k = 0, 1, \ldots$ and adding over k, we get, on simplification, for $0 < \rho < 1$,

$$V(s) = \frac{\{1 - K'(1)\} (1 - s) K(s)}{K(s) - s}$$

(see Example 7 (a), Ch. 3)

Putting $K'(1) = \rho$ we get

$$V(s) = \frac{(1-\rho)(1-s)K(s)}{K(s)-s} \tag{7.7}$$

$$= \frac{(1-\rho)(1-s)B^*(\lambda - \lambda s)}{B^*(\lambda - \lambda s)-s} \tag{7.7a}$$

This is known as *Pollaczek[†]-Khinchine (P.K.) formula.*

Remarks :

(1) Consider the situation when the server is intermittently available such that he starts service on completion of a service even when there may be none waiting (e.g. an automatic traffic signal) – there is then no idle period for such a server. The relation (7.1) will hold good in that case also ; so also relation (7.7) i.e., $V(s)$ will be of the same form.

(2) **Poisson Arrivals See Time Averages (PASTA)**

In a system with Poisson input, the fraction of arrivals who find n customers in the system is equal to the fraction of time with n customers (this property is called *PASTA* ; for proof, see Wolff (1982)).

(3) For the model $M/G/1$, the process $\{N(t)\}$ giving the system size at time t is not a Markov process. By extracting a process at departure points t_n, we get an associated process $X_n = N(t_n + 0)$, which is a Markov chain. The transition probabilities, p_{ij} of the Markov chain, give departure point probabilities (probabilities of the departure epoch system size), while

$$p_{in}(t) = \Pr\{N(t) = n \mid N(0) = i\}$$

give the probabilities of the general time system size. When the Markov chain is ergodic (the condition for which is $\rho < 1$), the limiting probabilities $v_n = \lim p_{in}^{(m)}$, as $m \to \infty$, exist and can be obtained by applying the ergodic theorem of Markov chain. When the steady state exists (the condition for which is also $\rho < 1$), the limiting probabilities $p_n = \lim p_{in}(t)$, as $t \to \infty$, exist. It can be shown that (for proof, see Cohen, Gross and Harris) when $\rho < 1$, then $v_n = p_n$ for all n, i.e. the limiting departure point probabilities are equal to the corresponding steady state probabilities of the system size. Thus ergodicity of the Markov chain is equivalent to the existence of steady state, the condition for each to hold good being $\rho < 1$.

It may also be noted that another interesting stochastic process can be extracted. The process $Y(t) = X_n, t_n \le t < t_{n+1}$ gives the number of units left behind in the system by the most recent departure. Clearly, $\{X_n, t_n\}$ is a Markov renewal process; and $\{Y(t), t \ge 0\}$ is a semi-Markov process, X_n being its embedded Markov chain and the waiting time in a state being distributed as the service time v. We shall consider such semi-Markov processes later.

[†]For a bibliography of the works of Felix Pollaczek (1892-1981), see *J. Appl. Prob* **18** (1981) 958-963.

Example 7(a). *The Model* $M/E_k/1$. Consider that the service time v is Erlang-k (with mean $1/\mu$) having p.d.f.

$$b(t) = \frac{(\mu k)^k t^{k-1} e^{-k\mu t}}{\Gamma(k)}, 0 < t < \infty.$$

Its L.T. is $\qquad B*(s) = \left(\frac{\mu k}{s + \mu k}\right)^k \quad$ (see Sec. 1.3.4.6)

From (7.5) and (7.7) we get

$$K(s) = B*(\lambda - \lambda s) = \left(\frac{\mu k}{\lambda - \lambda s + \mu k}\right)^k$$

$$= \left(1 + \frac{\rho(1-s)}{k}\right)^{-k}, \quad \rho = \lambda/\mu$$

and $\qquad V(s) = \frac{(1-\rho)(1-s)}{1 - s\{1 + \rho(1-s)/k\}^k}.$

Expanding in powers of s, and comparing coefficients of s^n from both sides, we get v_n, and equivalently p_n. We can find the moments of $\{p_n\}$ from $V(s)$.

This model could also be considered as non-birth and death Markovian; by considering that the service comprises of k exponential phases each with mean $1/(k\mu)$, the C.K. equations can be written and solutions obtained therefrom.

Special case: *The model* $M/M/1$. Putting $k = 1$, we at once get,

$$V(s) = \frac{(1-\rho)(1-s)}{(1 - s - s\rho(1-s))} = \frac{(1-\rho)}{(1-s\rho)}.$$

Example 7(b). *The model* $M/D/1$. As $k \to \infty$, the whole mass of the distribution E_k tends to concentrate at the point $1/\mu$, i.e. $E_k \equiv D$, as $k \to \infty$. We have then

$$B*(s) \to e^{-s/\mu},$$

so that

$$K(s) = B*(\lambda - \lambda s) \to e^{-\rho(1-s)}$$

and

$$V(s) = \frac{(1-\rho)(1-s)}{1 - s e^{\rho(1-s)}}.$$

Expanding the r.h.s. in powers of s, and after some algebraic manipulations, we get v_n, the coefficient of s^n, $n = 0, 1, 2, \ldots$, as follows:

$$v_0 = 1 - \rho$$

$$v_1 = (1 - \rho)(e^\rho - 1)$$

...   .
...   .

$$v_n = (1 - \rho)\left[e^{n\rho} + \sum_{r=1}^{n-1}(-1)^{n-r}e^{r\rho}\left\{\frac{(r\rho)^{n-r}}{(n-r)!} + \frac{(r\rho)^{n-r-1}}{(n-r-1)!}\right\}\right], \quad n \geq 2.$$

10.7.2.1 Waiting Time: Pollaczek-Khinchine Formula

Assume that the steady state exists ($\rho < 1$).

Let W and W_Q denote the waiting times in the system and in the queue respectively and $W(t)$ and $W_Q(t)$ denote their distribution functions. We have

$$p_n = v_n = \Pr \{\text{departing unit leaves } n \text{ in the system}\}$$

$$= \int_0^\infty \Pr \{\text{departing unit leaves } n \text{ in the system} \mid t \leq W < t + dt\}$$

$$\times \Pr \{t \leq W < t + dt\}$$

$$\simeq \int_0^\infty \frac{e^{-\lambda t}(\lambda t)^n}{n!}\, dW(t), \quad n = 0, 1, 2, \ldots . \tag{7.8}$$

Thus

$$V(s) = \sum_{n=0}^\infty v_n s^n$$

$$= \int_0^\infty \{e^{-\lambda t}\, dW(t)\}\left\{\sum_{n=0}^\infty \frac{(\lambda t)^n}{n!} s^n\right\}$$

$$= \int_0^\infty e^{-\lambda t(1-s)}\, dW(t)$$

$$= W^*(\lambda - \lambda s), \tag{7.9}$$

where

$$W^*(\theta) = \int_0^\infty e^{-\theta t}\, dW(t).$$

Now,

$$P(s) = \sum p_n s^n = \sum v_n s^n = V(s)$$

$$= W^*(\lambda - \lambda s). \tag{7.9a}$$

From (7.7), (7.9) and (7.6) we get

$$W^*(\lambda - \lambda s) = \frac{(1-\rho)(1-s)B^*(\lambda - \lambda s)}{B^*(\lambda - \lambda s) - s}.$$

Writing α for $\lambda - \lambda s$, we get

$$W^*(\alpha) = \frac{(1-\rho)(\alpha/\lambda)B^*(\alpha)}{B^*(\alpha) - (1-\alpha/\lambda)}$$

$$= \frac{\alpha(1-\rho)B^*(\alpha)}{\alpha - \lambda(1-B^*(\alpha))}, \qquad (7.10)$$

or

$$W^*(s) = \frac{s(1-\rho)B^*(s)}{s - \lambda(1-B^*(s))}. \qquad (7.10a)$$

This relation connects the L.T.'s of the distributions of the service time v and the waiting time W in the system.

Now, $W = W_Q + v$, so that $W^*(s) = W_Q^*(s) B^*(s)$, W_Q^* being the L.T. of W_Q. Thus, we have

$$W_Q^*(s) = \frac{s(1-\rho)}{s - \lambda(1-B^*(s))}. \qquad (7.11)$$

The relation (7.10a) (as well as (7.11)) is also known as Pollaczek-Khinchine (P.K.) formula. To distinguish it from the other P.K. formula (7.7), it is called *P.K. transform formula*. The formula has been rediscovered in special contexts by different researchers. There are many interesting direct proofs of the formula (see also Runnenberg (1960)).

Remark :

Waiting time distribution in terms of residual service time

Let X denote the residual service time X of unit receiving service at time t the test unit arrives. We have, for large t

$$G(x) = \Pr\{X \le x\} = \frac{1}{E(v)} \int_0^x [1 - B(y)]dy, \quad x > 0$$

(relation (7.4) Chapter 6), and

$$G^*(s) = \text{L.S.T of } G(x) = \frac{1-B^*(s)}{sE(v)}$$

$$= \frac{\mu[1 - B^*(s)]}{s}.$$

Then from (7.11), we have

$$W^*_Q(s) = \frac{1 - \rho}{1 - \frac{\lambda}{\mu}\left[\frac{\mu}{s}\{1 - B^*(s)\}\right]}$$

$$= \frac{1 - \rho}{1 - \rho G^*(s)}. \qquad (7.11a)$$

Thus the waiting time W_Q involves residual service time X (of the unit in service at the instant the test unit arrives), as is to be expected.

Expanding (7.11a) we get

$$W^*_Q(s) = (1 - \rho)\left[1 + \sum_{k=1}^{\infty} \rho^k \{G^*(s)\}^k\right]$$

$$= (1 - \rho) + (1 - \rho) \sum_{k=1}^{\infty} \rho^k \{G^*(s)\}^k. \qquad (7.11b)$$

Inversion of (7.11b) gives the p.d.f. $w_q(x)$ of W_Q. We get

$$w_q(x) = (1 - \rho)\, \delta(x) + (1 - \rho) \sum_{k=1}^{\infty} \rho^k g^{k*}(x), x \geq 0 \qquad (7.12)$$

$$(\delta(x) = 1, \quad x = 0$$

$$= 0, \quad x \neq 0)$$

where g denotes the p.d.f. of X and g^{k*} denotes the k-fold convolution of g with itself.

Kleinrock observes that, no satisfactory explanation has been found for the above form of w_q in terms of the weighted (geometric) sum of convolved p.d.f.'s of X. See Cooper & Niu., *J. App. Prob.* **23** (1986) 550.

*Particular case: M/M/*1. Here X has the same distribution as v, the service time, with mean $1/\mu$, so that k-fold convolution of X is a gamma variable with

$$g^{k*}(x) = \frac{\mu^k x^{k-1} e^{-\mu x}}{\Gamma(k)}, \quad x \geq 0$$

Thus from (7.12)

$$w_q(x) = (1 - \rho)\, \delta(x) + (1 - \rho)e^{-\mu(1-\rho)x}, \quad x \geq 0$$

as obtained earlier.

10.7.2.2 Expected Waiting Time and Expected Number in the System

Using (7.7) and (7.11) moments of $E\{N\}$ and $E\{W_Q\}$ can be obtained.

We have
$$B^{*(k)}(0) \equiv \frac{d^k}{ds^k}B^*(s)\Big|_{s=0} = (-1)^k E(v^k), \quad k = 1, 2, \dots.$$

From (7.11), we get

$$\frac{W_Q^{*(1)}(s)}{1-\rho} = \frac{s - \lambda + \lambda B^*(s) - s\{1 + \lambda B^{*(1)}(s)\}}{\{s - \lambda + \lambda B^*(s)\}^2}$$

which takes the form $\dfrac{0}{0}$ as $s \to 0$. We can use L'Hopital's rule to obtain its value as $s \to 0$. As $s \to 0$, the r.h.s. becomes

$$\frac{(1 + \lambda B^{*(1)}(s)) - (1 + \lambda B^{*(1)}(s)) - s(\lambda B^{*(2)}(s))}{2(s - \lambda + \lambda B^*(s))(1 + \lambda B^{*(1)}(s))} = \frac{-\lambda B^{*(2)}(0)}{2\{1 + \lambda B^{*(1)}(0)\}^2}$$

$$= \frac{\lambda E(v^2)}{2(1 - \lambda/\mu)^2}.$$

Thus,
$$E(W_Q) = W_Q^{*(1)}(0) = \frac{\lambda}{2(1-\rho)}E(v^2)$$

$$= \frac{\lambda}{2(1-\rho)}\left(\sigma_v^2 + \frac{1}{\mu^2}\right) \qquad (7.13)$$

$$= \frac{\rho}{2\mu(1-\rho)}(1 + c_v^2)$$

where σ_v^2 is the variance and $c_v = \mu\sigma_v$ is the coefficient of variation of the service time v.

By differentiating $W_Q^*(s)$, k times w.r.t. s and putting $s = 0$, we can obtain the k th moment of W_Q. The first $(k + 1)$ moments of the service time v determine the first k moments of W_Q and *vice-versa*. It can be seen that

$$\text{var}\{W_Q\} = \frac{\lambda}{12(1-\rho)^2}[4(1-\rho)E(v^3) + 3\lambda\{E(v^2)\}^2]. \qquad (7.14)$$

Note: Since $E(X) = \rho E(R) = \rho E(v^2)/2E(v)$, $R \equiv$ forward recurrence time of v

we get
$$E(W_Q) \equiv E(X)/(1-\rho) = [P/(1-\rho)]E(R) \qquad (7.13b)$$

It can also be shown that (Sphicas *et al.*, 1978)

$$c^2 \geq (4 - \rho)/3\rho \geq 1,$$

where $c = $ s.d. $\{W_Q\}/E\{W_Q\}$, is the coefficient of variation of W_Q.

The mean waiting time in the system is

$$E\{W\} = E\{W_Q + v\} = E(v) + E\{W_Q\}$$

$$= \frac{1}{\mu} + \frac{\lambda}{2(1 - \rho)} E^{\cdot}(v^2). \qquad (7.15)$$

Using Little's formula, we can obtain $E\{N\}$ from (7.15) and $E\{Q\}$ form (7.13). We have

$$E\{N\} = \lambda E\{W\} = \rho + \frac{\lambda^2}{2(1 - \rho)} E(v^2) \qquad (7.16)$$

$$= \rho + \frac{\rho^2}{1 - \rho} \{1 + c_v^2\}$$

and
$$E\{Q\} = \lambda E\{W_Q\} = \frac{\lambda^2}{2(1 - \rho)} E(v^2) = \frac{\lambda^2 \sigma_v^2 + \rho^2}{2(1 - \rho)}.$$

$E\{N\}$, $E\{N^2\}$, . . . can be obtained directly from (7.7). The formula (7.16) is also sometimes called P.K. formula; following Kleinrock (1976) we may call it *P.K. mean value formula*, to distinguish it from *P.K. transform formula*.

From (7.13) and (7.15) it becomes at once clear that for an *M/G/*1 queue with parameters λ, μ, expected waiting time, whether in the queue or in the system, is the least when $\sigma_v = 0$, i.e. service time distribution is constant ($= 1/\mu$). The model then becomes *M/D/*1.

Thus, *for all single server models with Poisson input, the one with constant service time has the least expected waiting time and hence the least expected queue and system size.*

Note: (1) By utilising a relationship between single server queues and random walks and without using P.K. formula, Lemoine (1976) obtains the moments of waiting time of *M/G/*1 queue.

(2) For an alternative derivation of P.K. mean value theorem, see Fakinos (1982).

(3) $E(W_Q)[M/G/1] = c_v^2 E(W_Q)[M/M/1] + (1 - c_v^2) E(W_Q)[M/D/1]$

10.7.3 Busy Period

We discussed the busy period for the $M/M/1$ model in Sec 10.3; here we consider it for the $M/G/1$ model through an alternative approach.

The expected duration of the busy period T (initiated by a single customer) follows immediately from the result noted in Example 10(a), Ch. 6. For $\rho < 1$, we have

$$p_0 = \lim_{t \to \infty} \{N(t) = 0\} = \frac{E(I)}{E(I) + E(T)}.$$

Clearly, the idle period I here is exponential with mean $1/\lambda$, since the interarrival distribution is so. As stated in the remarks noted earlier in this section, $p_n = v_n$ for all n. From equation (7.7) we find that v_0, the constant term in $V(s)$, is given by $v_0 = V(0) = 1 - \rho$, so that

$$p_0 = 1 - \rho = \frac{1/\lambda}{1/\lambda + E(T)}$$

whence $$E(T) = \frac{1}{\mu - \lambda} = \frac{E(v)}{1 - \rho}, \quad (v \text{ being the service time}).$$

$$(7.17)$$

Note that $E(T)$ for an $M/G/1$ queue has the same form as that for an $M/M/1$ queue (see (3.24)).

Thus, *given the mean arrival and service rates, the expected duration of a busy period in a queue with Poisson input is independent of the form of the distribution of the service time.*

We derive below an integral equation satisfied by the d.f. of a busy period, initiated by a single unit or customer.

10.7.3.1 Takacs Integral Equation

Let v be the service time,

$$B(t) = \Pr\{v \le t\}, \quad B^*(s) = \int_0^\infty e^{-st} \, dB(t),$$

and T be the duration of the busy period,

$$G(t) = \Pr\{T \le t\}, \quad G^*(s) = \int_0^\infty e^{-st} \, dG(t).$$

Let G^{n*} denote the n-fold convolution of G with itself. Let $A(x)$ denote the number of arrivals in $[0, x]$.

By conditioning on the duration of the service time v of the initial customer inaugurating the busy period and on the number A of arrivals during his service time, one gets,

$$G(t) = e^{-\lambda t} B(t) + \sum_{n=1}^{\infty} \int_0^t \Pr\{T \le t \mid v = x, A(x) = n\} \, dB(x) \cdot \Pr\{A(x) = n\};$$

the first term corresponds to the case that none arrives during the service period of the initial customer (and then $x = t$). Now let T_i be the total service time of the ith descendant[†] and of all his progeny. Then T_i's are i.i.d. and have the same distribution as T.

We have

$$\Pr\{T \le t \mid v = x, A(x) = n\} = \Pr\{T_1 + \ldots + T_n \le t - x\}$$

$$= G^{n*}(t - x), n \ge 1.$$

Writing $G^{n*} = 1$, for $n = 0$, and noting that then $x = t$, we get,

$$G(t) = \sum_{n=0}^{\infty} \int_0^t \frac{e^{-\lambda x}(\lambda x)^n}{n!} G^{n*}(t - x) \, dB(x).$$

This equation is known as *Takacs integral equation*.

Putting the above expression for $G(t)$ on the r.h.s. of

$$G^*(s) = \int_0^{\infty} e^{-st} dG(t),$$

changing the order of integration in the repeated integral (on the r.h.s.) and on simplification, we get

$$G^*(s) = B^*(s + \lambda - \lambda G^*(s)), \tag{7.18}$$

which is a functional equation in terms of the L.T.'s.   ▲

In view of its importance, we give an alternative derivation of the above equation.

Given that $v = x$, i.e. the initial customer ("ancestor") inaugurating the busy period departs at x, the number of ("descendants" i.e.) persons N arriving by time x has Poisson distribution with mean λx. Thus given $v = x$, the duration of the busy period T equals $x + S_N$, where

---

[†]In branching process terminology, the initial customer may be called "ancestor", those who arrive during his service time may be called (direct) "descendant" and the descendants of the descendants and so on, may be called "progenies".

$$S_N = \sum_{i=1}^{N} T_i.$$

Since the L.T. of T is $G^*(s)$, the L.T. of S_N is

$$\exp\{-\lambda x + \lambda x G^*(s)\} \qquad \text{(by Theorem 1.11)}$$

Thus the L.T. of the r.v. T, given $v = x$, is

$$E\{e^{-sT} \mid v = x\} = \exp(-sx) \exp\{-\lambda x + \lambda x G^*(s)\}$$
$$= \exp\{-x(s + \lambda - \lambda G^*(s))\}.$$

Finally,

$$G^*(s) = \text{L.T. of } T = E\{e^{-sT}\}$$

$$= \int_0^\infty E\{e^{-sT} \mid v = x\}\, dB(x)$$

$$= \int_0^\infty \exp\{-x(s + \lambda - \lambda G^*(s))\}\, dB(x)$$

$$= B^*(s + \lambda - \lambda G^*(s)). \qquad \blacktriangle$$

As can be imagined, such an equation would occur in branching processes. The reduction of the busy period to branching processes is due first to Good, whose heuristic arguments leading to the equation were given formal content by Takacs. We state the following proposition:

Let $H(s)$ be the L.T. of a distribution with mean b and let $\lambda > 0$, then the equation

$$K(s) = H(s + \lambda - \lambda K(s))$$

has a unique root $K(s) \leq 1$; further, $K(s)$ is the L.T. of a proper probability distribution *iff* $\lambda b \leq 1$.

(For a proof of the above proposition, see Feller, Vol. II. (1966))

From the above proposition it is clear that $G^*(s)$ is the L.T. of a proper probability distribution if $\rho = \lambda/\mu \leq 1$. If $\rho > 1$, there is a non-zero probability of a never-ending busy period, the probability being equal to the smallest positive root of the equation $\zeta = B^*(\lambda - \lambda \zeta)$.

Takacs obtains the inversion of $G^*(s)$ in the form

$$G(t) = \sum_{n=1}^{\infty} \int_0^t e^{-\lambda s} \frac{(\lambda t)^{n-1}}{n!}\, dB^{n^*}(t).$$

Even without finding $G(t)$, one can find the moments of T from the functional equation in $G^*(\cdot)$. As can be easily verified

$$E(T) = \frac{E(v)}{1-\rho}$$

and

$$\mathrm{var}(T) = \frac{E(v^2)}{(1-\rho)^3} - \frac{(E(v))^2}{(1-\rho)^2}. \qquad (7.19)$$

As a special case, we can obtain $G^*(s)$ for $M/M/1$ model from the functional equation by putting $B^*(s) = \mu/(s+\mu)$. The equation then reduces to a quadratic in $G^*(s)$, whose root is seen to be β/ρ (as given in equation (3.23)).

Remark: Alternative formula for $G^*(s)$ by a martingale method (Rosenkrantz, 1983).

Rosenkrantz's formula for $G^*(s)$ in terms of $B^*(s)$ is of independent interest. Denote

$$f(s) = \lambda[B^*(s) - 1] + s$$

and $f^{-1}(s)$ as the inverse of $f(s)$, i.e.

$$f(f^{-1}(s)) = f^{-1}(f(s)) = s,$$

(assuming the existence $f^{-1}(s)$ in the neighbourhood of 0). He shows that

$$G^*(f(s)) = B^*(s)$$

or, taking inverses

$$G^*(s) = B^*(f^{-1}(s))$$

One can find the moments of T from the above functional equation.

10.7.3.2 M/G/1 Queue under N-policy

Heyman (1968) considers a model such that the idle server starts service only when N are present in the queue and once he starts serving, goes on serving till the system size becomes 0 (*exhaustive service*). This is called N-policy (a *control operating policy*). It is shown that a queue with N-policy possesses certain optimal properties. The system is also known as a system with *removable server* (or with *server vacation*).

Consider a system with Poisson input, general service time with exhaustive service and under N-policy: Denote the system by $M_N/G/1$. For $N = 1$, we get $M/G/1$

queue. The idle period I (of the server) is the time for arrival of N units and is a gamma variate with $E(I) = N/\lambda$. The busy period B_N which starts with N units is the convolution of N busy periods of a standard $M/G/1$ queue, so that

$$E(B_N) = \frac{N}{\mu(1-\rho)}, \quad N \geq 1,$$

and the expected number of units served during a busy period is $N/(1-\rho)$.

Now, arrivals are Poisson and the busy and idle periods together constitute an alternating renewal process. The probability that the server is busy is given by

$$\text{Pr(server is busy)} = \frac{E(B_N)}{E(I) + E(B_N)}$$

$$= \frac{\lambda}{\mu} = \rho$$

Thus the fraction of time the server is busy is ρ and is independent of N, as should be intuitively clear.

More general models have been considered by Baker (*Infor.* 1973, 71-72), Borthakur *et al.* (1987), Lee and Srinivasan (1989) and Medhi and Templeton (1992).

10.7.3.3 *Number served During a Busy period in an M/G/1 Queue*

Denote by $N(T)$, the number of customers served during a busy period T initiated by a single customer ("ancestor") and by $P(z)$, its p.g.f. Denote by $N(T_j)$, the number served during the total service time T_j of the jth direct descendant and of all his progeny (i.e. $N(T_j)$ equals 1 plus the number of progeny of the jth direct descendant). Clearly

$$N(T) = 1 + \sum_j N(T_j)$$

($N(T)$ gives the total number of persons *ever born* in the family including the ancestor before the family becomes extinct).

Given that the initial customer departs at x, the number $A(x)$ of arrivals (direct "descendants") is a Poisson variate with mean λx. Denoting the p.g.f. of $N(t)$ by $P(z) = \xi$, we get

$$E\{z^{N(T_j)} \mid v = x\} = e^{-\lambda x + \lambda x \xi}.$$

Hence

$$E\{z^{N(T)} \mid v = x\} = ze^{-x(\lambda - \lambda \xi)}.$$

Thus we get

$$\xi = E\{z^{N(T)}\} = \int\limits_0^\infty E\{z^{N(T)} \mid v = x\} \, dB(x)$$

$$= z \int\limits_0^\infty e^{-x(\lambda - \lambda\xi)} \, dB(x),$$

i.e.
$$\xi = zB^*(\lambda - \lambda\xi). \tag{7.20}$$

Thus ξ satisfies the above functional equation. It can be shown that the equation has a unique root $\xi(z)$ such that $\xi(0) = 0$.

10.7.3.4 Moments of the Number Served during a Busy Period

The moments of N, the number served during a busy period (initiated by a single customer) of an $M/G/1$ queue can be obtained from the functional equation

$$P(z) = zB^*(\lambda - \lambda P(z)).$$

Denote by $B^{*(k)}$, the k the derivative of $B^*(\cdot)$. We have

$$B^*(0) = 1$$

$$B^{*(1)}(0) = -1/\mu$$

$$B^{*(2)}(0) = v_2 = E(v^2).$$

Differentiating once, we get

$$P'(z) = B^*(\lambda - \lambda p(z)) + zB^{*(1)}(\lambda - \lambda P(z))[-\lambda P'(z)]$$

Putting $z = 1$, and simplifying we get

$$E(N) = P'(1) = \frac{1}{1 - \rho}.$$

Differentiating twice, we get

$$P''(z) = 2B^{*(1)}(\lambda - \lambda P(z))[-\lambda P'(z)]$$
$$+ zB^{*(2)}(\lambda - \lambda P(z))[-\lambda P'(z)]^2$$
$$+ zB^{*(1)}(\lambda - \lambda P(z))[-\lambda P''(z)]$$

whence, putting $z = 1$ and simplifying, we get

$$P''(1) = \frac{2\rho(1-\rho) + \lambda^2 v_2}{(1-\rho)^3}$$

Thus

$$\text{var}(N) = P''(1) + P'(1) - [P'(1)]^2$$

$$= \frac{\rho(1-\rho) + \lambda^2 v_2}{(1-\rho)^3}.$$

Particular case. M/M/1 : Here $B^*(s) = \mu/(s + \mu)$. Thus

$$P(z) = \frac{z\mu}{\mu + \lambda - \lambda P(z)}$$

or

$$\lambda P^2(z) - (\mu + \lambda)P(z) + \mu z = 0$$

whence

$$P(z) = \frac{(\mu + \lambda) \pm \sqrt{(\mu + \lambda)^2 - 4\lambda \mu z}}{2\lambda} = \frac{1+\rho}{2\rho}\left[1 \pm \sqrt{\left\{1 - \frac{4\rho z}{(1-\rho)^2}\right\}}\right].$$

Since $P(1) = 1$, we take the root with negative sign before the radical sign and get

$$P(z) = \frac{1+\rho}{2\rho}\left[1 - \left\{1 - \frac{4\rho z}{(1-\rho)^2}\right\}^{1/2}\right].$$

Expanding the r.h.s. as a power series in z and comparing coefficients of z^n, we get

$$f_n = \Pr[N = n] = \frac{1+\rho}{2\rho}\left[\frac{\frac{1}{2}\left(\frac{1}{2} - 1\right) \text{ to } n \text{ terms}}{2^n(n!)}\left\{\frac{4\rho}{(1+\rho)^2}\right\}^n\right]$$

which, on simplification, yields

$$f_n = \frac{1}{n}\binom{2n-2}{n-1}\frac{\rho^{n-1}}{(1+\rho)^{2n-1}}, \quad n = 1, 2, \ldots \tag{7.21}$$

The above gives the distribution of the number served during a busy period in an *M/M/1* queue.

Remarks :

(1) Prabhu (1960,1965) considered the joint distribution of T and N; Enns (1969) considered the joint distribution of T, N and M (the maximum queue length during a busy period). Scott and Ulmer, Jr. (1972) and Tackacs (1976) discussed some related

joint distributions.

(2) *Delay Busy Period*. It is often useful to consider a more general kind of busy period when the server is required to perform some initial task (as soon as a unit arrives to an empty system and before starting service). Conway *et al* (1967) discussed this. Medhi and Templeton (1992) discuss a system under control operating policy and with start up time (time taken by than server to complete an initial task before starting servicing of units).

(3) Conway *et al* (1967) discuss how one can go from the analysis of busy period to waiting time distribution.

10.7.4 The Model M/G/∞: Transient Solution

Suppose that the input is Poisson with intensity λ, the service time v is general with distribution function B having mean $1/\mu$ and that there is an infinite number of channels so that an arriving unit is immediately taken into service whatever be the number present in the system. Let $N(t)$ denote the number present in the system (or number of busy channels) at epoch t, and $p_n(t) = \Pr\{N(t) = n\}$. Assume that initially the system is empty, i.e. $p_0(0) = 1$, $p_n(0) = 0$ $n \neq 0$. If $X(t)$ denotes the number of arrivals by time t, then $\Pr\{X(t) = k\} = e^{-\lambda t}(\lambda t)^k/k!$. If $Y(t)$ denote the number of departures by time t, then $X(t) = N(t) + Y(t)$. By conditioning on the number of arrivals $X(t)$, we get

$$p_n(t) = \Pr\{N(t) = n\} = \sum_{k=n}^{\infty} \Pr\{N(t) = n \mid X(t) = k\} \cdot \Pr\{X(t) = k\}.$$

Let $a \equiv a(t) = \Pr\{$an arbitrary unit arriving during $[0, t]$ is still in service at epoch $t\}$. Then conditioning on the epoch of arrival, we get

$$a = \int_0^t \Pr\{\text{service time of the arriving unit exceeds } t - x \mid \text{arrival occurs at epoch } x\} \cdot \Pr\{\text{arrival occurs at } x\}\, dx.$$

Now, arrival is in accordance with a Poisson process and so, given that an arrival occurs in $[0, t]$, the distribution of the arrival time is uniform in $[0, t]$ (Theorem 4.5). Thus,

$$a = \int_0^t [\Pr\{v > t - x\}\,(1/t)]\, dx$$

$$= (1/t) \int_0^t \{1 - B(t - x)\}\, dx = (1/t) \int_0^t \{1 - B(u)\}\, du$$

and

$$1 - a = (1/t) \int_0^t B(u)\, du.$$

Now, $\quad \Pr\{N(t) = n \mid X(t) = k\} = \Pr\{$given that k arrived in $[0, t]$, the number in service at epoch t is $n\}$

$$= \binom{k}{n} a^n (1-a)^{k-n}, \quad k \geq n.$$

Thus

$$P_n(t) = \sum_{k=n}^{\infty} \binom{k}{n} a^n (1-a)^{k-n} e^{-\lambda t} \frac{(\lambda t)^k}{k!}$$

$$= \left[\frac{e^{-\lambda a t}(\lambda a t)^n}{n!} \right] \sum_{k=n}^{\infty} \frac{[\lambda t(1-a)]^{k-n}}{(k-n)!}$$

$$= \frac{e^{-\lambda a t}(\lambda a t)^n}{n!}, \quad n \in I = \{0, 1, 2, ...\}$$

$$= \exp\left\{ -\lambda \int_0^t \{1 - B(u)\} \, du \right\} \left\{ \left[\lambda \int_0^t \{1 - B(u)\} \, du \right]^n \bigg/ n! \right\}.$$

That is, the distribution of $N(t)$ is Poisson with mean

$$\lambda \, at = \lambda \int_0^t \{1 - B(u)\} \, du.$$

Similarly, it can be shown that the distribution of the departure process $\{Y(t)\}$ is also Poisson with mean

$$\lambda(1-a)t = \lambda \int_0^t B(u) \, du.$$

Note: (1) *Limiting case: steady-state distribution*

As $t \to \infty$, $\qquad at = \int_0^t \{1 - B(u)\} \, du \to \frac{1}{\mu},$

so that, $\lambda \, at \to \lambda/\mu = \rho$. Then the steady-state distribution of the number in the system in $M/G/\infty$ model is Poisson with mean $\rho = \lambda/\mu$, irrespective of the value of ρ.

(2) $M/M/\infty$ in transient state with $p_0(0) = 1$. Here $B(t) = 1 - e^{-\mu t}$ so that $\lambda \, at = (\lambda/\mu)\{1 - e^{-\mu t}\}$. thus, for $n \in I$

$$p_n(t) = \left[\exp\left\{ -\frac{\lambda}{\mu}(1 - e^{-\mu t}) \right\} \right] \cdot \left\{ \frac{\lambda}{\mu}(1 - e^{-\mu t}) \right\}^n \frac{1}{n!}. \tag{7.22}$$

As $t \to \infty, p_n(t) \to e^{-\rho} \rho^n / n!, n \in I$.

Thus, the steady-state distribution of the number in the system in a model with Poisson input is, irrespective of the form of the distribution of the service time as well as of the value of ρ, always Poisson in case of infinite channel model. As mentioned in Sec. 10.4.4, the distribution is always truncated Poisson in case of a s-channel loss system. See also Exercise 10.13 and the note thereto.

10.7.4.1. An Application in Reliability : A spares inventory model (Barlow and Proschan, 1975).

Consider a maintenance depot with a large number of repairmen. Suppose that the failure of units occur in accordance with a Poisson process with rate λ. Failed units are brought to the depot for repair, the repair time v being i.i.d. random variables having common distribution function $B(t) = \Pr\{v \le t\}$ and mean $1/\mu$. Suppose that repair times are also independent of the arrival times and that the repaired units become part of the spares inventory, which consists initially (at $t = 0$), of certain number of spares. Let $N(t)$ denote the number of repairmen engaged at repair at epoch t and let $N(0) = 0$.

Then we have an $M/G/\infty$ model for the spares inventory and the distribution of $N(t)$ is Poisson with mean $\lambda \int_0^t \{1 - B(u)\} \, du$.

If n is the number of spares in the inventory at $t = 0$, then the expected number of customers waiting for spares at epoch t is given by

$$E[\max\{(N(t) - n), 0\}]$$

$$= \sum_{k=n+1}^{\infty} (k - n) \cdot \Pr\{N(t) = k\}.$$

As $t \to \infty$, this expected number tends to

$$\sum_{k=n+1}^{\infty} (k - n) \frac{e^{-\rho} \rho^k}{k!}.$$

10.8 THE MODEL $GI/M/1$

Here we assume that the service time distribution is exponential with mean $1/\mu$ and that the interarrival time is a random variable u, having an arbitrary (general) distribution with mean $1/\lambda$. Denote the d.f. of u by $A(t)$ and its p.d.f., when it exists, by $a(t)$; its L.T. is given by

$$A^*(s) = \int_0^{\infty} e^{-st} \, dA(t) \tag{8.1}$$

so that
$$A^{*(k)}(0) \equiv \frac{d^k}{ds^k} A^*(s) \bigg|_{s=0} = (-1)^k E(u^k) \tag{8.2}$$

and for $k = 1$, $A^{*(1)}(0) = -1/\lambda$.

Let t_n, $n = 1, 2, \ldots$, $(t_0 = 0)$ be the epoch at which the nth arrival occurs (or nth unit arrives). The process $N(t_n - 0) = Y_n$, $n = 0, 1, 2, \ldots$ gives the number in the system immediately before the arrival of the nth unit. Then

$$Y_{n+1} = Y_n + 1 - B_{n+1}, \quad \text{if} \quad Y_n \geq 0, \quad B_{n+1} \leq Y_{n+1},$$

where B_{n+1} is the number of units served during $(t_{n+1} - t_n)$, i.e. the interarrival time between the nth and $(n+1)$st unit. As B_n is independent of n, i.e. $B_n = B$ for all n, its distribution is given by

$$g_r = \Pr\{B = r\} = \int_0^\infty \frac{e^{-\mu t}(\mu t)^r}{r!} \, dA(t), \quad r = 0, 1, 2, \ldots \tag{8.3}$$

10.8.1 Steady State Distribution

The arrival point conditional probabilities (probabilities of arrival epoch system size)

$$p_{ij} = \Pr\{Y_{n+1} = j \mid Y_n = i\}$$

are given by

$$p_{ij} = g_{i+1-j}, \quad i+1 \geq j \geq 1, i \geq 0 \tag{8.4}$$

$$= 0, \quad i+1 < j,$$

and

$$\sum_{j=0}^{i+1} p_{ij} = 1.$$

Thus
$$p_{i0} = 1 - \sum_{j=1}^{i+1} p_{ij} = 1 - \sum_{j=1}^{i+1} g_{i+1-j}$$

$$= 1 - \sum_{r=0}^{i} g_r = h_i \text{ (say)}, i \geq 0. \tag{8.5}$$

Since all p_{ij}'s depend only on i and j, $\{Y_n, n \geq 0\}$ is a Markov chain having t.p.m.

$$P = (p_{ij}) = \begin{pmatrix} h_0 & g_0 & 0 & 0 & 0 & \dots \\ h_1 & g_1 & g_0 & 0 & 0 & \dots \\ h_2 & g_2 & g_1 & g_0 & 0 & \dots \\ \dots & & \dots & & & \dots \\ \dots & & \dots & & & \dots \end{pmatrix}. \tag{8.6}$$

As $g_r > 0$, the chain is irreducible and aperiodic. It can also be shown that it is ergodic, persistent non-null when $\rho < 1$. Thus, when $\rho < 1$, i.e. in the ergodic case, the limiting arrival point system size probabilities

$$v_j = \lim_{n \to \infty} p_{ij}^{(n)}$$

exist and are given as the unique solution of the system of equations

$$V = VP, \tag{8.7}$$

where
$$V = (v_0, v_1, \dots), \sum v_j = 1.$$

We now proceed to find V.

Let $G(s) = \sum_{r=0}^{\infty} g_r s^r$ be the p.g.f. of $\{g_r\}$, i.e. of the r.v. B, the number of units served during an interarrival interval. We have

$$G(s) = \sum_{r=0}^{\infty} g_r s^r$$

$$= \sum_{r=0}^{\infty} s^r \int_0^{\infty} \frac{e^{-\mu t}(\mu t)^r}{r!} \, dA(t)$$

$$= \int_0^{\infty} e^{-\mu(1-s)t} \, dA(t)$$

$$= A^*(\mu(1-s)).$$

Also
$$E(B) = G'(1) = -\mu A^{*(1)}(0)$$

$$= \mu/\lambda = 1/\rho.$$

The equations (8.7) can be written as

$$v_0 = \sum_{r=0}^{\infty} v_r h_r \tag{8.8}$$

$$v_j = \sum_{r=0}^{\infty} v_{r+j-1} g_r, \quad j \geq 1. \tag{8.9}$$

Denoting the displacement operator by $\mathbf{E}$ (so that $\mathbf{E}^r (v_k) = v_{k+r}$, etc.), we can write (8.9) as a difference equation

$$\mathbf{E}(v_{j-1}) = v_j = \sum_{r=0}^{\infty} g_r \, \mathbf{E}^r(v_{j-1}), \quad j \geq 1$$

or,

$$\left\{ \mathbf{E} - \sum_{r=0}^{\infty} g_r \, \mathbf{E}^r \right\} (v_{j-1}) = 0$$

or,

$$\{ \mathbf{E} - G \, (\mathbf{E}) \} v_{j-1} = 0, \quad j-1 \geq 0. \tag{8.10}$$

This equation can be solved by the method given in Appendix Sec. A. 2. The characteristic equation of the difference equation is

$$r \, (z) = z - G \, (z) \equiv z - A^* (\mu - \mu z) = 0.$$

It can be shown that when $G'(1) = 1/\rho > 1$, i.e. $\rho < 1$, $r \, (z)$ has only one zero inside $|z| = 1$. Assume that $\rho < 1$, then, if the root of $r \, (z) = 0$ inside $|z| = 1$ is denoted by r_0 and the roots on and outside $|z| = 1$ are denoted by $s_0, s_1, s_2, \ldots$ then the solution of (8.10) is

$$v_j = C_0 r_0^j + \sum_i D_i s_i^j, \quad j \geq 0,$$

where C_0, D_i are constants.

But since $\sum_j v_j = 1, D_i = 0$ for all i. Thus, when $\rho < 1$, we get, $v_j = C_0 \, r_0^j, j \geq 0$ and from $\sum_{j=0}^{\infty} v_j = 1$, we get $C_0 = 1 - r_0$, so that

$$v_j = (1 - r_0) r_0^j, j \geq 0. \tag{8.11}$$

The steady-state arrival point system size has a geometric distribution with mean $r_0/(1 - r_0)$, r_0 being the unique root of $r \, (z) = 0$ lying inside $|z| = 1$.

Note: Neuts (*Opns. Res.* **14** (1966) 313–317) gives an analytical proof of (8.11) using Weiner-Hopf technique.

10.8.2 Waiting Time Distribution

As the distribution of $\{v_n\}$ is geometric and the service time is exponential, the waiting time in the queue W_Q, being random geometric sum of exponential variables, is *modified* exponential.

We have,

$$\Pr(W_Q = 0) = 1 - r_0$$

and
$$\Pr\{W_Q \le t\} = 1 - r_0 e^{-\mu(1-r_0)t}, \quad t > 0. \tag{8.12}$$

We have $E(W_Q) = r_0/[\mu(1-r_0)]$.
The conditional distribution of the waiting time in the queue W_Q of those units that have to wait, is exponential with mean $1/\mu(1 - r_0)$.

Note: (1) The unique root r_0 $(0 < r_0 < 1)$ of $r(z) \equiv z - G(z) = 0$ can be obtained numerically by employing Newton-Raphson or other standard method.

(2) Kingman (1963) gives an explanation for the occurrence of geometric distribution as steady state solution of the $GI/M/1$ queue and discusses about the close formal connection between the theory of this queueing model and that of branching processes.

Remark: A question arises: for which type of arrival process (if any) is the average waiting time (delay) minimum? The answer is contained in what is known as **Folk's Theorem**, which is as follows. (The theorem is due to Hajek and Humblet).

The arrival process for which a queue having a single server with exponential service time has minimum average delay (as well as higher moments of W_Q and other related quantities) is the process with *constant interarrival time*, that is, a queue with deterministic arrival process (i.e. $D/M/1$ queue), has the minimum average waiting time among all queues having different arrival processes.

Example 8(a). *The model $E_k/M/1$.* Here the distribution of the interarrival interval is Erlang-k with mean $1/\lambda$ and L.T. $A^*(s) = \left(\dfrac{\lambda k}{s + \lambda k}\right)^k$.

Thus, the characteristic equation becomes

$$z - G(z) \equiv z - A^*(\mu - \mu z) = z - \left(\frac{\lambda k}{\mu - \mu z + \lambda k}\right)^k = 0.$$

If the unique zero of $z - G(z)$ lying inside $|z| = 1$ is denoted by r_0, then

$$v_n = (1 - r_0) r_0^n, \quad n \ge 0.$$

Note: As in the case of $M/E_k/1$ model, the queueing process in $E_k/M/1$ could also be viewed as non-birth-death Markovian and the solution obtained from the corresponding C.K. equations.

Particular cases: (1) $M/M/1$: When $k = 1$,

$$z - K(z) = 0 \quad \text{gives} \quad (z - 1)(z - \rho) = 0.$$

Thus when $\rho < 1$, the required root r_0 is ρ and we get

$$v_n = (1-\rho)\,\rho^n, \quad n \ge 0.$$

In this case $v_n = p_n$ for all n ($v_n = p_n$ only when $E_k \equiv M$, i.e. $k = 1$).

(2) $D/M/1$: When $k \to \infty, A*(s) \to e^{-s/\lambda}$

and
$$z - G(z) = 0 \text{ implies } ze^{(1-z)/\rho} = 1.$$

If the unique root inside $|z| = 1$ is denoted by ξ, then

$$u_n = (1-\xi)\xi^n, \quad n \ge 0.$$

Remark: We take as regeneration points the arrival epochs in case of $GI/M/1$ model and the departure epochs in case of $M/G/1$ model. The queueing processes occurring in the two models $M/G/1$ and $GI/M/1$ are non-Markovian; however by extracting processes at these regeneration points, it has been possible to obtain (embedded) Markov chains from those processes. The embedded Markov chains indicate *system state at regeneration* points. What one needs also is the distribution of *general (or random) time system state*.

Let us now consider the queueing process of $GI/M/1$, having for its embedded Markov chain $\{Y_n\}$, where Y_n is the system size immediately prior to nth arrival. We have obtained $\{v_j\}$, the limit distribution of $\{Y_n\}$. Define $Z(t) = Y_n, t_n \le t < t_{n+1}$. Then $\{Z(t)\}$, where $Z(t)$ gives the system size at the most recent arrival, is a semi-Markov process having $\{Y_n\}$ for its embedded Markov chain. $\{Y_n, t_n\}$ is an irreducible Markov renewal process.

Let $N(t)$ denote the system size at an arbitrary (general) time t. Denote

$$f_{ij}(t) = \Pr\{Z(t) = j \mid Z(0) = i\}$$

$$p_{ij}(t) = \Pr\{N(t) = j \mid N(0) = i\};$$

then $f_j = \lim_{t \to \infty} f_{ij}(t)$ gives the limiting probability that the system size of the s – M.P.

$Z(t)$ is j, whereas $p_j = \lim p_{ij}(t)$ as $t \to \infty$, gives the limiting probability that the system size of the general time process $N(t)$ is j.

We have to look for relationship, if any, existing between the three limit distributions $\{v_j\}$, $\{f_j\}$ and $\{p_j\}$.

From Theorem 7.1, we find that v_j and f_j are related as follows:

$$f_j = \frac{v_j M_j}{\sum_i v_i M_i}, \tag{8.13}$$

where M_i is the expected time spent in the state i during each visit.

The transitions, in case of this model, occur at the arrival points which are its regeneration points. Thus M_i, the expected time spent in a state i during each visit is the expected interarrival time. In other words, $M_i = 1/\lambda$ for all i. Putting the value of M_i in (8.13), we get

$$f_j = \frac{v_j/\lambda}{\sum_i v_i/\lambda} = \frac{v_j}{\sum_i v_i} = v_j \quad \text{for all } j,$$

$$= (1 - r_0)\, r_0^j \quad \text{(from (8.11))}.$$

In order to obtain p_j in terms of f_j (and in terms of v_j) we have to make use of the relationship (obtained by Fabens, 1961) between f_j and p_j.

It is shown that for this particular model,

$$p_j = (\lambda/\mu) f_j / r_0,$$

$$= (\lambda/\mu)(1 - r_0) r_0^{j-1} = \rho v_{j-1}, \quad j \geq 1$$

and

$$p_0 = 1 - \lambda/\mu.$$

We get

mean $= \rho/(1 - r_0)$ and variance $= \rho\,(1 - \rho + r_0)/(1 - r_0)^2$.

Note: Ross (1980) gives a simple probabilistic argument for the validity of the above relations. Fakinos (*Opns. Res.* **30** (1982) 1014–1017) also gives a derivation of the results.

10.9 THE MODEL $M/G(a, b)/1$

Model with Poisson input, general service time and under general bulk service rule. We easily see that the departure point process $\{X_n = N\,(t_n + 0)\}$ is a Markov chain.

Denote $q_r \equiv \Pr \{\text{Number of arrivals during a service time} = r\}$

$$= \int_0^\infty \frac{e^{-\lambda t}(\lambda t)^r}{r!}\, dB\,(t).$$

We have $Q(s) = \sum_r q_r s^r = B^*(\lambda - \lambda s)$, where $B^*(s) = \int_0^\infty e^{-st}\, dB(t)$. The transition probabilities $p_{ij} = \Pr\{X_{n+1} = j \mid X_n = i\}$ of the chain $\{X_n, n \geq 0\}$ are given by

$$
\begin{aligned}
p_{ij} &= q_{j-i+b}, & i \geq b, & \qquad j \geq i - b \\
&= 0, & i \geq b, & \qquad j < i - b \\
&= q_j, & a \leq i \leq b, & \qquad j \geq 0 \\
\text{and} \quad p_{ij} &= q_j, & 0 \leq i \leq a - 1, & \qquad j \geq 0.
\end{aligned}
$$

It is clear that the p_{ij}'s are *independent* of a; they are also independent of i for $0 \le i \le b$ and thus have the same values for all $i = 0, 1, \ldots, b$. The first $(b + 1)$ rows of the t.p.m. are identical. The t.p.m. $P = (p_{ij})$ is given by

$$
\begin{array}{c}
 \\
0 \\
1 \\
2 \\
\cdot \\
\cdot \\
\cdot \\
b \\
b+1 \\
b+2 \\
\cdot \\
\cdot \\
\cdot
\end{array}
\begin{array}{cccccc}
0 & 1 & 2 & 3 & \cdots & \\
\left[\begin{array}{ccccc}
q_0 & q_1 & q_2 & q_3 & \cdots \\
q_0 & q_1 & q_2 & q_3 & \cdots \\
q_0 & q_1 & q_2 & q_3 & \cdots \\
& \cdots & & \cdots & \\
& & & & \\
q_0 & q_1 & q_2 & q_3 & \cdots \\
0 & q_0 & q_1 & q_2 & \cdots \\
0 & 0 & q_0 & q_1 & \cdots \\
& \cdots & & \cdots & \\
& & & & \\
& \cdots & & \cdots &
\end{array}\right]
\end{array}
$$

The chain is irreducible and aperiodic. When the traffic intensity ρ (defined by $\rho = \lambda / b\mu$) is less than unity, the chain is ergodic. Then the limiting or steady-state departure point probabilities

$$
v_j = \lim_{n \to \infty} p_{ij}^{(n)}, j = 0, 1, 2, \ldots
$$

exist and $V = (v_0, v_1, \ldots)$ is given as the unique solution of $V = VP$. Thus, we have, for $j = 0, 1, 2, \ldots$

$$
v_j = (v_0 + v_1 + \ldots + v_{b-1}) q_j + v_b q_j + v_{b+1} q_{j-1} + v_{b+2} q_{j-2} + \ldots + v_{b+j} q_0
$$

$$
= \left(\sum_{r=0}^{b-1} v_r \right) q_j + \left(\sum_{r=0}^{j} q_{j-r} v_{b+r} \right).
$$

The p.g.f. of $\{v_j\}$ is given by

$$
V(s) = \sum_{j=0}^{\infty} v_j s^j
$$

$$
= \left(\sum_{r=0}^{b-1} v_r \right) \left(\sum_{j=0}^{\infty} q_j s^j \right) + \sum_{j=0}^{\infty} \left(\sum_{r=0}^{j} q_{j-r} s^{j-r} v_{b+r} s^r \right).
$$

The first term on the r.h.s. is $\left(\sum_{r=0}^{b-1} v_r \right) Q(s)$; the second term is

$$
\sum_j \{ (q_j s^j)(v_b) + (q_{j-1} s^{j-1})(s v_{b+1}) + \cdots + (q_0)(s^j v_{b+j}) \}
$$

$$= Q(s)\{v_b + sv_{b+1} + \cdots\}$$

$$= Q(s)\left\{\frac{1}{s^b}\sum_{r=b}^{\infty} v_r s^r\right\}$$

$$= Q(s)(1/s^b)\left\{V(s) - \sum_{r=0}^{b-1} v_r s^r\right\}.$$

Thus, $$(1 - Q(s)/s^b)V(s) = (Q(s)/s^b)\left\{\sum_{r=0}^{b-1}(s^b - s^r)v_r\right\}$$

or, $$V(s) = \frac{\sum_{r=0}^{b-1}(s^b - s^r)v_r}{s^b - Q(s)} \cdot Q(s)$$

$$= \frac{\sum_{r=0}^{b-1}(s^b - s^r)v_r}{\{s^b/Q(s)\} - 1}. \tag{9.1}$$

Putting $B^*(\lambda - \lambda s)$ for $Q(s)$, we also get the following expression for $V(s)$.

$$V(s) = \frac{\sum_{r=0}^{b-1}(s^b - s^r)v_r}{s^b - B^*(\lambda - \lambda s)} \cdot B^*(\lambda - \lambda s) \tag{9.2}$$

$$= \frac{\sum_{r=0}^{b-1}(s^b - s^r)v_r}{\{s^b/B^*(\lambda - \lambda s)\} - 1}. \tag{9.2a}$$

Particular cases : (1) *M/G/*1. Let $b = 1$, then $a = 1$. Noting that $v_0 = 1 - \rho$, and putting $a = b = 1$, we find that the result (7.7) at once follows from (9.1).

(2) *M/M(a, b)/*1 : We have $Q(z) = B^*(\lambda - \lambda z) = \frac{\mu}{\lambda - \lambda z + \mu}$. The distribution $\{v_k\}$ of the *system size at departure instants* in steady state (for $a = 0, 1, \ldots, b$) has the p.g.f.

$$V(z) = \sum_{k=0}^{\infty} v_k z^k$$

$$= \frac{\sum_{k=0}^{b-1}(z^b - z^k)v_k}{z^b\{1 + (\lambda/\mu)(1 - z)\} - 1}. \tag{9.3}$$

Now the denominator of $V(z)$ has $(b - 1)$ simple zeros $z = R_1, \ldots, R_{b-1}$ inside the unit circle $|z| = 1$. The b factors $(z - 1) \prod_{i=1}^{b-1}(z - R_i)$, common to both the numerator and the denominator cancel out, leaving out in the denominator a factor $(z - R_0)$, R_0 being the zero outside $|z| = 1$ of $z^b\{1 + (\lambda/\mu)(1 - z)\} - 1$ occurring in the denominator.

Thus, $$V(z) = \frac{C}{z - R_0} \quad \text{and} \quad V(1) = 1 \text{ gives } C = 1 - R_0$$

so that $$V(z) = \frac{1 - R_0}{z - R_0} = \frac{1 - r_0}{1 - r_0 z},$$

where $r_0 = 1/R_0$ is the unique root inside $|z| = 1$ of the equation whose roots are the reciprocals of those of $z^b\{1 + \lambda/\mu)(1 - z)\} - 1 = 0$, i.e. r_0 is the unique root *inside* $|z| = 1$ of $\lambda z^{b+1} - (\lambda + \mu)z + \mu = 0$.

Thus, for all $a = 0, 1, \ldots b$, we get, from $V(z) = (1 - r_0)(1 - r_0 z)^{-1}$,

$$v_k = (1 - r_0)r_0^k, \quad k = 0, 1, 2, \ldots$$

For example, for $b = 2$, $a = 0, 1, 2$, we get $r_0 = (2\lambda/\mu)(1 + \sqrt{(1 + 4\lambda/\mu)}$ and $v_k = (1 - r_0) r_0^k$; for $b = 1$, $a = 0, 1$ we get $r_0 = \rho$, so that $v_k = (1 - \rho) \rho^k$.

Remarks: (1) From (9.1), it is clear that the p.g.f. $V(s)$ which is a function of b, is independent of a. Thus given b, one gets the same steady state departure point distribution of system size under the general bulk service rule for all values of $a (= 1, 2, \ldots, b)$, and also for $a = 0$, (i.e., intermittently available server).

(2) To obtain the steady-state distribution $\{\pi_j\}$ of the general time system size $N(t)$, one can proceed as follows:

First consider the embedded Markov chain $\{X_n\}$ at departure points, i.e. $X_n = N(t_n + 0)$ and obtain its steady state distribution $\{v_j\}$ (this has already been considered in the example).

Secondly, consider the embedded Markov chain $\{Y_n\}$ not only at its instants of departure, but also at instants when $N(t)$ reaches the level a from below and the service commences. This will imply that transitions of Y_n will occur at all instants of departure as well as *all* instants of commencement of service (instants of departure and commencement of service will coincide when $N(t_i + 0) \geq a$). Deduce the steady-state distribution $\{r_j\}$ of $\{Y_n\}$ from that of $\{v_j\}$.

Thirdly, construct a semi-Markov process $Z(t)$, having $\{Y_n\}$ as its embedded Markov chain. Obtain the steady-state distribution $\{f_j\}$ of $Z(t)$ in terms of $\{r_j\}$ using Theorem 7.2.

Finally, obtain the steady-state distribution $\{\pi_j\}$ of $N(t)$ by conditioning $N(t)$ with $Z(t)$, (with the help of a theorem of Fabens (1961)–based on an elegant probabilistic argument concerning the number of units arriving between the instant of the last transition before t and the instant t itself).

Note that the distribution $\{r_j\}$ and therefore $\{\pi_j\}$ will also depend on a, although that of $\{v_j\}$ *does* not.

See Fabens (1961) and for more details, Neuts (1967), Teghem *et al.* (1969). For an alternative approach using the technique of inclusion of supplementary variables, see Borthakur and Medhi (1974); see also Chaudhry and Templeton (1981), who obtain the p.g.f.'s of the queue size at random epochs and at departure epochs and a relationship between the two p.g.f.'s (for the $M/G(1, b)/1$ model).

10.10. SOME GENERAL OBSERVATIONS

10.10.1 *GI/G/1* Model

The model with general interarrival and service time distributions has received a good deal of attention. Kingman [*Biometrika* **49** (1962) 315-324, *J.R.S.S.* **B** 32 (1970) 102 – 110] was the first to obtain bounds and inequalities for the *GI/G/1* queues. Marshall [*Opns. Res.* **16** (1968) 666-668 & 841-848, *SIAM J. Appl. Math.* **16** (1968) 324-327], Marchal [*Opns. Res.* **26** (1978) 1083-1088], Ross (1974), Schassberger [*Ann. Math. Statist.* **41** (1970) 182-187], Whitt [*Opns. Res.* **32** (1984) 41-51], Roseberg [*J.Appl. Prob.* **24** (1987) 749-757], among others, further consider waiting time distribution. Lindley [*P.C.P.S.* **48** (1952) 277–289] obtained an integral equation for waiting time. Marshall gives a generalisation of Pollackzek-Khinchine formula. Approximations for the mean waiting time have been examined, for example, by Kimura [*Mgmt. Sci.* **32** (1986) 751–763 for multiple server], Jagerman [*Queueing systems* **2** (1987) 351–362]. Whitt (1989) examines approximations for the mean work load. He also considers output process of such a queue [*Math. Opns. Res.* **9** (1984) 534–544]. He shows that the output process in a large class of *GI/G/c* queues in equilibrium, is, under certain conditions, approximately Poisson. Jain and Grassman [*Comp. & Opns. Res.* **15** (1989) 293–296] obtain numerical solution for the departure process. Many other interesting aspects have been studied.

We state below some results on bounds of the mean waiting time $E(W)$ in a *GI/G/1* queue with interarrival time u and service time v. Kingman's upper bound is given by

$$E(W) \leq \frac{\lambda(\sigma_u^2 + \sigma_v^2)}{2(1-\rho)}$$

where σ_u, σ_v are the *s.d.'s* of u and v respectively.

The equality holds for a *D/D/1* queue.
Kingman's lower bound is given by

$$E(W) \geq \frac{\lambda}{2(1-\rho)} E\left[\{\max(0, v-u)\}^2\right]$$

and Marchal's lower bound by

$$E(W) \geq \frac{\lambda^2 \sigma_v^2 + \rho(\rho - 2)}{2\lambda(1-\rho)}$$

which holds only if $\lambda\mu\sigma_v^2 \geq 2 - \rho$.

For details and other results, readers may see the papers mentioned above and also the text books, for example, by Cohen (1982), Gross & Harris (1985), Medhi (1991), as well as the references cited therein.

10.10.2 Queues with Vacations

We have assumed so far that the server is always available, even when the queue is empty after a busy period. There may be situations where as soon as the busy period ends, the server may shut down the service facility (may be engaged in other work or may just go away and may not be waiting) and thus the server may not be available when the next customer arrives to an empty system. This is known as a *queue with vacation* (earlier called *queue with removable server*). The period for which the server thus remains unavailable is known as *server vacation period*. Two cases may arise: the server may be away for a fixed period of time T (this is known as T- policy) or a random period. Let us consider that the server vacation period is a r.v.v. If at the end of a vacation, the server is available even when there is none waiting, then the system is called a *single-vacation system*. If however, none is waiting when the server returns after a vacation, he takes a further vacations, till at least there is one unit waiting, then the system is known as a *multiple-vacation system*.

Suppose that the number of arrivals during a server vacation is a r.v.X. The N-policy, where the server waits and begins serving only after a *specified number N* of arrivals, can also be considered as a vacation system; the period of vacation is then the sum of N interarrival times.

It is assumed that once the server starts serving, he goes on serving till none is left in the system; this is known as *exhaustive service discipline*.

Queues with vacation can be used to model production, communication and computer systems. Queues with vacations have been studied, among others, by Gaver [*J.R.S.S.* B 24 (1962) 73–90], Cooper [*Bell Syst. Tech. J.* 49 (1970) 399–413], Gelenbe and Iasnogorodski [*Ann. I.H.P.* XVI (1980) 63–73]. Fuhrmann [*Opns. Res.* 32 (1984) 1368–1373], Fuhrmann and Cooper [Opns. Res. 33 (1985) 1117–1129], Ali & Neuts [*J. Appl. Prob.* 21 (1984) 404–423]; Doshi [*Queueing Systems* 1 (1986) 29–66] gives a survey. Medhi (1991) also gives an account and includes some references.

Queues with server vacation are found to exhibit a very interesting property called *stochastic decomposition property*. It has been shown that under certain conditions on the r.v.v (vacation period), the waiting time in a *GI/G/*1 queue in steady state is the sum of (or may be decomposed into) two independent r.v.'s:

(i) the waiting time in the standard *GI/G/*1 queue without vacation (vacation independent r.v.) and (ii) a random variable related to the vacation period v (vacation-dependent r.v). Further, for Poisson input queues with vacation, the *decomposition property* is exhibited by the departure-epoch queue-length distribution as well (Fuhrmann and Cooper).

For an *M/G/*1 queue with server vacation, the queue-length at a departure epoch is equal to the sum of two r.v.'s:

(i) queue-length at a departure-epoch of an *M/G/*1 queue without vacation (a vacation-independent r.v.)

and (ii) queue-length at a random epoch given that the server is on vacation (a vacation-related r.v.).

The queues with server vacation have been pursued at great length. The results have been extended to cover variations in the vacation model as well as to cover non-exhaustive service disciplines.

Interested readers may see the works mentioned above and the references cited therein. See also H. Takagi's book *Queueing Analysis*, Vol. 1 (1991) North Holland.

10.10.3 Concluding Remarks

The pace at which research in queueing theory has progressed amply demonstrates its fertility. It is clear that the theory is moving forward and that its applicability is being enhanced. The extensive literature has mainly been concerned with describing the stochastic processes of interest in various queueing models and their applications. See Prabhu (1986) for an account of analytic innovations.

With a view to simplify the existing theory, various methods have been advanced. For example, random walk approach has been used to provide simple proof of some well-known results as well as to obtain new results on a variety of queueing problems. Reference may also be made to Wiener-Hopf techniques in queueing theory (Smith and Spitzer). While examining extensions of these techniques to continuous time, ladder processes have been investigated (Prabhu, 1970, 1980), Prabhu *et al.* (1971, 1973), and it is shown (Prabhu, 1974, 1980), how the techniques could be employed to simplify existing theory and to generate new results. Mention may also be made of the general definition and properties of the Markov renewal branching processes which have been shown to be useful in the study several queueing models (Neuts, 1974).

One topic of considerable mathematical and practical interest is the heavy traffic research, whose principal objectives are: (i) to describe unstable queueing systems (having traffic intensity greater than unity), and (ii) to approximate highly saturated stable queueing systems (having traffic intensity less than but close to unity). The term heavy traffic is due to Kingman who initiated work in this topic. Beginning with Iglehart's work (*J.Appl.Prob.* **2**. (1965) 429–441) there has been emphasis on approximation under heavy traffic conditions through diffusion process analysis. Mention may be made, for example, of the works of Gaver (*J.Appl. Prob.* **5** (1968) 607–623), Kobayashi (*J. Ass. Comp. Mach.* **21** (1974) 316–328 & 459–469), Halachmi and Franta (*Mgmt. Sci.* **24** (1978) 522–529 & 1448), Gelenbe (*Acta Informatica* **12** (1979) 285–303), Kimura (*Opns. Res.* **31** (1983) 304–321), Yao (*Opns. Res.* **33** (1985) 1266–1277). Whitt's (1974) paper contains a survey and a list of references; Reiser's (*Proc. IEEE* **70** (1982) 171–196) survey (containing a list of 142 references) indicates applications in computer modelling. For an account of diffusion process approximation, reference may be made, for example, to Kleinrock, II (1976), Gelenbe & Mitrani (1980), Heyman & Sobel (1982), Newell (1982), Medhi (1991) and so on.

The central theme for heavy traffic is limit theorems. Further areas of research are concerned with the rates of convergence, approximations for multi-channel queues, finite queues, control etc.

As for statistical analysis in queueing theory, pioneering work was done by Clarke, who began by obtaining the maximum likelihood estimators for the arrival and service parameters of the simple queue. Important contribution on estimation of parameters and confidence statements about them have been made by Cox (1965), Wolff (1965), Lilliefors (1966), Wolff extended the work of Billingsley on likelihood hypothesis testing and made use of the same to obtain results for birth and death queueing models.

The research in control and optimal operation of queueing systems is centered around optimal design and control of a model with a cost structure (for a review, see Stidham and Prabhu (1974), Sobel (1974) also Serfozo *{Adv. Appl. Prob.* **13**, (1981) 61–83); for a classified bibliography with over 150 titles, see Crabill *et al.* (1977). The decision variables considered are arrival and service parameters, queue parameters, operating time parameters; the objective functions considered are concerned with average or discounted costs. As for applications, a class of design problems is addressed to optimal allocation of servers as well as of vehicles in police control, hospital ambulance service and in transportation, in general. Mention may be made of the works, for example, of Chaiken & Larson (*Mgmt. Sci.* **19** (1972) 110–130), Parikh (*Mgmt. Sci.* **23** (1972) 972–977), Berman *et al.* (*Opns. Res.* **33** (1985) 746–771), Powell (1985), Powell and Humblet (1986), Shanthikumar & Yao (*Mgnt. Sci.* **33** (1987) 1173–1191; *Opns. Res.* **36** (1988) 333–342), Stidham, S. (*IEEE Trans. Auto. Cont.*, 30 (1986); IMA **10** (1988) 529).

The theory of Markov and semi-Markov decision processes have been used in various studies. Martingale approach has also been made in the formulation and solution of control problem in queues (Martins-Neto and Wong, 1976). Brémaud (1981) discusses the importance of martingale approach in queueing theory and deals with the subject in details. The traditional methods of Markov and semi-Markov processes as well as of renewal theory are useful in design; with these as tools one can compute the essential features and assess the performance measures of different policies and strategies with regard to queueing models. The martingale approach could be used in problems of dynamical nature, with a view to identify the best among those policies. Complete solution of some of the traditional topics could also be found by using martingale approach.

Approximations, algorithms, asymptotic methods, numerical and computational analyses, in general, have also been engaging a good deal of attention. These methods are useful to practioners in view of the fact that no tractable solution can be found nor could be exploited, even when found, for models of many real-life systems, encountered in practice. These aspects and methods counstitute a subject in itself. For an account of some of these aspects, reference may be made, for example, to Van Hoorn (1983), Borovkov (1984), Tijms (1986) and Bhat & Basawa (*Queueing and Related Models*, O.U.P., 1992).

These would throw a light on how queueing theory continues to grow with strong interplay between theory and applications.

10.11 RELIABILITY

10.11.1 Introduction.

Investigation in reliability began in the 1950's, and the growth of the theory began

---

It may be mentioned in passing that some researchers (for example, R. Larson in *Sloan Management Review*, MIT) are going beyond mathematical and stochastic analyses of queues, focussing attention on psychological reactions of those who have to wait in line.

to gain momentum after a decade or so (some standard works and survey papers are listed in the References).

Reliability theory is mainly concerned with the determination of the probability that a system, consisting possibly of several components, will operate adequately for a given period of time in its intended application.

Suppose that a component i of a system is capable of just two modes of performance—functioning and not functioning, which may be denoted by 1 and 0 respectively. The mode of performance of the component i may be represented by a Bernoulli random variable X_i, with values 1 and 0 depending on whether the component is functioning or not functioning. Suppose that

$$p_i = \Pr\{X_i = 1\} = 1 - \Pr\{X_i = 0\};$$

then p_i is the reliability of the i th component.

Suppose that a system has some finite number n, of independent components labelled $1, 2, \ldots, n$, and that the system is capable of just two modes of performances. Represent the mode of performance of the system by the Bernoulli r.v. X. Suppose that, given the structure of a system, the knowledge of its performance can be determined from that of its components. The system structures generally considered, are described below.

(i) *Series system*: The system functions *iff* all the n components function. We have $X = \min(X_1, \ldots, X_n)$; the reliability of n components is given by

$$P = \Pr\{X = 1\} = \Pr\{\min(X_1, \ldots, X_n) = 1\}$$

$$= \Pr\{X_1 = 1, X_2 = 1, \ldots, X_n = 1\} \qquad (11.1)$$

$$= \prod_{i=1}^{n} p_i.$$

(ii) *Parallel system* : The system functions *iff* at least one of the n components functions. We have $X = \max(X_1, \ldots, X_n)$. The system reliability is given by

$$P = \Pr\{X = 1\} = \Pr\{\max(X_1, \ldots, X_n) = 1\}$$

$$= 1 - \Pr\{X_1 = 0, X_2 = 0, \ldots, X_n = 0\} \qquad (11.2)$$

$$= 1 - \prod_{i=1}^{n}(1 - p_i).$$

(iii) *k-out of n-system*: The system functions *iff* at least k $(1 \le k \le n)$ of the n components function. As particular cases, we get the series system for $k = n$ and the parallel system for $k = 1$. Suppose, for simplicity, that $p_i = p$ for all i, then X is a binomial variate with parameters n, p. Thus

$$P = \Pr\{X = 1\}$$

$$= \Pr\{X \geq k\} = \sum_{i=k}^{n} \binom{n}{i} p^i (1-p)^{n-i}. \tag{11.3}$$

Lifetime distributions: Let us now consider time. Assume that a device functions for some time after which it fails, and then remains in that non-functioning state. The duration of time T, for which it functions is its lifetime. A number of failure distributions have been suggested as models to describe the lifetime of a device.

Exponential distribution, which has the memoryless property, characterises the phenomenon of no wear. Future lifetime of a device remains the same irrespective of its previous use, *iff*, its lifetime distribution is exponential. But the phenomenon of wearout has also to be taken into account in certain cases. In an attempt to find out distributions which describe diverse phenomena associated with a device, certain characteristics like hazard function for lifetime distribution have been evolved. These have been discussed in Section 1.4.

Definition: The *reliability* (or the *survival function*) of a device at time t (given that it started working at $t = 0$) is defined by

$$R(t) = \Pr\{T \geq t\} = 1 - F(t), \quad t \geq 0 \tag{11.4}$$

$$= 1 \qquad t < 0,.$$

where F is the distribution function of lifetime T.

Denote by $f(t)$, the p.d.f. of T, when it exists. Denote
$X(t) = 1$, if the device is working at epoch t
$= 0$, if the device is not working at epoch t
Then $\{X(t), t \geq 0\}$ is a continuous time process with state space $S = \{0, 1\}$ (two-state process). The reliability of the device is also defined by

$$R(t) = \Pr\{X(u) = 1, \quad 0 \leq u \leq t \mid X(0) = 1\}$$

It is a measure that the device is working during the interval $(0, t]$.
The *availability* $A(t)$ of the device is defined by

$$A(t) = \Pr\{X(t) = 1 \mid X(0) = 1\}$$

The availability refers to the behaviour (that it is working) at *an epoch of time* while reliability to that *during an interval*. Reliability indicates *failure-free* behaviour during an interval; availability does not have this restriction (the device may be down at some point of the interval and again working (after repairs) at the epoch t. Limiting (or stationary) availability is defined by $A = \lim_{t \to \infty} A(t)$, when the limit exists.

As $t \to \infty$, while $R(t) \to 0$, $A(t) \to A$ (a non-zero quantity in general)

10.11.2 System Reliability

10.11.2.1 System without Repair

Suppose that a system consists of n independent components 1, 2, ... i, ... n, and that the lifetime of the system is T, while that of ith component is T_i. The system reliability (or survival function) at time t is $R(t) = \Pr(T > t)$, and the reliability of ith component at time t is $R_i(t) = \Pr(T_i > t)$.

Expected lifetime or mean time to failure (MTTF) of the system is given by

$$E(T) = \int_0^\infty t\left(-\frac{dR}{dt}\right)dt$$

$$= -\int_0^\infty t\,dR(t) = \int_0^\infty R(t)\,dt.$$

The k th moment, when it exists, is given by

$$E(T^k) = k \int_0^\infty t^{k-1}R(t)dt, \quad k = 1, 2, \ldots \tag{11.5}$$

Thus

$$\text{var}(T) = 2 \int_0^\infty tR(t)dt - \left[\int_0^\infty R(t)dt\right]^2.$$

(i) *Series system* : We have $T = \min(T_1, \ldots, T_n)$.

Thus, $R(t) = \Pr\{T > t\} = \Pr\{\min(T_1, ..., T_n) > t\}$

$$= \prod_{i=1}^{n} \Pr\{T_i > t\} = \prod_{i=1}^{n} R_i(t). \tag{11.6}$$

The hazard rate function of the system is

$$r(t) = -\frac{R'(t)}{R(t)} = \frac{\sum_{i=1}^{n}(-R_i'(t))\prod_{j \neq 1}^{n} R_j(t)}{\prod_{j=1}^{n} R_j(t)}$$

$$= \sum_{i=1}^{n} \frac{-R_i'(t)}{R_i(t)} = \sum_{i=1}^{n} r_i(t),$$

where $r_i(t)$ is the hazard rate function of the component i. Thus, if the lifetime distribution of each component is an IFR, then that of the series system is also so. The IFR system is closed under the formation of series systems.

(ii) *Parallel system* : We have $T = \max (T_1, \ldots, T_n)$. Thus,

$$R(t) = \Pr \{T > t\} = \Pr \{\max (T_1, ..., T_n) > t\}$$

$$= 1 - \prod_{i=1}^{n} \Pr \{T_i \leq t\} = 1 - \prod_{i=1}^{n} (1 - R_i(t)). \quad (11.7)$$

For parallel system, the lifetime distribution of the system is not necessarily IFR, even if that of each component is.

From (11.6), since $0 \leq R_i (t) \leq 1$, we get

$$R(t) = \prod_i R_i(t) \leq \min_i R_i(t)$$

Thus

$$E(T) = \int_0^\infty R(t)dt$$

$$\leq \min_i \left\{ \int_0^\infty R_i(t)dt \right\}$$

$$= \min_i \{E(T_i)\}$$

$$= \min \{E(T_1), ..., E(T_n)\}.$$

Form (11.7), since $0 \leq R_i (t) \leq 1$, we get

$$R(t) \geq 1 - [1 - R_i(t)] \quad \text{for all } i$$

so that

$$E(T) \geq \max_i \{E(T_i)\}.$$

(iii) *k-out of n-system* : The system lifetime, is the corresponding order statistics

$$T^{(1)}, T^{(2)}, ..., T^{(k)}, ..., T^{(n)},$$

formed from $T_1, \ldots, T_n$ for $k = 1, 2, \ldots, n$.

Example 11 (a). Consider a system consisting of n components such that the failure of the ith component occurs in accordance with a Poisson process of intensity a_i. To find the reliability of the system under different system structures.

Here T_i is exponential with mean $1/a_i$, and $R_i(t) = e^{-a_i t}$.

(i) *Series system*

$$R(t) = \Pi R_i(t) = e^{-(\Sigma a i)t}.$$

T is exponential with mean $1/\Sigma a_i$.

(ii) *Parallel system*

$$R(t) = 1 - \prod_{i=1}^{n}\left(1 - e^{-a_i t}\right)$$

$$= \Sigma e^{-a_i t} - \Sigma e^{-(a_i + a_j)t} + \cdots \pm e^{-(\Sigma a_i)t},$$

and

$$E(T) = \Sigma \frac{1}{a_i} - \Sigma \frac{1}{(a_i + a_j)} + \cdots \pm \frac{1}{\Sigma a_i}.$$

In particular, if $a_i = a$ for all i, then

$$R(t) = 1 - (1 - e^{-at})^n = \sum_{r=1}^{n} (-1)^{r-1} \binom{n}{r} e^{-rat};$$

also

$$R(t) = \sum_{r=1}^{n} \binom{n}{r} (e^{-at})^r (1 - e^{-at})^{n-r},$$

and

$$E(T) = \sum_{r=1}^{n} (-1)^{r-1} \binom{n}{r} \Big/ (ar).$$

By integrating in $(0, 1)$, both sides of the identity

$$\sum_{r=0}^{n-1} (1 - t)^r = \{1 - (1 - t)^n\}/t,$$

we find that

$$E(T) = \frac{1}{a} \sum_{r=1}^{n} \frac{1}{r}.$$

For large n, $E(T) \to \frac{1}{a}(\log n + \gamma)$, where $\gamma = 0.577...$, is Euler's constant.

(iii) *k-out of n-system.* Suppose that $a_i = a$ for all i. We have

$$R(t) = \Pr\{T > t\}$$

$$= \sum_{r=k}^{n} \binom{n}{r} (e^{-at})^r (1 - e^{-at})^{n-r}$$

and

$$E(T) = \int_{0}^{\infty} R(t)dt = \sum_{r=k}^{n} \binom{n}{r} \int_{0}^{\infty} e^{-atr}(1 - e^{-at})^{n-r} \, dt$$

$$= \sum_{r=k}^{n} \binom{n}{r} \int_{0}^{1} y^{r-1}(1-y)^{n-r} \frac{dy}{a} \quad \text{(putting } y = e^{-at}\text{)}$$

$$= \frac{1}{a} \sum_{r=k}^{n} \binom{n}{r} \beta(r, n-r+1) = \frac{1}{a} \sum_{r=k}^{n} \frac{n!}{r!(n-r)!} \frac{\Gamma(r)\Gamma(n-r+1)}{\Gamma(n+1)}$$

$$= \frac{1}{a} \sum_{r=k}^{n} \frac{1}{r}.$$

Example 11(b). Suppose that a system consists of n components $1, 2, \ldots, i, \ldots, n$; the first r of them, $1, 2, \ldots, r$ are all of different kinds and the remaining $(n - r)$ are similar but are of another kind. The system functions *iff* each of the r different components functions and at least one of the remaining $(n - r)$ similar components function. Suppose that the failures of the (ith) components occur in accordance with independent Poisson processes, with parameter a_i for $i = 1, 2, \ldots, r$, and with parameter b for $i = r + 1, \ldots, n$.

Then

$$R_i(t) = e^{-a_i t}, \text{ for } i = 1, ..., r$$

$$= e^{-bt}, \text{ for } i = r + 1, ..., n$$

so that

$$R(t) = \left\{ \prod_{i=1}^{n} R_i(t) \right\} \left\{ 1 - \prod_{i=r+1}^{n} (1 - R_j(t)) \right\}$$

$$= \prod_{i=1}^{n} e^{-a_i t} \{ 1 - (1 - e^{-bt})^{n-r} \}.$$

10.11.2.2 *System with Repairs or Replacements*

Suppose that a system is either operating or under repair. Denote by $X(t)$, the state of the system at epoch t; $X(t) = 1$, if it is working and $X(t) = 0$, if it is under repairs. Assume that repairs (or replacements) on failures as well as instalments after repairs (or replacements) are instantly undertaken and that the system after repairs (replacements of components) behaves as new. Supose that the successive operating periods (lifetimes or failure-times) are *i.i.d.* r.v's $T_1, T_2 ...$ with $T_i \equiv T$ and that the successive failure times (repairtimes) are *i.i.d* r.v.'s $V_1, V_2...$, with $V_i \equiv V$. Then $\{T_i\}$ and $\{V_i\}$ constitute an alternating renewal process. $\{Z_i = T_i + V_i\}$ is also a renewal process;

Let $h(t)$ be the renewal density of $\{Z_i\}$, then $h(x) dx$ is the probability of one or more repairs in $(x, x + dx)$.

We have, if $X(0) = 1$, then the availabilit y $A(t)$ is given by

$$A(t) \doteq \Pr\{T > t\} + \int_0^t h(x) \Pr\{T > t - x\} dx$$

$$= R(t) + \int_0^t R(t - x) h(x) dx.$$

The first term on the *r.h.s.* corresponds to the event that the system is operating (without repair) in the interval $(0,t)$ and the second term to the event that the last repair (or renewal) occurred at instant x, $0 < x < t$ and that the system has been operating since that time up to instant t.

Note that $A(t)$ exceeds $R(t)$ (in general).

We have, from renewal theory,

$$A = \lim_{t \to \infty} A(t) = \frac{E(T)}{E(T) + E(V)}.$$

The limiting availability A depends on the means and not on the forms of distributions of T and V. For example, consider that the lifetimes and repair-times have independent exponential distributions with parameters λ and μ respectively. Then the process $\{X(t), t \geq 0\}$ is a two-state Markov process. The transition probabilities.

$$p_{ij}(t) = \Pr\{X(t) = j \mid X(0) = i\}, i, j = 1, 2$$

can be easily found (Example 5(*b*) Ch. 4); so also the limits $p_{ij}(t)$ as $t \to \infty$.

The limiting probability that the system is working after a very long period is given by $\lim_{t \to \infty} p_{01}(t) = \lim_{t \to \infty} p_{11}(t) = \dfrac{\mu}{\lambda + \mu}$, which is equal to $A = \dfrac{E(T)}{E(T) + E(V)}$.

10.11.3 Markovian Models in Reliability Theory

In Example 11(a), the failure process for individual components is assumed to be Markovian. The assumption that the failure process and the repair process, whenever repairs are considered, are Markovian is very often made; in such cases it is possible to write down at once the Chapman-Kolmogorov (C.K.) equation of the Markov process of the resulting system. From the solutions of these equations, one can find the system reliability. The following example will illustrate the method.

Example 11(c). To find the reliability of a system consisting of n identical components connected in parallel, such that the failure of a component occurs independently of the others and in accordance with a Poisson process with intensity a. (This reduces to Example 11(a) (ii) with $a_i = a$.)

The operating process of the system may be represented by a Markov process $\{X(t), t \geq 0\}$ with $(n+1)$ states $0, 1, \ldots, i \ldots, n$, the state i denoting that there are i failed components at the instant t. The state n is an absorbing state. The transition to state n denotes its first passage; and transitions from any of the other states to some other states avoid the state n.

Let $P_i(t) = \Pr\{X(t) = i\}$ and the initial conditions be $p_0(0) = 1$, $p_i(0) = 0$, $i \neq 0$. The reliability $R(t)$ of the system at time t is given by

$$R(t) = p_0(t) + p_1(t) + \cdots + p_{n-1}(t) = 1 - p_n(t).$$

It is clear that the process $\{X(t), t \geq 0\}$ is a birth process with state dependent rates λ_i given by $\lambda_i = (n-i)a$, $i = 0, 1, \ldots, n$, n being an absorbing state.

The C.K. equations of the system are thus given by

$$p_0'(t) = -n\,a\,p_0(t)$$

$$p_i'(t) = -(n-i)a p_i(t) + (n-i+1)a p_{i-1}(t), \quad i = 1, 2, \ldots, n-1$$

$$p_n'(t) = a p_{n-1}(t).$$

The first equation can be solved at once and the other equations (which are linear) can then be solved recursively. Alternatively, the equations can be solved by considering their Laplace transforms. Denote the L.T. of $p_i(t)$ by $P_i(s)$. By taking the L.T. of the above equations and making use of the initial conditions, one gets

$$sP_0(s) = -na\,P_0(s)$$

$$sP_i(s) = -(n-i)aP_i(s) + (n-i+1)aP_{i-1}(s), \quad i = 1, 2, \ldots, n-1$$

$$sP_n(s) = aP_{n-1}(s).$$

For the particular case, $n = 2$, we have

$$P_0(s) = \frac{1}{s+2a}$$

and

$$P_1(s) = \frac{2a}{s+a}P_0(s) = \frac{2a}{(s+a)(s+2a)} = 2\left(\frac{1}{s+a} - \frac{1}{s+2a}\right).$$

Taking the inverse L.T., we get

$$p_0(t) = e^{-2at}$$

and

$$p_1(t) = 2(e^{-at} - e^{-2at}).$$

Hence

$$R(t) = p_0(t) + p_1(t) = 2e^{-at} - e^{-2at}.$$

In the general case, we get

$$P_0(s) = \frac{1}{s + na}$$

$$P_i(s) = \frac{(n - i + 1)a}{s + (n - i)a} P_{i-1}(a), \quad i = 1, 2, ..., n - 1$$

$$P_n(s) = \frac{a}{s} P_{n-1}(s).$$

Thus,

$$P_n(s) = \frac{a}{s} P_{n-1}(s) = \frac{a}{s} \cdot \frac{2a}{s + a} P_{n-2}(s)$$

$$\cdots \qquad \cdots \qquad \cdots$$

$$= \frac{a(2a)...(na)}{s(s + a)...(s + na)}.$$

Resolving into partial fractions, we get

$$\frac{a(2a)...(na)}{s(s + a)...(s + na)} = \frac{A_0}{s} + \frac{A_1}{s + a} + \cdots + \frac{A_n}{s + na},$$

where A_i's are constants, which can be determined as follows. Multiplying both sides by s and letting $s \to 0$, we get $A_0 = 1$.

Multiplying both sides by $(s + ia)$ and letting $s \to -ia$, we get

$$A_i = \frac{a(2a)...(na)}{(-ia)(-(i - 1)a)...(-a)(a)(2a)...(n - i)a}$$

$$= (-1)^i \frac{(n - i + 1)(n - i + 2)...n}{i!} = (-1)^i \binom{n}{i}.$$

Thus,

$$P_n(s) = \frac{1}{s} - \frac{\binom{n}{1}}{s + a} + \cdots + (-1)^i \frac{\binom{n}{i}}{s + ia} + \cdots \pm \frac{\binom{n}{n}}{s + na}.$$

Inverting the L.T., we get

$$p_n(t) = 1 - \binom{n}{1} e^{-at} + \cdots \pm \binom{n}{n} e^{-nat}$$

$$= (1 - e^{-at})^n.$$

Thus, $$R(t) = 1 - p_n(t) = 1 - (1 - e^{-at})^n.$$

Note : This result was obtained in Example 11(a) by the straightforward method which was much simpler. This alternative method is given only to illustrate the application of the theory of Markov process.

We now make the assumption that a failed component can be repaired and renewed. The system can then be described by a queueing model, and the methods of queueing theory can be applied in its study. To find the reliability R (t) at time t, one needs the transient state probabilities. We consider an example below. For more general discussion refer to the standard texts on Reliability theory, such as Gnedenko *et al.*

Example 11(d). *Parallel system with maintenance.* Consider a two-component parallel system, with a single repairman available for maintenance . Suppose that the failure and repair processes are both Markovian with rates a and b respectively.

Here the process $\{X (t), t \geq 0\}$, has 3 states $i = 0, 1, 2$, the state i denoting the number of failed units at the instant t. The state 2 is an absorbing state. This may be considered as a machine interference problem (see Sec 10.4); the process may be represented by a birth and death process having 3 states 0, 1, 2, the state 2 being an absorbing state. Let $p_i = \Pr \{X (t) = i\}$; the C.K. equations are given by

$$p_0'(t) = -2a p_0(t) + b p_1(t)$$

$$p_1'(t) = -(a + b)p_1(t) + 2a p_0(t)$$

$$p_2'(t) = a p_1(t).$$

Let the initial conditions be $p_0 (0) = 1, p_i (0) = 0, i = 1, 2.$

The above equations can be easily solved by taking their Laplace transforms. Denote the L.T. of p (t) by P (s). Taking L.T and using initial conditions, the first two equations can be written as

$$(s + 2a)P_0(s) - bP_1(s) = 1$$

$$-2aP_0(s) + (a + b + s)P_1(s) = 0.$$

Solving, we get

$$P_1(s) = \frac{2a}{s^2 + (3a + b)s + 2a^2}, \quad P_0(s) = \frac{a + b + s}{s^2 + (3a + b)s + 2a^2}.$$

$P_1(s)$ can be written as

$$P_1(s) = \frac{2a}{(s - s_1)(s - s_2)}$$

where

$$\frac{s_1}{s_2} = \frac{-(3a + b) \pm \sqrt{\{(a + b)^2 + 4ab\}}}{2}$$

Thus,

$$P_1(s) = \frac{2a}{s_1 - s_2}\left(\frac{1}{s - s_1} - \frac{1}{s - s_2}\right)$$

and its inverse is

$$p_1(t) = \frac{2a}{s_1 - s_2}\left(e^{s_1 t} - e^{s_2 t}\right).$$

Again

$$P_0(s) = \frac{a + b + s}{(s - s_1)(s - s_2)}$$

$$= \frac{1}{(s_1 - s_2)}\left(\frac{a + b + s_1}{s - s_1} - \frac{a + b + s_2}{s - s_2}\right)$$

and its inverse is

$$p_0(t) = \frac{(a + b + s_1)e^{s_1 t} - (a + b + s_2)e^{s_2 t}}{s_1 - s_2}.$$

The system reliability is given by

$$R(t) = p_0(t) + p_1(t)$$

$$= \frac{(3a + b + s_1)e^{s_1 t} - (3a + b + s_2)e^{s_2 t}}{(s_1 - s_2)}.$$

Using the relation $s^2 + (3a + b)s + 2a^2 = (s - s_1)(s - s_2)$, $R(t)$ can also be expressed as

$$R(t) = \frac{s_1 e^{s_2 t} - s_2 e^{s_1 t}}{(s_1 - s_2)}.$$

10.11.4. Shock Models and Wear Processes

We consider here a model described by Cox, Esary, Epstein, Gaver and later exhaustively examined by Esary, Marshall and Proschan (1973) for the lifetime of a device. Here we give some of the interesting results obtained by Esary *et al.* to which references may be made for proof of the theorems stated. Some of the results of Esary *et al.* have been generalised to cover the multivariate case by Marshall and Shaked (1979). For a survey paper on replacement of systems subject to shocks, see Valdez-Flores and Fieldman (1989).

Suppose that a device is subjected to shocks which occur in accordance with a Poisson process with parameter a. Denote the probability that the device survives the first k shocks, $(k = 0, 1, 2, \ldots)$ by q_k. Clearly, the reliability of the device at time t is

$$R(t) = \Pr\{T > t\} = \sum_{k=0}^{\infty} q_k e^{-at}(at)^k \big/ k!, \quad t \geq 0 \qquad (11.8)$$

$$= 1, \qquad\qquad t < 0$$

Now, q_k being the probability of surviving k shocks, we have

$$1 \geq q_0 \geq q_1 \geq q_2, \ldots \qquad (11.9)$$

and p_k, the probability of failure on the kth shock, is given by

$$p_k = q_{k-1} - q_k, \quad k = 1, 2, \ldots \qquad (11.10)$$

and

$$p_0 = 1 - \sum_{k=1}^{\infty} p_k = 1 - q_0.$$

The p.d.f. $f(t)$ of the lifetime T is given by

$$f(t) = -R'(t) = a \sum_{k=1}^{\infty} p_k e^{-at}(at)^{k-1} \big/ (k-1)!, t > 0 \qquad (11.11)$$

and

$$\Pr\{T = 0\} = 1 - \lim_{\epsilon \to 0} \int_{\epsilon}^{\infty} f(t)dt = p_0, \qquad t = 0.$$

Its hazard rate function $r(t)$ is given by

$$r(t) = \frac{f(t)}{R(t)}$$

$$= a\left[1 - \left\{ \sum_{k=0}^{\infty} q_{k+1}(at)^k \big/ k! \right\} \Big/ \left\{ \sum_{k=0}^{\infty} q_k(at)^k / k! \right\} \right], \quad t > 0$$

so that $r(t) \leq a$.

The equality holds if $q_k = 0$ for all $k > 0$, i.e. if $q_0 = 1$ and then T has exponential distribution.

One can express q_k and p_k in terms of derivatives of $R(t)$ as follows: Denoting by $R^{(k)}(t)$, the k th derivative of $R(t)$, we have

$$q_k = \sum_{j=0}^{k} \binom{k}{j} R^{(j)}(0) \Big/ a^j,$$

$$p_k = 1 - \sum_{j=1}^{k} \binom{k-1}{j-1} R^{(j)}(0) \Big/ a^j.$$

The j th moment $\mu_j, j = 1, 2, \ldots$ of T is given by

$$\mu_j = \frac{j!}{a^j} \sum_{k=0}^{\infty} \binom{k+j-1}{k} q_k,$$

provided the series on the r.h.s converges. Some of the properties of q_k, of interest in reliability, carry over to $R(t)$. One such property is given below (without proof).

Theorem 10.1. $R(t)$ is IHR if q_k/q_{k-1} is decreasing in $k = 1, 2, \ldots$.

Models for q_k: Consider that damage is caused to the device by a shock. Suppose that the ith shock causes a random damage X_i and that damages are additive, so that the damage caused by the first k shocks is $X_1 + X_2 + \cdots + X_k$. Suppose that there is a capacity or threshold which the device can endure. The threshold may be considered to be a constant or to be a random variable ξ. We assume here that the capacity or threshold has a fixed value x. The device survives the kth shock if $X_1 + \cdots + X_k$ does not exceed the threshold value x, so that

$$q_k = \Pr\{X_1 + \cdots + X_k \leq x\}.$$

Now the non-negative random variables X_i may be (i) i.i.d., (ii) independent but not identical (as successive shocks could cause increasing damage) and (iii) neither independent nor identical (as accumulation of damage may result in loss of resistance to further damage).

If $X_i, i = 1, 2, \ldots$ are independent but not identical (and have F_i for their d.f.), then q_k is given by their convolution

$$q_k = F_1 * \ldots * F_k.$$

If X_i are i.i.d. with d.f.F, then the d.f. of $X_1 + \cdots + X_k$ is the kth convolution F^{k*} of $F, k = 1, 2, \ldots$. Then

$$q_k = F^{k^*}(x), (x \text{ fixed}), k = 0, 1, 2, \ldots$$

and
$$R(t) = \sum_{k=0}^{\infty} e^{-at} \frac{(at)^k}{k!} F^{k^*}(x).$$

The process represented by $R(t) \equiv R_x(t)$, which has x as a parameter, is a compound Poisson process.

The stochastic process $\{Z(t), t \geq 0\}$ representing the accumulated damage or wear in $(0, t)$ is called the *wear process*. For each t,

$$\Pr\{Z(t) \leq x\} = R_x(t) = R(t).$$

We state (without proof) below an important theorem concerning $Z(t)$.

Theorem 10.2. Assume that $Z(t)$ satisfies the following conditions:
(i) $\{Z(t), t \geq 0\}$ is a Markov process,
(ii) $Z(0) = 0$, $Z(t+h) - Z(t) \geq 0$ for all $t, h \geq 0$, with probability one, and
(iii) $\Pr\{Z(t+h) - Z(t) \leq u \mid Z(t) = u\}$ is decreasing both in Z and t for $t \geq 0$, $h \geq 0$, $u \geq 0$.
Then the first passage time $T_x = \inf\{t : Z(t) > x\}$ of the Markov process $Z(t)$, is IHRA

Note: (1) The condition (ii) means that the device enters service at time 0 with no accumulated wear, and that wear is non-negative and the condition (iii) implies that, for fixed t, an accumulation to wear can only weaken a device and make it more prone to further wear and that given equal amounts of accumulated wear, a device older in age is more prone to additional wear than one younger in age. The three conditions seem to be fairly realistic.

(2) Compound Poisson processes are included amongst the Markov processes satisfying the conditions (ii) and (iii).

(3) When the threshold has a fixed value x, the first passage time T_x of the Markov process $\{Z(t)\}$ is equivalent to the lifetime T of the device.

Example 11(e). If

$$q_k = 1, \quad k \leq n$$
$$= 0, \quad k > n,$$

then from (11.10) we get

$$p_k = 1, \quad k = n+1$$
$$= 0, \quad \text{otherwise};$$

from (11.11) we get

$$f(t) = \frac{a e^{-at}(at)^n}{n!} \, ,$$

so that T has a gamma distribution; in particular, if $n = 1$, i.e. $q_0 = 1$, $q_k = 0$, $k = 1, 2$, ..., then T has an exponential distribution.

Abdel-Hameed and Proschan (1973, 1975) have extended these shock models and wear processes to cover the cases in which shocks occur in accordance to non-homogeneous Poisson process and more general birth processes. Abdel-Hameed (1975) discusses preservation of life distributions under a gamma wear process.

The discussions in this section would give an indication of the role of stochastic processes in reliability. There are various other models in reliability where interesting stochastic processes arise. Cinlar (1975) shows how the theory of Markov renewal processes could be applied in reliability. Osaki (*J. Opns. Res. Japan* **12** (1970) 127) and Nakagawa and Osaki (*Micro Rel.* **15** (1976) 633) also deal with application of MRP in reliability analysis. Carraveta and R. De Dominics (*RAIRO, Automatique*, **15** (1981) 361) deal with application of semi-Markov processes in reliability analysis. See also Osaki (1985).

10.11.5 Current Trends: Some Surveys

The qualitative concepts of reliability are somewhat old but its quantitative aspects have been developed over the past two decades or so. The developments are quite varied; here we mention only a few recent survey papers of interest. All of them contain exhaustive and useful lists of uptodate references.

Barlow and Proschan (1976) indicate some directions of current academic research in reliability. See Osaki & Nakagawa (1976) for references.

Reliability, which has long been considered as a measure of system effectiveness, is now considered as an incomplete measure. *Availability*, which is a combined measure of reliability and maintainability is now receiving wider attention and usage as a measure of maintained system of effectiveness. An elaborate survey of the literature with a classification of the problems is given by Lie *et al.* (1977).

Considerable attention has been paid to reliability optimisation. A special issue of *IEEE Transactions*, Vol. R-26 (1977) is devoted to this topic. Tillman *et al.* (1977) present a review of the theoretical optimisation techniques used in reliability, and provide an extensive bibliography.

Statistical analysis has also been receiving a great deal of attention. See Barlow and Proschan (1975), Bain (1975), Kalbfleisch and Prentice (1980) and Lawless (1982), and references cited therein.

EXERCISES

10.1 Consider an $M/M/1$ queue and let $N(t)$ denote the system size at time t. Find the transition probabilities of the Markov process $\{N(t), t \geq 0\}$:

$$p_{ij}(\Delta t) = \Pr \{N(t + \Delta t) = j \mid N(t) = i\}.$$

Write down the rate matrix and hence derive the Kolmogorov equations.

10.2. Find the distribution of the number (Q) in the queue in steady state in an $M/M/1$ queue with rates λ, μ. Find $E(Q)$ and var (Q). Verify that $E(Q) = \lambda E(W_Q)$. Find also var (W_Q), and verify that var (W_Q) + var (v) = var (W_S), where v is the service time.

10.3. Show that the expected number of busy servers in an $M/M/c$ queue in steady state is $c\rho = \lambda/\mu$, and that the expected number of idle servers is $c(1-\rho)$.

10.4. Using Little's formula obtain the average waiting time in the system for $M/M/1/K$ model.

10.5. Centralisation of counters

Consider that two identical $M/M/1$ queueing systems with the same rates λ, μ (intensity $\rho = \lambda/\mu$) are in operation. Show that the steady state distribution of the number N in the two systems combined is given by

$$\Pr\{N = n\} = (n+1)(1-\rho)^2 \rho^n, n \geq 0.$$

Now suppose that it is decided to form a single queue instead of two queues but to retain the two servers. Show that the steady state distribution of the number M in the system is given by

$$\Pr\{M = n\} = \frac{2(1-\rho)\rho^n}{1+\rho}, \quad n \geq 1$$

$$= \frac{1-\rho}{1+\rho}, \quad n = 0.$$

Compare some measures of effectiveness and performance of the two models.

10.6. Output process of a Markovian queueing system

Show that for an $M/M/1$ queueing system in steady state, the intervals between successive departures are i.i.d. random variables having the same distribution as the interarrival interval. Show that the output distribution is identical to the input distribution also for the $M/M/s$ system in steady state. (In other words, the *output process* is identical to the *input Poisson process* for an $M/M/s$ ($s \geq 1$) queueing system in steady state) (Burke, 1968).

10.7. Let

$$p(k, r) = \frac{e^{-r} r^k}{k!},$$

$$E(s, r) = \sum_{k=0}^{s-1} p(k, r) = \Pr\{X \leq s-1\}$$

where X is a Poisson variate with mean r.

(Tables of these functions are available.)

Show that $E\{W_Q\}$ for an $M/M/s$ queue in steady state can be expressed as

$$E\{W_Q\} = \left[\frac{1}{s\mu(1-\rho)} \left\{ 1 + (1-\rho)\frac{E(s,r)}{p(s,r)} \right\}^{-1} \right], \rho = \lambda/\mu s = r/s.$$

10.8. Shonick (1970) refers to some studies on hospital administration and reports the following conclusion generally arrived at in regard to occupancy of hospital beds:

When the number of beds is such that all the beds are rarely filled up (such a situation arises in emergency wards for non-scheduled and emergency admissions), then one may expect the distribution of the number of occupied beds to be Poisson ; on the other hand, when the beds are often all filled, Poisson does not give a good fit.

Assuming that demand for beds occurs in accordance with a Poisson process and that occupancy times are exponentially distributed, give a plausible explanation for the occurrence of the observed phenomena.

10.9. *Combination of a loss and a delay (waiting) system* (Shonick, 1970).

Consider that two types of patients arrive at a hospital having a fixed number of beds, say s. These are (i) elective customers (or physician generated) and (ii) emergency customers (nature generated), their arrivals being in accordance with Poisson processes with rates a_1 and a_2 per day respectively. Assume that the duration of hospitalisation of both types of patients has exponential distribution with mean $1/b$ day. Elective customers who do not find a bed vacant wait and form a queue, while emergency customers who do not find a vacant bed leave and are lost.

Show that, in steady state, (when $a_1 < bs$), the distribution of the number in the system N is given by

$$\Pr(N = n) = p_n = \frac{(a/b)^n}{n!} p_0 , \qquad n = 1, 2, ..., s$$

$$= (a/b)^s \frac{(a_1/b)^{n-s}}{s^{n-s}(s!)} p_0 , \qquad n = s+1, s+2, ...$$

where

$$p_0 = \left\{ \sum_{r=0}^{s-1} \frac{(a/b)^r}{r!} + \frac{(a/b)^s}{s!} \left(1 - \frac{a_1}{bs} \right) \right\}^{-1} , \qquad a = a_1 + a_2.$$

Find the expected number of patients (i) hospitalised and (ii) in the queue. Obtain the corresponding results for $M/M/s/\infty$ and $M/M/s/s$ models as particular cases.

10.10. Consider an $M/G/s/s$-loss system in equilibrium. If the random variable N denotes the number of busy channels and $B(s) \equiv B(s, \lambda/\mu) = \Pr\{N = s\}$, then show that

$$E(N) = (\lambda/\mu)(1 - B(s)).$$

Suppose that x_i is the indicator variable of the ith randomly chosen channel; $x_i = 1$ or 0 according as the ith randomly chosen channel is busy or idle; and that $p_s(A)$ denotes the probability of the event A (with reference to a $M/G/s/s$-loss system in equilibrium). Then show that

$$p_s(x_1 = 1) = (\lambda/\mu)(1 - B(s))/s$$

$$p_s\{x_1 = 1, x_2 = 1, ..., x_k = 1\}$$

$$= B(s)/B(s - k), \quad 1 \le k \le s$$

$$p_s\{x_{k+1} = 1 \mid x_1 = 1, x_2 = 1, ..., x_k = 1\}$$

$$= (\lambda/\mu)[1 - B(s - k)] \big/ (s - k)$$

$$= p_{s-k}(x_1 = 1).$$

(Kaufman, 1979).

10.11. Machine interference with m machines and s repairman. (a) Show that the mean arrival rate λ' is given by

$$\lambda' = \sum_{n=0}^{s} (m - n)\lambda p_n = \lambda(m - L)$$

and that

$$W = \frac{L}{\lambda'} = \frac{L}{\lambda(m - L)}.$$

(b) Denote by X the expected number of machines in working order. Show that, when s is close to m, then

$$X \cong \frac{m}{1 + \lambda/\mu}$$

and when s is small, then

$$X \cong \frac{s}{(\lambda/\mu)}.$$

(c) Special case: $s = 1$. Show that the probability p_r that the number of machines in working order is r is given by

$$p_r = \frac{p(r, 1/\rho)}{1 - E(m + 1, 1/\rho)}, \quad 1 \leq r \leq m.$$

where $p(r, a) = (e^{-a}a^r)/r!$.

10.12. A mechanic looks after 8 automatic machines, a machine breaks down, independently of others, in accordance with a Poisson process, the average length of time for which a machine remains in working order being 12 hours. The duration of time required for repair of a machine has an exponential distribution with mean 1 hour. Find:

(a) the probability that 3 or more machines will remain out of order at the same time ;

(b) the average number of machines in working order ;

(c) for what fraction of time, on the average, the machanic will be idle ; and

(d) the average duration of time for which a machine is not in working order.

10.13. *Multistage Poisson queueing system with infinite number of servers*: Jackson *et al* (1980) have considered a finite number k of $M/M/\infty$ queues *in tandem* as a mathematical model for patient progress in a certain disease. Let the arrival to the first stage be Poisson with mean rate λ, the service time at ith stage be exponential with mean $1/\mu_i$; let the probability that at time t there are n_i customers in stage i ($i = 1, 2, \ldots, k$) be $P(n_1, n_2, \ldots, n_k, t)$. Show that the differential equation of the system state is given by

$$\frac{\partial P(n_1, n_2, \ldots n_k, t)}{\partial t}$$

$$= -\left(\lambda + \sum_{i=1}^{k} n_i \mu_i\right) P(n_1, n_2, \ldots, n_k, t)$$

$$+ \delta(n_1)\lambda P(n_1 - 1, n_2, \ldots, n_k, t)$$

$$+ \sum_{i=1}^{k-1} \delta(n_{i+1})(n_i + 1)\mu_i P(n_1, \ldots, n_i + 1, n_{i+1} - 1, \ldots, n_k, t)$$

$$+ (n_k + 1)\mu_k P(n_1, n_2, \ldots, n_k + 1, t),$$

where
$$\delta(n) = 1, \quad n > 0$$
$$= 0, \quad n = 0.$$

In particular, when $k = 2$, show that the equation admits of the solution

$$P(n_1, n_2, t) = \left\{\frac{e^{-a_1}a_1^{n_1}}{n_1!}\right\}\left\{\frac{e^{-a_2}a_2^{n_2}}{n_2!}\right\},$$

where
$$a_1 = (\lambda/\mu)(1 - e^{-\mu t})$$
$$a_2 = (\lambda/\mu)(1 - e^{-\mu t} - \mu t\, e^{-\mu t}).$$

Deduce the steady state results.

Note: See Harrison and A.J. Lemoine (*J. Appl. Prob.*, **18** (1981) 561–567) who consider an $M/G/\infty$ system in tandem. It has been shown that the steady state queue length distribution depends only on λ and μ_j (for a Poisson input queue) and not on the form of service time distributions.

10.14. *Transient distribution of the waiting time*

Denote by $w_n(v)$, the waiting time of the nth arrival when the arrival at epoch $t = 0$ finds a residual service of load v. When $v = 0$, the arrival at epoch $t = 0$ is immediately taken into service.

(a) Show that for an $M/M/1$ system (with $\rho < 1$)

$$\Pr\{w_n(v) = 0\} = \sum_{k=1}^{n} \frac{k}{2n-k} \binom{2n-k}{k} \frac{\rho^k}{(1+\rho)^{2n-k}}$$

$$\times \sum_{j=0}^{k-1} \frac{(\lambda v)^j}{j!} e^{-\lambda v}, \quad v \neq 0$$

and

$$\Pr\{w_n(0) = 0\} = \frac{1}{(1+\rho)^n} \left\{ \sum_{i=0}^{n-1} \left(\frac{\rho}{1+\rho}\right)^i \binom{-n}{i} - \frac{\rho}{1+\rho} \sum_{i=0}^{n-2} \left(\frac{\rho}{1+\rho}\right)^i \binom{-n-1}{i} \right\}$$

and that $\Pr\{w_n(0) =, 0\} \to 1 - \rho$ as $n \to \infty$.

(b) Show that for an $M/D/1$ system (with $\rho < 1$)

$$\Pr\{w_n(v) = 0\} = E_{n-1}(\lambda v + n\rho) - \rho E_{n-2}(\lambda v + n\rho)$$

where

$$E_k(\alpha) = \sum_{j=0}^{k} \frac{\alpha^j}{j!} e^{-\alpha}$$

and that

$$\Pr\{w_n(v) = 0\} \to 1 - \rho \quad \text{as} \quad n \to \infty \quad \text{(for each } v\text{)}.$$

[Mori, *J. Opns. Res. Soc. Japan* 19(1976) 14 - 31]

10.15. $M/M(a, b)/s$ queue (s-channel Markovian queue under the general bulk service rule).

Let $P_{r,n}$ denote the steady state probability that r channels are busy and n units are waiting in the queue. Find the difference equations of the system in steady state. Show that

$$P_{0,n} = A r_0^n, \quad n \geq 0,$$

where A is a constant and r_0 is the unique root in $(0, 1)$ of the equation

$$\lambda - (\lambda + s\mu)x + s\mu x^{b+1} = 0.$$

10.16. The system $M/G^a/1$ (single server system with Poisson input with service in batches of fixed size a). Let X_n, Y_n, $Z(t)$ denote the same processes as indicated in Sec. 10.9.

The Markov chain $\{X_n\}$ has the steady state distribution $\{v_k\}$ having p.g.f. as given by (9.2) with $b = a$.

Show that the Markov chain $\{Y_n\}$ has steady state distribution $\{r_k\}$,

where
$$
\begin{aligned}
r_k &= cv_k, & i < a \\
&= c(v + v_k), & k = a \\
&= cv_k, & k > a
\end{aligned}
$$

and
$$v = \sum_{j=0}^{a-1} v_j, \qquad c = 1/(1 + v).$$

The semi-Markov process $\{Z(t)\}$ has steady state distribution

$$f_j = \lim_{t \to \infty} \Pr\{Z(t) = j\} = \mu_j r_j \Big/ \sum \mu_i r_i,$$

μ_j being the mean waiting time in the state j (Fabens, 1961).

10.17. $M^{(X)}/G(a, b)/1$ model: Suppose that customers arrive in batches of size $r \geq 1$, the probability that a batch of r arrives in an infinitesimal interval of length dt being $\lambda c_r dt$ ($\sum_r c_r = 1$). Suppose that service is in groups of size k ($a \leq k \leq b$), and that $\eta(x)\,dx$ is the conditional probability that the service will be completed in $(x, x + dx)$, given that it has not been completed till time x. Let

$V_r(t) = $ Probability that at the instant 't', the queue length is equal to r and the service channel is idle (it is non-zero only when $0 \leq r \leq a - 1$);

$U_n(t) = $ Probability that at instant 't', the queue length is equal to n and the service channel is busy ($n \geq 0$);

$U_n(x, t)\,dx = $ Probability that at the instant 't' the queue length is equal to 'n' and a batch is being serviced with elapsed service time lying between x and $x + dx$ ($n \geq 0$), so that

$$U_n(t) = \int_0^\infty U_n(x, t)\,dx.$$

Obtain the differential-difference equations governing the system in terms of the above. Deduce the steady state relations. [Borthakur and Medhi (1974)].

10.18. *Average cost in a batch service queue*

Suppose that units arrive at a service facility according to a Poisson process and that the single server is an infinite capacity batch server. When the server is free to serve, and there are fewer than q (≥ 1) units waiting, he does not start service; but if there are q or more waiting, the service of all those waiting is initiated in a batch.

Let
λ = rate of arrival,
$\quad v$ = service time of a batch,
$\quad k$ = fixed cost of each service,
$\quad h$ = waiting cost of a unit per unit time,
$\quad X_n$ = length of time between the start of nth and $(n+1)$st batch services, and
$\quad Y_n$ = cost incurred during nth service.

Show that $\{X_n, Y_n\}$ is a renewal reward process and that the expected long run expected cost $\zeta(q)$ per unit time is given by

$$\zeta(q) = \frac{E(Y)}{E(X)} = \frac{k + \lambda h E(v^2) + (h/2\lambda) \sum\limits_{n=0}^{q-1} p_n(q^2 - q - n^2 + n)}{E(v) + \sum\limits_{n=0}^{q-1} p_n(q - n)/\lambda},$$

where p_n is the probability that exactly n units arrive during a service period. Deduce the expected waiting time per customer (Weiss, 1979, Medhi, 1984a).

10.19. Consider a k out of n-system, each of whose components has an exponential life time distribution with mean $(1/a)$. Show that the variance of the life time of the system is

$$(1/a^2) \sum\limits_{r=k}^{n} (1/r^2).$$

10.20. The lifetime of the ith component of an n-component system has an uniform distribution in $(0, t)$. Find the reliability of the system when the components are connected (i) in series, and (ii) in parallel.

10.21. Consider an equipment consisting of two components connected in parallel. Each component has a failure rate a when both the components are operative and has a higher failure rate ca ($c > 1$) when either of the components has failed. Find the reliability of the system, and also its mean lifetime.

10.22. Two-state process (see also Example 5(b), Ch. 4). Consider a system consisting of a single component that can be in two states, functioning and non functioning. Suppose that the failure and the repair processes are both Poissonian with rates a and b respectively. Find the probability that the system is in an acceptable state at the instant t given that the system was fully operative at $t = 0$.

This probability indicates the *availability* of the system, and is denoted by $A(t)$.

10.23. System reliability in a random environment

Consider a parallel system of n components, situated in a random environment such that

(i) shocks occur according to a Poisson process with parameter λ,

(ii) when a shock occurs, each component fails, independently of the others, with probability p $(=1 - q)$.

If the interval Z upto the first failure of the entire system has d.f. $H(t) = \Pr\{Z \le t\}$, then show that

$$\overline{H}(t) = 1 - H(t) = \sum_{r=1}^{n} (-1)^{r-1} \binom{n}{r} \exp\{-\lambda(1 - q^r)t\}.$$

Suppose that the condition (i) is replaced by (ia):

(ia) shocks occur according to a renewal process having interarrival distribution with d.f. $F(x)$, and with mean $1/\lambda$.

If $\overline{H}^*(s) = \int_0^\infty e^{-st}\{1 - H(t)\}\,dt$ and $F^*(s) = \int_0^\infty e^{-st}\,dF(t)$, then show that

$$\overline{H}^*(s) = \sum_{r=1}^{n} (-1)^{r-1} \binom{n}{r} \left\{ \frac{1 - F^*(s)}{s[1 - q^r F^*(s)]} \right\}.$$

Show that the mean time μ_n until failure of the system is given by

$$\mu_n = (1/\lambda) \sum_{r=1}^{n} (-1)^{r-1} \binom{n}{r} / (1 - q^r).$$

(Rade, 1976).

10.24. Consider the same situation as described in Exercise 10.23 but subject to the following replacement rule: The system is exchanged before failure (as a preventive measure before failure) if the total number of failed components exceeds k and is replaced upon failure *iff* all the components fail and otherwise it is left alone. A cost c_1 is incurred when the failed system is replaced and a cost c_2 $(< c_1)$ is incurred when the system is exchanged before failure.

Show that the meantime that the number of failed components exceeds k for the first time is

$$\frac{1}{\lambda} \sum_{r=0}^{k} (-1)^r \binom{n}{r} \sum_{i=0}^{k-r} \left\{ \binom{n-r}{i} \right\} / (1 - q^{n-i}).$$

Verify that this is equal to the meantime to failure of a $(k + 1)$-out-of-n system.

Show that the expected cost per unit of time for an infinite time span in the steady-state is

$$C(k) = \frac{c_2 + (c_1 - c_2) \sum_{r=0}^{k} (-1)^r \binom{n}{r} p^{n-r} \left[\sum_{i=0}^{r} \left\{ \binom{r}{i} (-1)^i \Big/ (1 - q^{n-i}) \right\} \right]}{(1/\lambda) \sum_{r=0}^{k} (-1)^r \binom{n}{r} \left[\sum_{r=0}^{k-r} \left\{ \binom{n-r}{i} \Big/ (1 - q^{n-i}) \right\} \right]}$$

$k = 0, 1, \ldots, n-1$ (Nakagawa, 1979a).

10.25. Consider a parallel system of n (≥ 2) components in a random environment which generates shocks at a mean interval of $(1/\lambda)$. Suppose that the damage caused to a component is cumulative and the probability that a component falls by ith shock is $p(i)$, $i = 1, 2, \ldots$

If $P(j) = \sum_{i=1}^{j} p(i)$ and $\overline{P}(j) = 1 - P(j)$, then show that the mean time until failure of the system is given by

$$\mu_n = \left(\frac{1}{\lambda} \right) \sum_{r=1}^{n} (-1)^{r-1} \binom{n}{r} \left\{ \sum_{j=0}^{\infty} [\overline{P}(j)]^r \right\}.$$

Obtain μ_n (of Exercise 23) as a particular case from the above.
[Hint: put $p(i) = q^{i-1} p, i = 1, 2, \ldots$] (Nakagawa, 1979b).

REFERENCES

J. Abate and W. Whitt. Transient behavior of *M/M/*1 queue starting at the origin, *Queueing Systems* **2**, 42–66 (1987).

———— Transient behavior of the *M/M/*1 queue via Laplace transforms. *Adv. Appl. Prob.* **20**, 145-178 (1988).

M. Abdel-Hameed. A gamma wear process. *IEEE Trans. on Rel.* **R 24**, 152–153 (1975).

———— and F. Proschan. Nonstationary shock models, *Stoch Pro & Appl.* **1**, 383–404 *(1973).*

———— Shock models with underlying birth process, *J. Appl. Prob.* **12**, 18–28 (1975).

S. Asmussen. *Applied Probability and Queueing*, Wiley, New York (1987),

N.T.J. Bailey. On queueing processes with bulk service, *J.R.S.S.B.*, **16**, 80-87 (1954).

———— A continuous time treatment of a simple queue using generating functions, *J.R.S.S.B.*, **16**, 288–291 (1954).

L.J. Bain. *Statistical Analysis of Reliability and Life-testing Models*, Marcel Dekker, New York (1978).

A A. Balkema and L. De Haan. Residual life time at great age, *Ann. Prob.*, **2**, 792–804 (1974).

D.Y. Barrer. Queueing with impatient customers and indifferent clerks, *Opns. Res.*, **5**, 650–656 (1957).

R.E. Barlow and F. Proschan. *Mathematical Theory of Reliability*, Wiley, New York (1965).

———— *Statistical Theory of Reliability and Life Testing*, Holt, Rinehart, New York (1975) [Reprint of the original edition with corrections (1981)].

———— Some current academic research in system Reliability theory, *IEEE Trans Reliability*, **R-25**, 3, 198–202 (1975).

F. Baskett, K. Chandy, R.R. Muntz and F. Palacios. Open, closed and mixed networks of queues with different classes of customers. *J. Ass. Comp. Mach.*, **22** 248–260 (1975).

D. Bertsekas and R. Gallager. *Data Networks*, Prentice-Hall, Englewood Cliffs, NJ (1987).

D. Bertsimas and X. Papaconstantinou. On the steady state solution of $M/C_2(a, b)/s$ queueing system, *Trans. Sci.*, **22**, 125–138 (1988).

A. Birolini. *On the Use of Stochastic Processes in Modelling Reliability Problems*, Springer-Verlag, Lecture Notes in Eco. & Math. Syst. No 252, Berlin (1985).

E. Borel. Sur l'emploi du theorème de Bernoulli pour faciliter le calcul d'une infinite de coefficients. Application au problème de l'attente à un guichet, *Comptes Rendus, Acad. Sci.*, Paris, **214**,452–456 (1942).

U.N. Bhat. Imbedded Markov chain analysis of bulk queues, *J. Aust. Math. Soc.*, **4**, 244–263 (1963).

———— *A Study of the Queueing Systems M/G/1 and GI/M/1*, Lecture Notes in Operations Research and Math. Eco., No. **2**, Springer-Verlag, (1968).

———— Sixty years of queueing theory, *Management Science*, **15B**, 280–292 (1969).

Z.W. Birnbaum, J.D. Esary, and A.W. Marshall A stochastic characterization of wear-out for components and systems, *Ann. Math. Stat.*, **37**, 816–825 (1966).

A.R. Bloemena. On a queueing process with a certain type of bulk service, *Bull. Inst. Statist.*, **37**, 219–227 (1960).

A.A. Borovkov. *Stochastic Processes in Queueing Theory*, Springer New York (1976).

———— *Asymptotic Methods in Queueing Theory*, Wiley, New York (1984).

A. Borthakur. On busy period of a bulk queueing system with a general bulk service rule, *Opsearch*, **12**, 40–46 (1975).

———— *A Study of some Queueing Models*, Ph. D. Dissertation, Gauhati University, India (1975).

A. Borthakur and J. Medhi. A queueing system with arrivals and service in batches of variable size, *Cah. du Centre d'etudes de Rech. Oper.* **16**, 117–126 (1974).

A. Borthakur, J. Medhi and R. Gohain. Poisson input queueing system with start up time and under control operating policy, *Comp. & Opns. Res.* **14**, 33–40 (1987).

O.J. Boxma. The cyclic queue with one general and one exponential server, *Adv. Appl. Prob.* **15**, 857–863 (1983).

P. Brémaud. *Point Processes and Queues: Martingale Dynamics*, Springer, New York (1981).

E. Brockmeyer. H.L. Halstorm, and A. Jenson. *The Life and Works of A.K. Erlang*, (Transl. of the Danish Academy of Sciences) No. **2** (1948).

S.L. Brumelle. A generalisation of $L = \lambda W$ to moments of queue length and waiting times, *Opns. Res.*, **20**, 1127–1136 (1972).

———— On the relation between customer and time averages in queues, *J. Appl. Prob.* **8**, 508–520(1971).

———— A generalization of Erlang's loss system to state dependent arrival and service rates, *Math. of Opns. Res.*, **3**, 10–16 (1978).

P.J. Burke. The output of a queueing system, *Opns. Res.*, **4**, 699–704 (1956).

———— The output process of a stationary $M/M/s$ queueing system, *Ann. Math. Stat.*, **39**, 1141–1152 (1968).

D.C. Champernowne. An elementary method of solution of the queueing problem with a single server and a constant parameter, *J.R.S.S.B.*, **18**, 125–128 (1956).

M.L. Chaudhry and J.G.C Templeton. The queueing system $M/G^B/1$ and its ramifications, *Euro. J. Opns Res.*, **6**, 56–60 (1981).

———— *A First Course in Bulk Queues*, Wiley, New York (1983).

———— A note on the distribution of a busy period for $M/M/c$ queueing system, *Math. Operations forsch u. Statist.* **4**, 75–79 (1973).

M.L. Chaudhry, J. Medhi, S.H. Sim and J.G.C. Templeton. On a two heterogeneous-server Markovian queue with general bulk service rules, *Sankhya*, **49**, **B**, 36–50 (1987).

M.L. Chaudhry, J.G.C. Templeton and J. Medhi. Computational results of multi-server bulk-arrival queues with constant service time $M^x/D/c$, *Opns. Res.* **40**, *Suppl.* **2** S 229–S 238 (1992).

A.B. Clarke. Maximum likelihood estimates in a simple Queue, *Ann. Math. Stat.*, **28**, 1036–1040 (1957).

———— (Ed.). *Mathematical Methods in Queueing Theory*, Lecture Notes in Eco. and Math. Syst., No. **98**, Springer-Verlag, New York, (1974).

J.W. Cohen. *The Single Server Queue*, 2nd ed., North Holland, Amsterdam, (1982).

B.W. Conolly. A difference equation technique applied to the simple queue with arbitrary arrival interval distribution, *J.R.S.S.B.*, **21**, 168–175 (1958).

———— *Lecture Notes on Queueing Systems*, E. Horwood, Chichester, (1975).

B.W. Conolly and J. Chan. Generalised birth and death queueing processes; recent results, *Adv. in Appl. Prob.*, **9**, 125–140 (1977).

R.W. Conway, W.L. Maxwell and L.W. Miller. *The Theory of Scheduling*, Addison- Wesley, Reading, MA ((1967).

R.B. Cooper. *Introduction to Queueing Theory*, 2nd ed., North-Holland, New York (1981).

D.R. Cox. The analysis of non-Markovian stochastic processes by the inclusion of supplementary variables, *Proc. Camb. Phil. Soc.* **51**, 433–441(1955).

———— Some problems of statistical analysis connected with congestion, in: *Proc. Symp on Queue Th.*, University of N. Carolina, Chapel Hill, (1965).

———— and W.L. Smith. *Queues*, Methuen, London (1961).

D.R. Cox and V. Isham. *Point Processes*, Chapman & Hall, London (1980).

T.B. Crabill, D. Gross, and M.G. Magazine. A classified bibliography of research on optimal design and control of queues, *Opns. Res.*, **25**, 219–232 (1977).

D.J. Daley. Queueing output processes, *Adv. Appl. Prob.*, **8**, 395–415 (1976).

A. Descloux. *Delay Tables for Finite and Infinite Source Systems*. McGraw-Hill, New York (1962).

R.L. Disney and D. König. Queueing networks. A survey of their random processes, *SIAM Review*, **27**, 335–403 (1885) [contains a list of 314 references].

B.T. Doshi. Queueing systems with vacations: a survey, *Queueing Systems* **1**, 29–66 (1986) [contains a list of 50 references].

F. Downton. Waiting time in bulk service queues, *J.R.S.S.B.*, **17**, 256–261 (1955).

———— On limiting distributions arising in bulk service, *J.R.S.S.B.*, **18**, 265–274 (1956).

E.G. Enns. The trivariate distribution of the maximum queue length, the number of customers served and the duration of the busy period for the $M/G/1$ Queueing system, *J. Appl. Prob.*, **6**, 154–161 (1969).

J.D. Esary, A.W. Marshall, and F. Proschan. Shock models and wear Processes, *Ann Prob.*, **1**, 627–649 (1973).

A.T. Fabens. The solution of Queueing and Inventory models by semi-Markov processes, *J.R.S.S.B.*, **23**, 113–127 (1961).

D. Fakinos. The $M/G/k$ blocking system with heterogeneous servers. *J. Opnl. Res. Soc.* **31**, 919–927 (1980).

R. Fortet. Random distributions with an application to Telphone Engineering, *Proc. Berk. Symp. Math. Stat. and Prob.*, Los Angeles, **2**, 81–88 (1956).

P. Franken, D. König, U. Arndt and V. Schmidt. *Queues and Point Processes*, Wiley, New York (1982).

E. Gelenbe and I. Mitrani. *Analysis and Synthesis of Computer Systems*, Academic Press, New York (1980).

E. Gelenbe and G. Pujolle. *Introduction to Queueing Networks*, Wiley, New York (1987).

P.M. Ghare. Multichannel queueing system with bulk service, *Opns. Res.*, **16**, 180–192 (1968).

P.W. Glynn and W. Whitt. A central-limit theorem version of $L = \lambda W$, *Queueing Systems* **1**, 191–215 (1986).

———— Extensions of the queueing relations $L = \lambda W$ and $H = \lambda G$, *Opns. Res.* **37**, 634–644 (1989).

B.V. Gnedenko, Yu.K. Belyayev and A.D. Solovyev. *Mathematical Methods of Reliability Theory*, Academic Press, New York (1969).

B.V. Gnedenko, and I.N. Kovalenko. *Introduction to Queueing Theory*, Israel Program for Scient. Tran. Ltd., Jérusalem, (1969).

W.K. Grassman. Transient solutions in Markovian queues: An algorithm for finding them and determining their waiting time, *Euro. J. Opnl. Res.* **1**, 396–402 (1977).

D. Gross, and C.M. Harris. *Fundamentals of Queueing Theory*, 2nd ed. Wiley, New York, (1985).

W.J. Gordon and G.P. Newell. Closed queueing systems with exponential servers, *Opns. Res.*, **15**, 254–265 (1967).

A. Harel. Sharp bounds and simple approximations for the Erlang delay and loss formulas, *Mgmt. Sc.* (to appear, 1991)

R. Harris. The expected number of idle servers in a queueing system, *Opns. Res.* **22**, 1258–1259 (1974).

J.M. Harrison. *Brownian Motion and Stochastic Flow Systems*, Wiley, New York (1985).

D.P. Heyman and M.J. Sobel. (Eds.) *Stochastic Models in Operations Research*, Vol. I, McGraw-Hill, New York (1982).

———— (Eds.). *Stochastic Models, Handbooks in Operations Research & Management Science*, Vol 2, North-Holland, Amsterdam (1990).

J.R. Hubbard, C.D. Pegden and M. Rosenshine. The departure process of an $M/M/1$ queue, *J. Appl. Prob.* **23**, 249–255 (1986).

J.R. Jackson. Networks of waiting lines, *Opns. Res.* **5**, 518–522 (1957).

———— Jobshop like queueing systems, *Mgmt. Sci.*, **10**, 131–142 (1963).

R.R.P. Jackson and J.C. Henderson. Time dependent solution to the many server Poisson queue, *Opns. Res.* **14**, 720–723 (1966).

R.R.P. Jackson and P. Aspden. A transient solution to the multistage Poisson queueing system with infinite servers, *Opns. Res.*, **28**, 618–622 (1980).

D.L. Jagerman. Some properties of the Erlang loss function, *Bell Syst. Tech. J.* **53**, 525–551 (1974).

N.K. Jaiswal. The time-dependent solution of the bulk-service queueing problem, *Opns. Res.*, **8**, 773–781 (1960).

———— *Priority Queues*, Academic Press, New York (1968).

A.K.S. Jardine. *Maintenance, Replacement and Reliability*, London (1973).

———— Solving industrial replacement problems, *Proc. of the 1979 Annual Reliability and Maintenance Symp.*, 136–141, *I.E.E.E.* New York (1979).

J.D. Kalbfleisch and R.L. Prentice. *The Statistical Analysis of Failure Time Data*, Wiley, New York (1980).

S. Karlin, and J.L. McGergor. The Classification of Birth and Death Processes, *Trans. Ann. Math. Soc.*, **86**, 366–400 (1957).

J.S. Kaufman. The busy probability in $M/G/N$-N loss systems, *Opns. Res.*, **27**, 204–206 (1979).

J. Keilson, and A. Kooharian. On Time-dependent Queueing Processes, *Ann. Math. Stat.*, **31**, 104–112 (1960).

F.P. Kelly. *Reversibility and Stochastic Networks*, Wiley, New York (1979).

———— Stochastic models of computer communication systems. *JRSS* **B47**, 379–395 (1985).

———— Loss networks, *Ann. Appl. Prob.*, **1**, 319–378 (1991).

D.G. Kendall. Some Problems in the Theory of Queues, *J.R.S.S.B.*, **13**, 151–185 (1951).

———— Stochastic Processes occurring in the Theory of Queues and their Analysis by the method of Markov chains, *Ann. Math. Stat.*, **24**, 338–354 (1953).

J.F.C. Kingman. The single server queue in heavy traffic, *Proc. Camb. Phil. Soc.*, **57**, 902–904 (1961).

———— Poisson counts for random sequence of events, *Ann. Math. Stat*, **34**, 1217–1232 (1963).

L. Kleinrock. A conservation law for a wide class of Queueing disciplines, *Nav. Res. Log. Quart.* **12**, 181–192 (1965).

———— *Queueing Systems*, Vol. I, Wiley, New York (1975).

———— *Queueing Systems*, Vol. II, Wiley, New York (1976).

E. Koenigsberg. Cyclic queues, *Oper. Res. Qrly.*, **9**, 22–35 (1958).

———— Twenty five years of cyclic queues and closed queue networks: a review. *J. Opnl. Res. Soc.* **33**, 605–619 (1982).

H. Kobayashi. *Modeling and Analysis: An Introduction to System Performance Evaluation Methodology*, Addison-Wesley, Reading MA (1978).

L. Kosten. *Stochastic Theory of Service Systems*, Pergamon, New York, (1973).

M. Krakowski. Conservation methods in Queueing Theory, *Rev. Fran. d'auto. Infor. Rech. Oper. (R.A.I.R.O.)* 7é annee, V-I, 3–20 (1973).

J.F. Lawless. *Statistical Models and Methods for Lifetime Data*, Wiley, New York (1982).

A. Lee. *Applied Queuing Theory*, Macmillan, Toronto, (1966).

H.S. Lee and M.M. Srinivasan. Control policies for the $M^x/G/1$ queueing system, *Mgmt. Sci.* **35**, 708–721 (1989).

A.J. Lemoine. On random walks and stable $GI/G/1$ queues, *Math. Opns. Res*, **1**, 159–164 (1976).

———— Network of queues – a survey of equilibrium analysis, *Mgmt. Sci.* **24**, 464–481 (1977).

———— Networks of queues – a survey of weak convergence results, *Mgmt. Sci.* **24**, 1175–1193 (1978).

D. Levhari, and E. Shesinski. A micro-economic production function, *Econometrika*, **38**, 559–573 (1970).

C.H. Lie, C.L. Hwang, and F.A. Tillman. Availability of Maintained systems: A state of the art Survey, *A.I.I.E. Trans.*, **9**, 247–259 (1977).

J.D.C. Little. A proof for the Queueing Formula $L = \lambda W$, *Opns. Res.* **9**, 383–387 (1961).

N.R. Mann, R.E. Schafer and N.D. Singapurwalla. *Methods for Statistical Analysis of Reliability and Lifetime Data*, Wiley, New York (1974).

A.W. Marshall and M. Shaked. Multivariate shock models for distributions with increasing hazard rate average, *Ann. Prob.*, **7**, 343–358 (1979).

J. Medhi. Waiting time distribution in a Poisson queue with a general bulk service rule, *Management Science*, **21**, 777–782 (1975).

———— Further results on the waiting time distribution of a Poisson queue with a general bulk service rule, *Cah. du Centre d' etudes de Rech. Oper.*, **21**, 183–191 (1979).

———— Bulk service queueing models and associated control problems IN: *Statistics : Applications and New Directions*, 369–377, Indian statistical Institute, Calcutta (1984a).

———— *Recent Developments in Bulk Queueing Models*, Wiley Eastern, New Delhi (1984b).

———— *Stochastic Models in Queueing Theory*, Academic Press, Boston & San Diego (1991).

J. Medhi and A. Borthapur. On a two server Markovian queue with a general bulk service rule. *Cah. du. Centre d' etudes de Rech. Oper.* **14** 151–158 (1972).

J. Medhi and J.G.C. Templeton. A Poisson input queue under N-policy and with a general start-up time, *Comp. & Opns. Res.* **19**, 35–41 (1992).

S.G. Mohanty and W. Panny. A discrete time analogue of the *M/M/*1 queue and the transient solution, *Sankhya* (1992) (to appear).

S.G. Mohanty, P.R. Parthasarathy and M.Sharafali. On the transient solution of discrete queue (to appear).

E.C. Molina. *Poisson's Exponential Binomial Limit*, Van Nostrand, New York, (1942).

P.M. Morse. *Queues, Inventories and Maintenance*, Wiley, New York, (1958).

T. Nakagawa. Replacement problem of a parallel system in random environment. *J. Appl. Prob.*, **16**, 203–205 (1979a).

———— Further results of replacement problem of a parallel system in random environment. *J. Appl. Prob.*, **16**, 923–926 (1979b).

P. Naor. On machine interference. *J.R.S.S.B.*, **18**, 280–287 (1956).

B. Natvig. On the input and output processes for a general birth and death queueing model, *Adv. Appl. Prob.* **7**, 576–592 (1975).

M.F. Neuts. The single server queue with Poisson input and semi-Markov service times, *J. Appl. Prob.*, **3**, 202–230 (1966).

———— A general class of bulk queues with Poisson input, *Ann. Math. Stat.*, **38**, 759–770 (1967).

———— Markov renewal branching process, in A.B., Clarke (ed.), *Op. cit.*, (1974).

———— *Algorithmic Methods in Probability Theory*, TIMS Studies in Management Science, Amsterdam 7, North-Holland (1977).

———— Queues solvable without Rouche's theorem, *Opns. Res.*, **27**, 767–781 (1979).

———— *Matrix-Geometrix Solutions in Stochastic Models*, Johns Hopkins Univ. Press, Baltimore, MD (1981).

———— *Structured Stochastic Matrices* of *M/G/*1 *Type and Their Applications*, Dekker, New York (1989).

G.F. Newell. *Applications of Queueing Theory*, 2nd ed. Chapman & Hall, New York (1982).

S. Osaki. *Stochastic System Reliability Modelling*, World Scientific Publ. Co., Singapore (1985).

———— and T. Nakagawa. Bibliography for reliability and availability of stochastic system, *IEEE Trans. Rel. Vol.* **R25**, 284–287 (1976).

G.C. Ovuworie. Multi-Channel Queues : a Survey and Bibliography, *Int. Stat. Review*, **48**, 49–71 (1980).

P.R. Parthasarathy. A transient solution to an *M/M/*1 queue: a simple approach, *Adv. Appl. Prob.* **19**, 997–998 (1987).

P.R. Parthasarathy and M. Sharafali. Transient solution to the many server Poisson queue. *J. Appl. Prob.*, **26**, 584–594 (1989).

W.B. Powell. Analysis of vehicle holding and cancellation strategies in bulk arrival, bulk service queues. *Transp. Sci.* **19**, 352–377 (1985).

———— and P.Humblet. The bulk service queue with a general control strategy: theoretical analysis and new computational procedure, *Opns. Res.* **34**, 267–275 (1986).

N.U. Prabhu. Some results for the Queue with Poisson arrivals, *J.R.S.S.B.*, **22**, 104–107 (1960).

———— *Queues and Inventories*, Wiley New York, (1965).

———— Ladder variables for continuous time stochastic processes, *Z. Wahr. Verw. Gleb*, **16**, 157–164 (1970).

———— Limit theorems for the single server queue with traffic intensity one, *J. Appl. Prob.*, **7**, 227–233 (1970).

———— *Stochastic Storage Processes*, Springer-Verlag, New York (1980).

———— Editorial introduction, *Queueing Systems* **1**, 1–4 (1986).

———— A bibliography of books and survey papers on queueing systems: theory and applications, *Queueing Systems* **2**, 393–398 (1987).

———— and U.N. Bhat. Further results for the queue with Poisson arrivals, *Opns. Res.*, **11**, 380–386 (1963).

———— and M. Rubinovitch. On a continuous time extension of Feller's lemma, *Z, Wahr. Verb. Gieb.*, **17**, 220–226 (1971).

L. Rade. Reliability systems in random environment, *J. Appl. Prob.*, **13**, 407–410 (1976).

N. Ravichandran. *Stochastic Methods in Reliability Theory*, Wiley Eastern, New Delhi (1990).

———— and A. Gravey. Some comments on the simple queue, *Adv. Appl. Prob.* **16**, 933–935 (1984).

E. Reich. Waiting times when queues are in tandem, *Ann. Math. Stat.* **28**, 768–773 (1957).

M. Reiser. Numerical methods in separable queueing networks IN: *Algorithmic Methods in Probability*, (Ed. M.F. Neuts) North Holland, Amsterdam (1977).

M.F. Romalhoto, J.A. Amaral and M.T. Cochito. A survey of J. Little's formula, *Int. Stat. Rev.*, **51**, 255–278 (1983).

S.M. Ross. *Introduction to Probability Models*, 2nd ed., Academic Press, New York (1980).

J.T. Runnenberg. Probabilistic interpretation of some formula in Queueing theory, *Bull d'Inst. Inter de Stat.*, **37**, 405–414 (1960).

T.L. Saaty. A.K. Erlang, *Opns. Res.* **5**, 293–294 (1957).

———— *Elements of Queueing Theory*, McGraw-Hill, New York (1961).

I. Sahin. *Regenerative Inventory Systems*, Springer-Verlag, New York (1990).

C.H. Sauer and K. Chandy. *Computer System and Performance Modelling*, Prentice-Hall Englewood Cliffs, NJ (1981).

M. Schwartz. *Computer Communication Network Designs and Analysis*, Prentice-Hall, Englewood Cliffs, NJ (1987).

M. Scott and M.B. Ulmer Jr. Some results for a simple queue with limited waiting room, *Zeit. f. Opns. Res.*, **16**, 199–204 (1972).

M. Sharafali and P.R. Parthasarathy. On the distribution of a busy period for the many server Poisson queue, *Opsearch* **26**, 125–132 (1989).

O.P. Sharma. *Markovian Queues*, Ellis Horwood, London (1990).

———— N. Ravichandran and J. Dass. Transient analysis of multiple reliability systems, *Nav. Res. Log. Orly.*, **36**, 255–264 (1989).

W. Shonick. A stochastic model for occupancy-related random variables in general acute hospitals, *J. Am. Stat. Ass.*, **65**, 1474–1500 (1970).

G.P. Sphicas, and D.G. Shimshak. Waiting time variability in some single server queueing systems, *J. Opnl. Res. Soc.*, **29**, 65–70 (1978).

M.J. Sobel. Optimal operation on Queues, IN A.B. Clarke (Ed.), *Op. cit.* (1974).

S. Stidham. A last word on $L = \lambda W$, *Opns. Res.*, **22**, 417–421 (1974).

S. Stidham and N.U. Prabhu. Optimal control in Queueing system, IN A. B Clarke (Ed.), *Op. cit.* (1974).

D. Stoyan. *Comparison Methods for Queues and other Stochastic Models* (English edition, edited and revised by D.J. Daley), Wiley, New York (1983).

R. Syski. *Introduction to Congestion Theory in Telephone Systems*, 2nd ed., North-Holland, Amsterdam (1986).

L. Takacs. *Introduction to the Theory of Queues*, Oxford University Press, Oxford (1962).

———— On Erlang's Formula, *Ann. Math. Stat.*, **40**, 71–78 (1969).

J. Teghem, J. Loris-Teghem and J.P. Lambotte. *Modeles d'attente M/G/1 et GI/M/1 à Arrivees et Services en Groupes*, Lecture Notes on O.R., **8**, Springer-Verlag, Berlin, (1969).

W.A. Thompson, Jr. On the foundations of reliability, *Technometrics*, **23**, 1–13 (1981).

H.C. Tijms. *Stochastic Modelling and Analysis: A Computational Approach*, Wiley, Chichester (1986).

F.A. Tillman, C.L. Wang, and W. Kuo. Optimization techniques for system reliability with redundancy: a review, *IEEE Trans. On Rel.*, **R-26**, 148–155 (1977).

———— *Optimization of System Reliability*, Marcel Dekker, New York (1980).

K.S. Trivedi. *Probability and Statistics with Reliability, Queuing and Computer Science Applications*, Prentice-Hall, Englewood Cliffs, NJ (1982).

C. Valdez-Flores and R.M. Fieldman. A survey of preventive maintenance models for stochastically deteriorating single unit systems, *Nav. Res. Log. Qrly.* **36**, 419–446 (1989) [Contains a list of 129 references].

M.H. Van Hoorn. *Algorithms and Approximations for Queueing Systems*, Math. Center, Amsterdam (1983).

J. Walrand. *Introduction to Queueing Networks*, Prentice-Hall, Englewood Cliffs, NJ (1988).

H.J. Weiss. The computation of optimal control limits for a queue with batch service, *Management Science*, **25**, 320–328 (1979).

W. Whitt. Heavy traffic limit theorems for queues: A survey IN: A.B. Clarke (Ed.), *Op. Cit.* (1974).

———— Untold horrrs of the waiting room: what the equilibrium distribution will never tell about the queue length process, *Mgmt. Sci* **29**, 395–408 (1983).

———— Comparing counting processes and queues, *Adv. Appl. Prob.*, **13**, 207–220 (1981).

———— Open and closed network of queues, *Bell Syst. Tech. J.* **63**, 1911–1979 (1984).

———— Deciding which queue to join : some counter examples, *Opns. Res.* **34**, 55–62 (1986).

———— An interpolation approximation for the mean workload in a $GI/G/1$ queue, *Opns. Res.* **37**, 936–952 (1989).

R.W. Wolff. Problems of statistical inference for birth and death queueing models, *Opns. Res.*, **13**, 343–357 (1965).

———— Poisson arrivals see time averages, *Opns. Res.* **30**, 223–231 (1982).

———— *Stochastic Modeling and the Theory of Queues*, Prentice-Hall, Englewood Cliffs, NJ (1989).

J.W. Wong. Response time distribution of the $M/M/m/N$ queueing model, *Opns. Res.* **27**, 1196–1202 (1979).

U. Narayan Bhat & Ishwar V. Basawa (Eds.). *Queueing and Related Models*, Clarendon Press, Oxford (1992).

Hideaki Takagi. *Queueing Analysis*, **Vol. 1**, *Vacation and Priority Systems*, North Holland, Amsterdam (1991). [indudes a Bibliography of all the Books on Queues and Teletraffic Engineering].

Some Basic Mathematical Results

A.1 IMPORTANT PROPERTIES OF LAPLACE TRANSFORMS

The Laplace transformation possesses some general properties; these enable one to find Laplace transforms of many functions in very simple ways. We discuss below some such properties. Denote Laplace transform of $f(t)$ by

$$L\{f(t)\} \equiv \bar{f}(s) = \int_0^\infty e^{-st} f(t).$$

1. *Linearity property*: The Laplace transformation is a linear operation, i.e. if a_i's are constants and $f_i(t)$'s, $i = 1, 2$, are functions whose Laplace transforms $L\{f_i(t)\}$ exist, then

$$L\{a_1 f_1(t) + a_2 f_2(t)\} = a_1 L\{f_1(t)\} + a_2 L\{f_2(t)\}$$

Proof: We have

$$L\{a_1 f_1(t) + a_2 f_2(t)\} = \int_0^\infty e^{-st} \{a_1 f_1(t) + a_2 f_2(t)\} \, dt$$

$$= a_1 \int_0^\infty e^{-st} f_1(t) dt + a_2 \int_0^\infty e^{-st} f_2(t) \, dt$$

$$= a_1 L\{f_1(t)\} + a_2 L\{f_2(t)\}$$

For example

$$L\{2t - 5\} = \frac{2}{s^2} - \frac{5}{s}.$$

The property holds for $i = 1, 2, \ldots, n$ where n is finite.

2. *First translation (or first shifting) property*:

If $\qquad L\{f(t)\} = \bar{f}(s) \quad$ exists for $s > \alpha$, then

$$L\{e^{at}f(t)\} = \bar{f}(s-a), s > \alpha + a$$

Proof: We have

$$\bar{f}(s-a) = \int_0^\infty e^{-(s-a)t} f(t)\, dt$$

$$= \int_0^\infty e^{-st} \{e^{at} f(t)\}\, dt$$

$$= L\{e^{at} f(t)\}.$$

For example,

$$L\{e^{-t} t\} = \frac{1}{(s+1)^2},$$

and

$$L\{e^{at} t^n\} = \frac{\Gamma(n+1)}{(s-a)^{n+1}}, n > -1.$$

3. *Second translation (or second shifting) property:*

If $L\{f(t)\} = f(s)$ exists and $g(t)$ is defined by

$$g(t) = \begin{cases} f(t-a), & t > a \\ 0, & t < a \end{cases}$$

then

$$L\{g(t)\} = e^{-as} \bar{f}(s).$$

Proof: We have

$$L\{g(t)\} = \int_0^\infty e^{-st} f(t-a)\, dt'$$

$$= \int_0^\infty e^{-s(t_1+a)} f(t_1) dt_1$$

$$= e^{-as} \bar{f}(s).$$

4. *Change of scale property*:

If $$L\{f(t)\} = \bar{f}(s) \quad \text{exists, then}$$

$$L\{f(at)\} = (1/a)\bar{f}(s/a), \, a \neq 0$$

Proof: We have

$$L\{f(at)\} = \int_0^\infty e^{-st} f(at)\, dt$$

$$= (1/a) \int_0^\infty e^{-(s/a)t_1} f(t_1)\, dt_1$$

$$= (1/a)\bar{f}(s/a).$$

For example,

$$L\{\sin at\} = (1/a)\frac{1}{(s/a)^2 + 1} = \frac{a}{s^2 + a^2}.$$

Example 1(a). For $n > -1$,

$$L\{e^{-at}(at)^n\} = (1/a)L\{e^{-t}t^n\}$$

$$= \frac{1}{a} \cdot \frac{\Gamma(n+1)}{(s/a+1)^{n+1}}$$

$$= \frac{a^n \Gamma(n+1)}{(s+a)^{n+1}}.$$

5. *Laplace transform of derivatives*:

First order derivative: Suppose that $f'(t)$ exists and is continuous on every finite interval in the range $t \geq 0$, then the Laplace transform of the derivative $f'(t)$ exists and

$$L\{f'(t)\} = s\, L\{f(t)\} - f(0)$$

$$= s\, \bar{f}(s) - f(0)$$

Proof: Integrating by parts, we get

$$L\{f'(t)\} = \int_0^\infty e^{-st} f'(t)\, dt$$

$$= e^{-st} f(t) \Big]_{t=0}^{\infty} + s \int_0^{\infty} e^{-st} f(t)\, dt$$

$$= s\bar{f}(s) - f(0).$$

Higher order derivatives: Suppose that $f''(t)$ exists and is continuous on every finite interval in the range $t \geq 0$ then $L\{f''(t)\}$ exists and is given by

$$L\{f''(t)\} = s^2 \bar{f}(s) - sf(0) - f'(0).$$

Proof: Applying the above result to the second order derivative, we get

$$L\{f''(t)\} = s L\{f'(t)\} - f'(0)$$

$$= s\{s\bar{f}(s) - f(0)\} - f'(0)$$

$$= s^2 \bar{f}(s) - sf(0) - f'(0).$$

Similarly

$$L\{f'''(t)\} = s^3 \bar{f}(s) - s^2 f(0) - sf'(0) - f''(0)$$

and more generally for nth order derivative

$$L\{f^n(t)\} = s^n \bar{f}(s) - s^{n-1} f(0) - s^{n-2} f'(0) - \cdots - f^{n-1}(0).$$

Example 1(b). Let $f(t) = t^2$. Then

$$L\{2\} = L\{f''(t)\} = s^2 L\{f(t)\} - sf(0) - f'(0)$$

$$= s^2 L\{t^2\}$$

$$\Rightarrow L\{t^2\} = \frac{L(2)}{s^2} = \frac{2/s}{s^2} = \frac{2}{s^3}.$$

Remarks: (1) Differentiation corresponds to multiplication by and addition of a constant.

(2) It is assumed that the function $f(t)$ and the derivatives satisfy a condition called piece-wise continuity. Since most of the functions considered here satisfy this condition, we donot discuss this here.

(3) The above results have important applications in differential equations. The

Laplace transformation reduces linear differential equations with constant coefficients to algebraic equations of the transform. For example, taking Laplace transform of

$$y''(t) + a^2 y(t) = r(t)$$

we get $$s^2 \bar{y}(s) - s y(0) - y'(0) + a^2 \bar{y}(s) = \bar{r}(s)$$

which is an algebraic equation yeilding

$$\bar{y}(s) = \frac{s y(0) + y'(0)}{s^2 + a^2} + \frac{\bar{r}(s)}{s^2 + a^2}$$

as its solution.

6. *Laplace transform of integrals* :

Under certain conditions satisfied by $f(t)$, we get

$$L\left\{ \int_0^t f(x)dx \right\} = \frac{\bar{f}(s)}{s} , \quad s > 0$$

Proof: Let $F(t) = \int_0^t f(x)dx$, so that $F'(t) = f(t)$.

Now $$L\{F'(t)\} = s\bar{F}(s) - F(0)$$

whence $$\bar{F}(s) = L\{F(t)\}$$

$$= \frac{\bar{f}(s)}{s} , \quad s > 0.$$

Example 1(c). Given that $\bar{g}(s) = \dfrac{1}{s + s^2}$ find $g(t)$

We have $$\bar{g}(s) = \frac{1}{s(s+1)} = \frac{\bar{f}(s)}{s}$$

where $\bar{f}(s) = \dfrac{1}{s+1}$, so that $f(t) = e^{-t}$. Thus

$$g(t) = \int_0^t f(x)dx = 1 - e^{-t}.$$

Note : Integration corresponds to division by s.

7. *Differentiation and integration of transforms* : The derivative of the transform. Let

$$\bar{f}(s) = L\{f(t)\} = \int_0^\infty e^{-st} f(t)dt$$

exist. Then under conditions the differentiation under the integral sign is valid and

$$\bar{f}'(s) = \int_0^\infty t\, e^{-st} f(t)\, dt$$

$$= L\{-tf(t)\}$$

or

$$L\{tf(t)\} = (-1)\bar{f}'(s).$$

Again

$$\int_s^\infty \bar{f}(z)dz = \int_s^\infty \int_0^\infty e^{-zt}\, dz\, f(t)\, dt$$

$$= \int_0^\infty \{f(t)dt\} \left\{ \int_s^\infty e^{-zt}\, dz \right\}$$

$$= \int_0^\infty \left\{ \frac{1}{t} f(t) \right\} e^{-st}\, dt$$

$$= L \left\{ \frac{f(t)}{t} \right\}.$$

Differentiation and integration of the transform of a function $f(t)$ corresponds respectively to the multiplication of the function $f(t)$ by $-t$ and the division of $f(t)$ by t.

It can be shown that

$$L\{t^2 f(t)\} = (-1)^2 \bar{f}''(s),$$

and in general, $L\{t^n f(t)\} = (-1)^n \bar{f}^n(s)$, $n = 1, 2, \ldots$

8. *Limit Property* : **Initial Value Theorem**

$$\lim_{t \to \infty} f(t) = \lim_{s \to 0} s\bar{f}(s)$$

whenever the limits exist.

For example, if $f(t) = e^{-t}$ so that $\bar{f}(s) = 1/(s+1)$, then

$$\lim_{t \to \infty} f(t) \quad = 0 = \lim_{s \to 0} s\bar{f}(s).$$

9. *Limit Property* : **Final Value Theorem**

$$\lim_{t \to 0} f(t) = \lim_{s \to \infty} s \, \bar{f}(s)$$

whenever the limits exist.

The properties (8) & (9) enable one to study the behaviour of $f(t)$ as $t \to \infty$ (or as $t \to 0$) from the behaviour of $\bar{f}(s)$ as $s \to 0$ (or as $s \to \infty$). There are other such results of more generality: these are covered under the name of Tauberian theorems.

10. *Integral Property*:

Taking limit, as $s \to 0$, of

$$\int_0^\infty e^{-st} f(t) dt = \bar{f}(s)$$

we get

$$\int_0^\infty f(t) dt = \bar{f}(0)$$

assuming the integral to be convergent. Further, for $a > 0$

$$\int_0^\infty e^{-at} f(t) dt = \bar{f}(a)$$

provided the integral converges.

A. 2 DIFFERENCE EQUATIONS

A.2.1 Introduction

Let $f(n)$ be a function defined only for non-negative *integral values* of the argument n. The first difference of $f(n)$ is defined by the increment of $f(n)$ and is denoted by $\Delta f(n)$, i.e.

$$\Delta f(n) = f(n+1) - f(n) \tag{2.1}$$

The second and the higher differences are defined by

$$\Delta^{k+1} f(n) = \Delta\{\Delta^k f(n)\}, \quad k > 0.$$

It can be verified that

$$\Delta^{k+1}f(n) = \Delta^k\{\Delta f(n)\} = \Delta^k\{f(n+1)-f(n)\}$$

$$= \Delta^k f(n+1) - \Delta^k f(n), \quad k > 0. \tag{2.2}$$

We denote $f(n+1)$ by $Ef(n)$ (E is here the displacement operator and is not to be confused with the symbol E for the expectation of a random variable): then

$$\Delta f(n) = Ef(n) - f(n) = (E-1)f(n).$$

Symbolically, $\qquad\qquad \Delta \equiv E - 1 \text{ or } E \equiv 1 + \Delta; \quad$ we have

$$E^k f(n) = f(n+k). \tag{2.3}$$

By a difference equation, we mean an equation involving a function evaluated at arguments which differ by any of a fixed number of values. Examples of such equations are:

$$f(n+2) - f(n+1) - f(n) = 0;$$

$$(\lambda + \mu_n)p_n = \mu_{n+1}p_{n+1} + \lambda p_{n-1}, \quad n \geq 1$$

$$p_n = qp_{n-1} - p(1-p_{n-1}) = 0; \tag{2.4}$$

$$Y_t = a(1+b)Y_{t-1} - abY_{t-2} + 1;$$

$$a_0 U_x + a_1 U_{x+1} + \cdots + a_k U_{x+k} = g(x).$$

In the first equation the argument runs from n to $(n + 2)$; this is a difference equation of order $(n + 2) - n = 2$. The last equation is of order $(x + k) - x = k$. The first two equations are homogeneous while the last three are not.

The equation $f(n+2) - f(n+1) - f(n) = 0$ may be written as

$$E^2 f(n) - Ef(n) - f(n) = 0, \quad \text{or} \quad r(E)f(n) = 0,$$

where $r(E) = (E^2 - E - 1)$.

A difference equation may be written as a function of the operator E in the same way that a differential equation can be written as a function of the operator $D \left(\equiv \frac{d}{dx}\right)$.

There is indeed a close similarity between the theory of differential equations and that of difference equations.

Given a function $f(n) = Ca^n$, we can easily get the difference equation $f(n+1) - af(n) = 0$. We shall generally be concerned with the inverse problem: given a difference equation, to find the function which satisfies the given difference equation.

A.2.2 Homogeneous Difference Equation with Constant Coefficients

Consider a homogeneous difference equation of order *two*

$$f(n+2)+af(n+1)+bf(n)=0 \qquad (2.5)$$

or $$r(E)f(n)=0, \text{ where } r(E)=E^2+aE+b.$$

Consider a trial solution $f(n)=m^n$ $(m \neq 0)$;
then $Ef(n)=m^{n+1}$, $E^2 f(n)=m^{n+2}$ so that (2.5) gives

$$(m^2+am+b)=0. \qquad (2.6)$$

This equation $r(m)=0$ is called the *characteristic* (or auxiliary or secular) *equation* of the difference equation (2.5).
(i) If the roots m_1, m_2 of the equation (2.6) are distinct, then m_1^n and m_2^n are two solutions of the equation (2.5). Since the equation is homogeneous $c_1 m_1^n$ and $c_2 m_2^n$ will also satisfy the equation and so also with their sum. In other words, if m_1, m_2 are the distinct roots of (2.6) and c_1, c_2 are arbitrary constants, then

$$f(n)=c_1 m_1^n+c_2 m_2^n \qquad (2.7)$$

will be the general solution of the equation (2.5).
(ii) In case the roots m_1, m_2 are equal, the general solution is given by

$$f(n)=(c_1+c_2 n)m_1^n. \qquad (2.8)$$

(iii) In case the roots are complex, then we can write

$$m_1=r(\cos\theta+i\sin\theta) \text{ and } m_2=r(\cos\theta-i\sin\theta)$$

and the general solution (2.7) will take form

$$f(n)=Ar^n\cos(n\theta+B), \qquad (2.9)$$

where A, B are arbitrary constants.

The above arguments can be carried over to homogeneous difference equations of any order k. We may state as follows the rule for the solution of the homogeneous difference equations:

$$f(n)+a_1 f(n+1)+\cdots+a_k f(n+k)=0 \qquad (2.10)$$

or, $$r(E)f(n)=0.$$

First, find the roots of the characteristic equation

$$r(m) \equiv 1 + a_1 m + \cdots + a_k m^k = 0.\tag{2.11}$$

Suppose that $m_1, m_2, \ldots, m_k$ are the real and distinct roots of (2.11), then the general solution of (2.10) will be given by

$$f(n) = c_1 m_1^n + c_2 m_2^n + \cdots + c_k m_k^n.\tag{2.12}$$

In case some of the real roots are equal, say $m_1 = m_2$, and others distinct then the solution will be

$$f(n) = (c_1 + c_2 n)m_1^n + c_3 m_3^n + \cdots + c_k m_k^n.\tag{2.13}$$

It is to be noted that the general solution of an equation of order k will, in general, contain k arbitrary constants $c_1, c_2, \ldots, c_k$. These can be obtained when k initial (or boundary) conditions are also given.

A.2.3. Linear Non-homogeneous Equation

Consider an equation of the form

$$f(n) + a_1 f(n+1) + \cdots + a_k f(n+k) = g(n)$$

$$\tag{2.14}$$

or $$r(E)\{f(n)\} = g(n).$$

Let u be the general solution of the corresponding homogeneous equation $r(E)$ $\{f(n)\} = 0$ and v be a particular solution of $r(E)\{f(n)\} = g(n)$.

Then the most general solution of (2.14) is given by $u + v$. Methods of obtaining particular solution of (2.14) depend largely on the form of the function $g(n)$. We consider below a particular form of $g(n)$. See Jordan (1947) for other forms.

Let $g(n) = AB^n$, where A and B are constants.

Case (i). Suppose that B is not a root of $r(x) = 0$, i.e. $r(B) \neq 0$.

Consider a trial solution $f(n) = CB^n$, $B \neq 0$, C is constant. Substitution in (2.14) gives

$$CB^n + a_1 CB^{n+1} + \cdots + a_k CB^{n+k} = AB^n$$

whence $$C = \frac{A}{r(B)}$$

and thus

$$v = \frac{AB^n}{r(B)}\tag{2.15}$$

is a particular solution of (2.14).

Case (ii). Suppose that B is a non-repeated root of $r(x) = 0$, i.e. $r(B) = 0$ but $r'(B) \neq 0$. Consider a trial solution $f(n) = n\, C\, B^n$. Substitution in (2.14) yields

$$C[nr(B) + Br'(B)] = A$$

whence
$$C = \frac{A}{Br'(B)}$$

and thus
$$v = \frac{AnB^n}{r'(B)} \tag{2.16}$$

is a particular solution of (2.14).

Case (iii). Suppose that $r(B) = r'(B) = \cdots = r^{(m-1)}(B) = 0$ but $r^{(m)}(B) \neq 0$.

This case can be similarly treated.
A particular solution will be

$$v = \frac{A n^{m-1} B^n}{r^m(B)}. \tag{2.17}$$

Particular Case: $g(n) = A$ (constant); we get the result by putting $B = 1$.

There are other methods for solving linear difference equations such as by using matrices, using Laplace transforms, using z-transforms (generating functions) and so on. See, Jordan (1947), Miller (1968).

Example 2(a). *Fibonacci numbers* are given by

$$0, 1, 1, 2, 3, 5, 8, 13, 21, \ldots;$$

these can be put as

$$f_0 = 0, \quad f_1 = 1, \quad f_n = f_{n-1} + f_{n-2} \ (n \geq 2). \tag{2.18}$$

The difference equation $f_n = f_{n-1} + f_{n-2}$ has the characteristic equation $m^2 = m + 1$, whose roots are $m_1 = (1 + \sqrt{5})/2$ and $m_2 = (1 - \sqrt{5})/2$. Hence $f_n = c_1 m_1^n + c_2 m_2^n$. Solving for c_1 and c_2 by using $f_0 = 0$, $f_1 = 1$, we get $c_1 = -c_2 = 1/\sqrt{5}$, so that

$$f_n = [\, \{(1 + \sqrt{5})/2\}^n - \{(1 - \sqrt{5})/2\}^n \,] \big/ \sqrt{5}, \quad n \geq 2 \tag{2.19}$$

Further, f_n can also be expressed as:

$$f_0 = 0 \quad \text{and} \quad f_n = \binom{n}{0} + \binom{n}{1} + \cdots + \binom{n-k}{k},$$

where $k = [n/2]$ is the largest integer $\leq n/2$, $n = 1, 2, \ldots$

For large n,
$$f_n \approx \frac{1}{\sqrt{5}} \left(\frac{1 + \sqrt{5}}{2} \right)^n = \frac{(1.618)^n}{2.236} \tag{2.20}$$

This gives quite close an approximation. Some approximations are given below.

n	approximation	actual value
5	4.96	5
8	21.007	21
10	54.995	55

For large n, $\dfrac{f_{n+1}}{f_n} = \dfrac{1}{2}(1+\sqrt{5}) = 1.618$;   this (and its reciprocal 0.6180) are known as golden section ratios.

Note. (1) Leonardo da Pisa, c. 1180–1250, better known as, Fibonacci (son of Bonacci), obtained this sequence of numbers which have found interesting applications (Jean, 1984, Vorobyov, 1963).

(2) Generalised Fibonacci numbers $F_{n,r}$ are given as solution of the difference equation

$$F_{n,,r} = F_{n-1,r} + F_{n-2,r} + \cdots + F_{n-r,r} \quad (r \geq 2).$$

For $r = 2$, we get $F_{n,r} = f_n$ (simple Fibonacci numbers).

A.2.4 Difference Equations in Probability Theory

Many problems in probability theory can be put down in terms of a difference equation involving the unknown probability. Such problems can then be solved completely by making use of the initial conditions supplied by the problem itself. This method is sometimes found to be very powerful.

Example 2(b). To solve the difference equation

$$U_n = \frac{1}{2}(U_{n+1} + U_{n-1}), \quad 1 \leq n \leq a-1$$

with the initial conditions

$$U_0 = 1 \text{ and } U_a = 0.$$

The difference equation has the characteristic equation $\frac{1}{2}m^2 - m + \frac{1}{2} = 0$, having two equal roots $m = 1, 1$. The solution of the difference equation is

$$U_n = (c_1 + c_2 n)(1)^n = c_1 + c_2 n.$$

The initial conditions give

$$1 = U_0 = c_1, 0 = U_a = c_1 + c_2 a, \quad \text{whence} \quad c_2 = -1/a.$$

The solution of the difference equation is thus

$$U_n = 1 + (-1/a)n = 1 - n/a.$$

The equation arises in the classical ruin problem, U_n being the probability of gambler's ruin in the symmetric case $p = q = \dfrac{1}{2}$.

In the study of stochastic processes, we frequently come across equations which are non-homogeneous and which have coefficients that are not constants but functions of the argument. The method of induction and the method of generating function could be used in such cases. These methods are illustrated by an example.

Example 2(c). The probability function

$$p_n = \Pr\{N = n\}, \quad n = 0, 1, 2, \ldots,$$

of a random variable N satisfies the difference equations

$$p_{n+1} - (1+a)p_n + ap_{n-1} = 0, \quad n \geq 1 \tag{2.21}$$

$$-p_1 + ap_0 = 0 \qquad (0 < a < 1). \tag{2.22}$$

Since p_n's are probabilities of the random variable N, we have also the condition

$$\sum_{n=0}^{\infty} p_n = 1. \tag{2.23}$$

We now proceed to find p_n, *i.e.* to solve the difference equation (2.21) given (2.22) and (2.23)

METHOD OF CHARACTERISTIC EQUATION

The second-order difference equation (2.21) which can also be written as

$$p_{n+2} - (1+a)p_{n+1} + ap_n = 0, n \geq 0 \tag{2.24}$$

has the characteristic equation $m^2 - (1 + a)m + a = 0$, whose roots are $m = 1$ and $m = a \, (< 1)$. So the equation (2.21) has the solution

$$p_n = A(1)^n + B(a)^n = A + Ba^n, n \geq 0. \tag{2.25}$$

We have from (2.24) and (2.22)

$$p_1 = A + Ba = ap_0$$

$$p_0 = A + B. \tag{2.26}$$

Solving in terms of p_0, we get $A = 0$, $B = p_0$, so that

$$p_n = p_0 a^n. \tag{2.27}$$

From (2.23) we get

$$1 = \Sigma p_n = \Sigma p_0 a^n = p_0 / (1-a) \quad \text{or} \quad p_0 = 1-a.$$

Hence the solution is

$$p_n = (1-a)a^n, \quad n \geq 0. \tag{2.28}$$

METHOD OF INDUCTION

The equation $p_{n+1} - (1+a)p_n + ap_{n-1} = 0$ can be written as

$$p_{n+1} - ap_n = p_n - ap_{n-1}, \quad n \geq 1.$$

Putting $(n-1)$ for n, we get

$$p_n - ap_{n-1} = p_{n-1} - ap_{n-2}$$

and so on. Hence

$$p_{n+1} - ap_n = p_n - ap_{n-1} = p_{n-1} - ap_{n-2}$$

$$\cdots \qquad \cdots$$

$$= p_1 - ap_0, \qquad n \geq 0 \tag{2.29}$$

But from (2.22) $p_1 - ap_0 = 0$. Hence

$$p_n = ap_{n-1} = a(ap_{n-2}) = \cdots = a^n p_0,$$

and

$$\sum_{n=0}^{\infty} p_n = 1 \text{ gives } p_0 = 1-a, \text{ so that } p_n = (1-a)a^n, n \geq 0.$$

METHOD OF GENERATING FUNCTIONS

Let

$$P(s) = \sum_{n=0}^{\infty} p_n s^n \tag{2.30}$$

be the p.g.f. of the distribution.

Multiplying equation (2.21) by s^n $(n \geq 1)$ and adding over $n = 1, 2, \ldots$, we get

$$\sum_{n=1}^{\infty} p_{n+1} s^n - (1+a) \sum_{n=1}^{\infty} p_n s^n + a \sum_{n=1}^{\infty} p_{n-1} s^n = 0. \qquad (2.31)$$

or, $(1/s)[P(s) - p_0 - p_1 s] - (1+a)[P(s) - p_0] + asP(s) = 0$

or, $[1/s - (1+a) + as]P(s) = \dfrac{p_0}{s} + p_1 - (1+a)p_0$

or, $\dfrac{(1-s)(1-as)}{s} P(s) = \dfrac{p_0(1-s)}{s}.$

Hence

$$P(s) = \frac{p_0}{1-as} = p_0(1-as)^{-1} = p_0(1 + as + a^2 s^2 + \cdots)$$

so that

$$p_n \equiv \text{coefficient of } s^n \text{ in } P(s) \qquad (2.32)$$

$$= p_0 a^n.$$

Finding p_0 as before from (2.23), we get

$$p_n = (1-a)a^n, \quad n \geq 0.$$

The method of Laplace transforms can also be used to solve a difference equation (Smith, Ch. 9).

Note. The generating function can be used to reduce a difference equation to an algebraic equation.

Example 2(d). To solve the difference equation

$$u_n = qu_{n-1} + p(1 - u_{n-1}), \quad n \geq 1, \quad p + q = 1, \qquad (2.33)$$

given the initial condition $u_0 = 1$. It is a non-homogeneous equation of order 1. Only one initial condition is needed to solve it completely.

The equation can be written as (denoting the displacement operator by E)

$$r(E)u_{n-1} = p, \quad r(E) = 1 + (p - q)E.$$

The general solution of the corresponding homogeneous equation
$$r(E) u_{n-1} = 0$$

is given by $u_n = A (q - p)^n$

and a particular solution is given by $v = p/r(1) = \dfrac{1}{2}$ so that

$$u_n = A (q - p)^n + \frac{1}{2}$$

$$= \frac{1}{2}[1 + (q - p)^n], \text{ since } u_0 = 1.$$

Alternatively: Let $A(s) = \sum\limits_{n=0}^{\infty} u_n s^n$ be the generating function of $\{u_n\}$. Multiplying

both sides of (2.33) by s^n and adding over $n = 1, 2, \ldots$, we get

$$\sum_{n=1}^{\infty} u_n s^n = (q - p) \sum_{n=1}^{\infty} u_{n-1} s^n + p \sum_{n=1}^{\infty} s^n$$

or,
$$(A(s) - u_0) = (q - p)s A(s) + p \frac{s}{1 - s}$$

or,
$$\{1 - (q - p)s\} A(s) = \frac{ps}{(1 - s)} + 1 \text{ (since } u_0 = 1) \tag{2.34}$$

or,
$$A(s) = \frac{ps}{(1 - s)\{1 - (q - p)s\}} + \frac{1}{1 - (q - p)s}$$

$$= \frac{1}{2}\left\{ \frac{1}{1 - s} - \frac{1}{1 - (q - p)s} \right\} + \frac{1}{1 - (q - p)s} \;,$$

(decomposing into partial fractions)

$$= \frac{1}{2}\left\{ (1 - s)^{-1} + [1 - (q - p)s]^{-1} \right\}.$$

Thus

$$u_n \equiv \text{coefficient of } s^n \text{ in } A(s) = \frac{1}{2}\{1 + (q - p)^n\}.$$

(Here u_n gives the probability that n Bernoulli trials result in an even number of successes. $\{u_n\}$ is not a probability distribution.)

The method of generating functions is very powerful. It can also be used when the coefficients of the equation are not constants, but are functions of the argument.

A.3 DIFFERENTIAL–DIFFERENCE EQUATIONS

Suppose that $u_n(t)$, $n = 1, 2, \ldots$, is a function of t having a derivative

$$\frac{d}{dt}u_n(t) = u_n'(t)$$

An equation involving $u_n'(t)$ and $u_n(t)$, $u_{n+1}(t)$ etc. is called a *differential-difference* (or *recurrence differential*) equation. For example,

$$p_n'(t) = -\lambda[p_n(t) - p_{n-1}(t)]$$

is a differential-difference equation (see Bellman and Cooke (1963) for a detailed account of this topic).

The existence of unique solution will depend on suitable initial conditions which are to be determined carefully. The solution of such equations can sometimes be found easily by the method of induction (see Example 2(c)). Besides, there are two powerful tools for the solution of such an equation: these are generating functions and Laplace transforms. By using generating functions, a differential-difference equation can often be reduced to an ordinary differential equation, while the use of Laplace transforms often enables one to reduce a differential-difference equation to a difference equation. A combination of the two tools sometimes facilitates the solution to a differential-difference equation. We illustrate the use of these tools by some examples.

Example 3(a). Solve the differential-difference equation

$$u'_n(t) = u_{n-1}(t), \quad t \geq 0, \quad n = 1, 2, \dots \tag{3.1}$$

given the initial conditions

$$u_0(t) = 1, \quad t \geq 0 \tag{3.2}$$

and

$$u_0(0) = 1, \quad u_n(0) = 0 \quad \text{for } n \neq 0 \tag{3.2a}$$

Let

$$G(s, t) = \sum_{n=0}^{\infty} u_n(t)s^n, \quad -s_0 < s < s_0 \tag{3.3}$$

be the generating function of the sequence $\{u_n(t)\}$. Term by term differentiation of the series is valid within its radius of convergence $(-s_0 < s < s_0)$; we thus get

$$\frac{\partial G}{\partial t} = \sum_{n=0}^{\infty} u'_n(t)s^n = u'_0(t) + \sum_{n=1}^{\infty} u'_n(t)s^n$$

$$= \sum_{n=1}^{\infty} u'_n(t)s^n, \text{ since } u_0'(t) = 0 \text{ from (3.2).}$$

Multiplying (3.1) by s^n and adding over for $n = 1, 2, \ldots$, we get

$$\sum_{n=1}^{\infty} u'_n(t)s^n = \sum_{n=1}^{\infty} u_{n-1}(t)s^n$$

or,

$$\frac{\partial G}{\partial t} = sG,$$

which is a partial differential equation. The solution is given by

$$G(s,t) = A \exp(st),$$

and putting $t = 0$, we get $G(s, 0) = A$.

Again, $G(s, 0) = \sum_{n=0}^{\infty} u_n(0)s^n = u_0(0) + \sum_{n=1}^{\infty} u_n(0)s^n = 1$, from (3.2a).

Hence $A = 1$ and $G(s, t) = \exp(st)$. Thus,

$$u_n(t) \equiv \text{coefficient of } s^n \text{ in } G(s,t)$$

$$= \text{coefficient of } s^n \text{ in } \exp(st)$$

$$= \frac{t^n}{n!}, \quad n = 0, 1, 2, \ldots.$$

Example 3(b). Consider the equations

$$p_n'(t) = -\lambda [p_n(t) - p_{n-1}(t)], n \geq 1, \quad \lambda > 0 \qquad (3.4)$$

$$p_0'(t) = -\lambda p_0(t) \qquad (3.5)$$

with the initial conditions $p_n(0) = 0$ for $n \neq 0$, $p_0(0) = 1$. $\qquad (3.6)$
Let

$$L\{p_n(t)\} = \int_0^{\infty} \exp(-st)p_n(t)dt \equiv \bar{p}_n(s) \qquad (3.7)$$

be the L.T. of $p_n(t)$; then

$$L\{p_n'(t)\} = s\bar{p}_n(s) - p_n(0),$$

$$= s\bar{p}_n(s), \quad n \geq 1$$

Hence taking the L.T. of both sides of (3.4), we have

$$s\overline{p}_n(s) = -\lambda[\overline{p}_n(s) - \overline{p}_{n-1}(s)] \tag{3.8}$$

or,

$$\overline{p}_n(s) = \left(\frac{\lambda}{\lambda+s}\right)\overline{p}_{n-1}(s), \quad n \geq 1$$

which is a difference equation in $\overline{p}_n(s)$. Its solution is given by

$$\overline{p}_{n-1}(s) = c\left(\frac{\lambda}{\lambda+s}\right)^{n-1}, n \geq 1. \tag{3.9}$$

By taking L.T. of (3.5) and using the initial condition $p_0(0) = 1$, we get

$$s\overline{p}_0(s) - 1 = -\lambda\overline{p}_0(s)$$

or,

$$\overline{p}_0(s) = 1/(s+\lambda). \tag{3.10}$$

The unknown constant c in (3.9) can be obtained by putting $n = 1$ and using (3.10). We have

$$\overline{p}_0(s) = c = 1/(s+\lambda)$$

Hence

$$\overline{p}_{n-1}(s) = \frac{\lambda^{n-1}}{(\lambda+s)^n}, \quad n \geq 1$$

or,

$$\overline{p}_n(s) = \frac{\lambda^n}{(\lambda+s)^{n+1}}, \quad n \geq 0. \tag{3.11}$$

Now, $p_n(t)$ can then be obtained by finding the inverse Laplace transform of $\overline{p}_n(s)$. We find that $\overline{p}_n(s)$ is the L.T. of

$$\frac{\exp(-\lambda t)(\lambda t)^n}{n!}.$$

Hence

$$p_n(t) = \frac{\exp(-\lambda t)(\lambda t)^n}{n!}, \quad n = 0, 1, 2, ... \tag{3.12}$$

The solutions of these equations governing Poisson processes have been obtained by an alternative method of generating function in Sec. 4.1 (Theorem 4.1).

Example 3(c). Consider the equations

$$p'_n(t) = -(a+b)p_n(t) + ap_{n-1}(t) + bp_{n+1}(t), \; n \geq 1 \tag{3.13}$$

$$p'_0(t) = -ap_0(t) + bp_1(t) \tag{3.14}$$

with initial conditions $p_n(0) = 0$, $n \neq 0$ and $p_0(0) = 1$. $\tag{3.15}$
Let

$$\bar{p}_n(s) = \int_0^\infty \exp(-st)p_n(t)dt \quad \text{be the L. T. of } p_n(t).$$

By taking L.T. of both sides of (3.13), we get the difference equation

$$s\bar{p}_n(s) - p_n(0) = -(a+b)\bar{p}_n(s) + a\bar{p}_{n-1}(s) + b\bar{p}_{n+1}(s), \; n \geq 1$$

or, $$\bar{p}_n(s) = \frac{a}{a+b+s}\bar{p}_{n-1}(s) + \frac{b}{a+b+s}\bar{p}_{n+1}(s). \tag{3.16}$$

Similarly from (3.14), we get

$$s\bar{p}_0(s) - p_0(0) = -a\bar{p}_0(s) + b\bar{p}_1(s)$$

or, $$\bar{p}_0(s) = \frac{1}{a+s} + \frac{b}{a+s}\bar{p}_1(s). \tag{3.17}$$

The L.T. of $p_n(t)$ (for $n \geq 1$) satisfies a homogeneous difference equation of order two. Its characteristic equation is

$$m = \frac{a}{a+b+s} + \frac{b}{a+b+s}m^2$$

or, $$bm^2 - (a+b+s)m + a = 0. \tag{3.18}$$

If m_1, m_2 are the two roots of this equation, then the solution of (3.16) is

$$\bar{p}_n(s) = Am_1^n + Bm_2^n, \quad n \geq 1 \tag{3.19}$$

where A, B are two constants. By putting the value of $\bar{p}_1(s)$, we get $\bar{p}_0(s)$ from (3.17).

Thus we get $\bar{p}_n(s)$, the L.T. of $p_n(t)$ $(n \geq 0)$, whence $p_n(t)$ can be determined. This technique leads to the solution of the $M/M/1$ queueing model (see Sec. 10.3).

A.4 MATRIX ANALYSIS

A.4.1 Introduction

Vector: A vector is an *ordered m*-tuple $x_1, \ldots, x_m$ and is denoted by

$$
\mathbf{x} = \begin{pmatrix} x_1 \\ \cdot \\ \cdot \\ \cdot \\ x_m \end{pmatrix};
$$

it is called a *column vector*. The transpose of x denoted by $\mathbf{x}' = (x_1, \ldots, x_m)$ is a *row vector*. When $x_1, \ldots, x_m$ are the probabilities of a r.v. then $\mathbf{x}'$ is known as *probability vector*.

Matrix: A matrix A is an ordered rectangular array of elements a_{ij}, $i = 1, 2, \ldots n$, $j = 1, 2, \ldots, m$

$$
A = \begin{pmatrix} a_{11} & \cdots & a_{1m} \\ \cdot & & \cdot \\ \cdot & & \cdot \\ \cdot & & \cdot \\ a_{n1} & \cdots & a_{nm} \end{pmatrix}
$$

A is written as $A = (a_{ij})$ $i = 1, 2, \ldots, n, j = 1, 2, \ldots, m$.
A is a matrix of n rows and m columns and is called a matrix of dimension $n \times m$. When $n = m$, one gets a square matrix of order n (or of dimension $n \times n$). When all the elements $a_{ij} = 0$, we get a zero matrix denoted by $\mathbf{0}$.

A column vector is a matrix of dimension $1 \times n$ and a row vector is a matrix of dimension $m \times 1$.

The transpose of a $n \times m$ matrix $A = (a_{ij})$ is a matrix A' of dimension $m \times n$. i.e.

$$
A' = \begin{pmatrix} a_{11} & \cdots & a_{n1} \\ \cdot & & \cdot \\ \cdot & & \cdot \\ \cdot & & \cdot \\ a_{1m} & \cdots & a_{nm} \end{pmatrix}.
$$

Obviously, $(A')' = A$.

Symmetric matrix: A square matrix is said to be *symmetric* if $A = A'$ i.e. if $a_{ij} = a_{ji}$ for all i, j, and is said to be *skew-symmetric* if $A = -A'$ i.e. $a_{ij} = -a_{ji}$.

Hermitian matrix: A square matrix $A = (a_{ij})$ with complex elements such that $a_{ji} = \overline{a}_{ij}$ $(\overline{a}_{ji}$ is the complex conjugate of $a_{ij})$ is called a Hermitian matrix. A real Hermitian matrix is a symmetric matrix.

Sum and product of matrices : The scalar product cA (where c is a scalar constant) is defined by $cA = (ca_{ij})$. For two matrices $A = (a_{ij})$ and $B = (b_{ij})$ of the same dimension $n \times m$, the sum $A + B = (a_{ij} + b_{ij})$ is a matrix of dimension $n \times m$. It follows that $(A + B)' = A' + B'$.

The matrix sum is commutative and associative.

The matrix product of $A = (a_{ij})$ (of dimension $n \times m$) and $B = (b_{ij})$ (of dimension $m \times r$) is a matrix $C = AB = (c_{ij})$ of dimension $n \times r$ where

$$c_{ij} = \sum_{k=1}^{m} a_{ik} b_{kj}, \quad i = 1, 2, ..., n, \quad j = 1, 2, ..., r.$$

The product is defined only if the number of columns of A is equal to the number of rows of B. Even when both AB and BA are defined, in general $AB \neq BA$. The matrix product, when defined, is distributive and associative but non commutative.

We have $\quad (AB)' = B'A'$

Diagonal matrix: A square matrix $A = (a_{ij})$ is said to be a *diagonal* matrix if $a_{ij} = a_{ji}$ $= 0$, $i \neq j$, i.e. if all of its off-diagonal elements are zero. The diagonal $m \times m$ matrix

$$I_m = \begin{pmatrix} 1 & 0 & ... & 0 \\ 0 & 1 & ... & 0 \\ 0 & 0 & ... & 1 \end{pmatrix}$$

is called the unit matrix of order m and is denoted by I ($\equiv I_m$). We have det $(I_m) = 1$.

Triangular matrix: A square matrix $A = (a_{ij})$ is said to be an upper triangular matrix if $a_{ij} = 0$ for all $j < i$. If $a_{ij} = 0$ for all $i < j$, then it is called a lower triangular matrix.

Non singular matrix: The determinant of a square matrix is the unique scalar quantity $|A|$ (or det A) equal to the determinant having (a_{ij}) for its elements. A square matrix is said to be *non-singular* if det $A \neq 0$; if det $A = 0$ it is said to be *singular*.

Inverse of a matrix: The inverse of a non-singular matrix A of order m is the *unique* matrix A^{-1} such that

$$A^{-1}A = A A^{-1} = I (\equiv I_m).$$

Orthogonal matrix: A real square matrix is said to be *orthogonal* if $AA' = A'A = I$.

For an orthogonal matrix A, det $A = 1$ or -1. A square matrix $A = (a_{ij})$ for which $A = \bar{A}^{-1}$ is called a *unitary* matrix. A real unitary matrix is an orthogonal matrix.

Sub-matrix: Any matrix obtained by omitting some rows and columns from a given matrix A (of dimension $n \times m$) is said to be a sub-matrix of A.

Trace of a matrix: The *trace* of a square matrix $A = (a_{ij})$ of order m is the sum of the diagonal elements and is denoted by $tr\,(A) = \sum\limits_{i=1}^{m} a_{ii}$. It follows that $tr\,(A) = tr\,(A')$, $tr\,(A + B) = tr\,(A) + tr\,(B)$ and that $tr\,(AB) = tr\,(BA)$, provided both AB and BA are defined.

Rank of a matrix: Consider a matrix A of dimension $n \times m$ and all its possible *square sub-matrices* (obtained by omitting rows and/or columns of A) of order $p = 1, 2, \ldots, k$, $k = \min\,(n, m)$. The matrix A is said to be of order r if it contains *at least* one r-rowed square matrix with a non-zero determinant, whereas the determinant of any square submatrix having $(r + 1)$ or more rows is zero. The rank of A of dimension $(n \times m)$ is less than or equal to $\min\,(n, m)$. The matrix having each of its elements equal to 0 is of rank zero.

A square matrix of order m has rank $r < m$ *iff* $\det A = 0$, i.e. if it is singular. The matrix has rank $r = m$ *iff* $\det A \neq 0$, i.e. if it is non-singular.

If A and B are square matrices of order m $(\neq 0)$ such that $AB = 0$, then both A and B are singular. If $AB = 0$ and A is non-singular, then $B = 0$.

If A is non-singular, then

$$AS = AT \quad \text{implies} \quad S = T.$$

Similar matrices: Two square matrices, A, B both of order m are said to be similar if there exists a mth order non-singular matrix R such that

$$R^{-1} A R = B$$

$$\Rightarrow A = RBR^{-1}.$$

If B is a diagonal matrix, then A is to be diagonalized.

A.4.2 Eigenvalues and Eigenvectors of a Square Matrix

Let $A = (a_{ij})$ be a square matrix of order m. The roots of the characteristic equation.

$$|A - tI| \equiv \det\,(A - tI) = 0, \, (t \text{ real})$$

are called the *eigenvalues (or spectral values or latent roots or characteristic roots)* of the matrix A. The characteristic equation is of degree m and has exactly m roots, which we denote by t_i, $i = 1, 2, \ldots, m$. The roots may not be all distinct and there may be one or more roots of order greater than one (so that the sum of the order of the roots is equal to m). The order of the root is the order (or the *algebraic multiplicity*) of the corresponding eigenvalue.

The set $T = \{t_i, i = 1, 2, \ldots, m\}$ of eigenvalues is called the *spectrum* of A. The largest of the absolute values $p\,(A) = \max\,\{\,|t_1|, |t_2|, \ldots |t_m|\,\}$ is called the *spectral radius* of A.

A non-zero column vector **x** which satisfies the vector equation

$$(A - t_i I) \quad \mathbf{x} = 0$$

is called a *right eigenvector* (or latent or characteristic vector) of the matrix A corresponding to (or associated with) its eigenvalue t_i.

A non-zero row vector $\mathbf{y}'$ which satisfies the vector equation

$$\mathbf{y}' (A - t_i I) = 0$$

is called a *left eigenvector* of A corresponding to its eigenvalue t_i.

Now $(A - t_i I) \mathbf{x} = 0$ has non-zero solution $\mathbf{x}$ *iff* $\det (A - t_i I) = 0$, i.e. *iff* t_i is an eigenvalue of the matrix A.

It may be noted that

$$\mathbf{y}' (A - tI) = 0 \text{ implies } (A - tI) \mathbf{y} = 0.$$

This can be proved by using $(AB)' = B' A'$ and $(A - B)' = A' - B'$.

An alternative way of defining eigenvalues and eigenvectors is as follows:

A number t is called an eigenvalue of the matrix A if there exists a non-zero column vector $\mathbf{x}$ such that

$$A\mathbf{x} = t\,\mathbf{x}.$$

The non-zero column vector $\mathbf{x}$ is called a *right eigenvector* corresponding to t.

The vector equation $A\mathbf{x} = t\mathbf{x}$ being homogeneous, the eigenvector $\mathbf{x}$ corresponding to an eigenvalue cannot be unique. Thus if $\mathbf{x}$ is an eigenvector then $k\mathbf{x}$ (where k is a scalar constant $\neq 0$) is also an eigenvector corresponding to the same eigenvalue.

Similarly, t is called an eigenvalue of A if there exists a non-zero row vector $\mathbf{y}'$ such that

$$\mathbf{y}' A = t\,\mathbf{y}'.$$

The non-zero row vector $\mathbf{y}'$ is called a left eigenvector corresponding to t. If $\mathbf{y}'$ is a left eigenvector, then $k\,\mathbf{y}'$ is also a left eigenvector. One might ask if there are any left eigenvectors corresponding to an eigenvalue t which are not multiples of $\mathbf{y}'$. To an eigenvalue, there *may* correspond several linearly independent eigenvectors. The number of independent (left) eigenvectors corresponding to an eigenvalue t is called the *multiplicity* (or *geometric multiplicity*) of the eigenvalue t.

The matrix $\begin{pmatrix} 1 & 0 \\ 3 & 1 \end{pmatrix}$ has the eigenvalue 1 of order 2 but of multiplicity 1, i.e. the eigenvalue 1 is of algebraic multiplicity 2 and of geometric multiplicity 1.

The unit matrix $\begin{pmatrix} 1 & 0 \\ 0 & 1 \end{pmatrix}$ has the eigenvalue 1 of order 2 and of multiplicity 2, i.e. the eigenvalue 1 has both geometric and algebraic multiplicities equal to 2. It can be shown that geometric multiplicity of an eigenvalue cannot exceed its algebraic multiplicity.

A.4.2.1 *Properties of eigenvalues and eigenvectors*

Let A be a $m \times m$ matrix with eigenvalues $t_1, \ldots, t_m$ (which may not be all distinct).
1. Det $(A) = 0$ *iff* A has a zero eigenvalue.
2. The matrix A and its transverse A' have the same eigenvalues.
3. Different matrices may have the same eigenvalues.
  For example, the matrices

$$\begin{pmatrix} 1 & 0 \\ 0 & 2 \end{pmatrix} \quad \text{and} \quad \begin{pmatrix} 2 & 0 \\ 3 & 1 \end{pmatrix}$$

have the same eigenvalues.
4. If A is of the form

$$\begin{pmatrix} A_{11} & 0 \\ B_{11} & A_{22} \end{pmatrix}$$

where A_{11}, A_{22} are square submatrices then t is an eigenvalue of A *iff* it is an eigenvalue of at least one of the matrices A_{11}, A_{22}.
5. Det (A) and $tr\,(A)$ are expressible in terms of eigenvalues as follows

$$\det(A) = \prod_{i=1}^{m} t_i, \quad tr\,(A) = \sum_{i=1}^{m} t_i.$$

6. If A is a triangular. matrix, then its eigenvalues are the elements of the principal diagonal. For example, the eigenvalues of

$$A = \begin{pmatrix} 2 & 0 & 0 \\ 1 & 3 & 0 \\ 4 & 2 & 1 \end{pmatrix}$$

are 2, 3, and 1.
In particular, the unit matrix has only one eigenvalue 1 of the same order as that of the unit matrix.
Since $I_m \mathbf{x} = \mathbf{x} = 1.\mathbf{x}$, any non-zero vector $\mathbf{x}' = (x_1, \ldots, x_m)$ is an eigenvector of I_m. The eigenvalue 1 which is a root of order m of the characteristic equation is of multiplicity m.
7. The eigenvalues of a real symmetric matrix are all real.
  The converse is not necessarily true. A real matrix which is not symmetric may have real eigenvalues.
8. The inverse of A exists *iff* $t_i \neq 0$ $(i = 1, 2, \ldots, m)$ and the inverse A^{-1} has the eigenvalues $1/t_i, i = 1, \ldots, m$.
9. If $f(A)$ is a scalar polynomial in A, then the eigenvalues of $f(A)$ are $f(t_i), i = 1, \ldots, m$. In particular, the eigenvalues of A^n are $(t_i)^n, i = 1, 2, \ldots, m$.
10. The eigenvalues form a *biorthogonal system*. Let t_i, t_j be two distinct eigenvalues with $\mathbf{x}_i, \mathbf{x}_j$ and $\mathbf{y}_j', \mathbf{y}_j'$ as the corresponding eigenvectors. We have

$$A\,\mathbf{x}_i = t_i\,\mathbf{x}_i \text{ and } \mathbf{y}_j{}'\,A = t_j\,\mathbf{y}_j{}'$$

and

$$t_j(\mathbf{y}_j{}'\,\mathbf{x}_i) = (t_j\,\mathbf{y}_j{}')\,\mathbf{x}_i = (\mathbf{y}_j{}'\,A)\mathbf{x}_i$$
$$= \mathbf{y}_j{}'\,(A\,\mathbf{x}_i) = \mathbf{y}_j{}'\,t_i\,\mathbf{x}_i$$
$$= t_i\,(\mathbf{y}_j{}'\,\mathbf{x}_i)\,.$$

Since $t_i \neq t_j$, the relation holds *iff* $\mathbf{y}_j{}'\,\mathbf{x}_i = 0$.

A.4.2.2 *Jordan canonical form*

A $(r \times r)$ square matrix $J_r\,(\lambda) = (a_{ij})$ is called a *Jordan block* of order r if

$$\begin{aligned}
a_{ij} &= \lambda\,, & i = j \\
&= 1\,, & i + 1 = j \\
&= 0\,, & \text{elsewhere}
\end{aligned}$$

A $(n \times n)$ square matrix A is said to be in *Jordan canonical form* if

$$A = \begin{pmatrix} J_1 & 0 & \dots & 0 \\ 0 & J_2 & \dots & 0 \\ \dots & \dots & \dots & \\ 0 & 0 & \dots & J_k \end{pmatrix}$$

where the sub-matrices $J_1, \dots, J_k$ are Jordan blocks.

Now $\qquad J_r\,(\lambda) = \lambda I + M$

where M is a $r \times r$ matrix with ones as the superdiagonal elements and zero elsewhere, i.e. $a_{i,j} = 1, j = i + 1, a_{ij} = 0, j \neq i + 1$ and

$$M = \begin{pmatrix} 0 & 1 & 0 & \dots & 0 \\ 0 & 0 & 1 & \dots & 0 \\ & \dots & & \dots & \\ 0 & 0 & 0 & \dots & 1 \\ 0 & 0 & 0 & \dots & 0 \end{pmatrix}$$

It can be seen that $M^r = 0$, so that for $m \geq r$

$$J_r^m(\lambda) = (\lambda I + M)^m$$

$$= \lambda^m I + \binom{m}{1} \lambda^{m-1} M + \cdots + \binom{m}{r-1} \lambda^{m-r+1} M^{r-1}$$

Thus, for $m \geq r$

$$A^m = \begin{pmatrix} J_1^m & 0 & \cdots & 0 \\ 0 & J_2^m & \cdots & 0 \\ \cdots & & \cdots & \\ 0 & 0 & \cdots & J_k^m \end{pmatrix}.$$

We have the following result

Theorem A.4.1. Suppose that the eigenvalues of the mth order matrix A are *not* all distinct. Let $\lambda_1, \ldots \lambda_k$ be distinct eigenvalues of algebraic multiplicity $m_1, \ldots, m_k$ respectively, so that $m_1 + \cdots + m_k = m$. Then there exists an mth order non-singular matrix R such that

$$R^{-1}AR = \begin{pmatrix} J_{n_1}(\lambda_1) & 0 & \cdots & 0 \\ 0 & J_{n_2}(\lambda_2) & \cdots & 0 \\ \cdot & \cdot & & \cdot \\ \cdot & \cdot & & \cdot \\ \cdot & \cdot & & \cdot \\ 0 & 0 & & J_{n_k}(\lambda_k) \end{pmatrix}.$$

This is known as the *Jordan canonical form* of A. The spectral representation is a special case of Jordan canonical form.

A.4.2.3 Determination of Eigenvalues

The problem of determination of eigenvalues and eigenvectors of a matrix is known as an *eigenvalue problem*. It has assumed importance because of the fact that such a problem arises in several physical and technical applications. Given a square matrix A of order, say m, its eigenvalues can be determined by solving the equation

$$\det (A - tI) = 0.$$

The roots can be determined by numerical methods, for instance, by Newton's method. However, when m is large, the determination of roots becomes a time consuming process. Thus it is necessary to consider more efficient means.

The problem has been studied from two different aspects:

(i) obtaining bounds for eigenvalues, and
(ii) computing approximate values for eigenvalues.

(i) We state the following result by Frobenius.

Theorem A.4.2. Let A be a square matrix whose elements are all positive. Then A has at least one real positive eigenvalue and the corresponding eigenvector can be so chosen that all its elements are positive.

From the above, we may derive the following result by Collatz.

Theorem A.4.3. Let A be a *positive* square matrix of order m. Let $\mathbf{x}$ be any vector whose elements $x_1, x_2, \ldots, x_m$ are positive and let $y_1, y_2, \ldots, y_m$ be the elements of the vector $\mathbf{y}$ such that $\mathbf{y} = A\,\mathbf{x}$.

If $q_i = y_i / x_i$, then the closed interval bounded by the smallest and the largest of m quotients $q_i, i = 1, 2, \ldots, m$ contains at least one eigenvalue of A.

(ii) Approximate determination of eigenvalues has been receiving attention. Several new methods have been recently developed and efficient computer software programmes have been put forward. We discuss an iteration method known for a long time. The method is general and holds for a square matrix whose elements are not necessarily positive.

Rayleigh's method : Let A be a square matrix of order m. Start with any vector $\mathbf{x}_0$ $(\neq 0)$ with m elements and compute successive $\mathbf{x}_i$'s as follows:
$$\mathbf{x}_1 = A\,\mathbf{x}_0, \ \mathbf{x}_2 = A\,\mathbf{x}_1, \ldots, \mathbf{x}_r = A\,\mathbf{x}_{r-1}.$$
Let $\mathbf{z} = A\,\mathbf{x}_r$, $m_0 = \mathbf{x}_r'\,\mathbf{x}_r$, $m_1 = \mathbf{x}_r\,\mathbf{z}$, $m_2 = \mathbf{z}'\mathbf{z}$.
Then the quotient $q = m_1/m_0$ is an approximation for an eigenvalue λ of A.

Further, if we write $q = \lambda + \varepsilon$, so that ε is the error, then

$$|\varepsilon| \leq \sqrt{\left(\frac{m_2}{m_0} - q^2\right)}.$$

Example 4(a). Consider the matrix

$$A = \begin{pmatrix} 1 & 2 & 1 \\ 2 & 1 & 3 \\ 3 & 2 & 1 \end{pmatrix}$$

and choose

$$\mathbf{x} = \begin{pmatrix} 1 \\ 1 \\ 1 \end{pmatrix}$$

Then

$$\mathbf{y} = A\,\mathbf{x} = \begin{pmatrix} 1 & 2 & 1 \\ 2 & 1 & 3 \\ 3 & 2 & 1 \end{pmatrix}\begin{pmatrix} 1 \\ 1 \\ 1 \end{pmatrix} = \begin{pmatrix} 4 \\ 6 \\ 6 \end{pmatrix}$$

Thus $q_1 = 4$, $q_2 = 6$, $q_3 = 6$.

By Theorem A 4.3, A has at least one eigenvalue λ in the interval $4 \le \lambda \le 6$.

Approximation

Start with $\mathbf{x}_0 = \begin{pmatrix} 1 \\ 1 \\ 1 \end{pmatrix}$

Computation gives

$$\mathbf{x}_1 = A\,\mathbf{x}_0 = \begin{pmatrix} 4 \\ 6 \\ 6 \end{pmatrix}$$

$$\mathbf{x}_2 = A\,\mathbf{x}_1 = \begin{pmatrix} 22 \\ 32 \\ 30 \end{pmatrix}$$

If we stop here (with $r = 2$) then

$$\mathbf{z} = A\,\mathbf{x}_2 = \begin{pmatrix} 116 \\ 166 \\ 160 \end{pmatrix}$$

$$m_0 = \mathbf{x}_2'\,\mathbf{x}_2 = 2408$$

$$m_1 = \mathbf{x}_2'\,\mathbf{z} = 12664$$

so that $\dfrac{m_1}{m_0} = 5.26$ is an approximation for an eigenvalue of A.

A.4.3 Non-negative Matrices

We shall now confine ourselves to square matrices with real elements. A matrix A order m is non-negative ($A \ge 0$) if all the elements of A are non-negative, i.e. $A \ge 0$ if $a_{ij} \ge 0$, and A is positive ($A > 0$) if $a_{ij} > 0$ for all $i, j = 1, 2, \ldots, m$.

Definition. A square matrix in which each row and each column has exactly one entry with value unity, and all others zero is called a *permutation matrix*.

Note that there are m ! permutation matrices of order m. An $m \times m$ matrix A is said to be *reducible* if there exists a permutation matrix P such that

$$PAP' = \begin{pmatrix} A_{11} & A_{12} \\ 0 & A_{22} \end{pmatrix}$$

where A_{11}, A_{22} are squares matrices of order r and $m - r$ respectively.

A is *irreducible* if no such permutation matrix P exists.

It is to be noted that a transformation of the matrix A to the form $P'AP$ has simply the effect of rearranging the elements of A. The reducibility of a matrix depends on the disposition of the zero and non-zero elements of A. *A matrix A with all positive real elements is necessarily irreducible*. Further, if the matrix is irreducible, all the off-diagonal entries of any row or column cannot vanish. We state below an elegant result concerning non-negative matrices due to Perron-Frobenius.

Theorem A.4.4. Let A be a non-negative and irreducible matrix. Then
  (1) A has a positive eigenvalue, r, equal to its spectral radius $p(A)$,
  (2) r is a simple eigenvalue of A, i.e. r is a simple root of the characteristic equation,
  (3) there is a positive right eigenvector associated with r, and
  (4) r increases when any entry of A increases.

Perron obtained the result in 1907 for positive matrices; it was later extended by Frobenius in 1912 to non-negative matrices.

If A is positive, (when it is necessarily irreducible) Perron showed that there is only one eigenvalue of modulus r. If A is non-negative and irreducible, then the number of eigenvalues having modulus equal to r $(= p(A))$ is h, where $h \geq 1$.

Corollary: We have

$$\min_i \sum_j a_{ij} \leq r \leq \max_i \sum_j a_{ij}$$

with equality in either side implying equality throughout.

A similar result holds for column sums also.

Remarks.

The theory of non-negative matrices has developed tremendously since the time of Perron and Frobenius. The theory has found application in several fields, such as probability theory, numerical analysis, demography, mathematical economics and in dynamic programming (see Seneta (1981) for an extensive discussion and application to Markov chain analysis).

Definition. An non-negative irreducible matrix is called *primitive*, if $h = 1$, and *imprimitive* (or *cyclic*) of index h, if $h > 1$.

A positive matrix is necessarily primitive.

A non-negative irreducible matrix A may be primitive or imprimitive. Suppose that A is cyclic of index h and t_i, $i = 1, 2, \ldots, h$ are the h eigenvalues of modulus r, i.e. $t_i = r = p(A)$, then $t_i = w_i r$, $i = 1, \ldots, h$, where w_i's are the h distinct roots of unity.

Note. A primitive matrix is used in demographic studies for population projection (see Keyfitz, 1968). The ergodic theory of demography can be explained with the help of properties of such a matrix.

A.4.4 Spectral Representation of a Matrix

Consider an $m \times m$ matrix A with m distinct *non-zero* eigenvalues t_i and $\mathbf{x}_i$, $\mathbf{y}_i'$, $i = 1, 2, \ldots, m$, as right and left eigenvectors, of A. Let $D = (d_{ij})$, $d_{ij} = 0$, $i \neq j$, $d_{ii} = t_i$

be a diagonal matrix and let $X = (\mathbf{x}_1, \ldots \mathbf{x}_m)$ and $Y = (\mathbf{y}_1, \ldots, \mathbf{y}_m)$ be the matrices formed with the right and left eigenvectors respectively of A. From $\mathbf{y}_j'A = t_j\mathbf{y}_j'$ and $A\mathbf{x}_j = t_j\mathbf{x}_j$, we get $Y'A = DY'$ and $AX = XD$ respectively. Now suppose X^{-1} exists, then $AX = XD$ implies $X^{-1}(AX)X^{-1} = X^{-1}(XD)X^{-1}$, or $X^{-1}A = DX^{-1}$. Thus, the rows of X^{-1} are the left eigenvectors of A.

Further, $X^{-1}AX = X^{-1}XD = D$.

Particular case: Suppose that $X^{-1} = Y'$; this implies that $XY' = I$, or $Y'X = I$. We may thus choose suitable multiplicative factors to get $\mathbf{y}'\mathbf{x} = 1$. Since $\mathbf{y}'\mathbf{x} = 0$, $i \neq j$, we may write $\mathbf{y}_j'\mathbf{x}_i = \delta_{ij}$. From $AX = XD$ we get

$A = XDX^{-1} = XDY'$. Further $A^2 = (XDX^{-1})(XDX^{-1}) = XD^2X^{-1} = XD^2Y'$

and $A^n = XD^nY'$ (for any positive integer $n \geq 1$), where D^n is a diagonal matrix with elements t_i^n, $i = 1, 2, \ldots, m$, along the main diagonal, the off-diagonal elements being all zero. On expanding, we get

$$A^n = \sum_{j=1}^{m} t_j^n \, \mathbf{x}_j\mathbf{y}_j'.$$

General case: Suppose that the eigenvectors $\mathbf{x}$ and $\mathbf{y}'$ are arbitrarily chosen; then we can write

$$\mathbf{y}_j'\mathbf{x}_j = \delta_{jj}/c_j, \quad \text{or} \quad c_j = 1/(\mathbf{y}_j'\mathbf{x}_j)$$

$$A^n = \sum_{j=1}^{m} t_j^n c_j\mathbf{x}_j\mathbf{y}_j', \quad n \geq 1.$$

The products $c_j\mathbf{x}_j\mathbf{y}_j' = B_j$ are matrices (called *constituent matrices*) and the above gives an expression of A^n as a sum of m such matrices.

We immediately find the following properties of constituent matrices:

      (i) $B_iB_j = 0$, for $i \neq j$;

      (ii) $B_j^2 = B_j$ (B_j's are idempotent);

and   (iii) $\sum_{j=1}^{m} B_j = I_m$.

The representation $A = XDY'$ is known as the *spectral (or canonical) representation* of A.

Example 4(b). Consider the matrix

$$A = \begin{pmatrix} 2 & 1 \\ 3 & 4 \end{pmatrix}$$

The roots of the equation $\det(A - tI) = 0$ are 1 and 5, i.e. $t_1 = 1$, $t_2 = 5$ are the eigenvalues. The right eigenvector $\mathbf{x}_1$ (where $\mathbf{x}_1' = (x_{11}, x_{12})$) corresponding to $t_1 = 1$ is given as solution of $A\mathbf{x}_1 = \mathbf{x}_1$, i.e.

$$2x_{11} + x_{12} = x_{11} \quad \text{and} \quad 3x_{11} + 4x_{12} = x_{12}.$$

We get, $x_{11} = -x_{12}$ and $\mathbf{x}_1 = \begin{pmatrix} x_{11} \\ x_{12} \end{pmatrix}$; thus we get $\begin{pmatrix} 1 \\ -1 \end{pmatrix}$ as a right eigenvector corresponding to the eigenvalue $t_1 = 1$.

Corresponding to $t_2 = 5$, right eigenvector $\mathbf{x}_2$ (where $\mathbf{x}'_2 = (x_{21}, x_{22})$) is given as solution of $A\mathbf{x}_2 = 5\mathbf{x}_2$, i.e.

$$2x_{21} + x_{22} = 5x_{21} \quad \text{and} \quad 3x_{21} + 4x_{22} = 5x_{22}.$$

We get, $3x_{21} = x_{22}$ and $\mathbf{x}_2 = \begin{pmatrix} x_{21} \\ x_{22} \end{pmatrix}$ and we get $\begin{pmatrix} 1 \\ 3 \end{pmatrix}$ as a right eigenvector corresponding to the eigenvalue $t_2 = 5$.

The left eigenvectors $\mathbf{y}'_i = (y_{i_1}, y_{i_2})$ can be similarly obtained as solutions of $\mathbf{y}'_i A = t_i \mathbf{y}'_i$, $i = 1, 2$. We get

$$\mathbf{y}'_1 = (y_{11}, y_{12}) = (-3, 1) \text{ as a left eigenvector corresponding to the eigenvalue } t_1 = 1 \text{ and}$$

$$\mathbf{y}'_2 = (y_{21}, y_{22}) = (1, 1) \text{ as a left eigenvector corresponding to the eigenvalue } t_2 = 5.$$

Now
$$c_1 = 1/(\mathbf{y}'_1 \mathbf{x}_1) = -\frac{1}{4}, c_2 = 1/(\mathbf{y}'_2 \mathbf{x}_2) = \frac{1}{4},$$

$$\mathbf{x}_1 \mathbf{y}'_1 = \begin{pmatrix} -3 & 1 \\ 3 & -1 \end{pmatrix}, \quad \mathbf{x}_2 \mathbf{y}'_2 = \begin{pmatrix} 1 & 1 \\ 3 & 3 \end{pmatrix};$$

$$A = \sum_{t=1}^{2} t_i B_i, A' = \sum_{i=1}^{2} t'_i B'_i \ (r \geq 1),$$

where $B_i = c_i \mathbf{x}_i \mathbf{y}'_i$.

Verify that   (i) $\mathbf{x}, \mathbf{y}$ are biorthogonal,

(ii) $B_i^2 = B_i$, $B_i B_j = 0, i \neq j$, $\sum_{i=1}^{2} B_i = I_2$.

(iii)   canonical representation of A is XDX^{-1}, where

$$X = (\mathbf{x}_1, \mathbf{x}_2) = \begin{pmatrix} 1 & 1 \\ -1 & 3 \end{pmatrix}.$$

Note: General forms of right and left eigenvectors are obtained by inclusion of multiplicative constants. By writing $a_i \mathbf{x}_i$ in place of $\mathbf{x}_i$ and $b_i \mathbf{y}'_i$ in place of $\mathbf{y}'_i$, we get $c_i/a_i b_i$ in place of c_i, but then B_i's remain unaltered (and so also A and A' in terms of B_i's).

A.4.5 Stochastic Matrix

Definition. A square matrix with non-negative elements and unit row sums is called a *stochastic* matrix. It is *doubly stochastic*, if, in addition, it has unit columns sums.

$P = (p_{ij})$ is stochastic *iff* $p_{ij} \geq 0$, $\sum_j p_{ij} = 1$ for all i, j;

P is doubly stochastic *iff* $p_{ij} \geq 0$, $\sum_j p_{ij} = 1, \sum_i p_{ij} = 1$ for all i, j.

A stochastic matrix P is called a *Markov matrix iff* $\mu(P) \equiv \max_j \{\min_i p_{ij}\} > 0$.

If P is a stochastic matrix, so also is P^n, n being a positive integer. Stochastic matrices (which are non-negative) have some very interesting properties which are useful in the study of Markov chains. Some of these are given below.
Note. A *Markov matrix* is a stochastic matrix having at least one column with positive elements.

Theorem A.4.5. The matrix D is stochastic *iff* it has an eigenvalue 1 and the corresponding right eigenvector **x** is a column vector with all the elements equal to unity, i.e. $\mathbf{x}' = (1, 1, \ldots, 1)$. Further $p(P) = 1$, i.e. the spectral radius of a stochastic matrix is 1.

If the matrix P is irreducible, then there can be one and only one eigenvalue equal to 1 (Theorem A.4.4). However, there can be h (> 1) eigenvalues of modulus 1 if P is cyclic of order h. If it is primitive $(h = 1)$, then there can be no other eigenvalue of modulus 1. From this we get that $\lim_{n \to \infty} P^n \neq 0$.

Theorem A.4.6. Suppose that P is an irreducible stochastic matrix. Then $\lim_{n \to \infty} P^n$ exists *iff* P is primitive.

In other words, a primitive matrix raised to a sufficiently high power becomes a positive matrix and *conversely*.
Remark: The hypothesis that P is irreducible is not necessary; the limit P^n exists when P is a primitive stochastic matrix.

These properties of stochastic matrices has been used in the study of Markov chains (Sec. 3.4).

EXERCISES

A. 1   Show that the difference equation

$$p_n = a p_{n-1} + b, a \neq 1, n = 2, 3, \ldots$$

has for solution

$$p_n = \left(p_1 - \frac{b}{1-a}\right) a^{n-1} + \frac{b}{1-a}.$$

A. 2   Solve the equation

$$U_n = p U_{n-1} + q U_{n-1}, 1 \leq n \leq a - 1, 0 < p = 1 - q < 1$$

given that

$$U_0 = 1, U_a = 0, (p \neq q).$$

By taking the limit of U_n as $p/q \to 1$, find the solution of

$$U_n = \frac{1}{2}(U_{n+1} + U_{n-1}).$$

A. 3   If $p_n(t)$, $n = 0, 1, 2, \ldots$ represents a probability distribution and $\overline{p}_n(s)$ is the L.T. of $p_n(t)$, then show that

$$\sum_{n=0}^{\infty} \overline{p}_n(s) = 1/s.$$

A. 4   Consider the differential-difference equations (3.13) and (3.14) with initial conditions (3.15).
If

$$G(z,t) = \sum_{n=0}^{\infty} p_n(t)z^n$$

show that

$$z \frac{\partial G(z,t)}{\partial t} = (1-z)\{(b-az)G(z,t) - b p_0(t)\}$$

If $\overline{p}_n(s) = L\{p_n(t)\}$ and $G^*(z,s) = L\{G(z.t)\}$,

then show that

$$G^*(z,s) = \frac{z - b(1-z)\overline{p}_n(s)}{sz - (1-z)(b-az)}.$$

(These relations occur in the study of transient solution of $M/M/1$ queue (Bailey, *JRSS* **B16**, 288–291, 1954).

A. 5   Suppose that $P_n(t)$ satisfies the differential-difference equations

$$P_n'(t) = -(a+b)P_n(t) + aP_{n-1}(t) + bP_{n+1}(t), \quad n \geq 2$$

$$P_1'(t) = -(a+b)P_1(t) + bP_2(t)$$

$$P_0'(t) = bP_1(t)$$

with initial conditions $P_1(0) = 1$, $P_n(0) = 0$ for $n \neq 1$.

Show that the generating function $G(z,t) = \sum_{n=0}^{\infty} P_n(t)z^n$ satisfies

$$z \frac{\partial G(z,t)}{\partial t} = (1-z)(b-az)\{G(z,t) - P_0(t)\};$$

show that L.T. $G^*(z, s)$ of $G(z, t)$ is given by

$$G^*(z,s) = \frac{z^2 - (1-z)(b-az)\overline{P}_0(s)}{sz - (1-z)(b-az)}.$$

$[\overline{P}_n(s) \equiv \text{L.T}\{P_n(t)\}]$.

(These relations lead to busy period distribution of $M/M/1$ queue Bailey (1954, *loc. cit*).

A. 6   Find the eigenvalues and the left eigenvectors of the matrix

$$P = \begin{pmatrix} 0.5 & 0.5 & 0 \\ 0.5 & 0 & 0.5 \\ 0 & 1 & 0 \end{pmatrix}.$$

A. 7   Examine whether the following matrices are primitive:

(i) $\begin{pmatrix} 0 & 1 & 0 \\ 0 & 0 & 1 \\ 1 & 0 & 0 \end{pmatrix}$ (ii) $\begin{pmatrix} 0 & 0 & 1 & 0 \\ 0 & 0 & 0 & 1 \\ 0 & 1 & 0 & 0 \\ 1 & 0 & 0 & 0 \end{pmatrix}$

(iii) $\begin{pmatrix} 0 & 1 & 0 & 0 \\ 0 & 0 & 1 & 0 \\ 0 & 0 & 0 & 1 \\ 1/2 & 1/2 & 0 & 0 \end{pmatrix}$ (iv) $\begin{pmatrix} 0 & 0 & 1/2 & 1/2 \\ 0 & 0 & 1/2 & 1/2 \\ 1/2 & 1/2 & 0 & 0 \\ 1/2 & 1/2 & 0 & 0 \end{pmatrix}$

(v) $\begin{pmatrix} 0 & 0 & 1 & 0 \\ 0 & 0 & 0 & 1 \\ 1 & 0 & 0 & 0 \\ 0 & 1 & 0 & 0 \end{pmatrix}$

Consider Markov chains having the above matrices as t.p.m.'s. Verify that: Chains with t.p.m.'s (i) to (iv) are irreducible; chain with t.p.m. (v) is reducible with 2 pairs of closed chains: identify them.

A. 8   If the two $m \times m$ matrices P and Q are stochastic, then show that the matrix

(i) PQ is stochastic

(ii) $aP + (1 - a)Q$ is stochastic, for $0 \le a \le 1$.

Is the product of two doubly stochastic matrices doubly stochastic?

A. 9   Probability vector: a vector $\mathbf{x}' = (x_1, \ldots, x_i, \ldots, x_m)$, such that

$$x_i \ge 0, \sum_{i=1}^{m} x_i = 1 \text{ is called a probability vector.}$$

Show that an $m \times m$ matrix M is stochastic *iff* $x'M$ is a probability vector whenever x' is a probability vector. (This is a characterization of stochastic matrices.)

A. 10 Prove that a doubly stochastic $m \times m$ matrix A may be written as

$$A = \Sigma C_r P_r, r = 1, ..., m!, c_r \geq 0, \Sigma c_r = 1,$$

where P_r are permutation matrices of order m.

REFERENCES

C.S. Beightler, L.G. Mitten and G.L. Nemhauser. A short table of z-transforms and generating functions, *ORSA* **9**, 574–578 (1961).

R. Bellman and K.L. Cooke. *Differential-Difference Equations*, Academic Press, New York (1963).

F.R. Gantmacher. *Applications of the Theory of Matrices*, Inter-Science, New York (1959).

G. Hadley. *Linear Algebra*, Addison-Wesley, Reading, MA (1964).

K. Hoffman and R. Kunze. *Linear Algebra*, Prentice-Hall, Englewood Cliffs, NJ (1965).

V.P. Jean. *Mathematical Approach to Pattern and Form in Plant Growth*, Wiley, New York (1984).

C. Jordan. *Calculus of Finite Differences*, 2nd ed., Chelsea, New York (1947).

P. Lancaster. *Theory of Matrices*, Academic Press, New York (1969).

E.B. McBride. *Obtaining Generating Functions*, Springer-Verlag, New York (1971).

K.S. Miller. *Linear Difference Equations*, W.A. Benjamin, New York (1968).

E.J. Muth. *Transform Methods with Applications to Engineering and Operations Research*, Prentice-Hall, Englewood Cliffs, NJ (1977).

E. Seneta. *Non-Negative Matrices and Markov Chains*, 2nd ed.; Springer-Verlag, New York (1981).

M.G. Smith. *Laplace Transform Theory*, Van Nostrand, London (1969).

N.N. Vorobyov. *The Fibonacci Numbers*, Heath, Boston (1963).

D.V. Widder. *The Laplace Transform*, Princeton Univ Press, Princeton (1941).

Answers to Exercises

CHAPTER 1

1.1 (a) $s(1-s-s^2)^{-1}$.

1.4 $rq/p, rq/p^2; \{p/(1-qs)\}^{mr}, mrq/p, mrq/p^2$.

1.5 (b) (i) $(e^a - 1)^{-2}(e^{as} - 1)^2$,

(ii) $(e^{a_1} - 1)^{-1}(e^{a_2} - 1)^{-1}(e^{a_1 s} - 1)(e^{a_2 s} - 1)$

(c) $(e^a - 1)^{-n}(e^{as} - 1)^n$;

$$\Pr(S_n = m) = [(e^a - 1)^{-n}] \sum_{r=0}^{n-1} (-1)^r \binom{n}{r} \frac{a^m(n-r)^m}{m!}$$

$$= \frac{n! a^m S(m,n)}{m!(e^a - 1)^n} \quad \text{for } m = n, n+1, \ldots,$$

where $S(m,n) = \sum_{r=1}^{n}(-1)^{n-r}\binom{n}{r}r^m/n!$ is the Stirling number of the second kind with arguments m and n.

$$E(S_n) = na\, e^a(e^a - 1)^{-1}.$$

1.7 (a) (i) $\exp\{(\lambda p/(1-qs) - 1\}$

(ii) $E(S_N) = aq/p;\ \text{var}(S_N) = (aq/p^2) + (aq^2/p^2)$.

1.9 $P(s_1, s_2) = \dfrac{p_1 p_2}{(1-q_1 s_1)(1-q_2 s_2)}.$

1.10 (i) $\dfrac{a}{s^2 + a^2}$,

(ii) $\dfrac{a}{s+2}\ (s > -2)$

(iii) $\dfrac{s^4 + 4s^2 + 24}{s^3}\ (s > 0)$,

(iv) $\dfrac{6}{(s+3)^4}$

(v) $\dfrac{s+1}{s^2+2s+5}$,

(vi) $\dfrac{\Gamma(n+1)}{(s-a)^{n+1}}$

(vii) $\dfrac{a^n}{(s+a)^{n+1}}$,

(viii) $e^{-a\sqrt{s}}$

(ix) $\exp(1/s)/s^{r+1}$,

(x) $\exp\!\left(\dfrac{-s}{s+1}\right)\Big/ (s+1)^{r+1}$

(xi) $\dfrac{\exp\{a\lambda/(\lambda+s)\}-1}{\exp(a)-1}$.

1.12   (i) $\lambda k e^{-\lambda x}$, (ii) $E(X)=k/\lambda$, (iii) $\lambda k/(s+\lambda)$.

1.14 $(1-e^{-s})^2/s^2$.

1.15(b) $E(X)=\dfrac{\Gamma(k+(1/c))}{\lambda\Gamma(k)}$; $var(X)=\dfrac{1}{\lambda^2}\left[\dfrac{\Gamma(k+(2/c))}{\Gamma(k)}-\left\{\dfrac{\Gamma(k+(1/c))}{\Gamma(k)}\right\}^2\right]$

Chapter 3

3.1 $p_{00} = p_{11}=p, \ p_{01}=p_{10}=q;$

$p_{00}^{(2)} = p_{11}^{(2)}=p^2+q^2, \ p_{01}^{(2)}=p_{10}^{(2)}=2pq.$

$p_{00}^{(3)} = p_{11}^{(3)}=p(p^2+3q^2), \ p_{01}^{(3)}=p_{10}^{(3)}=q(3p^2+q^2).$

$\Pr\{X_2=0\mid X_0=1\}=p_{10}^{(2)}, \Pr\{X_3=1\mid X_0=0\}=p_{01}^{(3)}.$

3.2   (i) 0.279, (ii) 0.0336 .

3.3 $p_{jj}=0, \ p_{jk}=\dfrac{1}{2}, k\neq j$

$\Pr\{X_2=1\mid X_0=1\}=p_{11}^{(2)}=\dfrac{1}{2}, \ \Pr\{X_2=2\mid X_0=3\}=\dfrac{3}{8}$

$\Pr\{X_2=3\mid X_0=2\}=\dfrac{3}{8}.$

When the number of children is $m(\geq 3)$, P will be an $m\times m$ matrix such that $p_{jj}=0, p_{jk}=1/(m-1), j\neq k$.

3.4 $P = \begin{pmatrix} p & q & 0 & 0 \\ 0 & 0 & p & q \\ p & q & 0 & 0 \\ 0 & 0 & p & q \end{pmatrix}$, $P^m = (p^2, pq, pq, q^2), m = 2, 3, \ldots$

3.5 P is a (6×6) triangular matrix (p_{jk}) such that

$$p_{jj} = j/6, \quad p_{jk} = 0, k < j, p_{jk} = 1/6, \quad k > j.$$

For $P^2, p_{jj}^{(2)} = (j/6)^2, p_{jk}^{(2)} = \sum_r p_{jr} p_{rk} = 0, \quad k < j$

$$p_{jk}^{(2)} = \sum_r p_{jr} p_{rk} = (2k - 1)/36$$

$$\Pr\{X_3 = 6\} = 91/216.$$

3.6 $P = (p_{jk})$ is a (3×3) matrix such that $p_{jj} = j/3$,

$p_{j,j+1} = (3 - j)/3, p_{j,k} = 0, k \neq j, j + 1$.

The eigenvalues of P are $t_j = j/3, j = 1, 2, 3$. Left eigenvectors corresponding to these eigenvalues are $(1, 0, 0), \left(1, \frac{1}{2}, 0\right)$ and $(1, 1, 1)$ respectively; the right eigenvectors are $(1, -2, 1)', (0, 1, -1)'$ and $(0, 0, 1)'$ respectively, so that $c_1 = 1, c_2 = 2, c_3 = 1$.
Thus

$$P = \sum_{j=1}^{3} c_j t_j B_j, \quad P^n = \sum_{j=1}^{3} c_j t_j^n B_j,$$

where $B_1 = \begin{pmatrix} 1 & -2 & 1 \\ 0 & 0 & 0 \\ 0 & 0 & 0 \end{pmatrix}$, $B_2 = \begin{pmatrix} 0 & 2 & -2 \\ 0 & 1 & -1 \\ 0 & 0 & 0 \end{pmatrix}$,

$$B_3 = \begin{pmatrix} 0 & 0 & 1 \\ 0 & 0 & 1 \\ 0 & 0 & 1 \end{pmatrix}$$

$$\lim P^n \to (0, 0, 1), \quad \text{as} \quad n \to \infty.$$

3.7   For r cells, $P = (p_{jk})$ is a $(r \times r)$ matrix such that

$$p_{jj} = j/r, p_{j,j+1} = (r - j)/r, p_{jk} = 0, k \neq j, j + 1.$$

Elements of P^2 are $(p_{jk}^{(2)})$,

where $\qquad p_{jj}^{(2)} = (j/r)^2, \quad p_{jj+1}^{(2)} = (r-j)(2j+1)/r^2,$

$$p_{jj+2}^{(2)} = (r-j)(r-j-1)/r^2,$$

$$p_{jk}^{(2)} = 0, k \neq j, j+1, j+2.$$

The eigenvalues of P are $t_j = j/r, j = 1, 2, ..., r$.

The right eigenvectors $\mathbf{x}_i$ of P are solutions of the equations

$$(P - t_j I)\, \mathbf{x}_i = 0, \qquad \mathbf{x}_i = (x_{i1}, ..., x_{is}, ..., x_{ir})'.$$

The solution of the homogeneous equations (which can be obtained by the method outlined in Jordan,[†] section 173) is given by

$$x_{is} = \binom{s-1}{i-1} \Big/ \binom{r-1}{i-1}, \qquad s \geq i$$

$$= 0, \qquad\qquad\qquad s < i$$

Similarly, $\quad y_{is} = (-1)^{i-s}\binom{r-s}{i-s}, \qquad s \leq i$

$$= 0, \qquad\qquad\qquad s > i$$

whence $1/c_s = 1 \Big/ \binom{r}{s}$. Thus P and P^n can be written down.

It follows that

$$\lim P^n \to (0, 0, ..., 1), \quad \text{as} \quad n \to \infty.$$

The interpretation of the limit is intuitively clear.

3.9 $\qquad p_{00}^{(2n)} = p\binom{2n}{n} p^n q^n.$

Using Stirling's approximation for $n!$ (n large), we get

$$p_{00}^{(2n)} \cong (4pq)^n / \sqrt{(n\pi)};$$

when $p = \dfrac{1}{2}, \sum p_{00}^{(2n)} = \sum_n \dfrac{1}{\sqrt{(n\pi)}}$ diverges but $p_{00}^{(2n)} \to 0$.

[†]*Calculus of Finite Differences.* Chelsea, New York (1950)

Hence the state 0 is persistent (recurrent) null if $p = \dfrac{1}{2}$.

When $p \neq \dfrac{1}{2}$, $p_{00}^{(2n)} < a^n$, $a = 4pq < 1$

and so $\Sigma p_{00}^{(2n)}$ converges; thus the state 0 is transient. Similarly for the state 1.

3.10   Let $f_{jk}^{(n)}$ be the probability that starting from state j, the state k is reached for the first time at nth step. We have

$$f_{00}^{(n)} = 1 - a_1, \qquad\qquad\qquad n = 0$$

$$= a_1 a_2 \ldots a_{n-1}(1 - a_n), \qquad n = 1, 2, \ldots$$

$$F_{00} = \Sigma f_{00}^{(n)} = (1 - a_1) + \sum_{n=1}^{\infty} a_1 \ldots a_{n-1}(1 - a_n)$$

$$= 1 - \lim (a_1 \ldots a_n).$$

Thus the state 0 is recurrent *iff* $\lim (a_1 \ldots a_n) = 0$.
We have

$$\mu_{00} = \Sigma(n + 1) f_{00}^{(n)} = 1 + \sum_{n=1}^{\infty} a_1 \ldots a_n .$$

Hence the state 0 is positive *iff* $\displaystyle\sum_{n=1} a_1 \ldots a_n$ is finite.

The other states $1, 2, \ldots$ will be recurrent and non-null (or null) according as the state 0 is.

3.11   To proceed as in **3.10**.

3.14   118/25.

Chapter 4

4.3 $\exp \{\lambda t(s - 1)/(1 - qs)\}$; $\lambda t/p$, $\lambda t(1 + q)/p^2$.

4.4   28; 42

4.8 $p'_n(t) = -n\mu p_n(t) + (n + 1)\mu p_{n+1}(t)$
$E\{N(t)\} = i \exp(-\lambda t)$
$\mathrm{var}\{N(t)\} = i \exp(-\lambda t)\{1 - \exp(-\lambda t)\}.$

4.9(b)   1, when $\mu > \lambda$

$$\left(\frac{\mu}{\lambda}\right)^i, \text{ when } \lambda > \mu.$$

4.10 $p'_n(t) = -\{(\lambda + \mu)n + a\} \, p_n(t) + \{\lambda(n-1) + a\} \, p_{n-1}(t) + (n+1)\mu p_{n+1}(t), n \geq 1$

$p'_0(t) = -ap_0(t) + \mu p_1(t)$

As t tends to ∞, $M(t)$ tends to $a / (\mu - \lambda)$ for $\mu > \lambda$.

4.12 $E(T_n) = \left(\dfrac{1}{\lambda}\right) \sum\limits_{k=1}^{n} (-1)^{k+1} \binom{n-1}{k} \dfrac{1}{k} = \dfrac{1}{\lambda} \sum\limits_{k=1}^{n-1} \dfrac{1}{k} .$

4.13 $E(Z_k) = kt/(n+1).$

4.14 (i) 0.458, (ii) 0.956.

4.16 $\Pr\{N = n\} = \int\limits_0^\infty e^{-\lambda t}(1 - e^{-\lambda t})^n \, \mu e^{-\mu t} \, dt = \dfrac{\mu}{\lambda}\beta\left(n+1, \dfrac{\mu}{\lambda}+1\right)$

(Note that N does not include the parent.)

4.18 Z is (negative) exponential with mean $1 / \sum \lambda_i$, $i = 1, 2, ...n.$

4.19 When $\sum \alpha_j < \infty$, $\{v_j\}$ defines a probability distribution.

When $\sum \alpha_j = \infty$, $v_j = 0$ for all j.

4.23
$$A = \begin{bmatrix} -2a & a & a & 0 \\ b_2 & -(a+b_2) & 0 & a \\ b_1 & 0 & -(a+b_1) & a \\ 0 & b_1 & b_2 & -(b_1+b_2) \end{bmatrix}$$

$V = (k \, b_1 \, b_2, \, kab_1, \, kab_2, \, ka^2),$

where $k^{-1} = (a + b_1) (a + b_2).$

4.24 $v_0 = k, v_1 = (a/b) k, v_2 = (a/b)^2 k$, where
$k = [1 + (a/b) + (a/b)^2]^{-1}$; reqd. prob. = $v_0 + v_1$.

Chapter 6

6.4 $L^*(s) = \{1/(s + B)^2\} [A + \{ab + A(2A + B)\}/s$

$$+ \{ab(4A + B)\}/s^2 + 2(ab)^2/s^3],$$

$A = pa + (1-p)b, B = (1-p)a + pb.$

$L(t) = \sum\limits_{i=1}^{4} K_i a_i(t),$

where

$K_1 = A, K_2 = ab + A (2A + B), K_3 = ab (4A + B), K_4 = 2 (ab)^2$

$a_1 = te^{-Bt}, a_2 = (1 - e^{-Bt})/B^2 - te^{-Bt}/B,$

$a_3 = t(1 + e^{-Bt})/B^2 - 2(1 - e^{-Bt})/B^3,$

$$a_4 = (t^2/2B^2) - t(2 + e^{-Bt})/B^3 + 3(1 - e^{-Bt})/B^4.$$

6.6 $M_1(t) = abt / (a + b) + \{a^2 / (a + b)^2\} \{1 - \exp(-(a + b)t)\}$
$M_0(t) = abt / (a + b) + \{ab / (a + b)^2\} \{\exp(-(a + b)t) - 1\}.$

6.7 $\dfrac{AM(T) + B}{T} \to \dfrac{A}{\mu} + \dfrac{1}{T}\left\{B + A\dfrac{\sigma^2 - \mu^2}{2\mu^2}\right\} + o(T^{-1})$

(μ, σ^2 are the mean and variance of the lifetime distribution).

Chapter 7

7.1 (i) $f_{jk}^{*}(s)$ can be found from

$$F^*(s) = [I - C^*(s)]^{-1} A^*(s);$$

$$f_{11}^{*}(s) = \frac{(3s + 7)(s + 2)^2}{s\Delta_1}$$

$$f_{12}^{*}(s) = \frac{(3s + 1)(s + 3)}{s\Delta_1}$$

$$f_{21}^{*}(s) = \frac{(3s + 7)(s + 4)}{s\Delta_1}$$

$$f_{22}^{*}(s) = \frac{(3s + 1)(s + 3)^2}{s\Delta_1}$$

where

$$\Delta_1 = 3s^3 + 22s^2 + 52s + 37.$$

On factorising Δ_1, then splitting into partial fractions, and finally inverting the transforms, expressions for $f_{ij}(t)$ can be obtained.
(ii) $f_1 = 28/37, f_2 = 9/37$.

7.2 (i) $g_{11}(t) = (2/3)e^{-t} + e^{-3t} + 2te^{-2t} - e^{-2t}$

$g_{12}(t) = e^{-2t} + 2te^{-2t}$

$g_{21}(t) = (3/4)e^{-3t} + (1/4)e^{-t/3}$

$g_{22}(t) = 2e^{-4t} + \left\{\dfrac{3e^{-2t}}{5} + \dfrac{3e^{-t/3}}{20} - \dfrac{3e^{-3t}}{4}\right\}.$

Using (5.6), expressions for g_{ij}^* (s) can be found.

Note that f_{ij}^* (s) can be found also by using (5.4) and (5.4a).

(ii) $\mu_{11} = 37/36,$ $\mu_{22} = 37/24$

(iii) $\mu_{12} = 7/3,$ $\mu_{21} = 3/4.$

7.3 (ii) $\mu_{01} = 7/6,$ $\mu_{10} = 23/7$

(iii) $f_0 = 49/187,$ $f_1 = 138/187.$

7.4 The relation can be found by simple probabilistic argument. Taking L.T. of the relation, and multiplying both sides by s,

$$\{sf_{ij}^*(s)\} = \{se_{ij}^*(s)\}\{h_j^*(s)\},$$

and then taking limits, as $s \to 0$, one gets the relation $e_j = f_j / M_j$.

7.6   var $(V_{12}) = 17/9,$ var $(V_{21}) = 13/16.$

7.10 $Q_{i,j}(t) = 0,$ $i \ge 1, j < i - 1$

$$= \int_0^t \frac{e^{-ay}(ay)^{j-i+1}}{(j-i+1)!} dB(y), \quad i \ge 1, j \ge i - 1$$

$$Q_{0,j}(t) = \int_0^t \{1 - e^{-a(t-y)}\} \frac{e^{-ay}(ay)^j}{j!} dB(y), \quad j \ge 0$$

7.11 $Q_{0,j}(t) = \int_0^t \{1 - e^{-a(t-y)}\} \frac{e^{-ay}(ay)^j}{j!} dB(y), \quad j \ge 0$

$$Q_{i,j}(t) = \int_0^t e^{-ay} \frac{(ay)^j}{j!} dB(y), \quad 1 \le i \le c, j \ge 0$$

$$= 0, \quad i \ge c, j < i - c$$

$$= \int_0^t \frac{e^{-ay}(ay)^{j-i+c}}{(j-i+c)!} dB(y), \quad i \ge c, j \ge i - c.$$

7.12 Equality holds in the former case and does not hold in case of the latter.

Chapter 9

9.2 $(\sqrt{5} - 1)/2$

9.6 $n^{r-1} / (n+1)^r, r = 1, 2, \ldots$

9.9 $H(s) = \dfrac{1 - \sqrt{(1 - 4pqs)}}{2q}$

9.10 $b_k = (1 - [q/p])(q/p)^{k-1}, \quad k = 1, 2, ..$

$B(s) = \sum\limits_{k=1}^{\infty} b_k s^k = \dfrac{(p - q)s}{p - qs}$

9.12   (i) $r_n = \dfrac{p^n q^{n-1}(q - p)^2}{(q^{n+1} - p^{n+1})(q^n - p^n)}, \quad p \neq q$

$= 1/n(n + 1), \qquad p = q, \qquad n = 1, 2, 3, \ldots$

(ii) $r_n = \dfrac{s_0(1 - s_0)m^{n-1}(m - 1)}{(m^{n-1} - s_0)(m^n - s_0)}, \qquad m \neq 1$

$= \dfrac{(1 - c)c}{\{1 + c(n - 1)\}\{1 + c(n - 2)\}}, \quad m = 1, n = 1, 2, 3, \ldots.$

9.14 $\text{Var}(X(t)) = \dfrac{u''(1) - u'(1)}{u'(t)} \exp\{u'(1)t\} [\exp\{u'(1)t\} - 1] \quad \text{if } u'(1) \neq 0$

$= u''(1)\, t \qquad\qquad\qquad\qquad\qquad\qquad\quad \text{if } u'(1) = 0.$

Appendix

A.2 $\dfrac{(q/p)^a - (q/p)^n}{(q/p)^a - 1}; \ 1 - n/a$

A.6   Three eigenvalues are $1, (-1 \pm\sqrt{5})/4$; corresponding left eigenvectors are $(1,1,1); ((-1 - \sqrt{5})/4, (-1 + \sqrt{5})/4, 1)$; $((-1 + \sqrt{5})/4, (-1 - \sqrt{5})/4, 1)$.

A.7   (i) Imprimitive (ii) Imprimitive
   (iii) Primitive   (iv) Imprimitive
   (v) Imprimitive

Abbreviations

Pr(A); $P(A)$	probability of the event A
r.v.	random variable
p.d.f. (pdf)	probability density function (or density function)
p.g.f. (pgf)	probability generating function
p.f.; p.m.f.	probability mass function
d.f.; c.d.f.	cumulative distribution function
iff	if and only if
i.i.d.	identically and independently distributed
$E(X)$	expectation of the r.v. X
$E[\mathbf{E}]$	also used as symbol for displacement operator
s.d.(X)	standard deviation of the r.v. X
var (X)	variance of the r.v. X
L.S.T. (LST)	Laplace-Stieltjes Transform
L.T. (LT)	Laplace transform
C.K.	Chapman-Kolmogorov
t.p.m.	transition probability matrix
i.t.p.m.	interval transition probability matrix
w.r.t.	with respect to
l.h.s. (r.h.s.)	left (right) hand side
~	asymptotic to
$X*Y$	convolution of two random variables X and Y
F_1*F_2	convolution of two distributions with d.f. F_1 and F_2
F^{n*}	convolution of n i.i.d. random variables having common d.f. F
$f^{(n)}(\cdot)$	n^{th} derivative of $f(\cdot)$
$\Phi(\cdot)$	distribution function of standard normal variate
$\mathbf{e}$	column vector with all elements equal to 1, i.e. $\mathbf{e}' = (1, 1, ...,1)$
$\mathbf{0}$ (0)	column vector with all elements equal to 0
▲	indicates completion of a proof or of an example

Glossary of Commonly Used Notations

Chapter 1

$P(s)$	probability generating function (p.g.f.) of a r.v.
S_N	sum of N i.i.d. random variables X_i, where N is a random variable independent of X_i
$G(s)$	p.g.f. of N
E_k	r.v. having Erlang-k distribution
$K_m\,(C_m)$	r.v. having Coxian distribution of order m
$F^*(s)$	$= \int\limits_0^\infty e^{-st}\, dF(x)$ (L.T. [L.S.T.] of a r.v. with d.f. F)
$\bar{f}(s)$	$= \int\limits_0^\infty e^{-st} f(t)dt$ [L.T. of $f(t)$]
$F(t)$	distribution function
$R(t)$	$= 1 - F(t)$
$h(t)$	hazard rate function
$r(t)$	residual mean-lifetime (or expected mean life).

Chapter 2

$\{X(t)\};\ \{X_t\}$	stochastic process in continuous time
$\{X_n\}$	stochastic process in discrete time
$C_{s,t};\ C(s, t)$	covariance function between $X(t)$ and $X(s)$

Chapter 3

p_{jk}	transition probability from state j to state k
$P = (p_{jk})$	t.p.m. with elements p_{jk}
$p_{jk}^{(m)}$	m-step transition probability from state j to state k
$P^{(m)}$	t.p.m. with elements $p_{jk}^{(m)}$
$f_{jk}^{(n)}$	probability that the system reaches the state k for the first time after n transitions, given that it started with state j
F_{jk}	$(= \sum\limits_n f_{jk}^{(n)}$, i.e. probability that starting with state j, the system will ever reach state k
μ_{jk}	expected first passage time from state j to state k
$P_{jj}(s),\ F_{jj}(s)$	p.g.f. of $\{p_{jj}^{(n)}\}$ and $\{f_{jj}^{(n)}\}$ respectively

t_i, $\mathbf{x}_i$, $\mathbf{y}_i'$	eigenvalue, right and left eigenvectors respectively of P
v_j	$=\lim p_{ij}^{(n)}$ as $n \rightarrow \infty$
$V(s)$	p.g.f. of $\{v_j\}$

Chapter 4

$N(t)$	number of occurrences of an event by time t (or upto epoch t)
$p_n(t)$	$\Pr\{N(t) = n \mid N(0) = .\}$
$P(s, t)$	p.g.f. of $\{N(t), t \geq 0\}$ or of $\{p_n(t)\}$
$M(t)$	compound Poisson process, i.e. $\sum_{i=1}^{N(t)} X_i$ where X_i are i.i.d. r.v's and $N(t)$ is a Poisson process
$Q(s, t)$	p.g.f. of a non-homogeneous Poisson process
$\{X(t)\}$	Markov process in continuous time with discrete state space or continuous-time Markov chain
$p_{ij}(t)$; $p_{i,j}(t)$	probability of transition from state i to state j by time t
$P(t)$	t.p.m. with elements $p_{ij}(t)$
v_j	$\lim p_{ij}(t)$ as $t \rightarrow \infty$
a_{ij}	transition density, i.e. $p'_{ij}(t)\mid_{t=0}$
A	rate matrix or transition density matrix whose elements are (a_{ij})

Chapter 5

$\{X(t)\}$	Wiener process
μ_t	mean-value function of $X(t)$
$\sigma^2(t)$	variance function of $X(t)$
$M(T)$	maximum of $X(t)$ in $0 \leq t \leq T$ $\left(\max_{0 \leq t \leq T} X(t)\right)$
T_a	first passage time to a fixed point a of a Wiener process with $X(0) = 0$, $\mu = 0$
$U(t)$	Ornstein-Uhlenbeck process (O-U.P)
$m(t)$	mean-value function of $U(t)$
$\sigma^2(t)$	variance-function of $U(t)$

Chapter 6

X_n	non-negative independent r.v. (interarrival time between $(n-1)$th and nth renewal) having mean μ
F	d.f. of X_n
S_n	nth renewal epoch, i.e. $S_n = \sum_{i=1}^{n} X_i$
$N(t)$	number of renewals in $[0, t]$ $(N(t) = \sup\{n: S_n \leq t\})$

$M(t)$	expected number of renewals in $[0, t]$; $M(t) = E\{N(t)\}]$
$m(t)\Delta t$	Pr {of one or more renewals in $(t, t + \Delta t)$}; $m(t) = M'(t)$
$M^*(s)$	L.T. of $M(t)$
$N_A(t, T)$ $(N_B(t, T))$	number of renewals in $[0, t]$ under age (block) replacement policy with replacement interval T
G	d.f. of initial distribution of a modified renewal process
$N_D(t)$ $(N_e(t))$	number of renewals in $[0, t]$ for modified (equilibrium) renewal process
$F_e(t)$	equilibrium distribution of F having d.f.

$$\frac{1}{\mu}\int_0^\infty \{1 - F(x)\}\, dx$$

$G(t, z)$	p.g.f. of $\{N(t)\}$
$G_D(t, z)$ $[G_e(t, z)]$	p.g.f. of $N_D(t)$ $[N_e(t)]$
$G^*(s, z)$	L.T. of $G(t, z)$
$M_i(t)$	renewal function of an alternating (or two-stage) renewal process $(i = 0, 1)$
$Y(t)$	residual or spent lifetime (*forward recurrence time*) at time (epoch) t, i.e. $Y(t) = S_{N(t)+1} - t$
$Z(t)$	age (*backward recurrence time*) at time t, i.e. $Z(t) = t - S_N(t)$
Y_n	reward or cost associated with nth renewal
$\{X_n, Y_n\}$	renewal reward process
$Y(t)$	total (cumulative) reward earned or cost incurred by time t
$C(t)$	expected cumulative reward (or cost) by time t
$A(\cdot)$	inter-demand time distribution function
Y_i	size of batch size of demand at ith occurrence of demand
$D(t)$	cumulative demand during $(u, u + t]$
$B_n(x)$	$= P\{Y_1 + \ldots + Y_n \leq x\}$
$R(x)$	$= \sum_{n=1}^\infty B_n(x)$
$I(t)$	inventory position at epoch t
$f(t, x)\, dx$	$= \Pr\{x \leq I(t) < x + dx\}$
$I[f(x)]$	asymptotic distribution of $I(t)$ $[f(t, x)]$ as $t \to \infty$

Chapter 7

$\{X(t)\}$	Markov process
t_n	nth transition epoch
$X_n = X(t_n +)$	state at transition occurring at epoch t_n
τ_n	time interval $(t_{n+1} - t_n)$ between nth and $(n + 1)$st transitions
$\{X_n, t_n\}$	Markov renewal process
Y_t	$(= X_n$ on $t_n \leq t < t_{n+1})$, i.e. state of the process at its most recent transition

$Q_{jk}(t); Q_{j,k}(t)$	$\Pr\{X_{n+1}=k, \tau_n \leq t \mid X_n = j\}$
p_{jk}	$\lim Q_{jk}(t)$ as $t \to \infty$
T_{jk}	(conditional) waiting time in state j given that the next transition is to state k
$W_{jk}(t)$	$(= Q_{jk}(t)/p_{jk})$, i.e. d.f. of T_{jk}
α_{jk}	expectation of T_{jk}
$\alpha_{jk}^{(r)}$	rth moment of T_{jk}
T_j	unconditional waiting time in state j
h_j	$\Pr\{T_j > t\}$
M_j	expectation of T_j
$M_j^{(r)}$	rth moment of T_j
S_n^j	successive epochs t_n for which $X_n = j$
$N_j(t)$	number of transitions to state j (renewals of state j) in $[0, t]$
$\mathbf{N}(t)$	$= (N_1(t), N_2(t), \ldots)$
V_{ij}	first passage time into state j from state i
G_{ij}	d.f. of V_{ij}
μ_{ij}	mean first passage time into state j from state i
$f_{ij}(t)$	conditional probability that state at epoch t is j given that initial state was i
$Q_{ij}^{(n)}(t)$	$= \Pr\{X_n = j, t_n \leq t \mid X_0 = i\}$
$M(i, j, t)$	Markov renewal function $(= E\{N_j(t) \mid X_0 = i\})$
$F(t)$	interval transition probability matrix (i.t.p.m) having elements $f_{ij}(t)$
$f_{ij}; f_j$	limit $f_{ij}(t)$ as $t \to \infty$
$f_{ij}^*(s)$	L.T. of $f_{ij}(t)$
$\{v_j\}$	stationary distribution of embedded Markov chain $\{X_n\}$
$U(t)$	$(= t - t_n)$ time since last transition (backward recurrence time)
$V(t)$	$(= t_{n+1} - t_n)$ time to next transition (forward recurrence time)

Chapter 8

$\{X_t\}$	wide-sense or covariance stationary process in discrete time
C_k	covariance function of X_t of lag k
ρ_k	correlation function of X_t of lag k
$\{e_t\}$	purely random process
B	backward shift operator (i.e. $B X_t = X_{t-1}$)
$f(.)$	spectral density function of X_t
$C(z)$	autocovariance generating function of $\{X_t\}$
$S(z)$	autocorrelation generating function of $\{X_t\}$
$X(t)$	stationary process in continuous time
$R(v) (\rho(v))$	covariance (correlation) function of $\{X(t)\}$

Chapter 9

$\{p_k\}$	offspring distribution
ξ_r	i.i.d. random variable with distribution $\{p_k\}$
$P(s)$	p.g.f. of offspring distribution $\{p_k\}$ and also r.v. ξ_r, $r = 1, 2, ...$
m, σ^2	mean and variance respectively of $\{p_k\}$
X_n	size of nth generation, $n = 1, 2, ...$
X_0	size of 0-th generation or ancestor
$\{X_n, n \geq 0\}$	Galton-Watson (G.W.) process
$P_n(s)$	p.g.f. of $\{X_n\}$
q	probability of extinction
Y_n	total number of progeny upto nth generation including the ancestor, i.e. $Y_n = 1 + Z_n$
$G_n(s)$	p.g.f. of $\{Z_n\}$, $Z_n = X_1 + ... + X_n$
$R_n(s)$	p.g.f. of $\{Y_n\}$
$H(s)$	$\lim R_n(s)$, as $n \to \infty$
b_{kn}	$\Pr\{X_n = k \mid X_n > 0\}$ for a subcritical G.W. process
b_k	$\lim b_{kn}$, as $n \to \infty$
$B_n(s)$	p.g.f. of $\{b_{kn}\}$
$B(s)$	p.g.f. of $\{b_k\}$
$\varphi(0)$	$\lim \Pr\{X_n > 0\}/m^n$, as $n \to \infty$
G.W.I. Process	Galton-Watson process with immigration
$h(s)$	p.g.f. of immigrant distribution
a	mean of immigrant distribution
$X_{(n)}$	size of nth generation of a G.W.I. process
$P_n(s)$	p.g.f. of $\{X_{(n)}\}$
$F(s)$	$\lim P_n(s)$, as $n \to \infty$
$p^{(i)}(r_1, r_2, ..., r_k)$	probability that an individual of type i produces r_j offspring of type j, $j = 1, 2, ..., k$ (in case of a multitype G.W. process)
$X_n = (X_n^{(1)}, ..., X_n^{(k)})$	population size of k types in the nth generation of a multitype G.W. process
m_{ij}	expected number of offspring type j produced by an object of type i
$X(t)$	continuous time branching process
$p_{ij}(t)$	$\Pr\{X(t) = j \mid X(0) = i\}$ for a Markov branching process
$F(t, s)$	p.g.f. of $\{p_{ij}(t)\}$
a_j	transition density of $\{X(t), t \geq 0\}$, $X(0) = 1$
$u(s)$	p.g.f. of $\{a_j\}$
q	$\lim p_{10}(t)$ as $t \to \infty$, i.e. (probability of extinction)
G	d.f. of lifetime of objects in an age-dependent branching process
$F(t, s)$	p.g.f. of an age dependent branching process $\{X(t)\}$ with $X(0) = 1$

Chapter 10

λ, μ	mean arrival and service rates respectively

ρ	utilization factor
$N(t)$	number in the system at epoch t
W_n	waiting time of nth arrival
$W(t)$	virtual waiting time, i.e. duration of time a unit would have to wait were it to arrive at epoch (instant) t
M	exponential (Markovian) distribution
G	arbitrary (general) distribution
$p_n(t)$ $(p_{in}(t))$	probability that the number in the system at t is n given that number at $t = 0$ is zero (is i)
p_n	$\lim p_n(t)$ as $t \rightarrow \infty$
$L(L_Q)$	expected number in the system (queue)
$W(W_Q)$	expected waiting time in the system (queue)
$w_s(w_q)$	p.d.f. of waiting time in the system (queue)
$\overline{p}_n(s)$	L.T. of $p_n(t)$
$P(z, t)$	p.g.f. of $\{p_n(t)\}$
T	busy period initiated by a single customer or unit
$N^*(t)$	number in the system at epoch t during a busy period
$M^{(x)}$	arrival streams follow exponential distribution, number arriving at an arrival instant being given by a r.v. X
$M(a,b)$	bulk service with exponential service time, under a general bulk service rule (i.e., a batch taking a minimum of a units and a maximum of b units)
$p_{1,n}(t)$ $(p_{0,n}(t))$	probability that the number in the queue at time t is n, the service channel being busy (idle) under general bulk service rule
$p_{1,n}(t)$ $(p_{0,n}(t))$	limiting value of $p_{1,n}(t)$ ($p_{0,n}(t)$) as $t \rightarrow \infty$
$P(n_1, ..., n_k)$	steady state probability that there are n_j in the jth node in a Jackson network of k nodes
α_j	total average rate of arrival of node j
M/G/1	
$B(t)$ $(b(t))$	d.f. (p.d.f.) of general or arbitrary service time v
$B^*(s)$	$\int_0^\infty e^{-st}\, dB(t)$
t_n	nth departure epoch
$X_n = N(t_n + 0)$	number in the system immediately after nth departure
k_r	probability that r units arrive during service time of a unit
p_{ij}	$\Pr\{X_{n+1} = j \mid X_n = i\}$
$\{v_j\}$	stationary distribution of $\{X_n\}$
$V(s)$	p.g.f. of $\{v_j\}$
$K(s)$	p.g.f. of $\{k_r\}$
$W(W_Q(t))$	d.f. of waiting time in the system (queue)
$W^*(s)$	$\int_0^\infty e^{-st}\, dW(t)$

GI/M/1

$A(t)$ $(a(t))$	d.f. (p.d.f.) of general or arbitrary independent interarrival time u
$A^*(s)$	$\int_0^\infty e^{-st}\, dA(t)$
t_n	nth arrival epoch
$Y_n = N(t_n - 0)$	number in the system immediately before the arrival of the nth unit
$Z(t)$	$(= Y_n, t_n \le t < t_{n+1})$ system size at the most recent arrival
p_{ij}	$= \Pr\{Y_{n+1} = j \mid Y_n = i\}$
g_r	probability that number served in the inter-arrival time between nth and $(n + 1)$st units is r
$G(s)$	p.g.f. of $\{g_r\}$
$\{v_j\}$	stationary distribution of $\{Y_n\}$
$r(z)$	$= z - A^*(-\mu z)$
r_0	unique root of $r(z) = 0$ lying inside $\mid z \mid = 1$
$N(t)$	system size at an arbitrary epoch t
$f_{ij}(t)$	$\Pr\{Z(t) = j \mid Z(0) = i\}$
$p_{ij}(t)$	$\Pr\{N(t) = j \mid N(0) = i\}$

Reliability

X_i	Bernoulli r.v. with p.f. $\Pr(X_i = 1) = p_i$, $\Pr(X_i = 0) = 1 - p_i$
X	$= \min(X_1, ..., X_n)$
T_i	lifetime of a device or equipment
$R_i(t)$	reliability of the device i at time t $(= \Pr\{T_i > t\})$
$r_i(t)$	hazard rate $(= -R_i'(t)/R(t))$
$R(t)$	reliability of a system (consisting of n independent components $1, 2, ..., n$) at time t
$A(t)$	availability of a system at time t
A	limiting or stationary reliability $(= \lim A(t)$ as $t \to \infty)$
$1/a_i$	mean of exponential distribution
$Z(t)$	accumulated damage by time t

Appendix A

t_i	eigenvalue of a matrix
$p(A)$	spectral radius of the matrix A
$\mathbf{x}(\mathbf{x}')$	column (row) vector
$\mathbf{x}$	right eigen vector
$\mathbf{y}'$	left eigen vector
P	stochastic matrix

Table D.1
Table of Laplace Transforms

Laplace transform of $f(t)$ is defined by

$$L\{f(t)\} = \int_0^\infty e^{-st} f(t) \, dt$$

$f(t)$	$L\{f(t)\}$
1. 1	$\dfrac{1}{s}$
2. t^{n-1}	$\dfrac{(n-1)!}{s^n}$, $n = 1, 2, 3, \ldots$
3. t^{a-1}	$\dfrac{\Gamma(a)}{s^a}$, $a > 0$
4. $\dfrac{1}{\sqrt{(\pi t)}}$	$\dfrac{1}{\sqrt{s}}$
5. e^{at}	$\dfrac{1}{s-a}$
6. $\dfrac{t^{n-1}e^{at}}{(n-1)!}$	$\dfrac{1}{(s-a)^n}$, $n = 1, 2, \ldots$
7. $(at)^n \, e^{-at}$	$\dfrac{a^n \Gamma(n+1)}{(s+a)^{n+1}}$, $n > -1$
8. $\sin \alpha t$	$\dfrac{\alpha}{s^2 + \alpha^2}$
9. $\cos \alpha t$	$\dfrac{s}{s^2 + \alpha^2}$
10. $\dfrac{e^{bt} \sin \alpha t}{\alpha}$	$\dfrac{s}{(s-b)^2 + \alpha^2}$
11. $f(t) = 0, 0 < t < a$ $\quad\quad = 1, \; a < t$ (Heaviside *unit function*)	$\dfrac{e^{-as}}{s}$

(contd.)

(Table D.1 Contd.)

12.	$\delta(t-a)$	e^{-as}
	(Dirac δ function located at a)	
13.	$\lambda e^{-\lambda t}(\lambda>0)$	$\dfrac{\lambda}{s+\lambda}$
14.	$\dfrac{\lambda^k t^{k-1} e^{-\lambda t}}{\Gamma(k)}$ $(\lambda>0, k>0)$	$\left[\dfrac{\lambda}{s+\lambda}\right]^k$
15.	$\dfrac{a\exp(-a^2/2t)}{\sqrt{(2\pi t^2)}}$	$\exp(-a\sqrt{(2s)})$
16.	$a^n I_n(at)$	$\dfrac{\left[s-(s^2-a^2)\right]^n}{\sqrt{(s^2-a^2)}}$, $n>-1$
17.	$\dfrac{n}{t} I_n(at)$	$\left[s-\sqrt{(s^2-a^2)}\right]^n$, $n=1,2,\dots$
18.	$\{\sqrt{(t^n)}\} I_n(2\sqrt{t})$	$\dfrac{e^{1/s}}{s^{n+1}}$, $n=1,2,\dots$
19.	$\dfrac{\sqrt{\pi}}{\Gamma(k)}\left(\dfrac{t}{2a}\right)^{k-\frac{1}{2}} I_{k-\frac{1}{2}}(at)$	$\dfrac{1}{(s^2-a^2)^k}$, $k>0$
20.	$\dfrac{1}{t}(e^{bt}-e^{at})$	$l_n\left(\dfrac{s-a}{s-b}\right)$ [$l_n \equiv$ log natural]

APPENDIX D
Table D.2
Important Probability Distributions and their Parameters, Moments, Generating Functions

Distribution (variable X)	Probability function $p_k = P(X = k)$: discrete $f(x)$: continuous	Parameter (s)	Mean $E(X)$	Variance $E[(X - E(X))^2]$	Probability generating function $P(s) = E(s^X)$	Laplace transform $f^*(s) = E[e^{-sX}]$	Characteristic function $\phi(t) = E[e^{itX}]$
1	2	3	4	5	6	7	8
Discrete distributions							
1. Bernoulli	$p_k = p, \quad k = 1$ $= 1 - p, k = 0$	$0 < p < 1$	p	$p(1 - p)$	$ps + (1 - p)$		
2. Binomial	$p_k = \binom{n}{k} p^k (1-p)^{n-k}$ $k = 0, 1, \ldots, n$	$0 < p < 1$ $n = 1, 2, \ldots$	np	$np(1 - p)$	$[ps + (1 - p)]^n$		
3. (a) Geometric	$p_k = p(1-p)^k$, $k = 0, 1, 2, \ldots$	$0 < p < 1$	$\dfrac{1-p}{p}$	$\dfrac{1-p}{p^2}$	$\dfrac{p}{1-(1-p)s}$		
(b) Decapitated geometric	$p_k = p(1-p)^{k-1}$ $k = 1, 2, \ldots$	$0 < p < 1$	$\dfrac{1}{p}$	$\dfrac{1-p}{p^2}$	$\dfrac{ps}{1-(1-p)s}$		

(contd.)

(Table D.2. Contd.)

1	2	3	4	5	6	7	8
4. (a) Poisson	$p_k = \dfrac{e^{-a}a^k}{k!}$ $k = 0,1,2,...$	$a > 0$	a	a	$e^{a(z-1)}$		
(b) Decapitated Poisson	$p_k = \dfrac{a^k}{(e^a-1)k!}$	$a > 0$	$\dfrac{ae^a}{e^a-1}$	$\dfrac{e^a[1-a/(e^a-1)]}{e^a-1}$	$\dfrac{e^{az}-1}{(e^a-1)}$		
5. Negative Binomial (Pascal when r is a positive integer)	$p_k = \binom{r+k-1}{k}$ $\times p^r(1-p)^k$ $k = 0,1,2,...$	$r > 0$ $0 < p < 1$	$\dfrac{r(1-p)}{p}$	$\dfrac{r(1-p)}{p^2}$	$\left[\dfrac{p}{1-(1-p)s}\right]^r$		
6. Logarithmic series	$p_k = \dfrac{(1-p)^k}{k\log(p^{-1})}$ $k = 1,2,3,...$	$0 < p < 1$	$\dfrac{1-p}{p(\log(p^{-1}))}$	$\dfrac{(1-p)\{1+(1-p)/\log p\}}{p^2(\log(p^{-1}))}$	$\dfrac{\log\{1-s\times(1-p)\}}{\log p}$		

Continuous distributions

1	2	3	4	5	6	7	8
7. Exponential distribution (negative exponential)	$f(x) = \lambda e^{-\lambda x}$ $0 \le x < \infty$	$\lambda > 0$	$\dfrac{1}{\lambda}$	$\dfrac{1}{\lambda^2}$		$f^*(s) = \dfrac{\lambda}{s+\lambda}$	

(contd.)

(Table D.2. Contd.)

1	2	3	4	5	6	7	8
8. Gamma distribution (2 parameter)	$f(x) = \dfrac{\lambda^k x^{k-1} e^{-\lambda x}}{\Gamma(k)}$ $0 \le x < \infty$	$\lambda > 0$ $k > 0$	$\dfrac{k}{\lambda}$	$\dfrac{k}{\lambda^2}$		$f^*(s) = \left(\dfrac{\lambda}{s+\lambda}\right)^k$	
9. Erlang-k distribution	$f(x) = \dfrac{(\lambda k)^k x^{k-1} e^{-\lambda k x}}{\Gamma(k)}$ $0 \le x < \infty$	$\lambda > 0$ $= 1, 2, \ldots$	$\dfrac{1}{\lambda}$	$\dfrac{1}{k\lambda^2}$		$f^*(s) = \left(\dfrac{\lambda k}{s+\lambda k}\right)^k$	
10. Uniform distribution	$f(x) = \dfrac{1}{b-a}$ $a \le x \le b$	a, b $(-\infty < a < b$ $< \infty)$	$\dfrac{a+b}{2}$	$\dfrac{(b-a)^2}{12}$		$f^*(s) = \dfrac{e^{-sa} - e^{-sb}}{s(b-a)}$	
11. Inverse Gaussian distribution	$f(x) = \dfrac{a}{\sigma\sqrt{(2\pi x^3)}} \times$ $\exp\left\{-\dfrac{(a-\mu x)^2}{2\sigma^2 x}\right\}$ $x > 0$	$a > 0$ $\mu > 0$ $\sigma > 0$	$\dfrac{a}{\mu}$	$\dfrac{a\sigma^2}{\mu^3}$		$f^*(s) = \exp[(a/\sigma^2) \times$ $\{\mu - \sqrt{(\mu^2 + 2s\sigma^2)}\}]$	
12. Normal distribution	$f(x) = \dfrac{1}{\sigma\sqrt{(2\pi)}} \times$ $\exp\left\{-\dfrac{1}{2}\left|\dfrac{x-\mu}{\sigma}\right|^2\right\}$ $-\infty < x < \infty$	$-\infty < \mu < \infty$ $\sigma > 0$	μ	σ^2			$\phi(t) = e^{it\mu - \frac{t^2\sigma^2}{2}}$

Table D.3

Performance Measures [Steady State] of Some Queueing Models

λ = Mean arrival rate ; μ = Mean Service Rate ; $a = \lambda/\mu \equiv$ traffic intensity

		M/M/1	M/M/s	M/G/1	M/D/1	M/M/∞ M/G/∞
		1	2	3	4	5
I	λ' [Effective arrival rate]	λ	λ	λ	λ	λ
II	ρ (Utilisation factor)	$\lambda/\mu = a$	$\lambda/s\mu = a/s$	$\lambda/\mu = a$	$\lambda/\mu = a$	
III	p_o [Prob. that system is empty]	$1-\rho$	$\left[\sum_{n=0}^{s-1}\dfrac{a^n}{n!}+\dfrac{a^s}{s!(1-\rho)}\right]^{-1}$	$1-\rho$	$1-\rho$	$e^{-\lambda/\mu}$
IV	p_n [Prob. that there are n in system] $n = 1, 2, \ldots$	$(1-\rho)\rho^n$	$\dfrac{a^n}{n!}\,p_o,\ n \le s$ $\dfrac{a^s}{s!}\left(\dfrac{a}{s}\right)^{n-s}p_o,\ n>s$	Given by P.K. formula	$(1-\rho)(e^{\rho}-1),\ n=1$ $(1-\rho)\sum_{j=1}^{n}\dfrac{(-1)^{n-j}}{(n-j)!}\times$ $(jp)^{n-j-1}(jp+n-j)e^{pj}$ $n = 2, 3, \ldots$	$\dfrac{e^{-a}a^n}{n!}$
V	$E(N_s)$ [expected no. in service]	$a\,(=\rho)$	$a\,(=s\rho)$	$a\,(=\rho)$	$a\,(=\rho)$	$a\,(=\rho)$

(contd.)

(Table D.3. Contd.)

	M/M/1	M/M/s	M/G/1	M/D/1	M/M/∞ M/G/∞
	1	2	3	4	5
VI $E(N_q)$ [Expected number in queue]	$\dfrac{\rho^2}{1-\rho}$	$\dfrac{\rho}{1-\rho}C(s,a)$ $[C(s,a)=P_r(N\geq s)]$	$\dfrac{\lambda^2}{\mu(1-\rho)}\cdot\dfrac{E(v^2)}{2E(v)}$	$\dfrac{\rho^2}{2(1-\rho)}$	0
VII $E(N)$ [Expected number in system]	$\dfrac{\rho}{1-\rho}$	$a+\dfrac{\rho}{1-\rho}C(s,a)$ $(=a+E(N_q))$	$a+E(N_q)$	$\dfrac{\rho(2-\rho)}{2(1-\rho)}$	a
VIII $E[W_Q]$ [Expected queueing time]	$\dfrac{\rho}{\mu(1-\rho)}$	$\dfrac{1}{s\mu(1-\rho)}C(s,a)$	$\dfrac{\rho}{1-\rho}\cdot\dfrac{E(v^2)}{2E(v)}$	$\dfrac{\rho}{2\mu(1-\rho)}$	0
IX $E(W)$ [Expected response time]	$\dfrac{1}{\mu(1-\rho)}$	$\dfrac{1}{\mu}+E(W_Q)$	$\dfrac{1}{\mu}+E(W_Q)$	$\dfrac{2-\rho}{2\mu(1-\rho)}$	$\dfrac{1}{\mu}$
X $E(T)$ [Expected length of busy period]	$\dfrac{1}{\mu(1-\rho)}$		$\dfrac{1}{\mu(1-\rho)}$	$\dfrac{1}{\mu(1-\rho)}$	

(contd.)

(Table D.3. Contd.)

	M/M/1/K	M/G//s/s	GI/M/1	GI/M/s	
	6	7	8	9	
I	λ^1	$\lambda(1-p_k)$	$\lambda(1-p_s)$	λ	λ
II	ρ_0	λ/μ $=a(1-p_k)$	$\lambda'/s\mu$ $=a(1-p_s)/s$	λ/μ	$\lambda/s\mu$ $(=a/s)$
III	p_0	$1-\rho$	$\left[\sum_{n=0}^{s}\dfrac{a^n}{n!}\right]^{-1}$	$1-\rho$	
IV	p_n $n=1,2,\ldots.$	$\dfrac{(1-a)a^n}{1-a^{k+1}}, \lambda\neq\mu$ $\dfrac{1}{k+1}, \lambda=\mu$	$\dfrac{a^n}{n!}\,p_0, n\leq s$	$\rho(1-r_o)r_o^{n-1}$	
V	$E(N_s)$	$a=\rho$ $=(\lambda/\mu)$	$s\rho(=\lambda'/\mu)$	$a(=\rho)$	$a=s\rho$

(contd.)

(*Table D.3. Contd.*)

		M/M/1/K	M/G/1/s/s	GI/M/1	GI/M/s
		6	7	8	9
VI	$E(N_q)$	$\dfrac{a}{1-a} - \dfrac{a(1+Ka^k)}{1-a^{k+1}}$ $\lambda \neq \mu$	0	$\dfrac{\rho r_o}{1-r_o}$	
VII	$E(N)$	$\dfrac{\lambda'}{\mu} + E(N_q)$	$\dfrac{\lambda'}{\mu}$	$\dfrac{\rho}{1-r_o}$	
VIII	$E(W_Q)$	$E(N_q)/\lambda'$	0	$\dfrac{r_o}{\mu(1-r_o)}$	
IX	$E(W)$	$\dfrac{1}{\mu} + E(W_Q)$	$\dfrac{1}{\mu}$	$\dfrac{1}{\mu(1-r_o)}$	
X	$E(T)$			$\dfrac{1}{\mu(1-r_o)}$	

Author Index

Subject Index